EXERCISE
and
CIRCULATION
IN HEALTH and DISEASE

Bengt Saltin, MD, PhD
Copenhagen Muscle Research Center, Copenhagen, Denmark

Robert Boushel, DSc
Institute for Sports Medicine Research, Copenhagen, Denmark

Niels Secher, MD, PhD
University of Copenhagen, Copenhagen, Denmark

Jere H. Mitchell, MD
Southwestern Medical Center, University of Texas, Dallas, Texas

Editors

Human Kinetics

Library of Congress Cataloging-in-Publication Data

Exercise & circulation in health & disease / edited by Bengt Saltin
  . . . [et al.].
      p.   cm.
   Includes bibliographical references and index.
   ISBN 0-88011-632-3
   1. Exercise--Physiological aspects.   2. Cardiovascular system-
-Physiology.   3. Blood--Circulation.   I. Saltin, Bengt, 1935-
II. Title: Exercise and circulation in health and disease.
   QP301.E9346   1999
   612'.044--DC21   99-24144
                                                      CIP

ISBN: 0-88011-632-3

Acquisitions Editor: Michael S. Bahrke, PhD; **Managing Editor:** Cynthia McEntire; **Assistant Editor:** John Wentworth; **Copyeditors:** Michele Sandifer and John Mulvihill; **Proofreader:** Kathy Bennett; **Indexer:** Marie Rizzo; **Graphic Designer:** Keith Blomberg; **Graphic Artist:** Denise Lowry; **Cover Designer:** Jack W. Davis; **Printer:** UG/Dekker

Printed in the United States of America   10   9   8   7   6   5   4   3   2   1

Human Kinetics
Web site: http://www.humankinetics.com/

United States: Human Kinetics, P.O. Box 5076, Champaign, IL 61825-5076
1-800-747-4457
e-mail: humank@hkusa.com

Canada: Human Kinetics, 475 Devonshire Road Unit 100, Windsor, ON N8Y 2L5
1-800-465-7301 (in Canada only)
e-mail: humank@hkcanada.com

Europe: Human Kinetics, P.O. Box IW14, Leeds LS16 6TR, United Kingdom
+44 (0)113-278 1708
e-mail: humank@hkeurope.com

Australia: Human Kinetics, 57A Price Avenue, Lower Mitcham, South Australia 5062
(08) 82771555
e-mail: humank@hkaustralia.com

New Zealand: Human Kinetics, P.O. Box 105-231, Aukland 1
09-523-3462
e-mail: humank@hknewz.com

Contents

Preface

In the fall of 1995, about one hundred scientists from around the world with expertise in cardiovascular regulatory physiology gathered at the Danish Academy of Science in Copenhagen for an intensive four-day meeting to discuss diverse aspects of circulatory control during exercise. The meeting was the first in an ongoing series of specialized symposia hosted by the Copenhagen Muscle Research Center to mark the centennial of the first publication on cardiovascular regulation by Erik Johansson in 1895. It was this meeting that inspired this book. While the original intent was to publish a collection of specialized papers in the format of current perspectives or proceedings, it was soon realized that the collective quality of authorship, as well as the breadth, scope, and integrated nature of the topical areas warranted the effort to synthesize the contents in the form of a textbook.

This text is intended for graduate students focusing on various aspects of cardiovascular regulation and circulatory physiology. It also has particular relevance for clinicians as a reference on cardiovascular function and regulation from an integrated, mechanistic viewpoint. Attention is on the physiological responses to exercise, since it is recognized that in this context cardiovascular regulation is most integrated. Also, the complexity of cardiovascular control can be appreciated as multiple systems are engaged and coordinated through the entire range of functional capacity, with regulatory signaling occurring from subcellular to systemic levels of organization. In this setting, the book provides a conceptual framework for understanding function in the healthy state, as well as analysis of altered cardiovascular control in disease states and in various physical and environmental conditions. Numerous topics are covered from a basic science approach with relevant implications for clinical and applied settings. Historical aspects are presented along with the progression of ideas toward current working hypotheses interpreted by many leading scientists in their respective disciplines. We believe that the aggregate contributions enclosed are of sufficient breadth to represent a comprehensive view of cardiovascular regulation while also serving as a stimulating reference for more advanced research specialties and future work.

We would like to express sincere thanks to Erik Kay-Hansen for his generous support for the symposium *Exercise and the Circulation in Health and Disease* held at the Danish Academy of Sciences in November 1995, in which most of the contributing authors participated and from which this book was inspired.

Introduction

Niels H. Secher and John Ludbrook

Over 100 years ago, studies about the integrated physiological responses during muscle contraction emerged in the scientific literature. Even in the earliest studies then, *exercise* was recognized as a valuable model for evaluating how the circulation and ventilation were controlled. These historical scientific reports about cardiovascular regulation established many of the seminal concepts from which much of our current understanding of autonomic regulation of the circulation in healthy and disease states extends.

Animal Studies

As noted by Hering in 1895, Asp initiated the evaluation of the neural control of cardiovascular responses. He demonstrated an increase in blood pressure and heart rate by stimulation of the proximal end of the sciatic nerve. This indicated that afferent neural signals influenced the cardiovascular response. He also showed that this response depended on efferent signals generated by the sympathetic nervous system. Zuntz and Geppert (1886, 1888) were interested in the control of ventilation during muscular exercise. They considered that either of two mechanisms enhanced the ventilatory response. The first involved a reflex from the working muscle that operated on the central nervous system directly or through receptors in the lungs. The second was a cortical control mechanism associated with the will to exercise. In the dog, exercise was associated with an increase in the arterial oxygen and a decrease in the carbon dioxide content secondary to the ventilatory response. Neither the ventilation nor the blood gases were affected by section of the spinal cord (T8 to T12) or by severing the nerves to the lungs, leaving the phrenic nerves intact. Thus, they felt that an unknown factor reaching the central nervous system through the blood generated the increase in ventilation upon commencement of exercise.

Johansson (1895) took up this line of reasoning (see figure 1) to include a discussion of the cardiovascular effects of exercise. In the rabbit, he demonstrated an increase in heart rate (from 254 to 302 beats/min), mean arterial pressure (from 126 to 156 mmHg), and ventilation (from 1.4 to 1.9 L/min) induced by voluntary movement (see figure 2). The increases in heart rate and ventilation with involuntary movement depended on the duration of the stimulus. Intermittent contractions (0.5 s, 2–5 s rest) induced similar responses. This was not replicated by passive movement of the limbs (to 256 beats/min, 148 mmHg, and 1.8 L/min, respectively), indicating the necessity of muscle contraction for the responses.

Figure 1 Erik Johan Johansson (1862–1938). Professor of physiology at the Karolinska Institute, Stockholm, Sweden (1901–1927) and head of the Nobel Committee (1918–1926).

Both in the rabbit and dog, sectioning of the spinal cord at L2 to L5 during electrically induced muscle contraction prevented the increase in blood pressure,

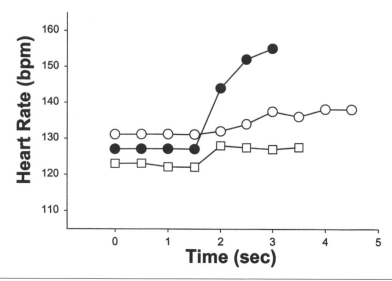

Figure 2   Heart rate response to spontaneous movement of a rabbit (filled symbol), to passive movement of the hind limbs (squares), and to electric stimulation of the legs with section of the spinal cord at L2 to LS (circles). (Data from Johansson 1895.)

attenuated the increase in heart rate (to 278 beats/min), and left the ventilatory response unaffected (1.8 L/min). This also occurred when the muscles were stimulated through the distal end of the cut spinal cord rather than transcutaneously to generate the tetanic contraction. In support of the latter observation, pinching the nose or the limbs of a dog did not affect either heart rate or blood pressure. Only if, in addition, the aorta and the vena cava were clamped were the increases in heart rate and ventilation to hind limb stimulation eliminated. Both variables increased as the clamps were removed and consequent reduction in blood pressure occurred. This implicated a role for hormonal influences in the control of the cardiovascular and ventilatory responses to muscle contraction.

Johansson (1893, 1895) also considered the role of ventilation on the heart rate response to exercise and showed that it was unaffected by maintaining the animal on a respirator. In addition, the fact that the blood pressure, but not the heart rate, response to electric stimulation was eliminated by severing the spinal cord at L2 to L5 argued against a coordinated cardiorespiratory response. This argument was further developed by showing that after section of the spinal cord at T1–T12, electric stimulation of the hind limb still caused blood pressure to increase, but heart rate fell. Furthermore, following electric stimulation, the increase in heart rate took longer (> 0.5 min) than expected by a reflex. However, with active or passive movement, it occurred within the first second (see figure 2). This response indicated the likelihood of blood-borne factors associated with contraction exerting control of blood pressure.

The role of the sympathetic and parasympathetic nervous system for the increase in heart rate during exercise was evaluated after the vagus and the stellate ganglia were eliminated in the dog with L1 or L5 spinal cord section. A small increase in heart rate still occurred. However, the response was too small and too late to account for the normal heart rate response to exercise. This added further support for the idea that the blood leaving the working muscles generates part of the exercise response.

Mansfeld (1910) tried to identify a substance in the blood released from the working muscle that increases heart rate. While the increase in heart rate could not be related to lactate, carbon dioxide, or blood obtained from another previously stimulated dog, blood temperature was of significance. Blood at the level of the heart increased 0.4–0.5 °C upon stimulation of the hind limb and was associated with a 23% to 90% increase in heart rate. Conversely, when the increase in blood temperature was restricted to 0.2–0.3 °C by cooling the vena cava, the heart rate response was reduced to 0.4%. In fact, intravenous infusion of heated saline gave rise to a 4–6 °C increase in blood temperature, and heart rate increased by 25–29%.

Hering (1895) conducted similar experiments in the rabbit in Prague. His interest was mainly about the effect of the vagus nerve on the heart. In the initial stage of the study, he described the reduction of heart rate in the playing dead reaction (from 237 to 157 beats/min) and demonstrated that it was vagally mediated. He also described marked bradycardia with inflation of smog in the nostrils, which also depended on the vagus. *Bewegung* (movement) was associated with an increase in heart rate from 205 to 324 beats/min and was similar to that obtained after vagotomy (321 beats/min). Also, running was not associated with any further increase in heart rate after the vagotomy. Yet, this lack of an increase in heart rate was explained by the higher initial value and also by the fact that the respiratory frequency was low (26 breaths/min versus 132 breaths/min). This respiratory frequency did not increase much during exercise (7 breaths/min), while the normal increase was about 51 breaths/min. This occurred because lung inflation itself increased the heart rate, e.g., from 220 to 277 beats/min, as did an increase in ventilatory frequency after curarization.

Elimination of the *beschleunigenden* (accelerating) heart nerves increased heart rate from 243 to 334 beats/min upon movement and to a similar value after an additional vagotomy when the resting heart rate was lower (269 beats/min). These observations led Hering (1895) to the conclusion that the increase in heart rate associated with exercise depends on the integrity of the sympathetic nervous system and that the vagus plays a supporting role through the effect from the lungs.

Human Studies

In 1869, Fraser used the Marey sphygmograph to record the pulsations of the radial artery before and after rowing. Before rowing, heart rate was 68 beats/min. It increased to 190 beats/min afterward. Also, the pulse wave was enlarged, and influences from the respiration disturbed the baseline. In both situations, Fraser noticed a clear dicrotic wave.

Athanasiu and Carvallo (1898) obtained similar observations in pulsatility by using plethysmography. In addition, they demonstrated a reduction of the forearm volume during static contraction of the fingers as blood was squeezed out of the muscle. A marked increase in the forearm volume was noted after the contraction and during rhythmic flexion of the finger. They also noted a small increase in the volume of the opposite resting arm followed by a decrease during the continued exertion and especially so after exercise.

A very important observation was made by Perthes (1895) and by Beecher, Field, and Krogh (1936) while working in August Krogh's laboratory. They described the fall in venous pressure at the ankle during exercise in the upright posture. Smirk (1936) later determined the decrease in venous pressure at the ankle to be about 60 mmHg. These observations pointed to the importance of translocation of blood from the lower limbs to the central circulation beds in contributing to the rapid increase in cardiac output at the onset of upright exercise.

Cortical Irradiation

In 1908, Aulo added pertinent information for interpreting the data of Johansson (1895). In a young male, he found the increase in heart rate during exercise was not developed by passive movement of the legs and that it increased only upon painful stimuli. On the other hand, a close relationship existed between the recovery heart rate and body temperature. He agreed with Johansson (1895) that an influence from the central nervous system dominated the rapid heart rate response to exercise. Aulo (1908) then introduced the term *cortical irradiation* (*Irradiation des motorischen Impulses nach den Zentren der Herznerven bedingst ist*) for this hypothesis.

In a follow-up study, Aulo (1911) focused on the influence of temperature on heart rate. At the onset of exercise, an increase in heart rate occurred despite a decrease in rectal temperature, which confirmed that another mechanism controlled the rapid changes in heart rate. Further evaluation demonstrated that a shortening of diastole caused exercise tachycardia. The duration of systole remained relatively constant and the duration of diastole recovered at the cessation of exercise. Since the vagus had been demonstrated to influence the duration of diastole more than systole while sympathetic activation influences both phases to an equal extent, Aulo (1911) concluded that vagal withdrawal dominated exercise tachycardia.

The more well-known studies by Krogh and Lindhard (1913, 1917) added significantly to our understanding of cardiovascular and ventilatory regulation. They studied the initial phases of ventilation and heart rate during exercise. In the first study, they were not able to record the ECG, so the heart rate was obtained in parallel experiments by Miss Buchanan in Oxford (see figure 3). They were impressed by the rapid increase in heart rate at the onset of exercise and even by the anticipatory response. They agreed with Johansson (1895) that irradiation of impulses from the motor cortex was the most probable explanation.

Krogh and Lindhard (1917) used the Bergonie apparatus to induce leg exercise by electric stimulation. A significant new finding showed that during evoked exercise, the increase in heart rate was delayed for one to several heart beats, and in one case, it decreased from 67 to 65 beats/min. Furthermore,

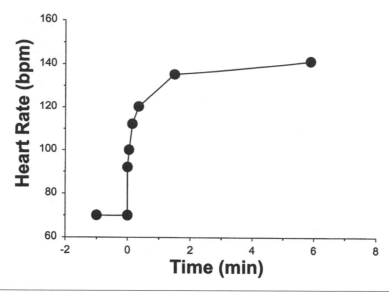

Figure 3   Heart rate response to cycling. (Data by Miss Buchanan from Krogh and Lindhard 1913.)

they established that during exercise, heart rate increases in proportion to oxygen uptake. Although the relation was significantly influenced by training, it was similar during voluntary and evoked exercise. They also demonstrated a direct relationship between cardiac output and oxygen uptake.

The Muscle Pressor Reflex

Alam and Smirk (1937, 1938a) published three papers in which they established the importance of neural input from exercising muscles to the increase in blood pressure and heart rate that occurs during dynamic exercise in humans. They also distinguished the effect of reflexes from muscle from the influence of voluntary effort. They used the simplest of experimental methods: a sphygmomanometer with which to measure blood pressure; a stopwatch to measure heart rate; home-made inflatable cuffs for arresting the circulation; and techniques for standardizing dynamic exercise of the forearm flexors and calf muscles. Most of the subjects studied were healthy volunteers. However, they also examined a patient with a curious spinal cord lesion that produced sensory paralysis.

At that time, researchers supposed that the hemodynamic changes of exercise could be accounted for by mental effort, by circulating metabolites, by an increase in circulating epinephrine, and/or by mechanical effects of increased venous return. Alam and Smirk (1937) showed that blood pressure increased during dynamic exercise of the calf or forearm flexor muscles, especially if the frequency of contraction was high. However, they also showed

that it increased much more if the exercise was done with occlusion of the limb circulation. Blood pressure remained elevated after exercise if circulatory arrest was maintained, but it fell dramatically as soon as the inflatable cuff was released (see figure 4). From this they concluded that the elevation of blood pressure was caused by a reflex, the afferent signal of which was transmitted by muscle afferent nerves stimulated by metabolites trapped in the muscles. They noted that the hypertensive effect of ischemic forearm exercise was comparable with that of ischemic exercise of the bulkier calf muscles. They also saw that even ischemic exercise of the flexors of the little finger caused a marked rise in blood pressure. In the ischemic forearm, they found that the rise in blood pressure was in proportion to the work done. They attributed the transient decline of blood pressure that occurred when exercise ceased (but with circulatory occlusion maintained) to the removal of mental effort. Neither pain nor the increase in peripheral resistance produced by reducing vascular area by cuff inflation were likely to account for the hypertensive effect of ischemic exercise.

They also addressed the question of the heart rate-raising effect of dynamic exercise during conditions of circulatory arrest (Alam and Smirk 1938a). While using the same experimental methods, they concluded that this, too, was accounted for chiefly by a muscle reflex. However, whereas ischemic exercise of even small muscle groups raised blood pressure, the heart rate-raising reflex seemed to depend on the bulk of the exercising muscle. As with blood pressure, when the circulation to the legs was arrested just prior to the end of exercise, heart

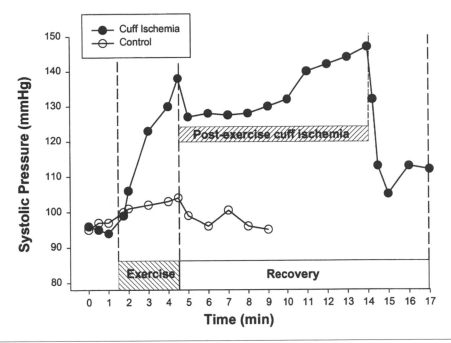

Figure 4 Contribution of a muscle reflex to the blood pressure-raising effect of dynamic exercise in humans. Exercise: A load of 500 g was lifted by the forearm flexors 180 times over 3 min. In one group, a period of circulatory arrest was created using an inflatable cuff on the upper arm (filled circles). Effects of the same exercise without circulatory arrest (open circles). (From Alam and Smirk 1937, with permission.)

rate remained elevated as long as the occlusion cuffs were maintained. However, an elevation of heart rate was not maintained when the same experiment was employed in the arms. While they proposed a muscle-mass effect for the reflex elevation of heart rate, they did not eliminate the possibility that lower-limb exercise was more potent in raising heart rate than upper-limb exercise.

Finally, they studied a subject with a spinal cord lesion in whom motor power in both legs was normal but sensation was absent below the knee in one leg (Alam and Smirk 1938b). During ischemic exercise of either calf, blood pressure and heart rate rose. However, in the case of the anesthetic leg, blood pressure and heart rate fell to normal levels when exercise ceased but circulatory arrest continued. They used this observation to reinforce their inference about the role of muscle afferent nerves in the postexercise cardiovascular changes. They explained the rise of blood pressure and heart rate during exercise of the anesthetic leg as being due to mental effort.

Thus, researchers defined the role of central command versus that of a reflex from the active muscle in regulation of the circulation during exercise before the Second World War. The numerous studies in both animals and humans since then have been devoted to defining the role of one or both of these influences during static and dynamic exercise and

their contribution when the mass of the engaged muscles is varied. Still, the contribution of cortical irradiation or central command relies on a negative hypothesis, i.e., that a given response is unlikely to be elicited by a reflex mechanism. Equally, researchers do not know what factor(s) in the muscle control blood flow or activation of the sympathetic nervous system during exercise.

The State of Integrative Cardiovascular Physiology

The contents of this book testify that the study of human exercise is integrative physiology par excellence and that it is flourishing. The following chapters present a comprehensive view of our current understanding of cardiovascular regulation during exercise in healthy adults as well as changes associated with training and disease. It is hard to imagine a physiological system that is not affected by or that does not affect muscular exercise. Volitional drive (mental effort, central command), together with neural input from exercising muscles and baroreflexes, initiates and maintains a pattern of muscular exercise designed to meet the goal of the exerciser. This immediately creates the problem of matching the abrupt increase in muscle O_2 consumption by a comparable increase in muscle blood flow. Just what local mechanisms account for the close coupling of blood flow and O_2 demand in an exercising muscle is not yet entirely clear.

One thing is certain, though. The increase in muscle blood flow requires a rapid increase in cardiac output. Sedentary office workers and highly trained athletes achieve this by somewhat different mechanisms. Neural mechanisms—the switching off of vagal and switching on of sympathetic nervous activity so that heart rate and ventricular contractility and thus cardiac output increase—are important in both. The trained athlete starts with the advantage of a low heart rate and near-maximal stroke volume, so these rapid, neural mechanisms are especially important. The untrained subject has to rely more on the slower process of increasing stroke volume. We are only just beginning to understand how training can remodel the left ventricle to provide for a large resting stroke volume. We are also starting to learn about how much the athlete's low heart rate is due to increased cardiac vagal drive and how much due to a reduction in the intrinsic heart rate. Much more can still be learned about the effects of the age at which training commences and the influences of heredity.

Humans are not the fastest-running mammals, but we do have one advantage over those with four legs. We have large, deep venous sinuses in our calves, and the muscles that surround them constitute a specialized musculovenous pump. At the onset of exercise, this delivers a bolus of blood to the right atrium, increasing cardiac preload. During the first few minutes of exercise in the erect or semierect posture, this pump also increases perfusion pressure in the muscles of the leg by increasing the arteriovenous pressure gradient.

What if cardiac output cannot match the vasodilatation in the exercising muscles? This question has taken a new twist with the recent discovery that maximal blood flow and conductance in an exercising muscle is much greater than had been thought. Therefore, if a substantial proportion of the body's muscle mass were to exercise, the demand for cardiac output could not possibly be satisfied. One solution would be to divert cardiac output away from nonexercising tissues in favor of the exercising muscles. This seems to occur in healthy humans but not in dogs at submaximal levels of exercise. To what extent it does so during periods of supramaximal exercise has not yet been adequately examined for technical reasons. It does occur in animals such as the rabbit that have a very small cardiac reserve. In them, the arterial baroreceptor reflex seems to play a key role in the diversion. In addition, redistribution seems to occur in humans with heart failure.

Another way of increasing O_2 delivery without increasing cardiac output would be to increase the O_2-carrying capacity of the blood. This strategy is employed in horses. In them, the hematocrit can be nearly doubled by the release of stored red cells from the spleen and other organs into the bloodstream. However, in humans and dogs, though splenic contraction can be detected, the red cell stores are so small that hematocrit rises only minimally. This does serve as a reminder, however, of the fascinating phylogenetic and comparative aspects of exercise physiology.

If steady-state exercise is to be maintained in the medium term, then the increased O_2 consumption by the exercising muscles must be matched by increased O_2 delivery. Important contributions are made by the increase in muscle blood flow and by increased extraction of O_2 from the arterial blood. However, this is not enough. Ventilation must, and does, increase to ensure full oxygenation of the returning venous blood. This, in turn, depends on an increased frequency and magnitude of neural drive from the respiratory oscillator in the brain stem. Researchers still debate the exact nature of the stimulus for this. Evidence shows a spillover of volitional drive to respiratory neurons. Other evidence indicates that a spillover of metabolites (especially K^+ ions) from the exercising muscles stimulates respiration. However, just how a precise balance is struck between ventilation and peripheral O_2 demand is not yet clear. Some evidence also shows that under extreme conditions of muscular exercise, O_2 transport from the atmosphere to the blood cannot keep up with the greatly increased pulmonary blood flow, that is, PO_2 falls. At high intensities of exercise, the O_2 demands of the respiratory muscles themselves also enter into the equation.

During long-term exercise, other problems arise. An important one is the great increase in metabolic heat production by the exercising muscles. The heat elimination mechanisms in humans are increased skin blood flow and sweating. These are linked, and the origins of the linkage are just beginning to be identified. Increased skin blood flow places additional demands on cardiac output. Increased sweating results in water and sodium loss that, if not replaced, leads to a contraction of the extracellular fluid space and blood volume, in particular. In extreme cases, these two consequences of heat stress can progress to the life-threatening condition of hyperthermic heat stroke. Is this last response merely a result of blood volume contraction and the competition of exercising muscles and the skin for cardiac output? Instead, do other, more sinister mechanisms come into play? There is just a hint, not yet substantiated, that redistribution of blood flow away from the splanchnic organs may allow bacterial endotoxins to enter the blood stream and trigger the cytokine cascade.

This brief and incomplete account of the integrative complexities of muscular exercise provides an overview of the fruitful integrative research performed over the last century and reviewed in this text. It also points to areas remaining to be explored.

Future Directions

A pervasive trend in human physiology and medical research is the application of molecular biology to the study of cellular and subcellular responses. To date, most of the related work in this area has focused on abnormalities of exercise performance rather than the normal and on pathological states rather than physiological. Considerable effort is being made to suggest how one or another cellular or subcellular mechanism might be responsible for the exercise intolerance that is part of common pathological conditions such as hypertension, heart failure, peripheral vascular disease, and diabetes mellitus.

Classical physiologists and those within many physiological societies express some concern about the threatened eclipse of integrative physiology by molecular and cellular biology (Boyd and Noble 1993; Folkow 1994; Jobe et al. 1994; Long Range Planning Committee 1990). The Nobel prize-winning pharmacologist, Sir James Black, is reported to have said when asked for his vision of the future of his discipline, "The progressive triumph of physiology over molecular biology" (Boyd and Noble 1993, ix). Those are harsh words.

Molecular biology unquestionably has a great deal to offer in the form of techniques with which the integrationists may be able to solve problems. What increases muscle mass during training? What increases capillary density under these circumstances? What is responsible for the cardiac hypertrophy and remodeling in the highly trained athlete? Exactly how do humans adapt to exercise at high altitude? What causes structural increases in the diameter of collateral arteries in the heart or skeletal muscle when the mainline arteries are blocked? What really causes intimal hyperplasia when atherosclerotic plaques are removed? Surely, answers to questions like these can be found only by employing molecular biology techniques. Thus, while the current emphasis in physiology is on understanding the molecular basis for function, the challenge for the future is for the integrationists to apply this knowledge to the whole organism. This will undoubtedly necessitate a deeper, more sophisticated integrative approach in order to incorporate knowledge of molecular processes to the understanding of systemic function. Perhaps this is what Sir James Black meant to say, and perhaps this is the message for physiologists of the future.

Acknowledgments

We wish to thank Frank Pott and Michael S. Kristensen for help in translation.

References

Alam, M., and F.H. Smirk. 1937. Observations in man upon a blood pressure raising reflex arising from the voluntary muscles. *Journal of Physiology London* 89:372–383.

Alam, M., and F.H. Smirk. 1938a. Observations in man on a pulse accelerating reflex from the voluntary muscles of the legs. *Journal of Physiology London* 92:167–177.

Alam, M., and F.H. Smirk. 1938b. Unilateral loss of a blood pressure raising, pulse accelerating, reflex from voluntary muscle due to a lesion of the spinal cord. *Clinical Science* 3:247–252.

Athanasiu, J., and J. Carvallo. 1898. Le travail musculaire et le rythme du coeur. *Archives de Physiologie* 30:347–362.

Aulo, T.A. 1908. Muskelarbeit und Pulsfrequenz: Untersuchungen am Menschen. *Skandinaviesches Archiv fiir Physiologie* 21:146–160.

Aulo, T.A. 1911. Weiteres iiber die Ursache der Herzbeschleunigung bei der Muskelarbeit. *Skandinaviesches Archiv fiir Physiologie* 25:347–360.

Beecher, H.K., M.E. Field, and A. Krogh. 1936. The effect of walking on the venous pressure at the ankle. *Skandinaviesches Archiv fiir Physiologie* 73:133–141.

Boyd, C.A.R., and D. Noble, eds. 1993. *The logic of life: A challenge of integrative physiology*. Oxford: Oxford University Press.

Folkow, B. 1994. Increasing importance of integrative physiology in the era of molecular biology. *News in Physiological Sciences* 9:93–95.

Fraser, T.R. 1869. The effects of rowing on the circulation, as shown by examination with sphygmograph. *Journal of Anatomy and Physiology* 3:127–130.

Hering, H.E. 1895. Ueber die Beziehung der extracardialen Herznerven zur Steigeruiig der

Herzschlagenzal bei Muskelthdtigkeit. *Archiv fiir die gesamte Physiologie* 60:429–492.

Jobe, P.C, L.E. Adams-Curtis, T.F. Burks, R.W. Fuller, C.C. Peck, R.R. Ruffolo, O.C. Snead, and R.L. Woosley. 1994. The essential role of integrative biomedical sciences in protecting and contributing to the health and well-being of our nation. *Physiologist* 37:79–84.

Johansson, J.E. 1893. Ueber die Einwirkung der Muskelthatigkeit auf die Athmung und Hertzhltigkeit. *Skandinaviesches Archiv fiir Physiologie* 5:20–66.

Johansson, J.E. 1895. Ueber die Einwirkung der Muskelthltigkeit auf die Athmung und die Hertzhiitigkeit. *Skandinaviesches Archiv fiir Physiologie* 5:20–66.

Krogh, A., and J. Lindhard. 1913. The regulation of respiration and circulation during the initial stages of muscular work. *Journal of Physiology London* 47:112–136.

Krogh, A., and J. Lindhard. 1917. A comparison between voluntary and electrically induced muscular work in man. *Journal of Physiology London* 51:182–201.

Long Range Planning Committee. 1990. What's past is prologue: A "white paper" on the future of physiology and the role of the American Physiological Society. *Physiologist* 33:161–180.

Mansfeld, G. 1910. Die Ursache der motorischen Acceleration des Herzens. *Archiv fiir die gesamte Physiologie* 134:598–626.

Perthes, G. 1895. Ueber die Operation der Unterschenkelvaricen nach Trendelenburg. *Deutsche Medizinische Wochenschrift* 21:253–257.

Smirk, F.H. 1936. Observations on the causes of oedema in congestive heart failure. *Clinical Science* 2:317–335.

Zuntz, N., and J. Geppert. 1886. Ueber die Natur der normalen Atemreize und den Ortihrer Wirkung. *Archiv fiir die gesamte Physiologie* 38:337–338.

Zuntz, N., and J. Geppert. 1888. Ueber die Regulation der Athmung. *Archiv fiir die gesamte Physiologie* 43:189–245.

PART I

Cardiovascular Control Mechanisms During Exercise

Chapter 1

Baroreceptor-Mediated Reflex Regulation of Blood Pressure During Exercise

Peter B. Raven, Jeffrey T. Potts, Xiangrong Shi, and James Pawelczyk

In 1863, Marey described in his book *The Physiology of the Circulation of the Blood* the inverse relationship that exists between heart rate and arterial pressure. Subsequently, in 1900, Pagano and Siciliano, two Italian physiologists, demonstrated that the pressure response to carotid artery occlusion depends upon the integrity of the nervous structures located in the carotid sinus and does not result from cerebral ischemia, as previously thought. However, it was not until after Hering (1927) and Koch and Meis (1929) described the afferent neural connections emanating from the carotid sinus bodies, which travel via the glossopharyngeal nerve to the central nervous system, and related these structures to the reflex nature of blood pressure regulation that their true function was enunciated. Subsequently, Heymans and Neil (1958) described the underlying mechanisms involved in the reflex control of the cardiovascular system at rest.

Over the next 70 years, the basic components of the baroreceptor reflex arc were identified. They include the sensory elements (receptors), the afferent nerves that travel to the brain stem, the central integrating nuclei, the sympathetic and parasympathetic efferent motor fibers, and the targeted organs (i.e., heart and blood vessels). There are two populations of arterial baroreceptors: the carotid sinus receptors and the aortic arch receptors. Both baroreceptor populations are implicated in the short-term control of systemic blood pressure. Carotid receptors are embedded in the wall of the internal carotid artery bilaterally (Sheehan, Mulholland, and Safiroff 1941). The Hering nerve, a branch of the glossopharyngeal nerve, carries impulses from these receptors centrally. The aortic arch receptors are found in the general vicinity of the aortic arch. However, these receptors have also been shown to be located at the origin of the right subclavian artery and in adjacent regions of the brachiocephalic artery in cats and dogs (Aumonier 1972; Boss and Green 1956). From the aortic arch, the afferent signals are carried centrally in small vagal branches that unite to form the aortic depressor nerves (Heymans and Neil 1958). It is not clear whether similar neural connections exist in humans.

In addition, differences have been reported in the thickness of the vascular smooth muscle content between the carotid sinus and the aortic arch. The muscle content of the aortic arch is greater than that of the carotid sinus. These differences in muscle content explain the greater compliance of the carotid sinus wall compared with that of the aortic arch wall (Rees and Jepson 1970). Finally, the carotid sinus apparently has two types of afferent fibers: large diameter A-fibers or Type I, and smaller A-fibers and nonmyelinated C-fibers or Type II. Type I and Type II carotid baroreceptor fibers have distinct firing patterns that enable the classification of the Type I fibers to be fast and the Type II fibers to be slow. In addition, the fast firing Type I fibers appear related to transient changes in dP/dt or pulse pressure, while the slow firing Type II fibers appear to be associated with changes in mean arterial pressure (Seagard, Gallenberg, et al. 1992; Seagard, Hopp, et al. 1992). Whether a similar distinction of afferent fiber types exists for the aortic baroreceptors has not been determined. However, research has shown that the aortic baroreceptor is insensitive to arterial pressure pulsation (Angell-James and Daly 1970).

In contrast with the aforementioned arterial baroreceptors, the cardiopulmonary receptors represent a diffuse population of mechanically sensitive receptors. They are localized in all four chambers of the heart, especially at the junctions of the great veins with the right atrium and the pulmonary veins with the left atrium (Coleridge et al. 1957; Miller and Kasahara 1964). These receptors are unencapsulated nerve endings (Nonidez 1941) subserved by myelinated vagal afferents. They carry signals centrally through small vagal branches, which join the main vagal trunk.

The first synapse for these baroreceptor afferents carried by the glossopharyngeal and vagal fibers lies within the nucleus of the tractus solitarius (NTS) in the medulla. Numerous techniques have been used to identify this site. First, baroreceptor denervation has resulted in degeneration of cells within the NTS (Cottle 1964). Second, nerve recordings from the NTS show pulse-synchronous activity similar to that seen in baroreceptor afferents (Koepchen et al. 1967). Third, lesions in this region produce lability of blood pressure control (Reis et al. 1977) similar to that produced following acute surgical removal of arterial baroreceptors. In addition, the NTS has extensive connections to higher centers of the nervous system, particularly the hypothalamus (Barman and Gebber 1982). The efferent outputs from the NTS include sympathetic preganglionic neurons in the intermediolateral cell columns of the thoracic spinal cord and in the nucleus ambiguus and dorsal motor nucleus of the vagus (Spyer 1981).

The overall performance of the arterial baroreflexes encompasses the ability to respond to an input stimulus (i.e., change in vascular distending pressure in the baroreceptor region) and alter systemic blood pressure (MAP) in the direction opposite to the input stimulus through changes in multiple reflex effector activities (Sagawa 1983). Arterial baroreflexes function in response to changes in arterial blood pressure (ABP) by affecting three physiological variables: heart rate (HR), stroke volume (SV), and total peripheral resistance (TPR). These relationships are summarized by the classic equation:

$$ABP = HR \times SV \times TPR$$

Consequently, techniques used to assess baroreflex function need to account for changes in each of the variables. Furthermore, although it is generally accepted that the carotid and aortic baroreceptors perform similar functions in providing efferent responses to changes in blood pressure, several factors have been established. First, the operating range of the aortic baroreceptor occurs over higher arterial pressures than that of the carotid baroreceptor (Sagawa 1983). Second, aortic baroreceptor reflex accounts for two-thirds and the carotid baroreceptor reflex accounts for one-third of the complete arterial reflex response (Ferguson, Abboud, and Mark 1985; Shi, Crandall, et al. 1993; Shi et al. 1995). Third, sustained selective stimulation of the carotid baroreceptor results in an attenuated reflex response because the aortic baroreceptor reflex opposes the carotid baroreceptor reflex response. Arterial baroreflex function and cardiopulmonary baroreflex function have been extensively studied at rest and during various perturbations of hypotension and hypertension (Secher,

Pawelczyk, and Ludbrook 1994). In this chapter, we will focus our attention on measuring arterial baroreflex function during exercise.

Quantification of Baroreceptor Reflex Function

The effect of exercise on the sensitivity (maximal gain) of the baroreceptor reflex has puzzled investigators for a number of years (see table 1.1). However, a large portion of this confusion is attributed to the methods used to quantify the operational characteristics of the reflex. In this section, we have limited our discussion to methods used to analyze the responses obtained from carotid baroreceptor stimulation in humans.

Probably the most utilized technique to characterize human carotid baroreceptor reflex function is the use of the neck collar developed by Ernsting and Parry in 1957. Since that time, the neck collar technique has been improved and perfected. Eckberg (Eckberg et al. 1975; Eckberg and Eckberg 1982; Sprenkle et al. 1986) developed the use of rapid, pulse-synchronous changes in neck pressure and suction, ranging between +40 torr neck pressure to –65 torr neck suction, to obtain an open-loop profile of carotid sinus baroreflex function at rest. The advantage of rapid changes in neck pressure and suction delivered to the surface of the neck is that the changes in carotid sinus transmural pressure closely mimic the physiological changes in carotid sinus pressure (Eckberg et al. 1975; Eckberg 1976; Eckberg and Eckberg 1982). However, asymmetry in the transmission of neck chamber pressure through the perivascular tissues overlying the carotid sinus region has been reported (Ludbrook 1983; Ludbrook et al. 1977a, 1977b).

The heart rate response following a change in neck chamber pressure immediately occurs by the first or second heartbeat. Therefore, the change in heart rate is generally believed to be unaffected by extracarotid baroreceptor populations. Also, the rapid change in heart rate is the primary function of the carotid-cardiac reflex change in cardiac output because stroke volume is unaffected (Levine et al. 1990). In addition, vasomotor changes of resistance are delayed to a time when the heart rate has returned to the prestimulus value (Potts and Raven 1995; Potts et al. 1998a, 1998b). The rapidity of heart rate responses to baroreceptor stimuli depends upon the short latency period of the carotid baroreceptor reflex and the baseline heart rate. Human carotid-cardiac baroreflex latency has been measured by Eckberg to average approximately 240 ms (Eckberg 1976), although values have been reported to vary

Table 1.1   Select References on Baroreflex Function During Exercise

Authors	Year	Model	Technique	Exercise	Baroflex Control	
					Heart rate	Blood pressure
Bevegard and Shephard	1966	Human	Neck collar	Dynamic	=	=
Robinson et al.	1966	Human	Pharm. block	Dynamic	=	NA
Bristow et al.	1971	Human	Pharm. agonist	Dynamic	↓	NA
Pickering et al.	1972	Human	Pharm. block	Dynamic	↓	NA
Cunningham et al.	1972	Human	Pharm. agonist	Static v. dynamic	↓	NA
Mancia et al.	1978	Human	Neck collar	Static	↓	NA
Ludbrook et al.	1978a	Human	Neck collar	Static	↓	=
Ludbrook et al.	1978b	Human	Isolated carot. sinus	Static	↓	NA
Melcher and Donald	1980	Dog	Isolated carot. sinus	Dynamic	=	=
Walgenback et al.	1983-84	Dog	Isolated carotid/aortic/CP	Dynamic	Reset	Reset
Daskalapoulos et al.	1984	Dog	Isolated carotid sinus/CP	Dynamic	NA	NA
Ebert	1983	Human	Neck collar	Static	=	=
Staessen et al.	1987	Human	Neck collar	Dynamic	↓	↓
Scher et al.	1991	Human	Pharm./MSNA	Static	Reset	Reset
Strange et al.	1990	Human	Neck suction	Dynamic	NA	Submax on, max off
DiCarlo and Bishop	1992	Rabbit	Pharm. depress.	Dynamic	Reset	Reset
Eiken et al.	1992	Human	Neck collar	Dynamic	↑	NA
Potts et al.	1993	Human	Neck collar	Dynamic	Reset	Reset
Papeller et al.	1994	Human	Neck collar	Dynamic	Reset	Reset
Potts and Mitchell	1998	Dog	Isolated carot. sinus	Static	Reset	Reset

= unchanged from rest to exercise; ↓ decrease gain from rest to exercise; ↑ increase gain from rest to exercise; NA Not measured.

between 200 and 500 ms (Sprenkle et al. 1986). In contrast, since the latency of the sympathetic branch of the baroreflex is one or two orders of magnitude larger (Rowell 1991), the reflex changes in blood pressure must be appropriately offset to account for the longer latency period needed to align the blood pressure response with the corresponding neck chamber pressure.

The second factor affecting how quickly heart rate can respond is the baseline heart rate (or R-R interval). Heart rate will respond to a change in neck chamber pressure within the first or second beat whenever the baseline heart rate (R-R interval) is significantly greater than the neural latency of the baroreflex arc. For example, at a heart rate of 60 beats/min, or an R-R interval of 1000 ms, heart rate will change by the next beat because there is sufficient time within the reflex arc to transduce a change in carotid transmural pressure and alter cardiac

parasympathetic efferent activity. However, when baseline R-R interval shortens and approaches a value similar to the neural latency of the baroreflex arc, as occurs during light-to-moderate exercise, heart rate is unable to change within the immediate heartbeat following a change in neck chamber pressure (figure 1.1).

This figure shows the heart rate responses obtained using the typical pulse-synchronous train of changes in neck pressure from a group of volunteer subjects at rest and during mild dynamic exercise at 75 watts (Potts and Raven 1995; Potts et al. 1998a, 1998b). Figure 1.1a shows the data typically plotted as a function of the estimated carotid sinus pressure. Two points should be stressed here. First, the tachycardia accompanying exercise is readily apparent by the upward shift of the heart rate stimulus-response curve during exercise. Second, the baroreflex threshold (i.e., the *minimal* input stimu-

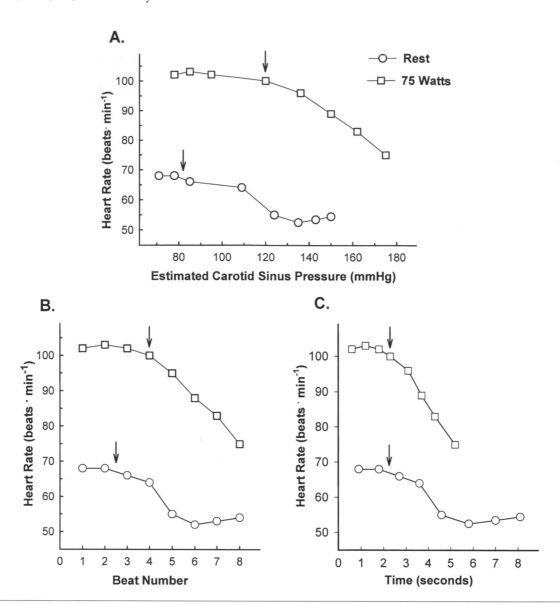

Figure 1.1   Data showing carotid baroreceptor reflex responses obtained from pulse-synchronous changes in neck pressure and suction from +40 torr to –65 torr in subjects during seated rest and upright cycling on an ergometer at 75 watts. *(a)* The baroreflex change in heart rate as a function of the estimated carotid sinus pressure. The same data plotted as a function of *(b)* beat number and *(c)* time. 1 = baroreflex threshold. (Adapted from unpublished personal data from Potts and Raven.)

lus necessary to evoke a change in heart rate) appears to have increased (shifted to the right) during exercise. However, the shift in baroreflex threshold can be completely explained by the method used to obtain the data in this experiment. The pulse-synchronous changes in neck chamber pressure used in this study are triggered by a discrete event in time (i.e., R wave of the EKG). Therefore, during exercise when the heart rate approaches 120 beats/min (R-R interval of 500 ms), the interbeat interval begins to approach the intrinsic latency of the baroreflex arc. As a result, the heart rate response begins to lag behind the sequence of neck chamber pressures. The outcome is the apparent increase in the thresh-

old pressure of the stimulus-response relationship. However, this is actually an artifact because the baseline interbeat interval has encroached on the neural latency of the baroreflex arc.

Figure 1.1b illustrates that during exercise, the reflex bradycardia does not occur until the fifth or sixth beat. At rest, however, the bradycardia begins as early as the second or third beat. Therefore, at higher heart rates (shorter R-R interval), a greater number of beats will occur during the train of neck chamber pressures (due to a shortened R-R interval), leading to an apparent delay in the reflex bradycardia. However, when the data are expressed as a function of the elapsed time per beat (see figure

1.1c) the onset of the bradycardia at rest and during exercise occurs at virtually the same time (~2 s).

What should be apparent from this discussion is that whenever the baseline interbeat interval approaches the neural latency of the reflex arc, the stimulus-response relationship for the carotid-cardiac baroreflex obtained from pulse-synchronous changes in neck pressure and suction will be incomplete during exercise. Therefore, a clear interpretation of the results obtained in this fashion will be very difficult. This has important ramifications for the use of the pulse-synchronous change in neck pressure protocol during exercise, especially when heart rate approaches or exceeds 120 beats/min. Under these conditions, the reflex cannot produce an immediate change in heart rate because the exercise heart rate is equal to, or shorter than, the latency of the neural arc. Furthermore, this situation is exacerbated as exercise intensity (and exercise heart rate) increase and the exercise interbeat interval shortens. In addition, many investigators have discussed the effect of baseline heart rate (R-R interval) on the calculated reflex changes in heart rate. They have noted that by reason of the mathematical relationship between heart rate and R-R intervals, comparing the reflex function between two different baseline heart rates is inappropriate.

In order to obtain an estimate of the open-loop characteristics of the carotid baroreflex when the baseline heart rate is reduced below the neural latency of the reflex arc, an alternative approach must be adopted. Potts, Shi, and Raven (1993) developed a method that uses brief (5 s) periods of neck pressure and suction (+40 torr to –80 torr) to stimulate carotid baroreceptors. They used the immediate reflex responses in heart rate and blood pressure to construct the baroreflex stimulus-response curves. In this procedure, the subject performed a breath-hold at the end-expired position. The breath-hold maneuver effectively minimized the possibility that respiration caused the recorded changes in heart rate and blood pressure. After the breath-hold was commenced and a stable heart rate was obtained, a discrete level of neck pressure or suction was precisely delivered 50 ms after the R wave and maintained for 5 s. The subject continued the breath-hold maneuver during a short recovery period (4–5 s) following the end of the baroreceptor stimulus so that the vascular response of the baroreflex (i.e., arterial blood pressure) was obtained. Under these conditions, we feel that the peak reflex change in heart rate and blood pressure elicited by the baroreceptor stimulus could be ascribed to a change in carotid baroreceptor afferent nerve activity. After multiple levels of neck pressure or suction had been randomly delivered, the reflex heart rate and blood pressure responses to each

stimuli were plotted as a function of the estimated change in carotid sinus transmural pressure to construct the open-loop stimulus-response curve for the carotid sinus baroreflex (Potts, Shi, and Raven 1993).

Using this approach to study carotid baroreflex function has several advantages. First, the heart rate response elicited by the baroreflex is extremely reproducible. Second, due to the brief nature of the stimulus and the short latency for the carotid-cardiac baroreflex, extracarotid (aortic or cardiopulmonary) baroreceptor reflexes do not affect reflex changes in heart rate. Third, since this approach plots the peak reflex response of heart rate and arterial pressure, which can occur at any time throughout the stimulation protocol, the baseline heart rate will not have an effect on this response. Fourth, as a breath-hold is performed, the reflex responses are not affected by respiration. However, this approach imposes several limitations. Due to the short 5-s period of neck pressure/suction, the absolute magnitude of the reflex change in heart rate and arterial pressure are likely attenuated. This will lead to an underestimation of the absolute baroreflex gain. In addition, there will be an exercise intensity beyond which the subject is unable to perform the breath-hold maneuver comfortably and thereby elicit chemo- and somato-pressor reflex responses.

Previous studies examining the effect of exercise on maximal baroreflex sensitivity utilized sustained levels of neck chamber pressure (between 20 s to 3 min) to activate carotid baroreceptors (Bevegard and Shepherd 1966; Papelier et al. 1994; Staessen et al. 1987; Stegemann, Busert, and Brock 1974). However, a continuous period of neck chamber pressure of sufficient magnitude to produce a reflex change in blood pressure will affect the gain of the carotid baroreflex by indirectly stimulating the aortic or cardiopulmonary baroreceptors. Regardless, the perturbation of a specific baroreceptor population, as presumed when using neck pressure and suction, may not always be the most appropriate experimental approach to use. However, with the immense complexity of the neural-humoral cardiovascular axis, studying and characterizing a single component of this large and complex system discretely is often desirable. The challenge then facing the investigator will be to determine how this one component functions when the system functions as a whole.

Role of the Arterial Baroreceptor Reflex

In this section, we will review the past and present work describing arterial baroreflex control of circulation during dynamic exercise. Those interested in the role of central command or the exercise pressor reflex in regulating sympathetic and parasympathetic efferent neural activity to the heart and peripheral blood vessels during exercise are referred to several excellent review articles that discuss these mechanisms in detail (Mitchell 1990; Rowell and O'Leary 1990) and to the proceeding chapters.

Animal Studies

Studies of the conscious dog have demonstrated that afferent information emanating from the aortic, carotid, and cardiopulmonary baroreceptors function to inhibit the vasomotor centers tonically to *minimum* blood pressure lability at rest. During exercise, however, only the arterial baroreflexes appear to function in this capacity. Melcher and Donald (1981) examined how the vascularly isolated carotid sinus baroreceptors regulate arterial pressure and heart rate during rest and treadmill exercise in the instrumented dog with intact aortic and cardiopulmonary baroreceptors. When the carotid sinuses were surgically isolated and perfused at pressures independent of the systemic circulation, they found an upward shift of the stimulus-response curve in heart rate and arterial pressure to carotid pressure. This was accompanied by no change in the functional characteristics of the reflex, i.e., maximal gain and carotid sinus pressure at maximal gain.

To investigate further the relative contribution of the carotid sinus baroreflex in cardiovascular control during exercise, they repeated their experiments following acute bilateral vagotomy to interrupt the buffering effects from the aortic and cardiopulmonary baroreceptors. Denervation caused an increase in carotid baroreflex gain and arterial pressure at rest. During exercise, however, maximal gain was not affected. When sinus pressure was altered during exercise, the reflex changes in heart rate and arterial pressure were still evident. Therefore, they concluded that the carotid baroreflex retained the ability to buffer disturbances in systemic pressure during exercise as effectively as at rest even in the absence of vagally mediated information from the aortic and cardiopulmonary baroreceptors.

Careful inspection of their data showed that although bilateral vagotomy acutely elevated resting arterial blood pressure, during the transition of rest to exercise the blood pressure was actually lower than the resting value. Furthermore, the arterial pressure stimulus-response curve was shifted downward and to the left following vagotomy, indicating that the carotid baroreflex was operating at a lower pressure range. This is opposite to their findings for the carotid baroreflex-blood pressure curve when the vagi were intact. Therefore, although not discussed by the authors, their results apparently show that the loss of vagal afferent input contributed to the downward shift in the arterial pressure stimulus-response curve during exercise. This suggests an important modulatory role by either cardiopulmonary or aortic baroreceptor afferents in altering carotid baroreflex function.

Stephenson and Donald (1980) developed the approach of surgically isolating the carotid sinus regions in the conscious dog. In order to obtain the open-loop stimulus-response relationships for the carotid baroreflex, they used 1–2 min of static pressure until achieving a steady-state response. Our approach of stimulating and quantifying the carotid baroreflex in humans (Potts, Shi, and Raven 1993) differs dramatically from the approach used by Melcher and Donald (1981). To determine if the mathematical model we applied to our data biases the results obtained from this type of experiment, we applied the logistic model of Kent et al. (1973) to data from Melcher and Donald's (1981) study.

As shown in figure 1.2, we reanalyzed Melcher and Donald's (1981) heart rate responses obtained at rest and during treadmill exercise at mild (5.5 km/h, 7% grade) and moderately severe (5.5 km/h, 21% grade) exercise. We found that although the absolute value for maximal gain at 7% grade was not the same as that reported by Melcher and Donald (1981) (–0.4 vs. –0.5 beats/min/mmHg), the gain values from the logistic model were very similar at rest and during exercise (–0.39, –0.409, –0.45 beats/min/mmHg). Therefore, except for the possible differences in the absolute value of baroreflex gain, the logistic model parameters can be applied to baroreflex responses obtained from at least two separate experimental approaches.

In a classic series of studies, Walgenbach and Donald (1983) examined the ability of the carotid baroreceptors to regulate blood pressure during graded exercise in the chronic absence of the aortic baroreflex in dogs. When the sinuses were vascularly isolated and unable to respond to the exercise-induced rise in arterial pressure, they found an exaggerated blood pressure response. However, the response of heart rate and cardiac output appeared to be similar to those produced when the carotid sinuses were permitted to respond to the changes in exercise blood pressure. They suggested that the exaggerated pressor response was produced by a powerful vasoconstrictor outflow that caused an

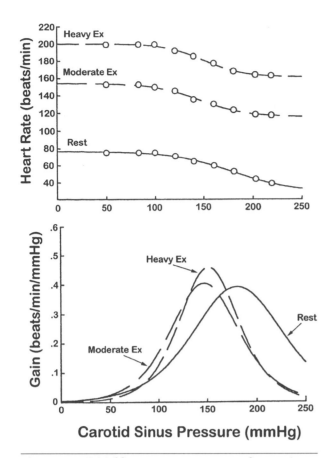

Figure 1.2 Initial heart rate responses to changes in carotid sinus pressure from four dogs at rest during mild (5.5 km/h, 7% grade) and moderately severe exercise (5.5 km/h, 21% grade). These were mathematically modeled with a logistic function described by Kent et al. (1973). The lower panel is the instantaneous rate of change of heart rate with respect to carotid sinus pressure derived from the first derivative of the logistic function. (Data modified from Melcher and Donald 1981 with permission of the American Physiological Society.)

increase in peripheral vascular resistance. In the absence of arterial baroreceptors, no mechanism senses and limits the rise in arterial pressure. Therefore, the increase in sympathetic outflow and the rise in blood pressure were exaggerated.

From these findings, the carotid sinus baroreceptor reflex may be thought of as a system designed to (1) fine-tune the effects of metabolic vasodilation in active skeletal muscle, and (2) increase sympathetic outflow to nonactive vascular beds. The advantage of such a system may be that it could operate as both an accelerator or a brake of vasoconstrictor outflow by permitting or limiting the rise in arterial pressure depending on the metabolic state of the exercising skeletal muscle. For the baroreflex to modulate sympathetic vasoconstriction, it must first be reset to a

higher arterial pressure. Strong, although indirect, evidence from animals shows that arterial baroreflexes continue to modulate sympathetic nerve activity and arterial blood pressure responses during dynamic exercise (DiCarlo and Bishop 1992; DiCarlo, Chen, and Collins 1996; DiCarlo, Stahl, and Bishop 1997; Scherrer et al. 1990; Sheriff et al. 1990). These findings support the idea that the arterial baroreflex resets during exercise.

In addition to limiting the increase in arterial pressure during exercise, resetting of the baroreflex may also facilitate the initial changes in autonomic outflow that occur immediately following the onset of exercise. Unfortunately, the speed at which the carotid baroreceptor reflex resets during dynamic exercise has not been experimentally determined. However, if the reflex resets immediately, then a *pressure error signal* would quickly arise, permitting sympathetic outflow to increase while decreasing the level of parasympathetic activity to the heart. Recently, DiCarlo and Bishop (1990) have provided indirect evidence suggesting that the resetting occurs at the same time as the onset of exercise. This suggests another role for central command. These investigators have also demonstrated that lumbar sympathetic nerve activity increases at the onset of dynamic exercise in the conscious rat. While the authors suggest that a central feed-forward mechanism mediates the increase in lumbar sympathetic nerve activity, it is difficult to eliminate the contribution of other sources of spinal afferent input (i.e., mechanically sensitive skeletal muscle afferents) that were also activated at exercise onset.

Recently, Potts et al. (1998a) demonstrated that sympathoexcitatory responses by carotid baroreceptors and skeletal muscle receptors interact in an inhibitory fashion. Moreover, this apparent inhibition was restricted to sites within the central nervous system. However, since only one point was examined on the baroreflex stimulus-response relationship, it was not known if changing the operating point would alter the interaction. Therefore, these investigators determined the effect of varying the level of carotid baroreceptor activity and, thus, the operating point on the reflex cardiovascular responses evoked by activation of skeletal muscle afferents (Potts and Li 1998c). In anesthetized dogs, an evoked contraction of the hind limb by ventral root stimulation was performed at varied levels of carotid sinus pressure. When sinus pressure was low and baroreceptor afferent activity was minimal, the cardiovascular response induced by muscle contraction was potentiated compared with when carotid sinus pressure was high (and central baroreceptor inhibition was near maximal). This finding suggests that the interaction between these two reflexes is not exclusively inhibitory. A facilitatory interaction may occur when carotid baroreceptor input is low.

Strong evidence now shows that the arterial baroreceptor reflex is reset during exercise to the prevailing level of arterial pressure. However, the mechanism(s) mediating baroreflex resetting remain in question. Rowell and O'Leary (1990) have proposed that neural signals by central command immediately reset the arterial baroreflex at exercise onset. In addition, others have reported that neural input from skeletal muscle receptors alters carotid baroreflex function. Since neural input from both central command and skeletal muscle receptors make synaptic connections in the lower brain stem, baroreflex resetting likely occurs within medullary regions. The controversy of baroreflex resetting is further complicated by the fact that central baroreceptor medullary neurons receive synaptic input from both spinal and supraspinal sources. Therefore, the *classical resetting* of the arterial baroreflex that has been reported during exercise (Papelier et al. 1994; Potts, Shi, and Raven 1993) likely results from the converging neural input from supraspinal and spinal sources.

In an attempt to separate the contribution of spinal and supraspinal neural input on baroreflex resetting, Potts and Mitchell (1998b) determined the effect of skeletal muscle receptor activation on the threshold pressure of the carotid baroreflex. In order to determine resetting, a slow linear increase in carotid sinus pressure was performed. The sinus pressure evoking a reflex bradycardia and hypotension was considered the threshold pressure. When the linear ramp of sinus pressure was accompanied by electrically induced contraction of the hind limb (ventral root stimulation), the threshold pressure was significantly increased. This represents the first evidence that neural input from contraction-sensitive skeletal muscle receptors resets the threshold pressure of the carotid baroreflex. Thus, the evidence suggests that in addition to supraspinal mechanisms (i.e., central command), neural input with a spinal origin (i.e., skeletal muscle afferents) may reset the arterial baroreceptor reflex.

Human Studies

The following describes the results of human studies about arterial baroreceptors. Specifically, it includes information about the carotid baroreceptor reflex, the aortic baroreceptor reflex, the cardiopulmonary baroreflex, and the interaction between the carotid and cardiopulmonary baroreceptors.

Carotid Baroreceptor Reflex

Information pertaining to carotid baroreceptor control of circulation in humans is much more limited largely due to the invasive procedures typically needed to study carotid baroreceptor function. As we indicated earlier in this chapter, the only approach available to characterize human carotid baroreflex function is the variable pressure neck collar. By utilizing this technique, Bevegard and Shepherd (1966) contrasted the effects of carotid baroreflex stimulation at rest and during supine cycle ergometry. They observed that increases in carotid sinus transmural pressure by neck suction produce similar changes in heart rate and blood pressure at rest and during exercise. The reflex decrease in blood pressure to neck suction was primarily mediated by changes in vascular resistance. This resulted from dilation of limb resistance vessels with only small decreases in cardiac output. This early study, conducted in 1966, supports the idea that the carotid baroreceptor reflex continues to regulate arterial blood pressure during dynamic exercise, a finding Melcher and Donald confirmed in the exercising dog in 1981.

Others have concluded that carotid baroreflex sensitivity is attenuated during exercise and does not contribute to the cardiovascular responses accompanying dynamic exercise (Bristow et al. 1971; Pickering et al. 1972; Staessen et al. 1987). Unfortunately, these studies used reflex changes in the R-R interval, instead of the heart rate, to make conclusions regarding the effect of exercise on baroreflex sensitivity. Using the R-R interval and heart rate interchangeably assumes that the relationship between these two variables is linear. However, over the physiological range for heart rate (50–200 beats/min), this relationship is nonlinear and will bias the results whenever the baseline heart rate changes (as occurs with mild-to-moderate exercise). By using reflex changes in the R-R interval, they made the erroneous conclusion that the sensitivity of the carotid baroreflex was attenuated during exercise. Researchers now widely accept that the nonlinear relationship between heart rate and R-R interval will bias the conclusions whenever the R-R interval is used to study baroreflex control of the heart during exercise. Another concern regarding comparisons between the baroreflex control of heart rate at rest and exercise is the recent finding of Yang, Senturia, and Levy (1994). They found an inhibition of vagal neural transmission as a result of an increase in sympathetic activation, which affected an attenuation of the bradycardiac response.

Recently, several studies conducted by independent investigators using somewhat different methodological approaches have reexamined carotid baroreflex function during dynamic exercise in humans (Eiken, Sun, and Mekjavic 1994; Papelier et al. 1994; Potts, Shi, and Raven 1993). Eiken and colleagues used pulse-synchronous changes in neck pressure and suction during bicycle exercise with and without lower-body positive pressure (LBPP +50 torr) (Eiken, Sun, and Mekjavic 1994). They wanted to determine the effect of exercise and exercise superimposed with a restriction of blood flow to the lower extremity on carotid baroreflex control of heart rate. They reported that the sensitivity of the carotid baroreflex increased during exercise and was further enhanced when exercise was superimposed with lower-body positive pressure. The increase of LBPP was intended to activate the muscle chemoreflex by restricting blood flow to the active muscle. These conclusions do support the idea that the carotid baroreflex continues to regulate heart rate during exercise. However, they are suspect due to the use of pulse-synchronous neck suction and pressure that, as discussed previously, should be restricted to conditions where the baseline heart rate remains constant.

To avoid the problems that arise when using pulse-synchronous changes in neck pressure and suction during exercise, Potts, Shi, and Raven (1993) used random-order presentation of discrete 5-s periods of sustained neck pressure. They used neck pressure/suction ranging between +40 torr and –80 torr to complete the stimulus-response curves for

heart rate and blood pressure. By doing so, they were able to minimize the effects emanating from the extracarotid baroreceptors as well as the respiratory influences on heart rate and blood pressure. By using this approach, they found that the carotid baroreflex continued to modulate heart rate and blood pressure during exercise at 25% and 50% $\dot{V}O_2$max with the same sensitivity as at rest (see figure 1.3). In addition, since they were able to demonstrate that the reflex responses had a clearly defined threshold and saturation point, they were able to quantify whether these points increased during exercise. If this occurred, it would support resetting of the baroreflex. The baroreflex threshold and saturation points for both heart rate and blood pressure were significantly increased, in an intensity-dependent manner, during exercise. This finding, along with no change in maximal baroreflex sensitivity, are the hallmark features of baroreflex resetting (Heesch and Carey 1987). This is the first clear evidence in humans that the carotid baroreflex resets during dynamic exercise. Potts, Shi, and Raven (1993) also found that the operating point (the intersection of steady-state heart rate and arterial pressure when neck chamber pressure equals 0 torr) for both the heart rate and blood pressure baroreflex curves shifted when the subject was no longer resting and started exercising. The operating point shifted away from the centering point of the curve (point of maximal gain) when the subject was rest-

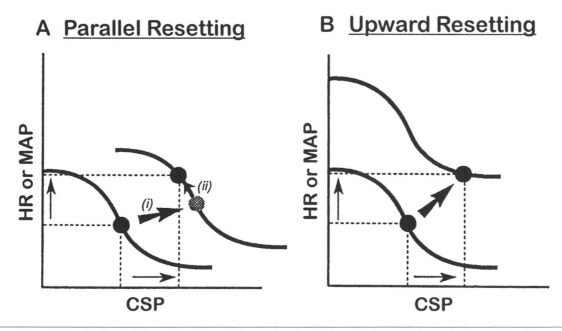

Figure 1.3 Stylized stimulus-response curves illustrating resetting of the carotid baroreceptor reflex during exercise. *(a) Parallel resetting* characterized by increases in threshold and saturation points and no change in the gain (or slope) of the curve. This figure illustrates a shift in the operating point toward threshold during exercise (ii). *(b) Upward resetting.* (Data modified from Potts, Shi, and Raven 1993 with permission of the American Physiological Society.)

ing and moved toward the threshold region of the curve during exercise (see figure 1.3).

The authors interpreted this shift in operating point during exercise as follows. At rest, with the operating point located at the centering point of the curve, the carotid baroreflex can alter the cardiac and vascular response when systemic pressure either falls or rises with equal sensitivity. In contrast, a shift in operating point toward threshold during exercise permits the reflex to respond preferentially to increases in systemic pressure and over a greater range of response. The muscle chemoreflex is a powerful pressure-raising reflex and is activated during moderate-to-severe intensities of exercise in humans. Therefore, the resetting and shifting of the operating point of the carotid baroreflex during exercise facilitates optimal buffering of the muscle chemoreflex.

Papelier and colleagues (Papelier et al. 1994) conducted a similar study in humans. They investigated carotid baroreflex control of heart rate and arterial pressure during graded bicycle exercise up to workloads that elicited approximately 75% $\dot{V}O_2$max. Although their approach to characterize the carotid baroreflex was similar to that used by Potts, Shi, and Raven (1993), there were distinct differences. First, they used 20 s of pulsatile neck pressure and suction between +60 torr and –80 torr. They plotted the reflex heart rate and blood pressure responses between 15 s and 20 s of the stimulus. Although pulsatile pressure delivered to the neck collar may be more physiological, the heart rate (and possibly the blood pressure) responses between the 15th and 20th second were likely affected by extracarotid baroreceptors. Second, subjects continued breathing during the 20 s of baroreceptor stimulation. Therefore, respiration also affected the changes in heart rate and blood pressure. Regardless of these differences in methodology, they found that the baroreflex curves for heart rate and blood pressure were shifted to higher pressures during exercise in an intensity-dependent manner with no change in gain. This agrees with the findings of Potts, Shi, and Raven (1993) and disagrees with the conclusions of Eiken and colleagues (Eiken, Sun, and Mekjavic 1994). One limitation of Papelier et al.'s (1994) work was its inability to demonstrate clearly a threshold and saturation point. Because it lacks this information, their data cannot be used to support the notion that the carotid baroreflex resets during exercise.

As mentioned previously, another function of the carotid baroreflex is to maintain a degree of vasoconstrictor tone to the vasculature supplying blood to active skeletal muscle during exercise. This vasoconstriction is necessary to limit the degree of meta-bolic vasodilation and thereby maintain arterial blood pressure during moderate-to-severe intensifies of exercise. This is needed since maximal metabolic vasodilation of the skeletal muscle can outstrip the cardiac pumping capacity of the heart. Strange et al. (1990) designed a study to address this issue specifically. They applied pulsatile neck suction of –50 torr for 30 s during graded bicycle exercise and measured femoral venous blood flow by constant-infusion thermodilution. At the lowest workload (40% $\dot{V}O_2$max), neck suction significantly reduced arterial pressure and heart rate and increased leg vascular conductance without affecting femoral venous blood flow. This finding suggests that during mild exercise, the net blood flow to active skeletal muscle results from the competing effects of metabolic vasodilation and sympathetic vasoconstrictor tone, which is under the direct control of the carotid baroreceptor reflex.

When exercise was performed at a high intensity (88% $\dot{V}O_2$max), however, leg vascular conductance was not altered when the carotid baroreflex was activated. Since no change in leg vascular conductance was observed during neck suction at the highest workload studied, the level of vasoconstrictor outflow to active skeletal muscle was not under the control of carotid baroreceptors. This, therefore, opens the possibility that some other mechanism(s) must be competing with (or overriding) the carotid baroreflex during higher intensities of exercise to produce vasoconstriction of active skeletal muscle. Another possible mechanism may be related to the shift in the operating point of the reflex at the higher workloads. This could result in the operating point being below the threshold of the reflex, thereby being inoperable. To date, the mechanism(s) limiting the effect of the carotid baroreflex on leg vascular conductance during severe exercise remains unknown.

From recent studies completed in this area, several definitive conclusions can be drawn regarding the function of the carotid baroreflex during exercise in humans. First, the maximal gain for the open-loop carotid-cardiac and carotid-vascular reflexes is not attenuated during mild-to-moderately severe *dynamic* exercise. Second, during graded exercise, the baroreflex curves reset in an intensity-dependent manner to higher levels of systemic pressure. Third, the operating point shifts toward the threshold region of the baroreflex reflex curve. Figure 1.3 illustrates acute resetting of the carotid baroreflex. The important question that has yet to be definitively answered is, "What mechanism(s) mediate these observable changes in carotid baroreflex function during exercise?"

The ability of the carotid baroreflex to regulate heart rate during periods of sympathetic activation (i.e., exercise) may depend, in part, on the neural latency of the reflex. The effect of exercise on baroreflex latency has recently been addressed. Potts and colleagues (Potts and Raven 1995; Potts et al. 1998a, 1998b) concluded that the peak neural effector response time (i.e., time from the onset of neck pressure/neck suction [NP/NS] to peak change in HR) during graded exercise was similar to resting conditions. In contrast, Sundblad and Linnarsson (1996) suggested that the latency of the carotid baroreflex lengthens during dynamic exercise. The discrepancy between these conclusions may be attributed to the method of baroreceptor activation (static versus pulsatile) and/or to the method used to determine baroreflex latency.

Aortic Baroreceptor Reflex

The work of Walgenbach and Donald (1983) made apparent that the aortic baroreflex is an important modulator of arterial blood pressure during exercise. However, because of a number of technical concerns, isolating the aortic baroreceptor reflex during exercise in humans was difficult. Initially, Mancia et al. (1977) used selective or nonselective stimulation of the carotid baroreceptors to calculate the contributions of the carotid and extracarotid baroreceptor control of heart rate at rest. In a unique modification of this experimental protocol, Ferguson, Abboud, and Mark (1985) were able to estimate the relative contribution of the aortic and carotid baroreflexes to the control of heart rate in the human at rest. Subsequently, Shi and colleagues (Shi, Andresen, et al. 1993; Shi, Crandall, et al. 1993) used a further modification of the protocol by Ferguson, Abboud, and Mark (1985) to describe differences in the relative contributions of aortic and carotid baroreflex control of heart rate in the trained and untrained subjects.

By utilizing the same modification of the aortic baroreceptor isolation technique, Shi et al. (1995) investigated the role of the isolated aortic-cardiac reflex during dynamic exercise. Aortic baroreflex function was assessed by the ratio change in mean arterial pressure (HR/MAP) during phenylephrine infusion (PE). The ratio obtained during PE was combined with low-level lower-body negative pressure (LBNP) and calculated neck pressure (NP). The result was used to assess the gain of the aortic-cardiac reflex during rest and 25% $\dot{V}O_2$max supine exercise. Table 1.2 presents the resultant data.

Shi et al. (1995) concluded from this data that the sensitivity (gain) of the aortic-cardiac reflex control

Table 1.2 Gain of Aortic-Cardiac Baroreflex at Rest and Exercise

PE	HR/MAP beats/min/mmHg
Rest	−1.10 ± 0.19
Exercise	−1.15 ± 0.15
P value	NS

PE + NP + LBNP	HR/MAP beats/min/mmHg
Rest	−0.74 ± 0.13
Exercise	−0.86 ± 0.12
P value	NS

PE = phenylephrine infusion; HR = heart rate; MAP = mean arterial pressure; NS = neck suction; NP = neck pressure; LBNP = lower-level lower-body negative pressure.

Data adapted from Shi, Andresen, et al. 1993 with permission of the American Physiological Society and X. Shi, C.G. Crandall, J.T. Potts, J.W. Williamson, B.H. Foresman, and P.B. Raven. 1993. A diminished aortic-cardiac reflex during hypotension in aerobically fit young men. *Medicine and Science in Sports and Exercise* 251: 1024–1030.

of heart rate was the same at rest as during exercise. However, the technique employed to isolate the aortic-cardiac reflex could not define the threshold or saturation pressures of the reflex. Hence, whether the classical resetting observed for the carotid-baroreceptor reflex during exercise was present for the aortic-baroreceptor reflex was unclear.

Cardiopulmonary Baroreflex

In the earlier work of Melcher and Donald (1981) using the dog model, their data clearly defined the role of the carotid baroreflex in maintaining arterial blood pressure during exercise. However, it also showed that cardiopulmonary baroreceptors were apparently necessary to obtain the normal expression of the reflex. More recently, DiCarlo and Bishop (1988, 1990) found that the renal vascular and mesentery vascular resistance of the exercising rabbit were significantly increased with anesthetic interruption of vagal afferent nerve traffic emanating from the heart wall.

In humans, a single investigation of the effect of cardiopulmonary baroreflex on the hemodynamic response to dynamic exercise has been reported. Mack, Nose, and Nadel (1988) found that stroke volume and cardiac output were significantly decreased and forearm vascular resistance increased in response to LBNP –10 and –20 torr during supine cycling at 77 watts without changing the MAP or HR. These investigations concluded that during exercise, the cardiopulmonary baroreflex was actively involved in maintaining systemic homeostasis in the same manner as that which occurs during rest. They also concluded that the elevated MAP during exercise was regulated and not merely the consequence of differential changes in cardiac output and total peripheral resistance.

Interaction Between Carotid and Cardiopulmonary Baroreceptors

As discussed earlier in this chapter, previous published research has been unable to provide definitive evidence to show a functional interaction between the cardiopulmonary and carotid baroreceptors in resting humans. However, the reported findings from several studies are rather compelling (Pawelczyk and Raven 1989; Shi, Potts, et al. 1993; Victor and Mark 1985). Some of the discrepancies may be due to (1) species differences, (2) different methodological approaches, and (3) rapid resetting of cardiopulmonary baroreceptors (Chapman and Parkhurst 1976). Also, in the upright human (as compared with the quadruped animal), the role of the cardiopulmonary baroreceptors have a different functional importance in cardiovascular regulation. They provide critical information regarding the "filling status" of the heart in humans.

If we agree that sufficient scientific evidence supports an interaction between these two baroreceptor populations at rest, then a similar interaction may be present during dynamic exercise is legitimate. Actually, in terms of overall cardiovascular regulation, this interaction may be of greater importance during exercise when thermoregulatory stresses are superimposed with the cardiovascular stresses accompanying exercise. The regulation of body core temperature during exercise is paramount. When sweating rate and cutaneous blood flow increase in response to elevations in body core temperature, the net effect is the "stealing" of blood volume away from the central circulation by the skin to cool the body. This can lead to a precipitous fall in venous return and cardiac filling pressure, which may jeopardize the maintenance of systemic blood pressure. Under these conditions, having an arterial baroreflex system with a *variable gain* would be advantageous. Such a system could be adjusted depending upon the level of cardiac filling pressure, which cardiopulmonary baroreceptors continuously monitor.

Eiken, Sun, and Mekjavic (1994) attempted to address this question in humans by applying –20 torr lower-body negative pressure at rest and during mild and moderate supine bicycle exercise. By using pulse-synchronous neck pressure and suction, they found that –20 torr lower-body negative pressure has no effect on the derived maximal gain of the carotid baroreflex. However, as discussed earlier, the interpretation of reflex changes of heart rate during exercise is difficult when the responses are obtained using trains of neck suction and pressure. Furthermore, without a direct measure of cardiac firing pressure, the degree of cardiopulmonary baroreceptor unloading at rest and during exercise is not known.

Recently, Potts, Shi, and Raven (1995) readdressed this issue using the technique outlined previously to stimulate carotid baroreceptors. Exercise was performed in a semirecumbent position on a modified cycle ergometer in a lower-body negative-pressure chamber. Central venous pressure was directly measured from a catheter placed into a forearm vein and advanced into the central circulation. This thereby allowed the investigators to control the level of central venous pressure precisely (and, by inference, the degree of cardiopulmonary baroreceptor unloading). In addition, to quantify the presence or absence of an interaction during exercise, Potts, Shi, and Raven (1995) used two approaches. First, an inhibitory interaction was confirmed if the combined effects of exercise and LBNP resulted in a greater increase in gain than the algebraic sum of the effects of exercise and LBNP alone on gain. Second, if the slope of the relationship between the increase in gain caused by a reduction in central venous pressure (CVP) was significant during exercise, this would indicate inhibitory interaction. Figure 1.4 illustrates the results.

Results from the two approaches used to show an interaction provided the same conclusion—unloading of cardiopulmonary baroreceptors during mild dynamic exercise increases the sensitivity of the carotid baroreceptor reflex to the control of heart rate. This finding supports the hypothesis that cardiopulmonary baroreceptors tonically inhibit carotid baroreceptor reflex control of heart rate during exercise so that a reduction in central filling pressure during exercise will increase the maximal sensitivity of the carotid baroreflex. This would produce a greater change in heart rate for a given change in systemic pressure sensed by carotid baroreceptors during exercise when filling pressure was reduced. It may also aid in the regulation of arterial blood pressure during exercise in hyperthermic conditions. Unfortunately, when the investigators

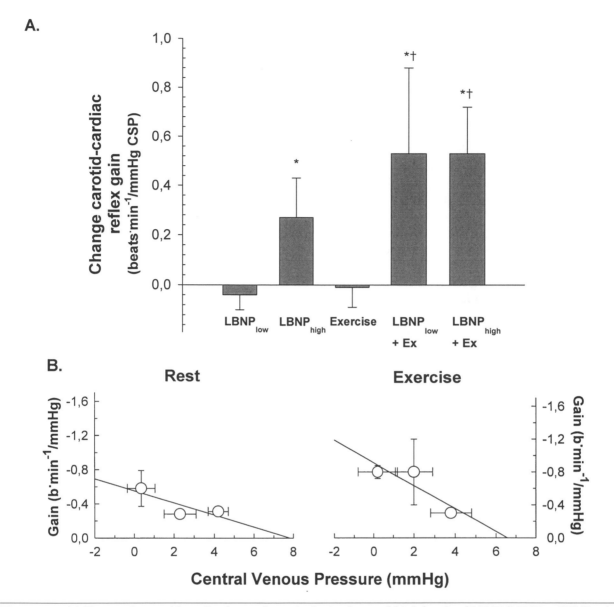

Figure 1.4 *(a)* Changes in carotid-cardiac baroreflex gain during low and high levels of LBNP (LBNP$_{low}$, LBNP$_{high}$), exercise (EX), and when LBNP was simultaneously imposed during exercise (LBNP$_{low}$ + EX, LBNP$_{high}$ + EX). The algebraic sum of the changes in gain during exercise and LBNP alone were significantly smaller than the change in gain when LBNP was performed during exercise. *(b)* Relationship between carotid-cardiac baroreflex gain and CVP at rest and during exercise. (Data adapted from Potts, Shi, and Raven 1995 with permission of the American Physiological Society.)

studied the vascular limb of the carotid baroreflex, they were unable to find a significant interaction between these two baroreceptor populations. A number of technical limitations may have contributed to their inability to demonstrate this effect.

Endurance Exercise Training

The following describes endurance exercise training and its relationship to arterial baroreceptors.

Specifically, it discusses the carotid baroreflex, cardiopulmonary baroreceptors, and the interaction between carotid and cardiopulmonary baroreceptors.

Carotid Baroreflex

In a NASA technical document, Luft et al. (1976) reported that endurance exercise-trained subjects were 42% less tolerant to lower-body negative pressure (LBNP) than untrained subjects. This study supported many anecdotal reports that endurance-trained athletes experienced syncopal symptoms during orthostasis. Raven, Rohm-Young, and Blomqvist (1984) investigated the hemodynamic responses of endurance-trained athletes (maximal aerobic capacity, $\dot{V}O_2 > 65$ ml O_2/kg/min) to LBNP that progressively increased to –50 torr. By using a ratio of the change in heart rate to a change in systolic blood pressure (HR/SBP) as a global index of the baroreflex control of heart rate, they found that the ratio was one-half that of untrained subjects ($\dot{V}O_2$max = 47 ml/kg/min). These data support the earlier work of Stegemann, Busert, and Brock (1974). Their work indicated that high-pressure arterial (carotid and aortic) baroreceptor-mediated reflexes of athletes were less responsive.

In a series of cross-sectional and longitudinal investigations, Raven and his colleagues (Pawelczyk 1989; Pawelczyk and Raven 1989; Raven and Pawelczyk 1993; Raven, Rohm-Young, and Blomqvist 1984; Shi, Andresen, et al. 1993; Shi, Crandall, et al. 1993; Smith et al. 1988) defined the effect of endurance training on the role of the arterial (carotid and aortic) and the cardiopulmonary baroreceptor-mediated control of arterial blood pressure. By using the pulse-synchronous neck pressure/neck suction (NP/NS) protocol of Eckberg (Eckberg and Eckberg 1982; Sprenkle et al. 1986) and of Pawelczyk (1989), Raven and Pawelczyk (1993) investigated the carotid baroreflex responsiveness of 24 young men. The subjects were matched for age, height, and lean body mass. They were assigned into three designated fitness groups with eight subjects in each group: low fit (LF; $\dot{V}O_2 < 40$ ml/kg/min), average fit (AF; $\dot{V}O_2 = 50.2$ ml/kg/min), or high fit (HF; $\dot{V}O_2 > 65$ ml/kg/min). Beat-to-beat changes in heart rate and blood pressure (direct radial artery pressures) were recorded in response to the discrete changes in carotid sinus transmural pressure described previously. Raven and Pawelczyk (1993) analyzed the modeled heart rate and blood pressure responses to the NP/NS protocol and found no differences in the responsiveness (maximal gain) between the three groups of subjects. This raised the question of whether endurance exercise training affects the high-pressure, baroreceptor-mediated reflex control of blood pressure.

Despite the negative findings of Pawelczyk, the cross-sectional data of Raven and colleagues (Raven, Rohm-Young, and Blomqvist 1984; Smith et al. 1988) and the longitudinal data of Stevens et al. (1992) consistently demonstrated an attenuated tachycardiac response to a systolic hypotensive response to LBNP (see figure 1.5).

These data indicate that an attenuated arterial baroreflex control of blood pressure of the endur-

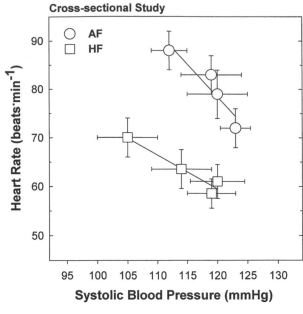

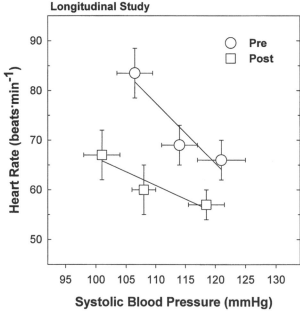

Figure 1.5   Illustrates tachycardiac responses to decreases in systolic blood pressure during lower-body negative pressure (0, –32, –40 and –50 torr in a cross-sectional study and 0, –35, and –45 torr in a longitudinal study). These data indicate that comparisons of the untrained average fit (AF) versus the exercise-trained high fit (HF) individuals or the pre- versus post-endurance exercise-trained individuals' heart rate reflex response during central hypovolemia-induced hypotension were less sensitive in exercise-trained individuals. (Data adapted from Stevens et al. 1992 and Raven, Rohm-Young, and Blomqvist 1984 with permission of the American Physiological Society.)

ance-trained individual was present during a global systolic hypotension. These findings focused the work of Shi and colleagues (Shi, Andresen, et al.

1993; Shi, Crandall, et al. 1993) on investigating the effect of endurance training on the discrete aortic baroreceptor-mediated reflex control of heart rate. They used a steady-state infusion of phenylephrine hydrochloride (PE) to raise systemic arterial blood pressure 15–18 mmHg or an infusion of sodium nitroprusside (SN) to lower systemic arterial pressure 12–14 mmHg. This allowed Shi and colleagues (Shi, Andresen, et al. 1993; Shi, Crandall, et al. 1993) to isolate the aortic baroreceptor reflex by using the appropriate neck pressure (PE protocol) and neck suction (SN protocol). They found an attenuated response range of the HF subjects compared with the LF subjects (see figure 1.6). By using a differential calculation in both drug infusion protocols, Shi and colleagues (Shi, Andresen, et al. 1993; Shi, Crandall, et al. 1993) reported no difference in the discrete carotid baroreflex control of heart rate. They also confirmed the NP/NS data of Pawelczyk (1989).

The data in figure 1.6 indicate that endurance exercise training selectively attenuates the aortic baroreflex control of heart rate. Interestingly, when subjects were aggressively untrained by placing them in a 6°, head-down, bed rest position for 15 days (Crandall et al. 1994), the selective carotid baroreceptor-mediated reflex control of HR was unchanged. However, the selective aortic baroreflex-mediated reflex control of HR was significantly augmented. The mechanism for the discrete effect of endurance exercise training on the aortic barore-ceptor-mediated reflex control of HR is unknown.

Recent work by DiCarlo and colleagues has suggested that daily exercise training attenuates arterial baroreflex of cardiovascular function. Previously, they reported that a regime of daily exercise attenuated the arterial baroreflex control of HR and RSNA in conscious rabbits (DiCarlo and Bishop 1988, 1990). However, the mechanism mediating the apparent inhibition of baroreflex function was not investigated. Recently, they have shown that the cardiac afferent blockade significantly increases the RSNA response evoked by treadmill exercise in rabbits (DiCarlo, Chen, and Collins 1996). This finding suggests that the attenuated sympathetic response evoked by dynamic exercise following daily exercise training is mediated, in part, by an augmentation of the inhibitory influence of cardiac afferents on the control of RSNA. In addition, DiCarlo, Stahl, and Bishop (1997) assessed the effect of daily exercise training on the sympathetic and parasympathetic components of the arterial baroreflex. They reported that daily exercise training in the rat attenuated baroreflex control of HR by differentially affecting the sympathetic component of the baroreflex. Therefore, these results combined with the findings from human studies provide rather

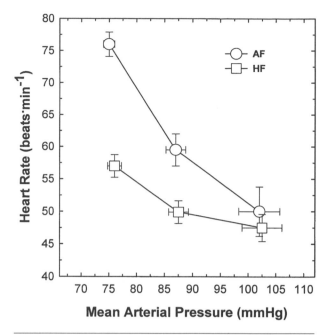

Figure 1.6 Illustrates the aortic baroreceptor-mediated heart rate reflex responses between untrained average fit (AF) and exercise-trained high fit (HF) individuals. Baseline heart rate was significantly lower in the HF group than in the AF group. However, arterial blood pressure was similar between the two groups. Similar changes occurred in mean arterial pressure in the two subject groups during phenylephrine infusion when combined with lower-body negative pressure and neck pressure. These similar changes also occurred during nitroprusside infusion when combined with neck suction in the two subject groups. During this time, both bradycardia and tachycardia were significantly less in the HF than in the AF individuals. This suggests that the aortic-cardiac reflex sensitivity significantly diminished with exercise training. (Data adapted from Shi, Andresen, et al. 1993 with permission of the American Physiological Society and X. Shi, C.G. Crandall, J.T. Potts, J.W. Williamson, B.H. Foresman, and P.B. Raven. 1993. A diminished aortic-cardiac reflex during hypotension in aerobically fit young men. *Medicine and Science in Sports and Exercise* 251:1024–1030.)

convincing evidence that daily exercise training attenuates arterial baroreflex control of the cardiovascular system.

Cardiopulmonary Baroreceptors

In 1984, Raven, Rohm-Young, and Blomqvist (1984) reported that endurance-trained subjects had 24% less of a vasoconstrictor response (change in peripheral vascular resistance) to changes in LBNP between –40 and –50 torr than their untrained counterparts. These data confirmed previous reports of Tipton, Mathis, and Bedford (1982) about the trained and untrained rat model. Further analysis of the data from both these investigations suggests high-pressure baroreceptor reflex regulation of vasomotion was attenuated by endurance exercise training. The mechanisms of this training effect have not been investigated.

In a cross-sectional comparison of fit (bicycle $\dot{V}O_2$ = 57 ± 2 ml/min/kg) and unfit (bicycle $\dot{V}O_2$ = 38.5 ± 3 ml/min/kg) subjects, Mack, Nose, and Nadel (1988) used LBNP from 0 to –20 torr. They plotted the changes in forearm vascular resistance (change in FVR) against changes in the estimated central venous pressure (change in estimated CVP) as an index of cardiopulmonary sensitivity. The linear relationship between the change in FVR and CVP, the gain of the cardiopulmonary baroreflex, was significantly decreased in the fit subjects (2.42 ± 0.6mmHg) compared with the unfit subjects (–5.15 ± 0.58 mmHg).

In a more comprehensive investigation of the effects of an endurance exercise training program, Mack, Convertino, and Nadel (1993) determined the change in FVR versus resting CVP in four groups of male subjects. The groups were divided into unfit individuals, fit subjects, persons before and after 10 weeks of endurance exercise training—with chronic blood volume expansion, and individuals before and after acute blood volume expansion. Table 1.3 presents the data obtained.

These data clearly indicate that the training-induced hypervolemia results in an attenuation of the cardiopulmonary baroreflex. Mack, Convertino, and Nadel (1993) suggest that the training-induced hypervolemia increased the tonic inhibitory influence of cardiac afferent nerve activity (i.e., activation of cardiopulmonary baroreceptors) because of the enlarged cardiac volumes resulting from the increased total blood volume. This suggestion is based upon the work of DiCarlo and Bishop (1988, 1990), which documents a significant interaction between cardiopulmonary baroreceptors and the arterial baroreceptor-mediated reflexes.

Interaction Between Carotid and Cardiopulmonary Baroreceptors

By using the NP/NS protocols for modeling the carotid baroreflex response previously discussed, Pawelczyk and Raven (1989) used progressive LBNP to –50 torr. They found that the carotid baroreflex gain of both HR and vasomotor reflexes linearly increased in direct relation to the decrease in CVP. At the same time DiCarlo and Bishop (1988, 1990) demonstrated that the inhibitory interaction between the cardiopulmonary baroreceptor and the reflex efferent sympathetic nerve activity was enhanced. In other words, endurance exercise-trained rabbits exhibited a greater inhibition of reflex sympathetic nerve activity to the vascular beds.

Subsequently, Pawelczyk (1989) used the same techniques as previously reported to evaluate the effect of endurance exercise training on the interaction between CVP and carotid baroreflex responsiveness (see figure 1.7).

These data clearly demonstrate that the interaction between the cardiopulmonary baroreflex and the carotid baroreflex was present in the low fit ($\dot{V}O_2$

Table 1.3   Changes in Forearm Vascular Resistance (ΔFVR) per Unit Change in Estimated Central Venous Pressure (est CVP) During 0 to –20 Torr Lower-Level Lower-Body Negative Pressure (LBNP)

Group	ΔFVR/est CVP (u/mmHg) Pre-	ΔFVR/est CVP (u/mmHg) Post-	Total blood volume (ml/kg) Pre-	Total blood volume (ml/kg) Post-
Unfit		–5.15		70.2
Fit		–2.421		91.01
Exercise trained	–5.96	–4.06**	63.5	68.0**
Acute volume expansion*	–4.24	–2.15**		

* Acute volume expansion accomplished by 8 ml/kg body weight infusion of 5% human serum albumin, which raised resting est CVP by 2 mmHg.
** Significantly different than pre- result in exercise trained group and unfit controls.

Data adapted from Raven, Rohm-Young, and Blomqvist 1984 with permission from the American Physiological Society and G.H.J. Stevens, B.H. Foresman, X. Shi, S.A. Stem, and P.B. Raven. 1992. Reduction in LBNP tolerance following prolonged endurance exercise training. *Medicine and Science in Sports and Exercise* 24:1235–1241.

A.

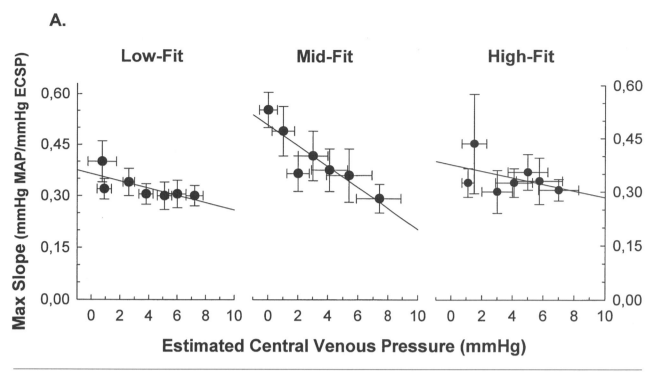

Figure 1.7a   These data represent the slope of the maximal gain of the carotid-vasomotor reflex relationship as compared with the decrease in CVP produced by LBNP.

B.

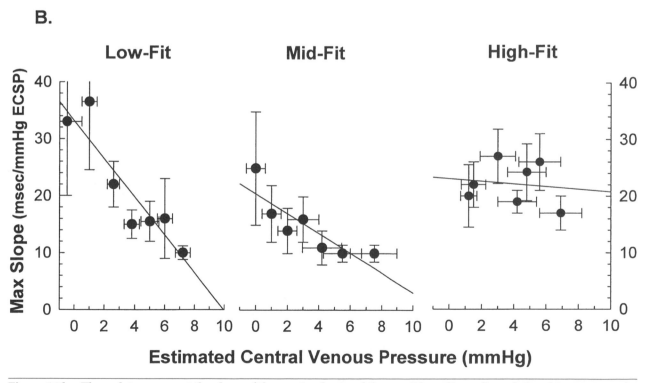

Figure 1.7b   These data represent the slope of the maximal gain of the carotid-cardiac reflex relationship as compared with the decrease in CVP produced by LBNP. In the carotid-cardiac relationship, a significantly greater slope occurred between fitness groups, LF > AF > HF. The HF relationship was not statistically significant from zero. The carotid-vascular relationship was not as consistently different, although the AF relationship was significantly greater than the HF. (Data adapted from Pawelczyk 1989.)

< 40ml/kg/min) and the average fit ($\dot{V}O_2 = 50 \pm$ 2ml/kg/min) subjects but was absent in the high fit subjects ($\dot{V}O_2 > 65$ml/kg/min). The mechanism for the lack of interaction between the cardiopulmonary baroreceptor and the carotid baroreflex gain of the endurance exercise-trained individuals may be related to increased inhibitory afferent nerve traffic from the cardiopulmonary baroreceptors, as suggested by DiCarlo and Bishop (1988, 1990).

Summary

The data of Potts, Shi, and Raven (1993, 1995) and Papelier et al. (1994) in conjunction with the classical animal model work of Donald and his colleagues (Melcher and Donald 1981; Stephenson and Donald 1980; Victor and Mark 1985) clearly suggest that arterial (carotid and aortic) and cardiopulmonary baroreflexes are fully operational during exercise. They also suggest that arterial and cardiopulmonary baroreflexes are reset to the prevailing arterial pressure and cardiac filling pressures established by the exercise. The mechanism of this resetting is probably an effect of central command, the role of which is modulated by the exercise pressor reflex.

The data of Raven and his colleagues (Crandall et al. 1994; Pawelczyk 1989; Raven, Rohm-Young, and Blomqvist 1984; Shi, Andresen, et al. 1993; Shi, Crandall, et al. 1993; Smith et al. 1988) support the concept that endurance exercise training modifies arterial baroreflex control of heart rate primarily by a down regulation of the aortic baroreceptor-mediated cardiac reflex. At this time, no animal data confirm these indications nor has a mechanism of effect been defined. Mack and colleague's data (Mack, Convertino, and Nadel 1993; Mack, Nose, and Nadel 1988) indicate that endurance exercise training attenuates the cardiopulmonary baroreflex control of forearm vascular resistance. Pawelczyk's data (1989), however, suggests that the cardiopulmonary carotid baroreflex interaction is attenuated by endurance exercise training. DiCarlo and Bishop (1988, 1990) suggest that the mechanism of effect of endurance exercise training on this interaction may be related to the training-induced increase in blood volume, cardiac hypertrophy, and increased cardiac afferent nerve traffic inhibiting central nervous system interaction pathways. It is important to note that these endurance exercise training effects on arterial and cardiopulmonary baroreflex control have been documented only during selective stimulation of the baroreceptors at rest and during simulated orthostasis. The effects of endurance exercise training on the reflex regulation of blood pressure during exercise have yet to be examined.

Acknowledgments

The authors acknowledge the support of the following specific grant awards in completing this work. NIH Grants HL 34397, 43202, 45547, and 07652; NASA RTOP NAS9-611; and NASA-NSCORT—NAGW 3853.

References

Angell-James, J.E., and M. Daly. 1970. Comparison of the reflex vasomotor responses to separate and combined stimulation of the carotid sinus and aortic arch baroreceptors by pulsatile and non-pulsatile pressure in the dog. *J. Physiol (London)* 209:257–293.

Aumonier, F.J. 1972. Histological observations on the distribution of the baroreceptors in the carotid and aortic regions of the rabbit, cat, and dog. *Acta. Anat.* 82:1–116.

Barman, S.M., and G.L. Gebber. 1982. Hypothalamic neurons with activity patterns related to sympathetic nerve discharge. *Am. J. Physiol.* 242:R34–R43.

Bevegard, B.S., and J.T. Shepherd. 1966. Circulatory effects of stimulating the carotid arterial stretch receptors in man at rest and during exercise. *J. Clin. Invest.* 45:132–142.

Boss, J., and J.H. Green. 1956. Histology of common carotid baroreceptor areas of the cat. *Circ. Res.* 4:12–17.

Bristow, J.D., E.B. Brown, D.J.C. Cunningham, M.G. Howson, E. Strange-Petersen, T.G. Pickering, and P. Sleight. 1971. Effect of bicycling on the baroreflex regulation of pulse interval. *Circ. Res.* 38:582–593.

Chapman, K.M., and J.J. Parkhurst. 1976. Strain sensitivity and directionality in cat atrial mechanoreceptors in vitro. *J. Physiol.* 259:405–426.

Coleridge, J.C., G. Hemingway, A. Holmes, and R.L. Linden. 1957. The location of atrial receptors in the dog: A physiological and histological study. *J. Physiol. (London)* 136:174–197.

Cottle, M.K. 1964. Degeneration studies of primary afferents of IXth and Xth cranial nerves in cats. *J. Comp. Neurol.* 122:329–345.

Crandall, C.G., K.A. Engelke, V.A. Convertino, and P.B. Raven. 1994. Aortic baroreflex control of heart rate after 15 days of simulated microgravity exposure. *J. Appl. Physiol.* 77:2134–2139.

DiCarlo, S.E., and V.S. Bishop. 1988. Exercise training attenuates baroreflex regulation of nerve activity in rabbits. *Am. J. Physiol.* 255 *(Heart Circ. Physiol.* 24):H974–H979.

DiCarlo, S.E., and V.S. Bishop. 1990. Regional vascular resistance during exercise: Role of cardiac afferents and exercise training. *Am. J. Physiol.* 258 *(Heart Circ. Physiol.* 2):H842–H847.

DiCarlo, S.E., and V.S. Bishop. 1992. Onset of exercise shifts operating point of arterial baroreflex to higher pressures. *Am. J. Physiol.* 262:H303–H307.

DiCarlo, S.E., C. Chen, and H.L. Collins. 1996. Onset of exercise increases lumbar sympathetic nerve activity in rats. *Med. Sci. Sports Ex.* 28:677–684.

DiCarlo, S.E., L.K. Stahl, and V.S. Bishop. 1997. Daily exercise attenuates the sympathetic nerve response to exercise by enhancing cardiac afferents. *Am. J. Physiol.* 273:H1606–H1610.

Eckberg, D.L. 1976. Temporal response patterns of the human sinus node to brief carotid baroreceptor stimuli. *J. Physiol.* 258:769–782.

Eckberg, D.L., S. Cavanaugh, A.L. Mark, and F.M. Abboud. 1975. A simplified neck suction device for activation of carotid baroreceptors. *J. Lab. Clin. Med.* 85:167–173.

Eckberg, D.L., and M.J. Eckberg. 1982. Human sinus node responses to repetitive, ramped carotid baroreceptor stimuli. *Am. J. Physiol.* 242:H638–H644.

Eiken, O., J.C.L. Sun, and I.B. Mekjavic. 1994. Effects of blood-volume distribution on the characteristics of carotid baroreflex in humans at rest and during exercise. *Acta Physiol. Scand.* 150:89–94.

Ernsting, J., and D.J. Parry. 1957. Some observations on the effects of stimulating the stretch receptors in the carotid artery in man. *J. Physiol. (London)* 137:44P–46P.

Ferguson, J.D.W., F.M. Abboud, and A.L. Mark. 1985. Relative contribution of aortic and carotid baroreflexes to heart rate control in man during steady state dynamic increases in arterial pressure. *J. Clin. Invest.* 76:2265–2274.

Heesch, C.M., and L.A. Carey. 1987. Acute resetting of arterial baroreflexes in hypertensive rats. *Am. J. Physiol.* 253:H974–H979.

Hering, H.E. 1927. *Die Karotissinus Reflex auf Herz und Gefasse.* Leipzig, Germany: Steinkopff.

Heymans, C., and E. Neil. 1958. *Reflexogenic Areas in the Cardiovascular System.* London: Churchill.

Interactions of the endocrine and cardiovascular systems in health and disease. 13th annual meeting, IUPS Commission on Gravitational Physiology. San Antonio, TX. September 29–October 3, 1991. *Physiologist* 34(4):239.

Kent, B.B., J.W. Drane, B. Blumenstein, and J.W. Manning. 1973. A mathematical model to assess changes in the baroreceptor reflex. *Cardiology* 57:295–310.

Koch, E., and H. Mies. 1929. Chronischer Arterieller Hochtruck Durch Experimentelle Dauerasschaltung der Blutdruckzugler. *Krankheitsforschung* 7:241–256.

Koepchen, H.P., P. Langhorst, H. Seller, J. Polster, and P.H. Wagner. 1967. Neuronale Aktivtat im Unteren I-Limstamm mit Beziechung zum Kreislauf. *Arch. Ges. Physiol.* 294:40–64.

Levine, B.D., J.A. Pawelczyk, J.C. Buckey, B.A. Parra, P.B. Raven, and C.G. Blomqvist. 1990. The effect of carotid baroreceptor stimulation on stroke volume. *Clin. Res.* 38:333A.

Ludbrook, J. 1983. Reflex control of blood pressure during exercise. *Ann. Rev. Physiol.* 45:155.

Ludbrook, J., G. Mancia, A. Ferrari, and A. Zanchetti. 1977a. Factors influencing the carotid baroceptor response to pressure changes in a neck chamber. *Clin. Sci. Mol. Med.* 51:347S–349S.

Ludbrook, J., G. Mancia, A. Ferrari, and A. Zanchetti. 1977b. The variable-pressure neck-chamber method for studying the carotid baroreflex in man. *Clin. Sci. Mol. Med.* 53:165–171.

Luft, U.C., L.G. Myhre, J.A. Loeppky, and M.D. Venters. 1976. *A Study of Factors Affecting Tolerance of Gravitational Stress Stimulated by Lower Body Negative Pressure. Contract NAS 9 14492.* Albuquerque, NM: Lovelace Foundation.

Mack, G.W., V.A. Convertino, and E.R. Nadel. 1993. Effect of exercise training on cardiopulmonary baroreflex control of forearm vascular-resistance in humans. *Med. Sci. Sports and Exerc.* 25:722–728.

Mack, G.W., H. Nose, and E.R. Nadel. 1988. Role of cardiopulmonary baroreflexes during dynamic exercise. *J. Appl. Physiol.* 65:1827–1832.

Mancia, G., A. Ferrari, L. Gregonni, R. Valintine, J. Ludbrook, and A. Zanchella. 1977. Circulatory reflexes from carotid and extracarotid baroceptor areas in man. *Circ. Res.* 44:309–315.

Marey, E.J. 1863. *Physiologie Medicale de la Circulation du Sang.* Paris: Delahaye.

Melcher, A., and D.E. Donald. 1981. Maintained ability of carotid baroreflex to regulate arterial pressure during exercise. *Am. J. Physiol.* 241:H838–H849.

Miller, M.R., and K. Kasahara. 1964. Studies on the nerve endings in the heart. *Am. J. Anat.* 115:217–234.

Mitchell, J.H. 1990. Neural control of the circulation during exercise. *Med. Sci. Sports Exerc.* 22:141–154.

Nonidez, J.F. 1941. Studies of the innervation of the heart II: Afferent nerve endings in the large arteries and veins. *Am. J. Anat.* 68:151–189.

Papelier, Y., P. Escourrou, J.P. Gauthier, and L.B. Rowell. 1994. Carotid baroreflex control of blood pressure and heart rate in men during dynamic exercise. *J. Appl. Physiol.* 77:502–506.

Pawelczyk, J.A. 1989. Interactions between carotid and cardiopulmonary baroreceptor populations in men with varied levels of maximal aerobic power. Ph.D. diss., University of North Texas, Denton.

Pawelczyk, J.A., and P.B. Raven. 1989. Reductions in central venous pressure improve carotid baroreflex responses in humans. *Am. J. Physiol.* 257:H1389–H1395.

Pickering, T.G., B. Gribbin, E. Strange-Petersen, D.J.C. Cunningham, and P. Sleight. 1972. Effects of autonomic blockade on the baroreflex in man at rest and during exercise. *Circ. Res.* 30:177–185.

Potts, J.T., G.A. Hand, J. Li, and J.H. Mitchell. 1998a. Central interaction between carotid baroreceptors and skeletal muscle receptors inhibits sympathoexcitation. *J. Appl. Physiol.* 84:1158–1165.

Potts, J.T., and J.H. Mitchell. 1998b. Rapid resetting of the carotid baroreceptor reflex is mediated by neural input from skeletal muscle receptors. *Am. J. Physiol. (Heart & Circ. Physiol.)* 275:H2000–2008.

Potts, J.T., and J. Li. Interaction between carotid baroreflex and exercise, pressor reflex depends on baroreceptor affernt input. *Am. J. Physiol. (Heart & Circ. Physiol.)* 274:H1841–H1847, 1998c.

Potts, J.T., and P.B. Raven. 1995. Effect of dynamic exercise on human carotid-cardiac baroreflex latency. *Am. J. Physiol. (Heart Circ. Physiol.)* 268(39):H1208–H1214.

Potts, J.T, X. Shi, and P.B. Raven. 1993. Carotid baroreflex responsiveness during dynamic exercise in man. *Am. J. Physiol. (Heart Circ. Physiol.)* 265(34):H1928—H1938.

Potts, J.T., X. Shi, and P.B. Raven. 1995. Cardiopulmonary baroreceptors modulate carotid sinus baroreflex control of heart rate during exercise in humans. *Am. J. Physiol.* 286:H1567–H1569.

Raven, P.B., and J.A. Pawelczyk. 1993. Chronic endurance exercise training: A condition of inadequate blood pressure regulation and reduced tolerance to LBNP. *Med. Sci. Sports Exerc.* 25:713–721.

Raven, P.B., D. Rohm-Young, and C.G. Blomqvist. 1984. Physical fitness and cardiovascular responses to lower body negative pressure. *J. Appl. Physiol.* 56:138–144.

Rees, P.M., and P. Jepson. 1970. Measurement of arterial geometry and wall composition in the carotid sinus baroreceptor region. *Circ. Res.* 26:461–467.

Reis, D.J., N. Doba, D.W. Synder. and M.A. Nathan. 1977. Brain lesions and hypertension: Chronic lability and elevation of arterial pressure produced by electrolytic lesions and 6-hydroxydopamine treatment of nucleus tractus solitari (NTS) in rat and cat. *Prog. Brain Res.* 47:169–188.

Rowell, L.B. 1991. Blood pressure regulation during exercise. *Ann. Med.* 23:329–333.

Rowell, L.B., and D.S. O'Leary. 1990. Reflex control of the circulation during exercise: Chemoreflexes and mechanoreflexes. *J. Appl. Physiol.* 69:407–418.

Sagawa, K. 1983. Baroreflex control of systemic arterial pressure and vascular bed. In *Line Handbook of Physiology. The Cardiovascular System. Peripheral Circulation and Organ Blood Flow.*, ed. R.M. Beme, Sect. 2, Vol. HI, Part 2, 453–496. Bethesda, MD: American Physiological Society.

Scherrer, U., S.L. Pryor, L.A. Bertocci, and R.G. Victor. 1990. Arterial baroreflex buffering of sympathetic activation during exercise-induced elevations in arterial pressure. *J. Clin. Invest.* 86:1855–1861.

Seagard, J.L., L.A. Gallenberg, F.A. Hopp, and C. Dean. 1992. Acute resetting in two functionally different types of carotid baroreceptors. *Circ. Res.* 70:559–565.

Seagard, J.L., F.A. Hopp, H.A. Drummond, and D.M. Van Wynskerghe. 1992. Selective contribution of two types of carotid sinus baroreceptors to the control of blood pressure. *Circ. Res.* 72:1011–1022.

Secher, N.H., J.A. Pawelczyk, and J. Ludbrook. 1994. *Blood Loss and Shock.* London: Edward Arnold Publishers.

Sheehan, D., J.H. Mulholland, and B. Safiroff. 1941. Surgical anatomy of the carotid sinus nerve. *Anat. Record* 80:431–442.

Sheriff, D.D., D.S. O'Leary, A.M. Scher, and L.B. Rowell. 1990. Baroreflex attenuates pressor response to graded muscle ischemia in exercising dogs. *Am. J. Physiol.* 258:H305–H310.

Shi, X., J.M. Andresen, J.T. Potts, B.H. Foresman, S.A. Stern, and P.B. Raven. 1993. Aortic baroreflex control of heart rate during hypertensive stimili. *J. Appl. Physiol.* 74:1555–1562.

Shi, X., C.G. Crandall, J.T. Potts., J.W. Williamson, B.H. Foresman, and P.B. Raven. 1993. A diminished aortic-cardiac reflex during hypotension in aerobically fit young men. *Med. Sci. Sports Exerc.* 251:1024–1030.

Shi, X., J.T. Potts, B.H. Foresman, and P.B. Raven. 1993. Carotid baroreflex responsiveness to lower body positive pressure-induced increases in central venous pressure. *Am. J. Physiol.* 265:H918–H922.

Shi, X., J.T. Potts, P.B. Raven, and B.H. Foresman. 1995. Aortic cardiac reflex during dynamic exercise. *J. Appl. Physiol.* 78:1569–1574.

Smith, M.L., H.M. Graitzer, D.L. Hudson, and P.B. Raven. 1988. Baroreflex function in endurance and static exercise trained men. *J. Appl. Physiol.* 65:1789–1795.

Sprenkle, J.M., D.L. Eckberg, R.L. Goble, J.J. Schelhom, and H.C. Halliday. 1986. Device for rapid quantification of human carotid baroreceptor-cardiac reflex responses. *J. Appl. Physiol.* 60:727–732.

Spyer, K.M. 1981. Neural organization and control of the baroreceptor reflex. *Rev. Physiol. Biochem. Pharmacol.* 88:123–124.

Staessen, J., R. Fiocchi, R. Fagard, P. Hespel, and A. Amery. 1987. Progressive attenuation of the carotid baroreflex control of blood pressure and heart rate during exercise. *Am. Heart J.* 114:765–772.

Stegemann, J., A. Busert, and D. Brock. 1974. Influence of fitness on the blood pressure control system in man. *Aerospace Med.* 41:45–48.

Stephenson, R.B., and E.E. Donald. 1980. Reversible vascular isolation of carotid sinuses in conscious dogs. *Am. J. Physiol.* 238:H809–H814.

Stevens, G.H.J., B.H. Foresman, X. Shi, S.A. Stem, and P.B. Raven. 1992. Reduction in LBNP tolerance following prolonged endurance exercise training. *Med. Sci. Sports Exerc.* 24:1235–1241.

Strange, S., L.B. Rowell, N.J. Christensen, and B. Saltin. 1990. Cardiovascular responses to carotid sinus baroreceptor stimulation during moderate to severe exercise in man. *Acta Physiol.* 138:145–153.

Sundblad, P., and D. Linnarsson. 1996. Slowing of carotid-cardiac baroreflex with standing and with isometric and dynamic exercise. *Am. J. Physiol.* 271:H1363–H1369.

Tipton, C.M., R.D. Mathis, and T.G. Bedford. 1982. Influence of training on the blood pressure changes during lower body negative pressure in rats. *Med. Sci. Sports Exerc.* 14:81–90.

Victor, R.G., and A.L. Mark. 1985. Interaction of cardiopulmonary and carotid baroreflex control of vascular resistance in humans. *J. Clin. Invest.* 76:1592–1598.

Walgenbach, S.C., and D.E. Donald. 1983. Inhibition by carotid baroreflex of exercise-induced increases in arterial pressure. *Clin. Res.* 52:253–262.

Yang, T., J.B. Senturia, and M.N. Levy. 1994. Antecedent sympathetic stimulation alters time course of chronotropic response to vagal stimulation in dogs. *Am. J. Physiol.* 266:H1339-1347.

Chapter 2

The Exercise Pressor Reflex: Afferent Mechanisms, Medullary Sites, and Efferent Sympathetic Responses

Jeffrey T. Potts, Lawrence I. Sinoway, and Jere H. Mitchell

Over 100 years ago, Johansson suggested two possible neurogenic mechanisms that control the cardiovascular response during exercise (Johansson 1893; see Introduction). In the first mechanism, signals from a central site that are responsible for recruiting motor units also activate the cardiovascular control areas in the ventrolateral medulla (Goodwin, McCloskey, and Mitchell 1972; Krogh and Lindhard 1913; see Introduction). In the second mechanism, signals from the contracting muscle activate the same control areas in the brain stem (Alam and Smirk 1937; Coote, Hilton, and Perez-Gonzalez 1971; McCloskey and Mitchell 1972; Mitchell, Kaufman, and Iwamoto 1983).

Alam and Smirk (1937) have been credited with providing the first experimental evidence that the cardiovascular responses during exercise are mediated, in part, by the activation of receptors located within the contracting skeletal muscle. They elegantly demonstrated that intermittent, high-intensity exercise during local circulatory occlusion increased arterial blood pressure which persisted after exercise until the vascular occlusion was removed.

These findings have been used as evidence that receptors in the skeletal muscle were activated by metabolites produced during muscle contraction. Neural signals from contracting skeletal muscle are one source of afferent input to the brain stem which control the sympathetic and parasympathetic nervous systems to increase arterial blood pressure, heart rate, and myocardial contractility. These cardiovascular responses have been reported in anesthetized animals, conscious cats, and in humans (Alam and Smirk 1937; Diepstra, Gonyea, and Mitchell 1980; Kaufman and Forster 1996; Lind 1983; Mark et al. 1985; Matsukawa et al. 1992; McAllen 1986a, 1986b; McCloskey and Mitchell 1972; Mitchell 1990; Seals, Chase, and Taylor 1988). It has been proposed that these signals are generated by activating mechanically- and metabolically-sensitive nerve endings (receptors) located in contracting skeletal muscle (Kaufman et al. 1984; Mitchell, Kaufman, and Iwamoto 1983). These neural signals are subsequently carried to the central nervous system by group III and group IV afferent fibers (McCloskey and Mitchell 1972). Together, activation of these receptors by muscle contraction along with the reflex cardiovascular responses has been termed the *exercise pressor reflex* (EPR) (Mitchell, Kaufman, and Iwamoto 1983; Mitchell 1990).

This concept was supported by studies which have found that skeletal muscle contraction reflexly increases arterial pressure and heart rate (Alam and Smirk 1937; Kaufman and Forster 1996; Lind 1983; McCloskey and Mitchell 1972; Mitchell, Kaufman, and Iwamoto 1983).

This chapter will summarize our current knowledge of the mechanisms which activate skeletal muscle afferents during static muscle contraction and the efferent sympathetic responses evoked following activation of muscle mechano- and metaboreceptors. The role of specific regions in the medulla and the associated neurochemical transmitters/modulators that are involved in mediating the cardiovascular responses to muscle contraction will also be discussed.

Afferent Mechanisms: Mechanically- and Metabolically-Sensitive Muscle Afferent Fibers and Their Activation

The reflex responses to static muscle contraction and passive muscle stretch are mediated by activation of group III (thinly myelinated AΔ) and group IV (unmyelinated C) afferent fibers (McCloskey and Mitchell 1972; Mitchell, Kaufman, and Iwamoto 1983). While group III and IV afferents mediate the reflex cardiovascular responses, their discharge patterns to an induced static muscle contraction differ considerably. The majority of group III afferents primarily exhibit mechanoreceptor properties and become activated abruptly at the onset of contraction (Kaufman et al. 1983, 1984; Kniffki, Mense, and Schmidt 1978; Mense and Stahnke 1983; Mitchell and Schmidt 1988).

By contrast, the majority of group IV afferents increase their activity approximately 15–20 sec following the onset of contraction (Kaufman et al. 1983, 1984; Kniffki, Mense, and Schmidt 1978; Mense and Stahnke 1983; Mitchell and Schmidt 1988). Furthermore, group IV afferent activity is potentiated by ischemia suggesting that group IV afferents are activated primarily by non-mechanical events such as an accumulation of metabolic by-products released by active muscle (Kaufman and Forster 1996; Kniffki, Mense, and Schmidt 1978, 1981; Mense and Stahnke 1983). Also, ischemia may potentiate the activation of group III fibers. Therefore, the reflex cardiovascular responses to muscle contraction appear to be mediated by both mechanically- (predominantly group III) and metabolically- (predominantly group IV) sensitive afferents while the reflex responses to passive muscle stretch are mediated by mechanically-sensitive fibers (Kaufman et al. 1983). This has been supported by the finding that anodal blockade prevented impulses from group I and II afferents (large, myelineated fibers with fast conduction velocity) from reaching the spinal cord, but failed to block the cardiovascular responses evoked by muscle contraction (McCloskey and Mitchell 1972).

However, a clear separation of the relative contribution of group III and IV afferents has been difficult to ascertain since these afferents exhibit polymodal firing behavior. At least a portion of the group IV muscle afferents are thought to transmit nociceptive input (Mense 1993). However, two recent studies have provided the first clear evidence that slowly-conducting muscle afferents transmit contraction-related sensory input. Pickar, Hill, and Kaufman (1994) and Adreani, Hill, and Kaufman (1997) reported that mild dynamic exercise, generating very low levels of muscle tension, was a sufficient stimulus to activate group III and IV muscle afferents. In these studies, rhythmical muscle contraction induced by stimulation of the mesencephalic locomotor region in unanesthetized decerebrate cats increased the firing of single-fiber group III and IV afferents. The low level of muscle tension generated during locomotion (i.e., approximately 350 gm) was considerably less that the tension development during tetanic muscle contraction (Kaufman et al. 1983; Kaufman and Forster 1996; Mense and Stahnke 1983).

Therefore, these studies indicate that group III and IV muscle afferents are stimulated by very low levels of muscle activity, thus strengthening the role of mechano- and metabosensitive muscle afferents

in transmiting sensory input associated with mild muscle contraction apparently distinct from nociception.

Metabolite Activation of Muscle Afferents

A considerable amount of attention has been placed upon which substances activate the metabolically-sensitive muscle afferent fibers during muscle contraction. Since the function of muscle afferent fibers in situ is to monitor the relationship between blood supply and the metabolic demand during muscle contraction, it follows that the generation of metabolites may activate muscle afferents. One approach has been to infuse specific metabolites into the arterial blood supply of a limb to examine the discharge activity of the muscle afferents, efferent sympathetic neural outflow, or the blood pressure response. Another method involves measuring these same responses during exercise with blockade or infusion of a metabolite, or in patients with metabolic or circulatory diseases (see chapter 20).

Infusion Studies

Arterial infusion of adenosine, lactic acid (but not sodium lactate, pH 7.4), diprotonated phosphate, potassium, bradykinin, and serotonin have been shown to evoke reflex sympathetic responses (Rybicki et al. 1984; Rotto and Kaufman 1988; Biaggioni et al. 1991; Kaufman and Forster 1996; Pan, Stebbins, and Longhurst 1993; Sinoway et al. 1993, 1994). Arterial infusion of the cyclooxygenase (and to a lesser extent the lipoxygenase) metabolites of arachidonic acid also elicit these responses but only during muscular contraction (Rotto and Kaufman 1988). Also, blockade of the cyclooxygenase metabolites by indomethacin attenuates the blood pressure response during static handgrip (Davy, Herbert, and Williams 1993). Thus, arachadonic acid and cyclooxygenase metabolites are thought to sensitize the discharge response of type III afferents to muscle contraction.

Adenosine infusion in the forearm increases muscle sympathetic nerve activity and blood pressure and elicits local sensations of pain (Biaggioni et al. 1991). However, it has been shown recently that the increase in sympathetic nerve activity in response to adenosine infusion may not be due to muscle metaboreceptor activation as previously suggested. MacLean et al. (1997) found that when adenosine was infused into the femoral artery, the increase in sympathetic nerve activity had a latency corresponding to the circulatory time for activation of either the central or peripheral chemoreceptors. In a subsequent trial, adenosine was infused, and cuff occlusion of the leg circulation was applied shortly after infusion to allow the infusate to distribute in the muscle in the vicinity of the afferent nerves but prevent recirculation centrally. Sympathetic nerve activity did not increase as long as the occlusion was maintained. When the cuff was released, there was again an increase in sympathetic nerve activity that corresponded with the time of recirculation necessary to activate the central or peripheral chemoreceptors. It was concluded that adenosine may not play a role in activating the muscle metaboreflex, but rather may act on peripheral or central chemoreceptors.

Exercise Studies

During exercise, an inherent difficulty in apportioning the reflex contribution to sympathetic nerve activity or blood pressure is that other cardiovascular regulatory mechanisms are redundantly activated. Early during exercise, blood pressure is increased by an elevation of cardiac output (Victor, Seals, and Taylor 1987; Mitchell et al. 1989a, 1989b) without concomitant increases in muscle sympathetic nerve activity and vascular resistance (Fernandes et al. 1990; Saito et al. 1993; Ray and Mark 1993; Toska and Eriksen 1994; Williamson et al. 1996). Yet, several metabolites, including pH, potassium, arachadonic acid, prostaglandins, bradykinin, and diprotonated phosphate, accumulating in muscle during contraction and measured either in the intramuscular compartment or in venous blood have been proposed to evoke the reflex elevation of sympathetic nerve activity and blood pressure (Victor et al. 1988; Sinoway, Prophet, Gorman, et al. 1988; Sinoway et al. 1993; Fallentin et al. 1992; Davy, Herbert, and Williams 1993; Fontana et al. 1994; Pan, Stebbins, and Longhurst 1993; Sinoway et al. 1994). Of these metabolites, pH, potassium, and diprotonated phosphate have been proposed as primary candidates for activating the muscle afferents. Still, there may not be a singularly important factor; rather the metaboreceptors may be redundantly activated.

It has been suggested that lactic acid plays an important role in activating muscle afferents during muscle contraction (Kaufman and Forster 1996; Mitchell, Kaufman, and Iwamoto 1983; Mitchell and Schmidt 1988). However, over the pH range that is typically found in muscle interstitium during exercise most of the generated lactic acid is in the form of lactate ion. Furthermore, a previous study has found that lactate is far less effective in evoking muscle reflex responses than lactic acid (Rotto and

Kaufman 1988). Therefore, it is likely that the acidification of some other metabolic product of muscle contraction may be responsible for sensitizing muscle afferent fibers during muscle contraction.

Muscle and venous pH fall in parallel with the increases in sympathetic nerve activity (Victor, Seals, and Taylor 1987), vascular resistance (Sinoway, Prophet, Gorman, et al. 1988), and discharge rates of group III and IV muscle afferents during exercise (Rotto and Kaufman 1988; Sinoway et al. 1993). Also, muscle sympathetic nerve activity, vascular resistance, and mean arterial pressure are elevated during postexercise muscle ischemia in association with reduced intramuscular and venous pH (Sinoway, Prophet, Gorman, et al. 1988; Victor et al. 1988; Boushel et al. 1998). This relation is also supported by the finding that during handgrip followed by postexercise muscle ischemia, dichloroacetate infusion attenuates the reflex increase in sympathetic nerve activity and blood pressure (Ettinger et al. 1991). Dichloroacetate decreases lactic acid formation for a given amount of pyruvate production by inhibiting pyruvate dehydrogenase kinase, which increases mitochondrial levels of pyruvate dehydrogenase. Similarly, the reflex increase in sympathetic nerve activity is attenuated during handgrip in patients with McArdle's disease who are incapable of lactic acid production (Pryor et al. 1990).

Recent experiments involving McArdle's patients dispute previous findings (Pryor et al. 1990) on the role of pH as the primary activator of the muscle metaboreflex. Vissing et al. (1998) recently found increases in muscle sympathetic nerve activity during moderate and high intensity handgrip in McArdle's patients in the absence of pH changes. The explanation given for the discrepancy between this finding and that of Pryor et al. is that in the study by Pryor et al., the maximal voluntary contractions were probably not truly maximal because of the concern for the development of contractures (personal communication). Thus, since the relative intensities of contractile force during exercise likely were modest and the duration of the exercise was brief, it is likely that metabolite accumulation in the interstitium was modest. The recent finding by Vissing et al. (1998) of increased sympathetic discharge during exercise in McArdle's patients is consistent with the observation that cardiac output and blood pressure responses during dynamic exercise are exaggerated in McArdle's patients. The stimulus for the higher cardiac output and blood pressure responses compared to healthy adults is thought to be of reflex origin, yet there is also no change in muscle pH (Lewis and Haller 1990). It is worth noting that in these patients, potassium and possibly adenosine concentrations are exaggerated compared to controls. Thus, these patients may have some compensatory metabolic responses which activate the muscle reflex.

The importance of pH as a direct stimulant of the afferent nerves has also been disputed by the observation that upon cessation of exercise, sympathetic nerve activity and MAP decrease rapidly while pH remains depressed and returns slowly to resting levels. It has been suggested that the active stimulant of the afferent nerves may not be the hydrogen ion itself but the diprotonated form of phosphate ($H_2PO_4^-$), which has also been linked to muscle fatigue and vasodilation (Hilton, Hudlicka, and Marshall 1978; Miller et al. 1988; Wilson et al. 1988; Sinoway et al. 1994). This idea is based on the finding that in animals arterial infusion of $H_2PO_4^-$ induces reflex increases in muscle sympathetic nerve activity, and in humans intracellular $H_2PO_4^-$ is closely correlated to sympathetic nerve activity during handgrip exercise, postexercise ischemia, and recovery (Sinoway et al. 1994). The acid dissociation constant (pKa) for the conversion of the monoprotonated to the diprotonated form of phosphate is ~6.8 (Wilson et al. 1988). During exercise, the concentration of $H_2PO_4^-$ rises significantly as pH drops and as Pi is generated. During recovery, $H_2PO_4^-$ decreases quickly due to the rapid resynthesis of PCr. pH displays a much slower recovery because the diminished production of cellular H^+ is offset by the release of H^+ as PCr is regenerated (Meyer 1988). Sinoway and colleagues (1994) investigated the role of diprotonated phosphate in evoking muscle reflex responses in both cats and humans (figure 2.1). They reported that during ischemic handgrip exercise in human subjects, the cellular concentration of $H_2PO_4^-$ correlated with muscle sympathetic nerve activity (MSNA) in seven of the eight subjects tested, while in only one subject was there a significant correlation between cellular pH (or H^+) and MSNA. Furthermore, the pressor response evoked by an intra-arterial injection of a solution of pH 6 phosphate (diprotonated phosphate) supplying the triceps surae of the cat was greater than the responses to injections of lactate (at the same pH) or a solution of pH 7.5 phosphate (predominantly monoprotonated phosphate). These results suggest that when the muscle becomes acidic during exercise, the generation of phosphate leads to an increased concentration of $H_2PO_4^-$, which activates muscle afferent fibers to evoke the reflex cardiovascular responses to muscle contraction.

The importance of potassium for evoking the reflex during exercise is indicated by potassium's close relationship to the increase in blood pressure during forearm handgrip since both blood pressure and venous potassium remain elevated during post-

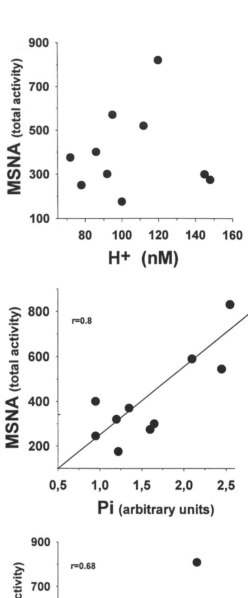

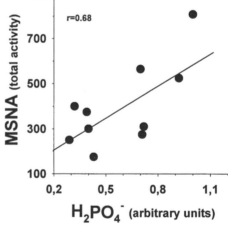

Figure 2.1 Representative example from one subject showing the relationship between muscle sympathetic nerve activity (MSNA) versus H+ ion in the upper panel, versus inorganic phosphate (Pi) in the middle panel, and versus diprotonated phosphate (H2PO4-) in the lower panel during ischemic handgrip exercise. Significant correlations were found for Pi and H2PO4- with MSNA, however, lack of a significant relationship was found between H+ and MSNA. (Adapted from Sinoway et al. 1994.)

exercise muscle ischemia (Fallentin et al. 1992). However, during constant-load exercise of moderate intensity, venous potassium reaches a plateau while mean arterial pressure continues to rise (Gaffney, Sjøgaard, and Saltin 1990; Fallentin et al. 1992; Boushel et al. 1998). Thus, potassium is likely to play some role in activating the muscle reflex, but other factors also appear to be involved.

In summary, several metabolites have been shown to activate the muscle afferents and be linked to the efferent responses to muscle reflex activation. Yet, clear evidence in support of a direct causal relationship between specific metabolites and the muscle reflex is limited by the fact that little is known about the interstitial metabolite concentrations where the afferent nerve endings are located. Presently, it is still uncertain what substances evoke the muscle reflex during exercise or the mechanisms whereby redundant metabolite activation may constitute the stimulus.

Muscle Fiber Type

Because many of the cardiovascularly-related afferents in skeletal muscle respond to metabolic signals, it is likely that the reflex cardiovascular responses are influenced by the fiber composition, and thus the type of muscle metabolism of the contracting muscle. This hypothesis was tested by Wilson et al. (1995). Following 21 days of continuous low-frequency electrical stimulation of the tibial nerve of one hind limb in rabbits, the gastrocnemius (predominantly a glycolytic muscle) was converted into a primarily oxidative muscle based on approximately a 300% increase in citrate synthase and succinate dehydrogenase activity. Accompanying this change in fiber composition of the chronically stimulated hind limb, the reflex cardiovascular response to electrically-induced static muscle contraction was significantly less than the response evoked by contraction of the control glycolytic hind limb. These results suggest that the magnitude of the pressor response to static contraction is influenced by the fiber composition of the contracting muscle in such a fashion that a primarily glycolytic muscle group will evoke larger reflex responses.

Neuropeptides and Neurotransmitters in the Dorsal Horn

The majority of group III and group IV skeletal muscle afferent fibers make their first synapse in the dorsal horn of the spinal cord (Kalia, Mei, and Kao 1981; Light and Perl 1979). Since the dorsal horn contains a large network of cell bodies and nerve fibers (Calaresu and Yardley 1988), it is not surprising that anatomical studies have identified an abundance of neurotransmitters and neuromodulators in this region (McNeill, Westlund, and Coggeshall 1989; Seybold and Elde 1980). A large concentration of substance P (SP) and somatostatin (SOM) has been found in this region, and the predominant source of these neuropeptides has been shown to be the primary afferents (Johansson, Hokfelt, and Elde 1984; McNeill, Westlund, and Coggeshall 1989; Pickel, Reis, and Leeman 1977).

Furthermore, previous studies have demonstrated that activation of primary afferents causes the spinal release of SP (Brodin et al. 1987; Go and Yaksh 1987; Kuraishi et al. 1985). Recently, Wilson and colleagues (Wilson et al. 1992, 1993; Wilson, Fuchs, and Mitchell 1993; Wilson, Wall, and Matsukawa 1995) conducted a series of studies that examined the modulatory role of these two peptides on neurotransmission of muscle afferent activity through the dorsal horn. Microinjections of antagonists to either SP or SOM into the dorsal horn attenuated the reflex MAP and HR response evoked by electrically-induced contraction of the triceps surae (Wilson et al. 1992), while coadministration of both antagonists was found to significantly reduce, but not obliterate, the reflex cardiovascular responses to muscle contraction (Wilson, Wall, and Matsukawa 1995).

A study by Wilson et al. (1993) found that activation of muscle afferents by static muscle contraction increased the extracellular concentration of SP in the dorsal horn utilizing in vivo microdialysis. In another study, Wilson, Fuchs, and Mitchell (1993) found that the extracellular concentration of SP was related to the level of developed muscle tension during static muscle contraction (figure 2.2). These findings strongly implicate the role of SP release in mediating the cardiovascular responses to muscle contraction at the level of the dorsal horn. SP was likely acting as a neuromodulator and not as a neurotransmitter in the dorsal horn as it was found that microinjections of SP into the dorsal horn when the hindlimb was quiescent failed to evoke a pressor response (Wilson et al. 1992). Therefore, the release of additional neuroactive substances are also involved in mediating the cardiovascular responses evoked by muscle contraction.

It has been shown that excitatory amino acids, such as L-glutamate, mediate a fast excitatory postsynaptic potential (EPSP) which accompanies the slower EPSP in second-order neurons in the dorsal horn of the spinal cord (Jessell, Yoshioka, and Jahr 1986). Furthermore, involvement of both SP (NK1) and N-methyl-D-aspartate (NMDA) receptor mechanisms have been demonstrated in spinal

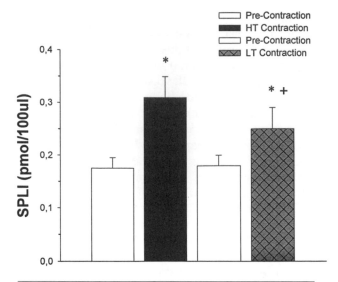

Figure 2.2 Concentration of substance P-like immunoreactivity before (open bars) and during high-tension (HT) contractions (filled bar) and low-tension (LT) contractions (hatched bar). Data are mean +SD (n=8). * Significant difference compared with precontraction values. + Significant difference compared with HT contractions. (From Wilson, Fuchs, and Mitchell 1993.)

facilitation of sensory transmission produced by repetitive stimulation of C-fiber afferents (Kellerstein et al. 1990; Liard, Hargreaves, and Hill 1993; Xu, Dalsgaard, and Wiesenfeld-Hallis 1991). In accordance with these findings, Hand et al. (1996a) reported that the extracellular concentration of L-glutamate was increased in the dorsal horn during static contraction. Furthermore, these same investigators also found that microdialyzing both a NMDA (AP-5) and a non-NMDA (CNQX) receptor antagonist into the dorsal horn of the spinal cord significantly attenuated the reflex cardiovascular responses to static muscle contraction (Hand et al. 1996b; Hand, Ordway, and Wilson 1996).

These findings show that the reflex pressor response to static muscle contraction is mediated by activation of both NMDA and non-NMDA receptors at the level of afferent fiber entry into the dorsal horn. In conjunction with previous studies demonstrating the neuromodulatory role of SP, these results implicate the involvement of both L-glutamate and SP in neurotransmission of muscle afferent activity at the level of the dorsal horn during muscle contraction.

Ascending Projections of Skeletal Muscle Afferents to the Brain Stem

From the dorsal horn, muscle afferents project to the brain stem via the spinal cord. Supraspinal centers

are essential for the complete expression of the exercise pressor reflex. Following transection of the spinal cord at the C1 level, most of the reflex cardiovascular responses to muscle contraction were eliminated (Iwamoto et al. 1985). Conversely, transection of the rostral portion of the medulla (5mm rostral to the obex) had only a minor effect on the magnitude of the pressor response (Iwamoto et al. 1985). Therefore, intact medullary sites are a prerequisite for the expression of the exercise pressor reflex. Although the precise pathways within the spinal cord from the dorsal horn to the brain stem remain unknown, the dorsolateral sulcus (Kozelka and Wurster 1985) and the ventral spinal cord (Iwamoto, Botterman, and Waldrop 1984) appear to carry the majority of the afferent projections from muscle to the brain stem.

Medullary Sites: Identification of Brain Stem Centers Involved in the Exercise Pressor Reflex by Expression of the Proto-Oncogene c-Fos

The tracing of CNS multisynaptic pathways has been made possible by a technique based on the observation that the cellular proto-oncogene c-Fos encoding for a nuclear phosphoprotein Fos (Dragunow and Faull 1989; Morgan and Curran 1991). Because c-Fos expression in neurons is increased following synaptic stimulation, immunohistochemical detection of the protein product Fos can serve as a marker of CNS polysynaptic pathways involved in specific physiological responses. A number of recent studies have taken advantage of this technology to map out the central neuronal pathways of the arterial baroreceptor reflex (Erickson and Millhorn 1991; Li and Dampney 1992; McKitrick, Krukoffm, and Calaresu 1992; Rutherford et al. 1992; Sved et al. 1994). Recently, a study by Iwamoto et al. (1996) measured the levels of Fos protein in the diencephalic and brain stem of rats following 45 minutes of treadmill running. Heavy labeling of Fos protein was found in the hypothalamic locomotor regions, periaqueductal grey matter, parabrachial complex, the cuneiform nucleus of the diencephalon, the nucleus tractus solitarius, and ventrolateral medulla. Furthermore, the cellular colocalization of Fos with specific markers of synthetic enzymes, neurotransmitters, and/or neuropeptides can be determined using double-labeling techniques (Murphy et al. 1994; Sved et al. 1994).

Recently, Li et al. (1997) examined the distribution of Fos protein in several brain stem sites in response to electrically-induced contraction of the triceps surae in anesthetized cats. The goal of this study was to identify medullary neurons which were activated during skeletal muscle contraction. Additionally, Fos-postive neurons were co-stained for phenylethanolamine-N-methyltransferase (PNMT) to determine if these neurons were catecholaminergic. Significant Fos-like immunoreactivity was found in several regions of the brain stem, including the rostral and caudal ventrolateral medulla and the nucleus of the solitary tract (figure 2.3). Within a region of the rostral ventrolateral medulla (subretrofacial nucleus) which is known to be critical in the regulation of cardiovascular function (Ciriello, Caverson, and Polosa 1986; Dampney et al. 1987; McAllen 1986a; Millhorn and Eldridge 1986; Spyer 1994), 66% of the Fos-positive neurons were found to contain PNMT. This finding suggests that adrenergic neurons of the rostral ventrolateral medulla may be involved in eliciting the pressor response to skeletal muscle contraction.

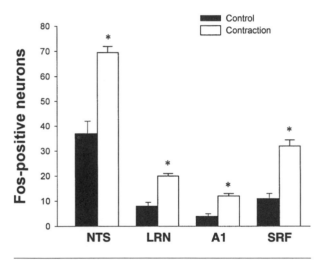

Figure 2.3 Histogram showing the mean number of Fos-positive staining neurons per section in the medulla in sham-operated control animals (solid bars) and in cats following 60 minutes of electrically-induced static muscle contraction (open bars). *Significant difference compared with sham-operated values (P<0.05). NTS = nuleostractus solitarius; LRN = lateral reticular nucleus; A1 = A1 region of the medulla; SRF = subretrofacial nucleus. (From Li et al. 1997.)

A potential limitation of using Fos-immunohistochemistry during muscle contraction is that multiple afferent pathways are activated. Since the central baroreceptor pathways include neuronal populations in the VLM and NTS, it is difficult to reconcile if Fos-labeled neurons in these regions were activated by synaptic input from skeletal muscle receptors and/or arterial baroreceptors. Recently, Li, Potts, and Mitchell (1998) completed a

study to determine if neurons in the VLM and NTS were activated by muscle contraction in the absence of baroreceptor activation. The findings demonstrated that some VLM and NTS neurons were activated during muscle contraction independent of arterial baroreceptor input. This finding supports the notion that neurons in these medullary regions may be involved in the central processing of neural input during muscle contraction.

Nucleus of the Solitary Tract

While the majority of muscle afferents synapse in the dorsal horn of the spinal cord, some evidence suggests that not all afferents make their first synapse at this location. Using transganglionic transport of horseradish peroxidase (HRP), Kalia, Mei, and Kao (1981) demonstrated that some muscle afferents project directly to the dorsal column nuclei and the nucleus tractus solitarius (NTS). This finding supports the notion that first-order afferent fibers from muscle project to a major sensory nucleus related to cardiovascular and respiratory control, and provides anatomical evidence for the short-latency reflex responses mediated by muscle afferents. Further evidence that muscle afferents project to the NTS has been shown by electrophysiological recordings from single cells in the NTS. Person (1989) electrically stimulated the peroneal (muscle), sural (cutaneous), and vagal (cardiopulmonary) nerves in the anesthetized cat and simultaneously recorded single-unit extracellular activity from neurons in several subnuclei of the NTS. A high degree of convergence of peroneal, sural, and vagal afferents on NTS neurons was observed. These inputs were capable of altering (inhibiting) the activity of NTS cells. Similarly, McMahon et al. (1992) determined the effect of electrical stimulation of the carotid sinus and peroneal nerves on NTS neuronal activity in the anesthetized cat. They found that 17 of 18 neurons that were activated by carotid sinus nerve stimulation were inhibited by peroneal nerve stimulation. Recently, Toney and Mifflin (1994) demonstrated a time-dependent inhibition of somatic nerve-evoked NTS cell discharge. These findings indicate that peripheral input from skeletal muscle is capable of modulating the excitability of NTS neurons. This raises the possibility that the NTS may represent a site of integration for the exercise pressor reflex.

Since earlier studies identified NTS neurons using whole nerve electrical stimulation, this may limit interpretation of these results to normal physiological conditions. Recently, Potts, Li, and Waldrop (1998) characterized the responsiveness of NTS neurons to sensory input from skeletal muscle receptors and arterial baroreceptors using an activation paradigm that more closely mimics the normal physiological state. Using this paradigm, skeletal muscle afferents were activated by induced contraction of the hind limb using graded electrical stimulation of the peripheral cut ends of L7-S1 ventral roots, while arterial baroreceptors are activated by mechanical means (i.e., graded inflation of an aortic balloon catheter). These studies were performed in anesthetized cats. Using this approach (instead of electrical stimulation of whole nerves), two distinct populations of NTS neurons have been identified: one population is barosenstive; a second population of NTS cells are sensitive to neural input from skeletal muscle receptors but appear to be insensitive to synaptic input from arterial baroerceptors. These findings suggest that the central processing of sensory input evoked by muscle contraction may be accomplished, in part, by two separate and distinct populations of barosensitive and contraction-sensitive NTS neurons. Further experimentation is needed to completely understand the neural circuitry of the NTS involved in the synaptic processing of baroreceptor and skeletal muscle receptor afferent input.

In addition to afferent projections from skeletal muscle, cells within the NTS receive arterial and cardiopulmonary-related afferent projections from the vagus and glossopharyngeal nerves, as well as gastrointestinal afferents (Jordan and Spyer 1986; Spyer 1994). The central baroreceptor pathway consists of: 1) neurons in the NTS that receive primary baroreceptor afferents (Ciriello 1983); 2) efferent projection from the NTS to neurons in the cVLM (Aicher et al. 1995); 3) direct inhibitory projection from the cVLM to sympathoexcitatory neurons in the C1 area of the rVLM (Jeske, Reis, and Milner 1995); and 4) direct projection from the rVLM to preganglionic sympathetic neurons in the spinal cord (Ciriello, Caverson, and Polosa 1986; Dampney et al. 1987). Furthermore, neurons in both the cVLM and rVLM that respond to muscle contraction also possess cardiovascular rhythmicity, some of which persists following barodenervation (Bauer et al. 1992; Iwamoto and Kaufman 1987; Iwamoto et al. 1989). This implies that VLM neurons receive input from both skeletal muscle and baroreceptor afferents. Therefore, this region of the medulla may represent the anatomical substrate which mediates the well-documented inhibitory interaction between the arterial baroreceptor reflex and the exercise pressor reflex (Sheriff et al. 1990; Waldrop and Mitchell 1985).

Ventrolateral Medulla

Considerable evidence has now accumulated indicating that the sites within the medulla, particularly the ventrolateral medulla (VLM), play an important role in the autonomic reflex arc of the exercise pressor reflex. Several lines of evidence support the role of the caudal (cVLM) and rostral (rVLM) ventrolateral medulla in the transmission of muscle afferent activity in the exercise pressor reflex arc. Ciriello and Calaresu (1977) found that bilateral lesions of the lateral recitular nucleus (LRN) in the cat, located within a region generally referred to as the caudal VLM, abolished the reflex pressor response to sciatic nerve stimulation. Similar results were found by Iwamoto et al. (1982) to electrically-induced static muscle contraction following LRN lesioning. To circumvent the limitations of electrolytic lesions, Bauer, Iwamoto, and Waldrop (1989) "chemically" lesioned the cVLM in cats using bilateral microinjections of kynurenic acid (a non-specific inotropic excitatory amino acid [EAA] receptor antagonist). They reported that the pressor response to static contraction of the hind limb was reversibly attenuated by kynurenic acid. Further evidence for the involvement of neurons in the cVLM in mediating afferent activity from skeletal muscle comes from studies recording the extracellular discharge of cells in this region. The discharge activity of neurons in this region has been shown to be altered by static muscle contraction (Bauer, Iwamoto, and Waldrop 1990; Iwamoto and Kaufman 1987). In addition, static muscle contraction increases the metabolic activity of cVLM cells (Iwamoto et al. 1984). Cells in the cVLM increased their uptake of radioactive glucose during electrically-induced static muscle contraction in the anesthetized cat. Interestingly, these cells were in the same area of the cVLM that attenuated the reflex cardiovascular responses to muscle contraction when microinjected with kynurenic acid (Bauer, Iwamoto, and Waldrop 1989). These studies demonstrate that neurons in the cVLM which utilize EAA receptors are activated by hind limb muscle contraction and appear to be involved in the neurotransmission of the exercise pressor reflex.

The rostral ventrolateral medulla (rVLM) also plays a role in transmission of muscle afferent activity (Kaufman and Forster 1996). Previous studies have demonstrated that pressor neurons within the rVLM play an important role in the tonic and phasic control of blood pressure (Calaresu and Yardley 1988; Ciriello, Caverson, and Polosa 1986;

Person 1989). These cells project directly to preganglionic sympathetic neurons in the intermediolateral (IML) cell column of the spinal cord (Ciriello, Caverson, and Polosa 1986; Dampney et al. 1987) and are responsible for basal sympathetic nerve discharge (Calaresu and Yardley 1988). In the cat, several lines of evidence indicate that neurons in the rVLM are pressor neurons that may be activated during static muscle contraction. First, microinjections of EAAs into this discrete region elicit large increases in blood pressure (McAllen 1986a, 1986b). Second, blocking EAA receptors in the rVLM by microinjecting kynurenic acid has been shown to attenuate the cardiovascular responses to hind limb muscle contraction (Bauer, Iwamoto, and Waldrop 1989). Furthermore, Kiely and Gordon (1993) have recently shown that bilateral microinjection of a specific non-NMDA receptor antagonist in the rVLM of the rat virtually attenuated the pressor response to electrical stimulation of the sciatic nerve. Finally, single-unit extracellular recording techniques have shown that the firing pattern of rVLM neurons is altered by muscle contraction. Bauer and colleagues (Bauer, Iwamoto, and Waldrop 1990; Bauer et al. 1992) have found cells within the rVLM which increase their discharge activity by 70% during electrically-induced static muscle contraction of the hind limb in anesthetized cats. These findings provide strong evidence that neurons in the rVLM play an important role in mediating the reflex sympathoexcitation following activation of mechano- and metaboreceptors during skeletal muscle contraction.

Role of Excitatory Amino Acids and Neuropeptides in the Brain Stem

Knowledge of the mammalian neurotransmitters/neuromodulators in the central nervous system has been well-established. Mounting evidence suggests that EAAs and their receptors play a pivotal role in neurotransmission in many regions of the brain stem, including the NTS, rVLM, and cVLM (Guynet, Filtz, and Donaldson 1987; Lawrence and Jarrott 1996; Sapru 1994). For example, blockade of EAA receptors in the cVLM attenuates the baroreflex response to an elevation in arterial pressure (Gordon 1987). Furthermore, administration of an antagonist to EAA receptors in the rVLM blunts the pressor response to static muscle contraction or high-frequency stimulation of vagal afferent fibers (Abrahams et al. 1994; Bauer, Iwamoto, and Waldrop 1989; Rutherford et al. 1992).

In addition, a number of neuropeptides, including substance P, calcitonin gene-related peptide, and 5-hydroxytryptamine, have been identified as mediating neurotransmission in brain stem regions (Batten 1995; Hokfelt et al. 1975; Lawrence and Jarrott 1996; Sapru 1994; Sykes, Spyer, and Izzo 1994). Previous studies using in vivo microdialysis have shown an increased release of substance P in the NTS in response to aortic depressor nerve stimulation and during hypoxia (Lindefors et al. 1986; Morilak, Morris, and Chalmers 1988; Srinivasan et al. 1991). Williams et al. (1994) have also detected immunoreactive substance P-like substances from cat brain stem sites during fatiguing isometric contractions. However, since arterial baroreceptor populations remained intact in this study, it was not possible to discern the source of afferent input responsible that increased substance P-like immunoreactivity during muscle contraction. Recently, Potts et al. (1999) completed a study to determine the relative contribution of these two sources of afferent input (i.e., arterial baroreceptor input vs. afferent input orginating from skeletal muscle receptors) on substance P release during induced muscle contraction (Potts et al. 1999). Utilizing the in vivo microdialysis technique, release of substance P from the NTS during muscle contraction in baroreceptor intact and barodenervated cats was determined. Activation of contraction-sensitive skeletal muscle receptors in baroden-ervated cats released substance P in the NTS independent of arterial baroreceptor input. However, when the relative contribution of baroreceptor and skeletal muscle receptor activation on substance P release was determined, a very interesting finding emerged. Muscle contraction alone, and baroreceptor activation alone, increased substance P-like immunoreactivity approximately 40%, respectively. However, when muscle contraction was performed in barointact animals, substance P-like immunoreactivity increased 150% (figure 2.4). This finding suggests that converging sensory input from arterial baroreceptors and skeletal muscle receptors facilitate the release of substance P in the NTS via a synergistic interaction. A differential distribution of neuropeptides in the central axons of sensory nuclei in the dorsal brain stem (Sykes, Spyer, and Izzo 1994), in conjunction with the viscerotropic organization of efferents from this region (Dampney and McAllen 1988), suggests that "chemical coding" of neural output may be involved in the central processing mechanisms intrinsic to the NTS.

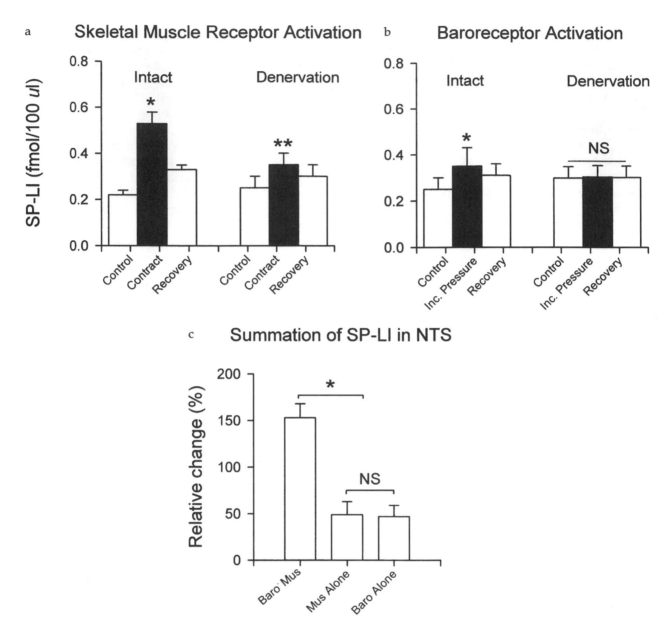

Figure 2.4 *(a)* Average changes in substance P-like immunoreactivity (SP-LI) during electrically-induced muscle contraction (contract) in baroreceptor intact and denervated animals. * Significant difference from control and recovery. ** Significant difference from control. *(b)* Average changes for SP-LI during baroreceptor activation (inc. pressure) in baro-intact and denervated animals. * Significant difference from control. *(c)* Relative change in SP-LI from control during the following conditions: contraction in intact animals (baro mus); contraction in denervated animals (mus alone); baroreceptor activation in intact animals (baro alone). * Significant difference between mus alone/baro alone and baro mus. (Adapted from Potts et al., 1999.)

Efferent Sympathetic Responses of the Exercise Pressor Reflex in Animals

It has been known for a number of years that sustained tetanic contraction of muscle by spinal ventral root stimulation increases arterial blood pressure, heart rate, left ventricular contractility, and cardiac output in both cats and dogs (McCloskey and Mitchell 1972; Mitchell, Kaufman, and Iwamoto 1983; Mitchell and Schmidt 1988). These reflex cardiovascular responses to muscle contraction are thought to be mediated by increases in efferent sympathetic nerve activity directed to the heart and regional vascular beds. Direct projections from the ventrolateral quadrant of the medullary reticular formation (rostral ventrolateral medulla) to the intermediolateral cell columns activate sympathetic preganglionic neurons that innervate the heart and regional vascular beds (Ciriello, Caverson, and Polosa 1986; Dampney and McAllen 1988). Matsukawa and colleagues (1992, 1994) have recorded postganglionic sympathetic nerve activity to both the heart and the kidney in response to static muscle contraction in the cat (figure 2.5). In the same animal they simultaneously recorded cardiac sympathetic nerve activity (CSNA) from a branch of the left inferior cardiac sympathetic nerve and renal sympathetic nerve activity (RSNA) from a branch of the renal nerve innervating the left kidney during electrically-induced static muscle contraction. They reported that both CSNA and RSNA increased immediately following the onset of muscle contraction indicating that activation of mechanically-sensitive muscle afferents increases sympathetic nerve activity. Furthermore, the elevation in both CSNA and RSNA persisted throughout the duration of the contraction despite falls in tension development, suggesting that activation of receptors other than mechanoreceptors (e.g., muscle metaboreceptors) may also evoke sympathoexcitation. Therefore, both

reflex mechanisms contribute to the increased inotropic and chronotropic states of the myocardium, as well as to the redistribution of cardiac output from the kidney and to the contracting muscle during static muscle contraction.

Sympathoadrenal activation also plays an important role in producing the cardiovascular and metabolic adjustments to exercise. To determine the effect of skeletal muscle contraction on adrenal sympathetic nerve activity (ASNA), Vissing et al. (1991) recorded ASNA in anesthetized rats during tibial nerve stimulation before and after neuromuscular blockade. During electrically-induced static contraction ASNA rapidly increased following the onset of contraction with a latency of <1 sec. This finding supports the hypothesis that neurogenic reflexes arising from contracting skeletal muscle may play a role in triggering the sympathoadrenal activation during exercise.

These findings indicate that a mechanism responsible for mediating the changes in efferent sympathetic activity during muscle contraction is activation of mechanically and metabolically sensitive muscle afferents. Recent studies by O'Hagan, Casey,

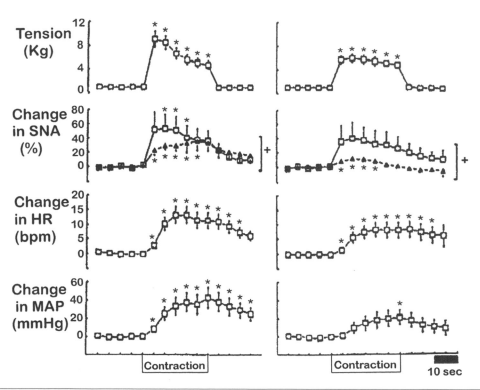

Figure 2.5 (a) Left panel shows the average changes in tension of triceps surae muscle, cardiac (▲) and renal (□) sympathetic nerve activity (SNA), heart rate (HR), and mean arterial pressure (MAP) during static muscle contraction induced by electrical stimulation of L7-S1 ventral roots at 40 Hz. * Significant difference between mean values of cardiac SNA and renal SNA during ventral root stimulation at a given time point. (b) Right panel shows the same responses as illustrated in (a) when the ventral roots were stimulated at 20 Hz. (Adapted from unpublished personal data from Matsukawa and Mitchell.)

and Clifford (1997) and Mittelstadt and colleagues (1996) have shown that increases in RSNA during dynamic exercise in conscious animals is mediated, in part, by activation of metabolically-sensitive skeletal muscle receptors. However, the targets for efferent sympathetic outflow during muscle contraction are not limited to the kidney. Hill, Adreani, and Kaufman (1996) reported that static contraction of the triceps surae increased the discharge of sympathetic efferents innervating the contralateral hindlimb in decerebrate cats. This would have likely resulted in vasoconstriction to the inactive skeletal muscle bed and a reduction in blood flow. Mittelstadt et al. (1994) also reported that activation of metabolically-sensitive skeletal muscle afferents by localized ischemia reduced vascular conductance in nonischemic exercising muscle. These findings suggest that in addition to the renal vasculature, nonactive skeletal muscle and metabolically-active skeletal muscle beds are also targets for sympathetic vasoconstriction. Since skeletal muscle has the capacity for enormous increases in blood flow during exercise, a tonic level of sympathetic vasconstriction may be necessary to prevent the exercise-related increases in vascular conductance from exceeding the capacity of the heart to pump blood. This hypothesis was recently confirmed by O'Leary, Robinson, and Butler (1997) who reported that sympathetic activity directed to hind limb skeletal muscle tonically opposed metabolic vasodilation during dynamic exercise in the conscious dog. Therefore, the control of efferent sympathetic nerve activity during muscle contraction appears to be exquisitely regulated to facilitate blood flow perferentially to contracting skeletal muscle while maintaining a critical level of systemic perfusion pressure.

Efferent Responses of the Exercise Pressor Reflex in Humans

Microneurographic techniques have been used to record sympathetic neural discharge of nerve fascicles to muscle. These studies have established that the exercise pressor reflex increases sympathetic neural discharge to skeletal muscle during exercise in man (Mark et al. 1985; Victor, Seals, and Taylor 1987; Victor et al. 1988; Victor and Seals 1989; Wallin, Victor, and Mark 1989; Pryor et al. 1990; Sinoway et al. 1993; Hansen et al. 1994, 1996; Ray, Secher, and Mark 1994; Sinoway et al. 1994). Sympathetic nerve activity increases during both static and dynamic exercise and remains elevated during postexercise muscle ischemia involving the forearm and leg (Mark et al. 1985; Sinoway, Wilson, Zelis, et al. 1988;

Victor et al. 1988; Wallin, Victor, and Mark 1989; Saito et al. 1993; Hansen et al. 1994; Ray 1993; Ray, Secher, and Mark 1994).

The link between sympathetic nerve activity and the exercise pressor reflex was established in studies that reported an increase in efferent neural discharge and blood pressure that coincided with an accumulation of muscle metabolites measured by 31P magnetic resonance spectroscopy or reflected in effluent venous blood (Sinoway, Prophet, Gorman, et al. 1988; Victor et al. 1988; Pryor et al. 1990; Fallentin et al. 1992; Sinoway et al. 1994). These findings have been supported by studies using selective blockade of specific metabolites produced during contraction. Further evidence has been gathered from studies involving patients with disorders of muscle metabolism in which sympathetic nerve activity is attenuated (Pryor et al. 1990; Ettinger et al. 1991; Davy, Herbert, and Williams 1993; Fontana et al. 1994).

Conclusions

Over the past century, major advances have been made in our understanding of the neural mechanisms regulating the cardiovascular system during exercise. This work has significantly increased our knowledge of the role of skeletal muscle afferents in mediating the cardiovascular responses to muscular exercise, as well as the mechanisms involved in neurotransmission of these signals at the level of both the spinal cord and the brain stem. Clearly, more research is needed to elucidate the neural circuitry, the neuroactive substances, and the neuronal mechanisms involved in transmission of these signals during exercise. Future direction for research in this field needs to focus on several areas. First, the role of the central nervous system in integrating the myriad of neural inputs that the brain continuously receives and processes during muscular exercise needs to be elucidated. Second, despite considerable work devoted to uncovering the metabolites that activate the exercise pressor reflex, it is still unclear which substances are the primary activators. Moreover, the neurophysiology of muscle afferent fiber activation and the potential for redundant or synergistic responses by different metabolites remain to be explored.

References

Abrahams, T.P., P.J. Hornby, K. Chen, A.M. Taveira Dasilva, and R.A. Gillis. 1994. The non-NMDA subtype of excitatory amino acid receptor plays the major role in control of cardiovascular function by the subretrofacial nucleus in cats. *J. Pharmacol. Expt. Therp.* 270(1):424–432.

Adreani, C.M., J.M. Hill, and M.P. Kaufman. 1997. Responses of group III and IV muscle afferents to dynamic exercise. *J. Appl. Physiol.* 82(6):1811–1817.

Aicher, S.A., O.S. Kurucz, D.J. Reis, and T.A. Milner. 1995. Nucleus tractus solitarius efferent terminals synapse on neurons in the caudal ventrolateral medulla that project to the rostral ventrolateral medulla. *Brain Res.* 693:51–63.

Alam, M., and F.H. Smirk. 1937. Observation in man upon a blood pressure raising reflex arising from the voluntary muscles. *J. Physiol. (Lond)* 89:372–383.

Batten, T.G.C. 1995. Immunolocalization of putative neurotransmitters innervating autonomic regulating neurones of the cat medulla. *Brain Res. Bull.* 37(5):487–506.

Bauer, R.M., G.A. Iwamoto, and T.G. Waldrop. 1989. Ventrolateral medullary neurons modulate the pressor reflex to muscular contraction. *Am. J. Physiol.* 257:R1154–R1161.

Bauer, R.M., G.A. Iwamoto, and T.G. Waldrop. 1990. Discharge patterns of ventrolateral medullary neurons during muscle contraction. *Am. J. Physiol.* 259:R606–R611.

Bauer, R.M., T.G. Waldrop, G.A. Iwamoto, and M.A. Holzwarth. 1992. Properties of ventrolateral medullary neurons that respond to muscular contraction. *Brain Res. Bull.* 28:167–178.

Biaggioni I., T.J. Killian, R. Mosqueda-Garcia, R.M. Robertson, and D. Robertson. 1991. Adenosine increases sympathetic nerve traffic in humans. *Circ.* 83:1668–1675.

Boushel, R., P. Madsen, B. Quistorff, and N.H. Secher. 1998. Contribution of pH, potassium and diprotonated phosphate for the reflex increase in blood pressure during handgrip. *Acta Physiol Scand.* 164:269–275.

Brodin, E., B. Linderoth, B. Gazelius, and U. Ungerstedt. 1987. In vivo release of substance P in the cat dorsal horn studies with microdialysis. *Neurosci. Lett.* 76:357–362.

Calaresu, F.R., and C.P. Yardley. 1988. Medullary basal sympathetic tone. *Ann. Rev. Physiol.* 50:511–524.

Ciriello, J. 1983. Brain stem projections of aortic baroreceptor afferent fibers in the rat. *Neurosci. Lett.* 36:37–42.

Ciriello, J., and F.R. Calaresu. 1977. Lateral reticular nucleus: A site of somatic and cardiovascular integration in the cat. *Am. J. Physiol.* 233:R100–R109.

Ciriello, J., M.M. Caverson, and C. Polosa. 1986. Function of the ventrolateral medulla in the control of the circulation. *Brain Res. Rev.* 11:359–391.

Coote, J.H., S.M. Hilton, and J.F. Perez-Gonzalez. 1971. The reflex nature of the pressor response to muscular exercise. *J. Physiol. (Lond)* 214:789–804.

Dampney, R.A.L., J. Czachurski, K. Dembowsky, A.K. Goodchild, and H. Seller. 1987. Afferent connections and spinal projections of the pressor region in the rostral ventrolateral medulla of the cat. *J. Auton. Nerv. Syst.* 20:73–86.

Dampney, R.A.L., and R.M. McAllen. 1988. Differential control of sympathetic fibers controlling hindlimb skin and muscle by subretrofacial neurones in the cat. *J. Physiol. (Lond)* 395:41–56.

Davy, K.P., W.G. Herbert, and J. Williams. 1993. Effect of indomethacin on the pressor responses to sustained isometric contraction in humans. *J. Appl. Physiol.* 75(1): 273–278.

Diepstra, G., W. Gonyea, and J.H. Mitchell. 1980. Cardiovascular response to static exercise during selective autonomic blockade in the conscious cat. *Circ. Res.* 47:530–535.

Dragunow, M., and R. Faull. 1989. The use of c-Fos as a metabolic marker in neuronal pathway tracing. *J. Neurosci. Methods* 29:261–265.

Erickson, J.T., and D.E. Millhorn. 1991. Fos-like protein is induced in neurons of the medulla oblongata after stimulation of the carotid sinus nerve in awake and anesthetized rats. *Brain Res.* 567:11–24.

Ettinger, S., K. Gray, S. Whisler, and L. Sinoway. 1991. Dichloroacetate reduces sympathetic nerve activity responses to static exercise. *Am. J. Physiol.* 261:H1653–H1658.

Fallentin, N., B.R. Jensen, S. Byström, and G. Sjøgaard. 1992. Role of potassium in the reflex regulation of blood pressure during static exercise in man. *J. Physiol.* 451:643–651.

Fernandes, A., H. Galbo, M. Kjaer, J.H. Mitchell, N.H. Secher, and S.N. Thomas. 1990. Cardiovascular and ventilatory responses to dynamic exercise during epidural anaesthesia in man. *J. Physiol.* 420:281–293.

Fontana, G.A., T. Pantaleo, F. Bongiani, F. Cresci, F. Lavorini, C. Guerra, and P. Panuccio. 1994. Prostaglandin synthesis blockade by ketoprufen attenuates respiratory and cardiovascular responses to static exercise. *J. Appl. Physiol.* 78:449–457.

Gaffney, F.A., G. Sjøgaard, and B. Saltin. 1990. Cardiovascular and metabolic responses to static contraction in man. *Acta Physiol. Scand.* 138: 249–258.

Go, V.L.W., and T.L. Yaksh. 1987. Release of substance P from the cat spinal cord. *J. Physiol. (Lond)* 391:141–167.

Goodwin, G.M., D.I. McCloskey, and J.H. Mitchell. 1972. Cardiovascular and respiratory responses to changes in central command during isometric exercise at constant muscle tension. *J. Physiol. (Lond.)* 226:173–190.

Gordon, F.J. 1987. Aortic baroreceptor reflexes are mediated by NMDA receptors in caudal ventrolateral medulla. *Am. J. Physiol.* 252:R628–R633.

Guynet, P.G., T.M. Filtz, and S.R. Donaldson. 1987. Role of excitatory amino acids in rat vagal and sympathetic baroreflexes. *Brain Res.* 407:272–284.

Hand, G.A., G.L. Kramer, F. Petty, G.A. Ordway, and L.B. Wilson. 1996a. Excitatory amino acid concentrations in the spinal dorsal horn of cats during muscle contraction. *J. Appl. Physiol.* 81:368–373.

Hand, G.A., A.F. Meintjes, A.W. Keister, A.A. Ally, and L.B. Wilson. 1996b. NMDA receptor blockade in cat dorsal horn blunts reflex pressor response to muscle contraction and stretch. *Am. J. Physiol.* 270:H500–H508.

Hand, G.A., G.A. Ordway, and L.B. Wilson. 1996. Microdialysis of a non-NMDA receptor antagonist into the L7 dorsal horn attenuates the pressor response to static muscle contraction but not passive stretch in cats. *Expt. Physiol.* 81:225–238.

Hansen, J., G.D. Thomas, S. Harris, W. Parsons, and R. Victor. 1996. Differential sympathetic neural control of oxygenation in resting and exercising human skeletal muscle. *J. Clin. Invest.* 98(2):584–596.

Hansen, J., G.D. Thomas, T.N. Jacobsen, and R.G. Victor. 1994. Muscle metaboreflex triggers parallel sympathetic activation in exercising and resting human skeletal muscle. *Am. J. Physiol.* 266:H2508–H2514.

Hill, J.M., C.M. Adreani, and M.P. Kaufman. 1996. Muscle reflex stimulates sympathetic postganglionic afferents innervating triceps surae muscle of cats. *Am. J. Physiol.* 271:H38–H43.

Hilton, S.M., O. Hudlicka, and J.M. Marshall. 1978. Possible mediators of functional hyperemia in skeletal muscle. *J. Physiol.* 282:131–147.

Hokfelt, T., J.O. Kellerth, G. Nilsson, and B. Perrow. 1975. Experimental immunohistochemical studies on the localization and distribution of substance P in cat primary sensory neurons. *Brain Res.* 100:235–252.

Iwamoto, G.A., B.R. Botterman, and T.G. Waldrop. 1984. The exercise pressor reflex: Evidence for an afferent pressor pathway outside the dorsolateral sulcus region. *Brain Res.* 292:160–164.

Iwamoto, G.A. and M.P. Kaufman. 1987. Caudal ventrolateral medullary cells responsive to static muscular contraction. *J. Appl. Physiol.* 62:149–157.

Iwamoto, G.A., M.P. Kaufman, B.R. Botterman, and J.H. Mitchell. 1982. Effects of lateral reticular nucleus lesions on the exercise pressor reflex in cats. *Circ. Res.* 51:400–403.

Iwamoto, G.A., J.G. Parnaveles, M.P. Kaufman, B.R. Botterman, and J.H. Mitchell. 1984. Activation of caudal brain stem cell groups during the exercise pressor reflex as elucidated by 2-[14C] deoxyglucose. *Brain Res.* 304:178–182.

Iwamoto, G.A., T.G. Waldrop, M.P. Kaufman, B.R. Botterman, K.J. Rybicki, and J.H. Mitchell. 1985. Pressor reflex evoked by muscular contraction: contributions by neuraxis levels. *J. Appl. Physiol.* 59:459–467.

Iwamoto, G.A., T.G. Waldrop, R.M. Bauer, and J.H. Mitchell. 1989. Pressor responses to muscular contraction in the cat: contributions by caudal and rostral ventrolateral medulla. In *Progress in Brain Research (Volume 81)* edited by J. Ciriello, M.M. Caverson, and C. Polosa. New York: Elsevier Science. 253–263.

Iwamoto, G.A., S.M. Wappel, G.M. Fox, K.A. Buetow, and T.G. Waldrop. 1996. Identification of diencephalic and brain stem cardiorespiratory areas activated during exercise. *Brain Res.* 726:109–122.

Jeske, I., D.J. Reis, and T.A. Milner. 1995. Neurons in the barosensory area of the caudal ventrolateral medulla project monosynaptically on sympathoexcitatory bulbospinal neurons in the rostral ventrolateral medulla. *Neurosci.* 65(2):343–353.

Jessell, T.M., K. Yoshioka, and C.E. Jahr. 1986. Amino acid receptor-mediated transmission at primary afferent synapses in rat spinal cord. *J. Exp. Biol.* 124:239–258.

Johansson, J.E. 1893. Uber die Einwirkung der Musdeltatigkeit auf die Atmun und die Herztatigeit. *Skand. Arch. Physiol.* 5:20–66.

Johansson, O., T. Hokfelt, and R.P. Elde. 1984. Immunohistochemical distribution of somatostatin-like immunoreactivity in the central nervous system of the adult rat. *Neurosci.* 13:265–339.

Jordan, D., and K.M. Spyer. 1986. Brain stem integration of cardiovascular and pulmonary afferent activity. In *Progress in Brain Research: Visceral Sensation (Volume 67)* edited by F. Cervero and J.F.B. Morrison. Amsterdam: Elsevier Science Publishers. 295–314.

Kalia, M., S.S. Mei, and F.F. Kao. 1981. Central projections from ergoreceptors (C fibers) in muscle involved in cardiopulmonary responses to static exercise. *Circ. Res.* 48:I48–I62.

Kaufman, M.P., and H.V. Forster. 1996. Reflexes controlling circulatory, ventilatory and airway responses to exercise. In *Handbook of Physiology, Section 12 Exercise: Regulation and Integration of Multiple Systems* edited by L.B. Rowell and J.T. Shepherd. Bethesda: American Physiological Society. 381–447.

Kaufman, M.P., J.C. Longhurst, K.J. Rybicki, J.H. Wallach, and J.H. Mitchell. 1983. Effects of static muscle contraction on impulse activity of groups III and IV afferents in cats. *J. Appl. Physiol.* 55:105–112.

Kaufman, M.P., T.G. Waldrop, K.J. Rybicki, G.A. Ordway, and J.H. Mitchell. 1984. Effects of static and rhythmic twitch contractions on the discharge of group III and IV muscle afferents. *Cardiovasc. Res.* 18:663–668.

Kellerstein, D.E., D.D. Price, R.L. Hayes, and D.J. Mayers. 1990. Evidence that substance P selectively modulates C-fiber-evoked discharges of dorsal horn nociceptive neurons. *Brain Res.* 526:291–298.

Kiely, J.M., and F.J. Gordon. 1993. Non-NMDA receptors in the rostral ventrolateral medulla mediate somtaosympathetic pressor responses. *J. Auton. Nerv. Syst.* 43:231–240.

Kniffki, K.D., S. Mense, and R.F. Schmidt. 1978. Responses of group IV afferent units from skeletal muscle to stretch, contraction, and chemical stimulation. *Exp. Brain Res.* 31:511–522.

Kniffki, K.D., S. Mense, and R.F. Schmidt. 1981. Muscle receptors with fine afferent fibers which may evoke circulatory reflexes. *Circ. Res.* 48:I25–I31.

Kozelka, J.W., and R.D. Wurster. 1985. Ascending spinal pathways for somato-autonomic reflexes in the anesthetized dog. *J. Appl. Physiol.* 58:1832–1839.

Krogh, A., and J. Lindhard. 1913. The regulation of respiration and circulation during the initial stages of muscle work. *J. Physiol. (Lond)* 47:112–136.

Kuraishi, Y., N. Hirota, Y. Sato, N.Y. Hino, M. Satan, and H. Takagi. 1985. Evidence that substance P and somatostatin transmit separate information related to pain in the spinal dorsal horn. *Brain Res.* 325:294–298.

Lawrence, A.J., and B. Jarrott. 1996. Neurochemical modulation of cardiovascular control in the nucleus tractus solitarius. *Prog. Neurobiol.* 48:21–53.

Lewis, S.F., and R.G. Haller. 1990. Disorders of muscle glycogenolysis/glycolysis: The consequences of substrate-limited oxidative metabolism in hu-

mans. In *Biochemistry of Exercise* edited by A.W. Taylor. Champaign, IL: Human Kinetics. 211–226.

Li, J., G.A. Hand, J.T. Potts, L.B. Wilson, and J.H. Mitchell. 1997. c-Fos expression in the medulla induced by static muscle contraction in cats. *Am. J. Physiol.* 272:H48–H56.

Li, J., J.T. Potts, and J.H. Mitchell. 1998. Effect of barodenervation on c-Fos expression in the medulla induced by static muscle contraction in cats. *Am. J. Physiol.* 274:H901–H908.

Li, Y-W., and R.A.L. Dampney. 1992. Expression of c-Fos protein in the medulla oblongata of conscious rabbits in response to baroreceptor activation. *Neurosci. Lett.* 144:70–74.

Liard, J.M.A., R.J. Hargreaves, and R.G. Hill. 1993. Effect of RP 67580, a non-peptide neurokinin receptor antagonist, on facilitation of a nociceptive spinal flexion reflex in the rat. *Br. J. Pharmacol.* 109:713–718.

Light, A.R., and E.R. Perl. 1979. Re-examination of the dorsal root projection to the spinal dorsal horn including observations on the differential termination of coarse and fine fibers. *J. Comp. Neurol.* 186:117–132.

Lind, A.R. 1983. Cardiovascular adjustments to isometric contractions: Static effort. In *Handbook of Physiology: The Cardiovascular System III* edited by R.M Berne, N. Sperelakis, and S.R. Geiger. Bethesda: American Physiological Society. 947–967.

Lindefors, N., Y. Yamamoto, T. Pantaleo, H. Lagercrantz, E. Brodin, and U. Ungerstedt. 1986. In vivo release of substance P in the nucleus tractus solitarii increases during hypoxia. *Neurosci. Lett.* 69:94–97.

MacLean, D.A., B. Saltin, G. Radegran, and L.I. Sinoway. 1997. Femoral arterial injection of adenosine in humans elevates MSNA via central but not peripheral mechansims. *J. Appl. Physiol.* 83(4):1045–1053.

Mark, A.L., R.G. Victor, C. Nerhed, and B.G. Wallin. 1985. Microneurographic studies of the mechanisms of sympathetic nerve responses to static exercise in humans. *Circ. Res.* 57:461–469.

Matsukawa, K., P.T. Wall, L.B. Wilson, and J.H. Mitchell. 1992. Neurally-mediated renal vasoconstriction during isometric muscle contraction in cats. *Am. J. Physiol. (Heart and Circ. Physiol.)* 31:H833–H838.

Matsukawa, K., P.T. Wall, L.B. Wilson, and J.H. Mitchell. 1994. Reflex stimulation of cardiac sympathetic nerve activity during static muscle contraction in cats. *Am. J. Physiol.* 267:H821–H827.

McAllen, R.M. 1986a. Identification and properties

of sub-retrofacial bulbospinal neurones: A descending cardiovascular pathway in the cat. *J. Auton. Nerv. Syst.* 17:151–164.

McAllen, R.M. 1986b. Location of neurons with cardiovascular and respiratory function at the ventral surface of the cats medulla. *Neurosci.* 18:43–49.

McCloskey, D.I., and J.H. Mitchell. 1972. Reflex cardiovascular and respiratory response originating in exercising muscle. *J. Physiol. (Lond)* 224:173–186.

McKitrick, D.J., T.L. Krukoffm, and F.R. Calaresu. 1992. Expression of c-Fos protein in the rat brain after electrical stimulation of the aortic depressor nerve. *Brain Res.* 599:215–222.

McMahon, S.E., P.N. McWilliams, J. Robertson, and J.C. Kaye. 1992. Inhibition of carotid sinus baroreceptor neurones in the nucleus tractus solitarius of the anesthetized cat by electrical stimulation of hindlimb afferents. *J. Physiol. (Lond)* 452:224P.

McNeill, D.L., K.N. Westlund, and R.E. Coggeshall. 1989. Peptide immunoreactivity of unmyelinated primary afferent axons in rat lumbar dorsal roots. *J. Histochem. Cytochem.* 37:1047–1052.

Mense, S. 1993. Nociception from skeletal muscle in relation to clinical muscle pain. *Pain* 54:241–289.

Mense, S., and S. Stahnke. 1983. Responses in muscle afferent fibers of slow conduction velocity to contractions and ischaemia in the cat. *J. Physiol. (Lond)* 342:383–397.

Meyer, R.A. 1988. A linear model of muscle respiration explains monoexponential phosphocreatine changes. *Am. J. Physiol.* 254:C548–C553.

Miller, R.G., M.D. Boska, R.S. Moussavi, P.J. Carson, and M.W. Weiner. 1988. ^{31}P Nuclear magnetic resonance studies of high energy phosphates and pH in human muscle fatigue. *J. Clin. Invest.* 81:1190–1196.

Millhorn, D.E., and F.L. Eldridge. 1986. Role of ventrolateral medulla in regulation of respiratory and cardiovascular systems. *J. Appl. Physiol.* 61:1249–1263.

Mitchell, J.H. 1990. Neural control of the circulation during exercise. *Med. Sci. Sports Exer.* 22:141–154.

Mitchell, J.H., M.P. Kaufman, and G.A. Iwamoto. 1983. The exercise pressor reflex: Its cardiovascular effects, afferent mechanisms, and central pathways. *Ann. Rev. Physiol.* 45:229–242.

Mitchell, J.H., D.R. Reeves, Jr., H.B. Rogers, and N.H. Secher. 1989a. Epidural anesthesia and cardiovascular responses to static exercise in man. *J. Physiol.* 417:13–24.

Mitchell, J.H., D.R. Reeves, Jr., H.B. Rogers, and N.H. Secher. 1989b. Autonomic blockade and cardiovascular responses to static exercise in partially curarized man. *J. Physiol.* 413:433–445.

Mitchell, J.H., and R.F. Schmidt. 1988. Cardiovascular reflex control by afferent fibers from skeletal muscle receptors. In *Handbook of Physiology: The Cardiovascular System III* edited by R.M. Berne, N. Sperelakis, and S.R. Geiger. Bethesda: American Physiological Society. 623–658.

Mittelstadt, S.W., L.B. Bell, K.P. O'Hagan, and P.S. Clifford. 1994. Muscle chemoreflex alters vascular conductance in nonischemic exercising skeletal muscle. *J. Appl. Physiol.* 77:2761–2766.

Mittelstadt, S.W., L.B. Bell, K.P. O'Hagan, J.E. Sulentic, and P.S. Clifford. 1996. Muscle chemoreflex causes renal vascular constriction. *Am. J. Physiol.* 270:H951–H956.

Morgan, J.I., and T. Curran. 1991. Stimulus-transcription coupling in the nervous system: Involvement of the inducible proto-oncogene Fos and jun. *Annu. Rev. Neurosci.* 14:421–451.

Morilak, D.A., M. Morris, and J. Chalmers. 1988. Release of substance P in the nucleus tractus solitarius measured by in vivo microdialysis: Response to stimulation of the aortic depressor nerves in rabbit. *Neurosci. Lett.* 94:131–137.

Murphy, A.Z., M. Ennis, M.T. Shipley, and M.M. Behbehani. 1994. Directionally specific changes in arterial pressure induce differential patterns of Fos expression in discrete areas of the rat brain stem: A double-labeling study for Fos and catecholamines. *J. Comp. Neurol.* 349:36–50.

O'Hagan, K.P., S.M. Casey, and P.S. Clifford. 1997. Muscle chemoreflex increases renal sympathetic nerve activity during exercise. *J. Appl. Physiol.* 82(6):1818–1825.

O'Leary, D.S., E.D. Robinson, and J.L. Butler. 1997. Is active skeletal muscle functionally vasoconstricted during dynamic exercise in conscious dogs? *Am. J. Physiol.* 272:R386–R391.

Pan, H.L., C.L. Stebbins, and J.C. Longhurst. 1993. Bradykinin contributes to the exercise pressor reflex: Mechanism of action. *J. Appl. Physiol.* 75:2061–2068.

Person, R.J. 1989. Somatic and vagal afferent convergence on solitary tract neurons in cat: Electrophysiological characteristics. *Neurosci.* 30(2):283–295.

Pickar, J.G., J.M. Hill, and M.P. Kaufman. 1994. Dynamic exercise stimulates group III muscle afferents. *J. Neurophysiol.* 71:753–760.

Pickel, V.M., D.J. Reis, and S.E. Leeman. 1977. Ultrastructural localization of substance P in neurons of the spinal cord. *Brain Res.* 122:534–540.

Potts, J.T., I.E. Fuchs, J. Li, B. Leshnower, and J.H. Mitchell. 1999. Skeletal muscle afferent fibres release substance P in the nucleus tractus solitarii of anaesthetized cats. *J. Physiol. (Lond)* 514(3):829–841.

Potts, J.T., J. Li, and T.G. Waldrop. 1998. Extracellular recordings from nucleus tractus solitarii neurons: Response to arterial baroreceptor and skeletal muscle receptor input. *FASEB J.* 12:A61.

Pryor, S.L., S.F. Lewis, R.G. Haller, L.A. Bertocci, and R.G. Victor. 1990. Impairment of sympathetic activation during static exercise in patients with muscle phosphorylase deficiency (McArdle's disease). *J. Clin. Invest.* 85:1444–1449.

Ray, C.A. 1993. Muscle sympathetic nerve responses to prolonged one-legged exercise. *J. Appl. Physiol.* 74(4):1719–1722.

Ray, C.A., and A. Mark. 1993. Augmentation of muscle sympathetic nerve activity during fatiguing isometric leg exercise. *J. Appl. Physiol.* 75(1):228–232.

Ray, C.A., N.H. Secher, and A.L. Mark. 1994. Modulation of sympathetic nerve activity during post-handgrip muscle ischemia in humans. *Am. J. Physiol.* 266:H79–H83.

Rotto, D.M., and M.P. Kaufman. 1988. Effect of metabolic products of muscular contraction on discharge of group III and IV afferents. *J. Appl. Physiol.* 64:2306–2313.

Rutherford, S.D., R.E. Widdop, F. Sannajust, W.J. Louis, and A.L. Gundlach. 1992. Expression of c-Fos and NGFI-A messenger RNA in the medulla oblongata of the anesthetized rat following stimulation of vagal and cardiovascular afferents. *Mol. Brain Res.* 13:301–312.

Rybicki, K.J., M.P. Kaufman, J.L. Kenyon, and J.H. Mitchell. 1984. Arterial pressure responses to increasing interstitial potassium in hindlimb muscles of dogs. *Am. J. Physiol.* 247:R717–R721.

Saito, M., A. Tsukanaka, D. Yanagihara, and T. Mano. 1993. Muscle sympathetic nerve responses to graded leg cycling. *J. Appl. Physiol.* 75:663–667.

Sapru, H.U. 1994. Transmitter/receptor mechanisms in cardiovascular control by the NTS: Excitatory amino acids, acetylcholine, and substance P. In *Nucleus of the Solitary Tract* edited by I.R.A. Barraco. Boca Raton, FL: CRC Press. 267–282.

Seals, D.R., P.B. Chase, and J.A. Taylor. 1988. Autonomic mediation of the pressor response to isometric exercise in humans. *J. Appl. Physiol.* 64:2190–2196.

Seybold, V.S., and R. Elde. 1980. Immunohistochemical studies of peptidergic neurons in the dorsal horn of the spinal cord. *J. Histochem. Cytochem.* 28:367–370.

Sheriff, D.D., D.S. O'Leary, A.M. Scher, and L.B. Rowell. 1990. Baroreflex attenuates pressor response to graded muscle ischemia in exercising dogs. *Am. J. Physiol.* 258:H305–H310.

Sinoway, L., J.M. Hill, J.G. Pickar, and M.P. Kaufman. 1993. Effects of contraction and lactic acid on the discharge of group III muscle afferents. *J. Neurophysiol.* 69: 1053–1059.

Sinoway, L., S. Prophet, I. Gorman, T. Mosher, J. Shenberger, M. Dolecki, R. Briggs, and R. Zelis. 1988. Muscle acidosis during static exercise is associated with calf vasoconstriction. *J. Appl. Physiol.* 66:429–436.

Sinoway, L., M.B. Smith, B. Enders, U. Leuenberger, T. Dzwonczyk, K. Gray, S. Whisler, and R.L. Moore. 1994. Role of diprotonated phosphate in evoking muscle reflex responses in cats and humans. *Am. J. Physiol.* 267:H770–H778.

Sinoway, L., J.S. Wilson, R. Zelis, J. Shenberger, D.P. McLaughlin, D.L. Morris, and F.P. Day. 1988. Sympathetic tone affects human limb vascular resistance during a maximal metabolic stimulus. *Am J. Physiol.* 255:H937–H946.

Spyer, K.M. 1994. Central nervous mechanisms contributing to cardiovascular control. *J. Physiol. (Lond)* 474(1):1–19.

Srinivasan, M., M. Goiny, T. Pantaleo, H. Lagercrantz, E. Brodin, M. Runold, and Y. Yamamoto. 1991. Enhanced in vivo release of substance P in the nucleus tractus solitarii during hypoxia in the rabbit: role of peripheral input. *Brain Res.* 546:211–216.

Sved, A.F., D.L. Mancini, J.C. Graham, A.M. Schreihofer, and G.E. Hoffman. 1994. PNMT-containing neurons of the C1 cell group express c-Fos in response to changes in baroreceptor input. *Am. J. Physiol.* 266:R361–367.

Sykes, R.M., K.M. Spyer, and P.N. Izzo. 1994. Central distribution of substance P, calcitonin gene-related peptide, and 5-hydroxytryptamine in vagal sensory afferents in the rat dorsal medulla. *Neurosci.* 59(1):195–210.

Toney, G.M., and S.W. Mifflin. 1994. Time-dependent inhibition of hindlimb somatic afferent inputs to nucleus tractus solitarius. *J. Neurophysiol.* 72(1):63–71.

Toska, K., and M. Eriksen. 1994. Peripheral vasoconstriction shortly after the onset of moderate exercise in humans. *J. Appl. Physiol.* 77:1519–1525.

Victor, R.G., L.A. Bertocci, S.L. Pryor, and R.L. Nunnally. 1988. Sympathetic nerve discharge is coupled to muscle cell pH during exercise in humans. *J. Clin. Invest.* 82:1301–1305.

Victor, R.G., and D.R. Seals. 1989. Reflex stimulation of sympathetic outflow during rhythmic exercise in humans. *Am. J. Physiol.* 257:H2017–2024.

Victor R.G., D.R. Seals, and A.J. Taylor. 1987. Differential control of heart rate and sympathetic nerve activity during dynamic exercise: Insight from direct intraneural recordings in humans. *J. Clin. Invest.* 79:508–516.

Vissing, J., S. Vissing, D. MacLean, B. Saltin, B. Quistorff, and R.G. Haller. 1998. Sympathetic activation in exercise is not dependent on muscle acidosis. *J. Clin. Invest.* 101(8):1654–1660.

Vissing, J., L.B. Wilson, J.H. Mitchell, and R.G. Victor. 1991. Static muscle contraction reflexly increases adrenal sympathetic nerve activity in rats. *Am. J. Physiol.* 261:R1307–R1312.

Waldrop, T.G., and J.H. Mitchell. 1985. Effects of barodenervation on cardiovascular response to static muscle contraction. *Am. J. Physiol.* 249:H710–H714.

Wallin, B.G., R.G. Victor, and A.L. Mark. 1989. Sympathetic outflow to resting muscles during static handgrip and postcontraction muscle ischemia. *Am. J. Physiol.* 256:H105–H110.

Williams, C.A., P.L. Brien, P.L. Nichols, and P. Gopalan. 1994. Detection of immunoreactive substance P-like substances from cat brain stem sites during fatiguing isometric contractions. *Neuropeptides* 26:319–327.

Williamson, J.W., H.L. Olesen, F. Pott, and N.H. Secher. 1996. Central command increases cardiac output during static exercise in humans. *Acta Physiol. Scand.* 156(4):429–434.

Wilson, J.R., K. McCully, D. Mancini, B. Boden, and B. Chance. 1988. Relationship of muscle fatigue to pH and diprotonated Pi in humans: A ^{31}P NMR study. *J. Appl. Physiol.* 64:2333–2339.

Wilson, L.B., C.K. Dyke, D. Parsons, P.T. Wall, J.A. Pawelczyk, R.S. Williams, and J.H. Mitchell. 1995. Effect of skeletal muscle fiber type on the pressor response evoked by static contraction in rabbits. *J. Appl. Physiol.* 79(5):1744–1752.

Wilson, L.B., I.E. Fuchs, K. Matsukawa, J.H. Mitchell, and P.T. Wall. 1993. Substance P release in the spinal cord during static muscle contraction in anesthetized cats. *J. Physiol. (Lond)* 460:79–90.

Wilson L.B., I.E. Fuchs, and J.H. Mitchell. 1993. Effects of graded muscle contractions on spinal cord substance P release, arterial blood pressure, and heart rate. *Cir. Res.* 73:1024–1031.

Wilson, L.B., P.T. Wall, and K. Matsukawa. 1995. Attenuation of the reflex responses to muscle contraction by the co-administration of antagonist to substance P and somatostatin into the dorsal horn. *Cardiovas. Res.* 29:379–384.

Wilson, L.B., P.T. Wall, K. Matsukawa, and J.H. Mitchell. 1992. Effect of spinal microinjections of an antagonist to substance P or somatostatin on the exercise pressor reflex. *Circ. Res.* 70:213–222.

Xu, J-X., C.J. Dalsgaard, and Z. Wiesenfeld-Hallis. 1991. Spinal substance P and N-methyl-D-aspartate receptors are coactivated in the induction of central sensitization of the nociceptive flexor reflex. *Neurosci.* 51:641–648.

Chapter 3

The Cardiopulmonary Baroreflex

Chester A. Ray and Mitsuru Saito

Mechanical stretch receptors in the heart and lungs responsive to changes in blood volume and pressure have important functions for autonomic neural control of cardiovascular adjustments for various conditions (Mark and Mancia 1983). These receptors, termed *cardiopulmonary baroreceptors*, elicit reflex alterations in sympathetic nervous activity to specific organs during maneuvers that alter cardiac and pulmonary filling volumes. These include postural change and exercise. This chapter is concerned with the function of the cardiopulmonary reflex in the regulation of sympathetic neural influences on cardiovascular responses at rest and during exercise. Although mounting evidence suggests the importance of the cardiopulmonary reflex in humans during exercise, its physiological role and the mechanism of its involvement remain inconclusive. Excellent reviews of the cardiopulmonary baroreflexes have been previously written. The reader is encouraged to seek these reviews for more detailed information (Hainsworth 1991; Mark and Mancia 1983).

Sympathetic Nerve Activity in Humans

The intraneural recording of postganglionic sympathetic nerve activity (SNA) in humans was developed by Vallbo and Hagbarth in the 1960s (Vallbo et al. 1979). This technique for studying sympathetic efferent activity is advantageous compared with indirect methods that assess responses in heart rate, blood pressure, and vascular resistance. This advantage occurs because the sympathetic effector end-organ response, for instance vascular resistance, is influenced by mechanisms other than altered nerve activity; i.e., by metabolic changes and circulatory vasoactive agents. Restricting assessment of SNA to one organ does not generally reflect a uniform, whole-body response because SNA is highly differentiated. For example, the spontaneous and grouped burst discharges of SNA in muscle (MSNA) differs from that in skin. The burst discharges of muscle SNA show an entrainment with heartbeat rhythm, enhancement with a drop in arterial blood pressure, and little reaction in the burst discharge to arousal stimuli. Skin SNA shows irregular bursts without heartbeat rhythm and sensitivity to arousal stimulus. Skin SNA also responds to atmospheric temperature changes and shows little reflex effect in response to baroreceptor input. The primary cardiovascular function of MSNA in humans is to mediate vasoconstriction. Thus, its main role is to maintain systemic blood pressure and to control muscle blood flow at rest and during exercise.

Cardiopulmonary Interventions at Rest

Postural changes that influence the distribution of blood volume activate the cardiopulmonary baroreflex. Roddie, Shepherd, and Whelan (1957) showed that passive lifting of the legs resulted in a vascular dilation of the forearm. Based on the finding that the increase in forearm blood flow was not correlated with an increase in arterial pressure and heart rate, they suggested that this response was mediated by low-pressure intrathoracic receptors.

Although the physiological response to postural change from a lying to an upright position is complex and appears to involve arterial as well as other physiological sensors, pooling of blood in the dependent portion of the body is a primary response. This causes a drop in central venous pressure (CVP) and subsequent unloading of cardiopulmonary receptors, which elicits reflex increases in MSNA (see figure 3.1). During graded head-up tilt, the MSNA burst rate increases linearly as a sine function of tilt angle (Iwase, Mano, and Saito 1987). In contrast, during head-down tilt, MSNA decreases to below that in the horizontal position (Iwase et al. 1987).

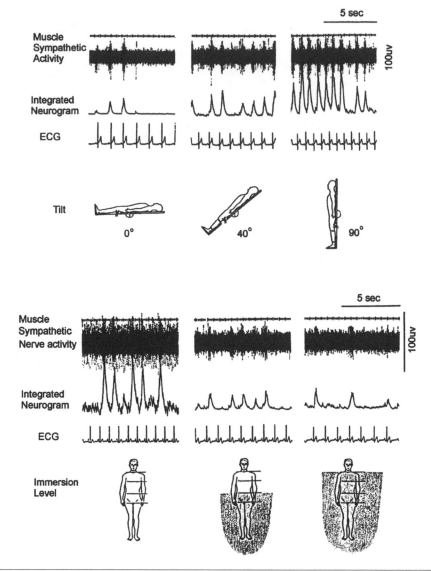

Figure 3.1   Muscle sympathetic nerve responses during head-up tilting and water immersion. (From Iwase et al. 1987.)

With the introduction of lower-body negative pressure (LBNP), human investigation of the physiological effects of the cardiopulmonary baroreflex progressed dramatically. LBNP is a maneuver designed to restrain the venous return to the heart and thus to reduce cardiac filling pressure. When the LBNP is less than –15 to –20 mmHg, CVP and cardiac filling pressure decrease with no detectable changes in arterial pulse pressure, mean pressure, or heart rate. Thus, researchers have suggested that the arterial baroreflexes are not engaged and the physiological response elicited by this maneuver is due to the cardiopulmonary reflex. This intervention elicits reflex vasoconstriction in skeletal muscle of the forearm and leg. It also elicits similar but smaller responses in the renal and splanchnic vasculature

regions (Abboud et al. 1979; Johnson et al. 1974; Sundlof and Wallin 1978; Vissing, Scherrer, and Victor 1989; Zoller et al. 1972). This response is consistent with increases in sympathetic neural outflow to skeletal muscle during LBNP, which produce decreases in CVP and atrial volume (see Figure 3.2). However, this response is absent in skin (Oren et al. 1993; Victor and Leimbach 1987; Vissing, Scherrer, and Victor 1994).

When the body is immersed in thermoneutral water in a standing position, intrathoracic blood volume increases due to the hydrostatic pressure effect. This maneuver stimulates the cardiopulmonary receptors, causes an increase in the inhibitory afferent activity, and subsequently suppresses MSNA (see figure 3.1) (Iwase et al. 1987). The sup-

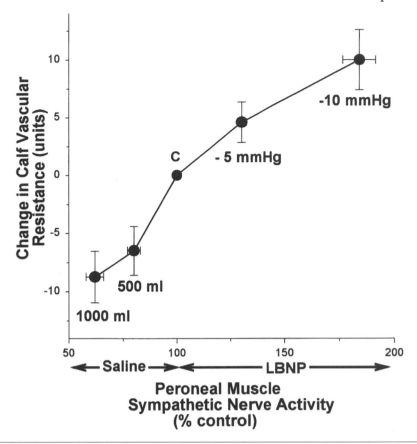

Figure 3.2   The relationship between leg muscle sympathetic nerve activity and calf vascular resistance during changes in central venous pressure elicited by volume loading (saline infusion) and lower-body negative pressure. (Adapted from Vissing, Scherrer, and Victor 1989.)

pression becomes stronger with water level and reaches a maximum with immersion up to the level of the neck.

Cardiopulmonary Reflex During Exercise

The primary mechanism responsible for activating sympathetic neural outflow during static and dynamic exercise is thought to be the skeletal muscle reflex (muscle metaboreflex) (Mitchell 1990). Afferent signals from skin nociceptors and peripheral chemoreceptors also have excitatory effects on MSNA, while central command exerts a minor excitatory influence. In contrast, both the arterial and cardiopulmonary baroreflexes are thought to have inhibitory effects on the sympathetic neural outflow during exercise.

Unlike investigations of other mechanisms that alter sympathetic activity and vascular resistance during exercise (i.e., muscle metaboreflex and arterial baroreflex), methodological difficulties make concrete conclusions regarding the importance of the cardiopulmonary baroreflex during exercise in

human investigation difficult. Walker et al. (1980) first investigated the possible role of the CPBR during exercise by measuring forearm vascular resistance during mild isometric handgrip with a low level of LBNP (–5 mmHg). Forearm vascular resistance during the exercise with LBNP was greater than the algebraic sum of each stressor performed alone. This finding prompted the idea that the cardiopulmonary baroreflex has a facilitative interaction with the muscle reflexes. Arrowood et al. (1993) have reproduced this finding. Yet, when examining muscle MSNA during higher levels of isometric handgrip, LBNP (–5 to –15 mmHg) does not facilitate muscle MSNA (Sanders, Ferguson, and Kempf 1988; Scherrer, Vissing, and Victor 1988; Seals 1988). Rather, LBNP equates a simple summation of each response when performed alone (LBNP alone plus handgrip alone). However, in these studies, the marked elevation in arterial pressure that occurs with high-intensity isometric handgrip may override the influence of the cardiopulmonary baroreflex. This is likely a much different response to that induced during mild static handgrip where increases in arterial pressure are modest.

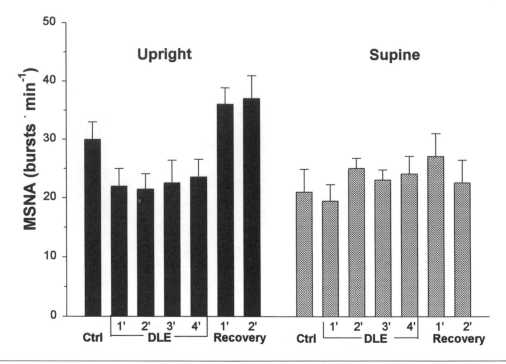

Figure 3.3   The effect of upright and supine one-legged cycling on muscle sympathetic nerve activity (MSNA). Exercise in the upright (sitting) position decreased MSNA, whereas MSNA remained unchanged during supine exercise. (Adapted from Ray et al. 1993.)

During dynamic exercise, a role for the cardiopulmonary reflex has been proposed because exercise increases preload, afterload, and cardiac contractility. All of these can stimulate the cardiopulmonary receptors and decrease firing frequency of the vagal afferent nerves. However, the results from animals and humans are not entirely consistent. The discrepancies may be due to differences in species, exercise conditions (i.e., intensity and duration), and/or the intervention.

Walgenbach and Donald (1983) and Daskalopoulos, Shepherd, and Walgenbach (1984) investigated the influence of the cardiopulmonary receptors on blood pressure and cardiac output during dynamic exercise in dogs. They found that blood pressure and cardiac output responses were similar before and after vagotomy but that total vascular conductance was reduced during heavy exercise. This suggested that the cardiopulmonary reflex has a role in inhibiting sympathetic vasoconstriction. Similarly, O'Hagan et al. (1994) found that cardiac afferents modulated the increase in renal sympathetic nerve activity during dynamic exercise in the rabbit. Collins and DiCarlo (1993) came to this conclusion in exercising rats. They found that cardiac afferents tonically inhibit the pressor response to reductions in terminal aortic blood flow during exercise. These studies support the notion that the cardiopulmonary baroreceptors are involved in sympa-

thetic and cardiovascular regulation during heavy exercise. In contrast, Ludbrook and Graham (1985) found that vagotomy reduced blood pressure during brief treadmill walking in rabbits. They concluded that input from cardiopulmonary receptors supports systemic vascular resistance at the onset of exercise and contributes to the rise of arterial pressure.

In human subjects, orthostatic stress leads to marked hemodynamic changes at rest and during exercise as compared with quadrupeds. For instance, 70% of the total blood volume is at or below the level of the heart in humans in the upright position. In the dog, however, 70% of the total blood volume is at or above the heart in the normal posture (Rowell 1986). Thus, the cardiopulmonary reflex may play a more important role in circulatory adjustments in humans than in animals. Mack, Nose, and Nadel (1988) investigated the role of the cardiopulmonary reflex during dynamic cycling exercise in humans. Cardiopulmonary receptor unloading by various levels of LBNP during cycling did not affect mean arterial blood pressure, pulse pressure, or heart rate. However, forearm vascular resistance had increased. This finding suggested that the cardiopulmonary reflex acted to buffer the expected arterial blood pressure change that would result from a change in cardiac filling pressure or cardiac output. Sprangers et al. (1991) reported similar conclusions in upright cycling exercise.

Orthostatic Effect on MSNA Response During Exercise

Ray et al. (1993) reported that sympathetic outflow to the quiescent leg muscles was depressed to below the resting control level during one-legged dynamic exercise in the upright sitting position (see figure 3.3). They did not observe any significant change in the supine position. A role for the cardiopulmonary baroreflex in the attenuation of sympathetic outflow is suggested by the finding that exercise induced the same increase in arterial blood pressure in both supine and upright positions. The decrease in MSNA during the upright exercise was related to an increase in central venous pressure. However, the researchers did not observe changes in central venous pressure during supine exercise. Saito et al. (1993) reported similar suppression of sympathetic outflow to the forearm muscles during low-intensity cycling exercise in the sitting position. These results suggest that orthostatic stress influences MSNA response during dynamic exercise. They also suggest that the cardiopulmonary baroreflex plays a role in regulating MSNA and vascular resistance, especially during mild-to-moderate exercise.

A different response is observed during static exercise. Recent investigations demonstrated the ineffectiveness of LBNP on MSNA responses during static exercise (Sanders, Ferguson, and Kempf 1988; Scherrer, Vissing, and Victor 1988; Seals 1988). This may be due to the more pronounced involvement of the arterial baroreflex accompanied by the marked pressor response. Nevertheless, MSNA at rest and during exercise is elevated during LBNP compared with those without LBNP. Analogous results were reported when blood volume expansion mediated by water immersion (Saito, Mano, et al. 1993) stimulated cardiopulmonary receptors. Altered central venous pressure with orthostatic stress may be the important factor determining cardiopulmonary afferent nerve activity during exercise.

Cardiopulmonary Reflex Influence on the Muscle Metaboreflex

The concept that cardiopulmonary vagal afferent input inhibits the sympathetic vasoconstrictor response to activation of muscle afferents was proposed from animal studies (Thames and Abboud 1979). Later studies in which cardiac efferent and afferent nerves were blocked by pharmacological agents reexamined this concept (Collins and DiCarlo 1993). Both cardiac efferent and afferent nerve block-

ades enhanced metaboreflex-mediated vasoconstriction during treadmill exercise. Blood volume expansion, however, showed an attenuating effect. This finding suggested that sympathetic outflow to the skeletal muscle vascular beds during exercise could be modified by activity of the afferent fibers from the cardiac receptors.

In humans, an attenuated increase in cardiac output produced by beta-blockade by metoprolol resulted in less leg blood flow during graded recumbent cycling exercise, while no significant change was observed in blood pressure (Pawelczyk, Warberg, and Secher 1992). Thus, less stimulation of cardiac receptors may have diminished afferent nerve activity with less inhibition of sympathetic vasoconstrictor outflow to the legs. In contrast, during isometric handgrip, cardiopulmonary receptor unloading with LBNP (a nonspecific cardiopulmonary intervention) did not modify metaboreflex increases in MSNA of nonexercising muscles (Sanders, Ferguson, and Kempf 1988; Scherrer, Vissing, and Victor 1988; Seals 1988). Again, this difference could be ascribed to the larger venous return and cardiac output during dynamic exercise with a large muscle mass. At the same time, the small change in preload and cardiac output during static handgrip exercise would exert less mechanical stretch on the heart.

Cardiopulmonary Baroreflexes and Exercise Training

The effect of exercise training on the cardiopulmonary baroreflex is equivocal. Training studies have indicated unchanged, augmented, and diminished cardiopulmonary baroreflex control of forearm vascular resistance. The number of studies addressing this issue is limited. Long-term longitudinal studies are lacking. Additionally, little discussion concerning the implication of altered cardiopulmonary baroreflex control with training have been undertaken.

Takeshita et al. (1986) reported augmented cardiopulmonary baroreflex control of forearm vascular resistance in young athletes compared with sedentary, age-matched controls. The same laboratory shortly thereafter reported augmented cardiopulmonary baroreflex control of forearm vascular resistance in middle-aged subjects (56 ± 7 years) after four months of exercise training (Jingu et al. 1988). The training regimen consisted of cycling at approximately 50% of VO_2max for 60 min/day, 3 days per week. However, Mack et al. (1987, 1991) found the opposite results. In a cross-sectional study of unfit and fit subjects (VO₂max = 39 and 57 ml/kg/min, respectively), fit subjects had less of an increase in forearm vascular resistance for a given decrease in peripheral venous pressure during LBNP. (Changes in peripheral venous pressure reflected changes in central venous pressure during LBNP.) These results were supported by findings in rabbits. DiCarlo and Bishop (1990) found increased inhibitory influences of cardiac afferents after training.

Mack et al. (1991) replicated these findings in a longitudinal training study. Fourteen subjects cycled for 30 min/day, 4 days/week for 10 weeks at 75% of VO₂max. Training elicited a marked increase in VO₂max from 38 to 64 ml/kg/min. As in their earlier study, forearm vascular resistance at a given central venous pressure was diminished after training. Seals and Chase (1989) and, most recently, Kouame et al. (1995) have also reported a diminished forearm vascular response to LBNP following

30 weeks of endurance training. They found an attenuated cardiopulmonary baroreflex control of forearm vascular resistance following training at 70% VO₂max but not at 50% VO₂max (see figure 3.4). Thus, these studies suggested an intensity-related training effect on the cardiopulmonary baroreflexes. Others have concluded that exercise training has no effect on cardiopulmonary baroreflex control during low levels of LBNP (Lightfoot et al. 1989).

Giannattasio et al. (1990, 1992) found that athletes with concentric cardiac hypertrophy had diminished cardiopulmonary baroreflex control of forearm vascular resistance. Later, this same group found that regression of the cardiac hypertrophy by detraining reversed the impairment of cardiopulmonary baroreflex of forearm vascular resistance (Giannattasio et al. 1992). These studies emphasized the importance of myocardial morphology and ventricular contractility in addition to altered

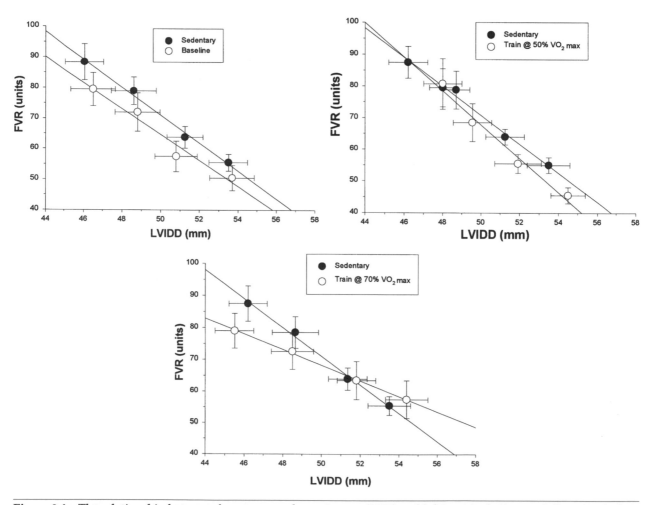

Figure 3.4   The relationship between forearm vascular resistance (FVR) and left ventricular internal diameter during diastole (LVIDD) before and after 10 weeks of exercise training (cycling) in (a) nontraining controls and at (b) 50% and (c) 70% of VO₂max. Changes in LVIDD were elicited by LBNP at –40 and –15 mmHg at rest and during leg raising (the four points from left to right on each graph). Exercise training at 50% VO₂max failed to alter the FVR-LVIDD relationship, whereas training at 70% VO₂max decreased the slope. (Adapted from Kouame et al. 1995.)

changes in central venous pressure in the activation of cardiac afferents during exercise.

Training studies examining cardiopulmonary baroreflex control of MSNA rather than vascular responsiveness may provide more consistent and definitive results. Alteration in forearm vascular responses after training represent possible changes not only in sympathetic neural activity. These altered responses also represent changes in local control systems (e.g., intrinsic changes in the vascular smooth muscle cells) and the effects of structural changes on the distribution of resistance throughout the microcirculation (Laughlin, McAllister, and Delp 1994).

Summary

Cardiopulmonary receptors subserving vagal afferents appear to have a strong influence on MSNA at rest. During exercise, both inhibitory and excitatory afferent activity from cardiopulmonary receptors and muscle metaboreceptors, respectively, show concomitant alterations acting to control sympathetic outflow. These patterns of afferent activity are differentiated depending on the exercise mode, load, and degree of orthostatic stress.

During mild upright dynamic exercise, the cardiopulmonary baroreflex reduces sympathetic outflow and vascular resistance. These responses are primarily evoked by an increase in central filling pressure produced by engagement of the muscle pump. The functional significance of this response at the onset of exercise remains to be determined. Animal (Peterson, Armstrong, and Laughlin 1988; Sheriff, Rowell, and Scher 1993) and human (Corcondilas, Koroxenidis, and Shepherd 1964; Joyner et al. 1990) studies indicate that the initial hyperemia at the onset of exercise does not depend upon the sympathetic nervous system. However, sympathetic activity modulates muscle blood flow during sustained exercise (Joyner et al. 1990; Peterson, Armstrong and Laughlin 1988; Sheriff, Rowell, and Scher 1993). Therefore, the decrease in muscle MSNA during upright exercise may facilitate the increase in muscle blood flow during steady-state exercise.

The significance of the cardiopulmonary baroreflex during moderate-to-heavy exercise is unclear. Its role is difficult to assess in humans because other regulators of MSNA and blood flow (i.e., muscle metaboreflex, arterial baroreflex, and local metabolic influences) are actively engaged. Research has shown that after the initial decrease in MSNA during upright leg exercise, MSNA begins to increase as the exercise duration (Ray 1993; Ray et al. 1993) and intensity progress (Saito et al. 1993). The cardiopulmonary baroreflex likely buffers increases in MSNA elicited by the muscle metaboreflex.

The cardiopulmonary baroreflex may serve an important function during prolonged exercise or exercise in the heat. Decreases in venous return, central venous pressure, and stroke volume are associated with prolonged exercise or exercise in the heat due to greater cutaneous blood flow and plasma loss (i.e., cardiovascular drift). The diminished activation of cardiopulmonary baroreceptors by the fall in venous return may serve to increase vascular resistance to defend arterial pressure. This may occur by directly increasing MSNA and indirectly increasing the sensitivity of the arterial baroreflex.

During static exercise, a role for the cardiopulmonary baroreflex is less apparent due to a more pronounced influence of the arterial baroreflex. With endurance training, the cardiopulmonary baroreflex appears to exert greater inhibition of sympathetic vasoconstriction due to enhanced venous return and cardiac filling. However, research has not demonstrated this consistently. This inhibitory influence may also be coupled to changes in the sensitivity of the arterial baroreflex.

References

Abboud, F.M., D.L Eckberg, U.J. Johaasen, and A.L. Mark. 1979. Carotid and cardiopulmonary baroreceptor control of splanchnic and forearm vascular resistance during venous pooling in man. *J. Physiol.* 286:173–184.

Arrowood, J.A., P.K. Mohanty, C. McNamara, and M.D. Thames. 1993. Cardiopulmonary reflexes do not modulate exercise pressor reflexes during isometric exercise in humans. *J. Appl. Physiol.* 74:2559–2565.

Collins, H.L., and S.E. DiCarlo. 1993. Cardiac afferents attenuate the muscle metaboreflex in the rat. *J. Appl. Physiol.* 74:114–120.

Corcondilas, A., G.T. Koroxenidis, and J.T. Shepherd. 1964. Effect of brief contraction of forearm muscles on forearm blood flow. *J. Appl. Physiol.* 19:142–146.

Daskalopoulos, D.A., J.T. Shepherd, and S.C. Walgenbach. 1984. Cardiopulmonary reflexes and blood pressure in exercising sinoaortic-denervated dogs. *J. Appl. Physiol.* 57:1417–1421.

DiCarlo, S.E., and V.S. Bishop. 1990. Exercise training enhances cardiac afferent inhibition of baroreflex function. *Am. J. Physiol.* 258 *(Heart Circ. Physiol.)* 27:H212–H220.

Giannattasio, C., G. Seravalle, G.B. Bolla, B.M. Cattaneo, J. Cleroux, C. Cuspidi, L. Sampieri, G. Grassi, and G. Mancia. 1990. Cardiopulmonary receptor reflexes in normotensive athletes with cardiac hypertrophy. *Circulation* 82:1222–1229.

Giannattasio, C., G. Seravalle, B.M. Cattaneo, J. Cleroux, C. Cuspidi, L. Sampieri, G.B. Bolla, G. Grassi, and G. Mancia. 1992. Effect of detraining on the cardiopulmonary reflex in professional runners and hammer throwers. *Am. J. Cardiol.* 69:677–680.

Hainsworth, R. 1991. Reflexes from the heart. *Physiol. Reviews* 71:617–658.

Iwase, S., T. Mano, and M. Saito. 1987. Effects of graded head-up tilting on muscle sympathetic activities in man. *The Physiologist* 30:S62–S63.

Iwase, S., T. Mano, M. Saito, K. Koga, H. Abe, T. Matsukawa, and H. Suzuki. 1987. Comparison of muscle sympathetic nerve activity in man during water immersion and during body tilting. *Environ. Med.* 31:33–42.

Jingu, S., A. Takeshita, T. Imaizumi, M. Nakamura, M. Shindo, and H. Tanaka. 1988. Exercise training augments cardiopulmonary baroreflex control of forearm vascular resistance in middle-aged subjects. *Japanese Circ. J.* 52:162–168.

Johnson, J.M., L.B. Rowell, M. Niederberger, and M.M. Eisman. 1974. Human splanchnic and foream vasoconstrictor responses to reductions of right atrial and aortic pressure. *Circ. Res.* 34:515–524.

Joyner, M.J., R.L. Lennon, D.J. Wedel, S.H. Rose, and J.T. Shepherd. 1990. Blood flow to contracting human muscles: Influence of increased sympathetic activity. *J. Appl. Physiol.* 68:1453–1457.

Kouame, N., A. Nadeau, Y. Lacourciere, and J. Cleroux. 1995. Effects of different training intensities on the cardiopulmonary baroreflex control of forearm vascular resistance in hypertensive subjects. *Hypertension* 25:391–398.

Laughlin, M.H., R.M. McAllister, and M.D. Delp. 1994. Physical activity and the microcirculation in cardiac and skeletal muscle. In *Physical activity, fitness, and health: International proceedings and consensus statement*, ed. C. Boushard. 302–319. Champaign, IL: Human Kinetics.

Lightfoot, S.T., R.P. Claytor, D.J. Torok, T.W. Journell, and S.M. Fortney. 1989. Ten weeks of aerobic training do not affect lower body negative pressure responses. *J. Appl. Physiol.* 67:894–901.

Ludbrook, J., and W.F. Graham. 1985. Circulatory responses to onset of exercise: role of arterial and cardiac baroreflexes. *J. Appl. Physiol.* 248:H457—H467.

Mack, G., H. Nose, and E.R. Nadel. 1988. Role of cardiopulmonary baroreflexes during dynamic exercise. *J. Appl. Physiol.* 65:1827–1832.

Mack, G., H. Shi, H. Nose, A. Tripathi, and E.R. Nadel. 1987. Diminished baroreflex control of forearm vascular resistance in physically fit humans. *J. Appl. Physiol.* 63:105–110.

Mack, G.W., C.A. Thompson, D.F. Doerr, E.R. Nadel, and V.A. Convertino. 1991. Diminished baroreflex control of forearm vascular resistance following training. *Med. Sci. Sports Exerc.* 23:1367–1374.

Mark, A.L., and G. Mancia. 1983. Cardiopulmonary baroreflexes in humans. In *Handbook of physiology, section 2: The cardiovascular system, peripheral circulation and organ blood flow*, eds. J.T. Shepherd and F.M. Abboud. Vol. III, 795–813. Bethesda, MD: American Physiological Society.

Mitchell, J.H. 1990. Neural control of the circulation during exercise. *Med. Sci. Sports Exerc.* 22:141–154.

O'Hagan, K.P., L.B. Bell, S.W. Mittelstadt, and P.S. Clifford. 1994. Cardiac receptors modulate the renal sympathetic response to dynamic exercise in rabbits. *J. Appl. Physiol.* 76:507–515.

Oren, R.M., H.P. Schobel, R.M. Weiss, W. Stanford, and D. W. Ferguson. 1993. Importance of left atrial baroreceptors in the cardiopulmonary baroreflex of normal humans. *J. Appl. Physiol.* 74:2672–2680.

Pawelczyk, J.A., J. Warberg, and N.H. Secher. 1992. Leg vasoconstriction during dynamic exercise with reduced cardiac output. *J. Appl. Physiol.* 73:1838–1846.

Peterson, D.F., R.B. Armstrong, and M.H. Laughlin. 1988. Sympathetic neural influences on muscle blood flow in rats during submaximal exercise. *J. Appl. Physiol.* 65:434–440.

Ray, C.A. 1993. Muscle sympathetic nerve responses to prolonged one-legged exercise. *J. Appl. Physiol.* 74:1719–1722.

Ray, C.A., R.F. Rea, M.P. Clary, and A.L Mark. 1993. Muscle sympathetic nerve responses to dynamic one-legged exercise: Effect of body posture. *Am. J. Physiol.* 264 (Heart Circ. Physiol 33):H1–H7.

Roddie, I.C., J.T. Shepherd, and R.F. Whelan. 1957. Reflex changes in vasoconstrictor tone in human skeletal muscle in response to stimulation of receptors in a low pressure area of the intrathoracic vascular bed. *J. Physiol. (London)* 139:369–376.

Rowell, L.B. 1986. *Human circulation regulation during physical stress.* New York: Oxford University Press.

Saito, M., T. Mano, S. Iwase, K. Koga, C. Miwa, and K. Inamura. 1993. Effect of water immersion on muscle sympathetic nerve response during static muscle contraction. *Jap. J. Aerospace & Environ. Med.* 30:63–69.

Saito, M., A. Tsukanaka, D. Yanagihara, and T. Mano. 1993. Muscle sympathetic nerve response to graded leg cycling. *J. Appl. Physiol.* 75:663–667.

Sanders, J.S., D.W. Ferguson, and J.L. Kempf. 1988. Cardiopulmonary baroreflexes fail to modulate sympathetic responses during isometric exercise in humans: Direct evidence from microneurographic studies. *J. Am. Coll. Cardiol.* 12:1241–1251.

Scherrer, U., S.F. Vissing, and R.G. Victor. 1988. Effects of lower body negative pressure on sympathetic nerve responses to static exercise in humans. *Circulation* 78:49–59.

Seals, D.R. 1988. Cardiopulmonary baroreflexes do not modulate sympathetic activity to muscle during isometric exercise in humans. *J. Appl. Physiol.* 64:2197–2203.

Seals, D.R., and P.B. Chase. 1989. Influence of physical training on heart rate variability and baroreflex circulatory control. *J. Appl. Physiol.* 66:1886–1895.

Sheriff, D.D., L.B. Rowell, and A.M. Scher. 1993. Is the rapid rise in vascular conductance at onset of dynamic exercise due to muscle pump? *Am. J. Physiol.* 265 *(Heart Circ. Physiol 34)*:H1227–H1234.

Sprangers, R.L.H., K.H. Wesseling, A.L.T. Imholz, B.P.M. Imholz, and W. Wieling. 1991. Initial blood pressure fall on stand up and exercise explained by changes in total peripheral resistance. *J. Appl. Physiol.* 70:523–530.

Sundlof, G., and B.G. Wallin. 1978. Effect of lower body negative pressure on human muscle nerve sympathetic activity. *J. Physiol. (London)* 278:525–532.

Takeshita, A., S. Jingu, T. Imaizumi, Y. Kunihilo, S. Koyangi, and M. Nakamura. 1986. Augmented cardiopulmonary baroreflex control of forearm vascular resistance in young athletes. *Circ. Res.* 39:43–48.

Thames, M.D., and F.M. Abboud. 1979. Interaction of somatic and cardiopulmonary receptors in control of renal circulation. *Am. J. Physiol.* 237 (Heart Circ. Physiol 6):H560–H565.

Vallbo, A.B., K.E. Hagbarth, H.E. Torebjørk, and B.G. Wallin. 1979. Somatosensory, proprioceptive, and sympathetic activity in human peripheral nerves. *Physiol. Rev.* 59:919–957.

Victor, R.G., and W.N. Leimbach. 1987. Effects of lower body negative pressure on sympathetic discharge to leg muscles in humans. *J. Appl. Physiol.* 63:2558–2562.

Vissing, S.F., U. Scherrer, and R.G. Victor. 1989. Relation between sympathetic outflow and vascular resistance in the calf during perturbations in central venous pressure: Evidence for cardiopulmonary afferent regulation of calf vascular resistance in humans. *Circ. Res.* 65:1710–1717.

Vissing, S.F., U. Scherrer, and R.G. Victor. 1994. Increase of sympathetic discharge to skeletal muscle but not to skin during mild lower body negative pressure in humans, *J. Physiol. (London)* 481:233–241.

Walgenbach, S.C., and D.E. Donald. 1983. Cardiopulmonary reflexes and arterial pressure during rest and exercise in dogs. *Am. J. Physiol.* 244 (Heart Circ. Physiol):H362–H369.

Walker, S.L., F.M. Abboud, A.L. Mark, and M.D. Thames. 1980. Interaction of cardiopulmonary and somatic reflexes in humans. *J. Clin. Invest.* 65:1491–1497.

Zoller, R.D., A.L. Mark, F.M. Abboud, P.G. Schmid, and D.D. Heistad. 1972. The role of low pressure baroreceptors in reflex vasoconstriction responses in man. *J. Clin. Invest.* 51:2967–2972.

Chapter 4

Control of Circulation and Respiration During Exercise: Central Neural Integration

Tony G. Waldrop and Jeffrey Kramer

Studies about the control of circulation and respiration are now into their second century of investigation. The pioneering theories of Zuntz and Geppert in 1886, Johansson in 1893, and Krough and Lindhard in 1913 have been tested and have led to important fields of research. These earlier investigators correctly identified possible control mechanisms responsible for regulating arterial pressure, heart rate, and breathing during exercise. Subsequent investigators have provided additional support for the role of two major neural mechanisms in exercise control: descending central command and a feedback reflex originating in contracting muscles.

Central command involves a parallel activation of the brain stem and the spinal circuits that control locomotor, cardiovascular, and respiratory activity during exercise (Waldrop et al. 1996). This concept implies that the brain sites that generate locomotor signals also send descending projections onto cardiovascular and respiratory neurons. They thereby provide a means for an immediate and coordinated adjustment of cardiorespiratory function to meet the elevated metabolic demands of exercise. Determination of the exact neuroanatomic loci at which the central command signal originates is a current area of investigation. However, several candidates have been identified. These include sites in the cerebral cortex and the so-called hypothalamic and mesencephalic locomotor regions (Bedford, Loi, and Crandall 1992; Eldridge et al. 1985; Green and Hoff 1937; Shik and Orlovski 1976; Waldrop, Bauer, and Iwamoto 1988).

A second control mechanism originates from metabolic and mechanical activation of nerve endings located in contracting skeletal muscles (Kaufman and Forster 1996). Signals from these nerve endings project to the dorsal horn of the spinal cord through group III and group IV afferent fibers and, ultimately, to supraspinal sites. The efferent limb of this reflex involves innervation of respiratory muscles as well as sympathetic and parasympathetic outflow to the heart, vasculature, and other target organs including the adrenal gland.

Central command and the muscle feedback reflex along with other cardiorespiratory reflexes are active during voluntary exercise. Numerous studies have demonstrated that either of these mechanisms acting alone can elicit a largely complete cardiorespiratory response to exercise (Eldridge and Waldrop 1991). However, both are activated during voluntary exercise. This simultaneous functioning of both control mechanisms necessitates central integration such that coordinated cardiorespiratory responses can be elicited to meet increasing metabolic demands. Despite the importance of central neural integration, only a few investigators have focused upon determining the level of integration that occurs during exercise and the brain sites involved in multiple control mechanisms.

Locomotor Regions and Central Command

Several known locomotor regions in the brain have been examined as possible locations of origin for the central command signals (Eldridge and Waldrop 1991; Waldrop et al. 1996) (see figure 4.1). Earlier studies utilized electric stimulation of the motor cortex to examine the possible role of this area in cardiorespiratory regulation (Clark, Smith, and Shearn 1968; Green and Hoff 1937; Hilton, Spyer, and Timms 1979). Even though some cardiovascular responses could be elicited, Hilton, Spyer, and Timms (1979) concluded that the observed responses are most likely due to factors other than activation of neurons in the motor cortex. A recent study by Fink et al. (1995), however, has provided evidence for a role of cortical sites in the central command mechanism. In this study, brain sites activated during exercise in humans were localized by combining

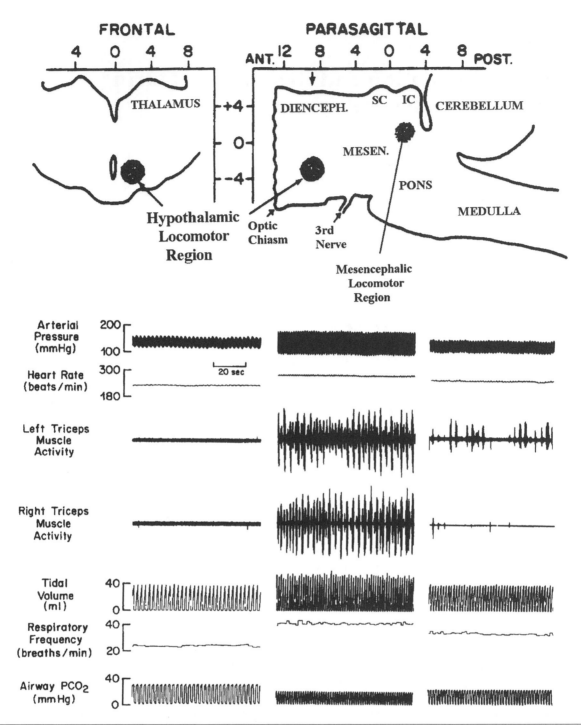

Figure 4.1 *(top)* Schematic representation of the location of the hypothalamic and mesencephalic locomotor regions. *(bottom)* Effects of microinjection of a GABA antagonist upon cardiovascular, locomotor, and respiratory activity in an anesthetized cat. Left panel: Control conditions. Center panel: After microinjection of the GABA antagonist, picrotoxin, into the hypothalamic locomotor region. Right panel: Reversal of the responses elicited by the GABA antagonist after microinjection of the GABA agonist muscimol. (Reprinted with permission from Waldrop, T.G., R.M. Bauer, and G.A. Iwamoto, "Microinjections of GABA antagonists into the posterior hypothalamus elicits locomotor activity and a cardiorespiratory activation," *Brain Research* 44⁴:84–94. Copyright 1988 American Association for the Advancement of Science.)

positron emission tomographic measurements of cerebral blood flow with magnetic resonance imaging for identifying neuroanatomic structures. This sophisticated combination of techniques demon-strated that several cortical sites, including the motor cortex, have increased activity during exercise.

Researchers have extensively studied the involve-ment of the hypothalamic locomotor region (HLR) in the central command mechanism (Eldridge et al. 1985; Shik and Orlovski 1976; Waldrop, Bauer, and Iwamoto 1988). Their studies have demonstrated that electric or chemical stimulation in the caudal hypothalamus produce running movements in the unanesthetized, decorticate, or anesthetized cat. These movements are accompanied by appropriate alterations in breathing, sympathetic nerve activity, blood pressure, and heart rate (see figure 4.1). Moreover, a redistribution of organ blood flows occurs, which is similar to that observed during voluntary exercise in the awake cat (Waldrop et al. 1986). This suggests that coordinated control of autonomic outflow is achieved during HLR stimulation. These hypothalamically evoked responses persisted even after muscular paralysis and in the absence of feedback changes from peripheral chemoreceptors, baroreceptors, and central chemoreceptors. Therefore, HLR stimulation can provide a parallel drive to descending motor as well as cardiorespiratory pathways independent of feedback from contracting muscles. These results have led to the conclusion that activation of the hypothalamic locomotor region provides a model for studying central command in the cat.

Bedford, Loi, and Crandall (1992) have studied the potential role of the mesencephalic locomotor region (MLR) in central command. Electric stimulation of the mesencephalic locomotor region in the decerebrate rat produced running movements and increases in blood pressure and heart rate. Muscular paralysis did not prevent the increases in blood pressure with stimulation. This suggests that the region also provides parallel activation of locomotor and cardiovascular areas in the brain. However, even though anatomic connections exist, possible interactions between the HLR and MLR, with respect to cardiorespiratory control, have yet to be determined.

Locomotor Regions and Feedback From Contracting Muscles

The hypothalamic locomotor region is also involved in the feedback reflex originating from activation of mechanical and metabolic receptors in contracting skeletal muscles. Waldrop and Stremel (1989) recorded from HLR neurons in anesthetized cats which contraction of hind limb muscles was evoked by stimulation of the cut peripheral ends of the L7 and S1 ventral roots (see figure 4.2). Either static or

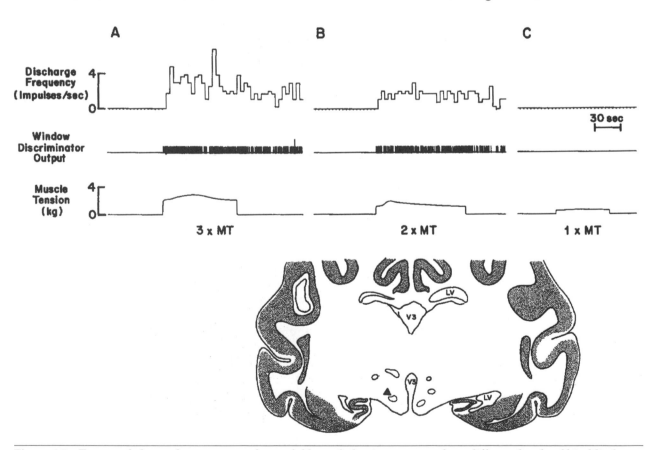

Figure 4.2 Top panel shows the responses of a caudal hypothalamic neuron to three different levels of hind limb muscle tension developed during stimulation of the ventral roots at 3, 2, and 1 times the motor threshold (MT). The line drawing depicts the location of the neuron. (From Waldrop and Stremel 1989.)

rhythmic contractions stimulated the majority of neurons in this hypothalamic region; neurons in this region possess basal discharge rhythms correlated temporally with the cardiac cycle and/or sympathetic nerve activity.

Waldrop et al. (1986) provided additional evidence for a role of the hypothalamic locomotor region in mediation of the cardiorespiratory responses evoked by muscular contraction. These investigators examined how placing bilateral lesions in the caudal hypothalamus affected the cardiovascular and respiratory responses evoked by static muscular contraction. Both the heart rate and respiratory frequency increases produced by contraction of hind limb muscles were attenuated after the lesions. They concluded that this region normally modulates some of the cardiorespiratory responses elicited by muscular contraction. However, compensation by other brain sites controlling the cardiorespiratory system during muscle contraction may have accounted for the lack of change in blood pressure responses before and after lesioning.

Another locomotor area that receives input from hind limb muscles is the mesencephalic locomotor region. Contraction of hind limb muscles altered the activity of neurons in this region of the brain (Plowey and Waldrop 1997). Since electric or chemical stimulation of this brain region can induce increases in blood pressure, these findings suggest a role in mediating cardiovascular responses generated by muscle contraction. Moreover, these data provide evidence that neurons in the MLR may integrate descending locomotor and muscle feedback mechanisms during exercise. Taken together, these studies suggest that central command pathways may not be exclusively separate from pathways serving muscle reflex responses.

Brain Stem Sites Involved in the Feedback Reflex Originating in Exercising Muscle

Afferents arising from exercising muscle are thought to contain their primary synapse in the central nervous system at the level of the dorsal horn of the spinal column (Mense and Craig 1988; Wilson et al. 1992). Research has shown that neurons in a similar area of the dorsal horn directly project to the nucleus tractus solitarius (NTS) in the brain stem. (Esteves, Lima, and Coimbra 1993). These studies provide neuroanatomic evidence suggesting that afferents from exercising muscle project through one or more synapses to neurons in the NTS. In this way, muscle afferents could modulate the activity of NTS neu-

rons. Moreover, the NTS has projections to other medullary sites, such as the ventrolateral medulla, that are involved in cardiovascular and respiratory regulation (Spyer 1994). However, because afferents from muscle receptors project to several brain stem sites, the pathways critical for expression of the reflex remain elusive.

The medial and ventral brain stem are complex structures that contain several nuclei related to cardiorespiratory regulation. These include ventrolateral medullary nuclei, nucleus ambiguus, lateral reticular nucleus, the lateral tegmental field, and nucleus paragigantocellularis. Much of the differentiation of these nuclei was first determined histologically. However, within the past 20 years, distinct functional differences have been discovered with regards to autonomic functioning. Moreover, researchers have investigated the role of these nuclei in regulating cardiorespiratory responses to muscle contraction (Waldrop et al. 1996).

The ventrolateral medulla (VLM) is important in the regulation of the cardiovascular and respiratory system. This region is postulated to be the site of the origin of respiratory drive as well as a major output area for sympathetic outflow to the spinal cord. In addition, several reflexes, including the baroreceptor reflex and chemoreceptor reflexes, involve neurons in the ventrolateral medulla (Millhorn and Eldridge 1986).

Bauer and colleagues (Bauer, Iwamoto, and Waldrop 1989, 1990; Bauer et al. 1992) have conducted several studies that provide strong support for areas in both the caudal and rostral VLM to mediate feedback reflexes from contracting muscles. An initial set of experiments examined the effects of blocking excitatory amino acid (EAA) transmission in the VLM upon the pressor response evoked by contraction of hind limb muscles (Bauer, Iwamoto, and Waldrop 1989). Bilateral microinjections of kynurenic acid (KYN) into the VLM of anesthetized cats attenuated the increase in arterial pressure evoked by muscular contraction (see figure 4.3) but did not alter the responses produced by stimulation in the caudal hypothalamus. Moreover, microinjection of an inactive analog of kynurenic acid, xanthurenic acid, had no effects on the pressor reflex. These findings suggest that an excitatory amino acid synapse in the ventrolateral medulla modulates the pressor response associated with muscular contraction.

Electrophysiological studies have provided additional support for a role of the caudal and rostral VLM in exercise regulation (Bauer, Iwamoto, and Waldrop 1990; Iwamoto and Kaufman 1987). In one study, contraction of hind limb muscles in anesthetized cats produced an increase in discharge fre-

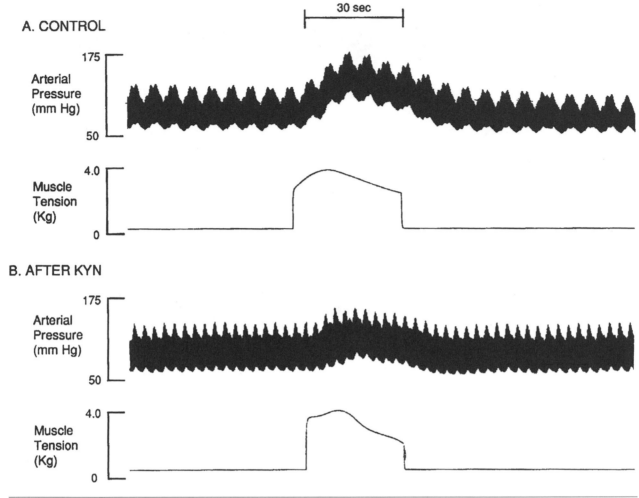

Figure 4.3 Microinjection of the excitatory amino acid antagonist, kynurenic acid, blunts the pressor response evoked by muscular contraction. *(a)* Control response. *(b)* After bilateral microinjection of kynurenic acid into the ventrolateral medulla. (From Bauer, Iwamoto, and Waldrop 1989.)

quency in a majority of the VLM neurons tested. Most of these neurons had a basal discharge frequency related to the cardiac cycle and/or sympathetic nerve discharge as revealed by spike-triggered averaging. Similar responses were obtained in peripherally barodenervated cats. Subsequent studies, utilizing antidromic activation, determined that many of the neurons that responded to muscular contraction had projections to the thoracic intermediolateral cell columns (IML) of the spinal cord (Bauer et al. 1992) (see figure 4.4). Comparable findings have been obtained for the somatic pressor reflex elicited by sciatic nerve stimulation in anesthetized rats (Morrison and Reis 1989). Therefore, these studies demonstrate that neurons in the ventrolateral medulla with a cardiovascular-, sympathetic-, or respiratory-related discharge, and that project to the intermediolateral columns in the spinal cord, are involved in the mediation of cardiorespiratory responses evoked by the muscle feedback reflex.

Several studies have suggested that glutamate is the primary neurotransmitter at the synapse of VLM neurons upon sympathetic preganglionic cells in the spinal cord (Morrison et al. 1991). Consistent with this pathway having a role in the mediation of the pressor reflex evoked by muscular contraction is the finding that glutamate immunoreactivity was observed in neurons in the same areas of the VLM that contain the reticulospinal neurons responsive to muscular contraction (Bauer et al. 1992). Moreover, subsequent experiments have documented that the cardiovascular responses associated with the muscle feedback reflex active during exercise involves an excitatory amino acid synapse in the thoracic spinal cord (Bauer et al. 1993). These experiments demonstrated that intrathecal administration of the EAA antagonist, kynurenic acid, into the thoracic spinal cord blunted the rise in arterial pressure observed during muscular contraction.

Iwamoto and Waldrop (1996) have investigated the possible role of neurons in the lateral tegmental

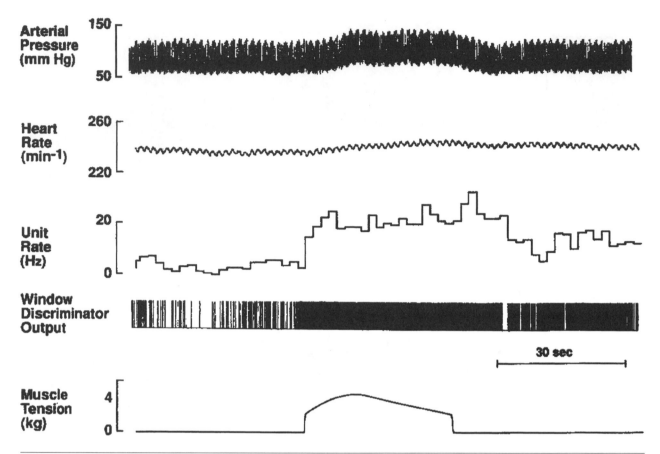

Figure 4.4 Contraction of hind limb muscles elicits an increase in the discharge rate of a neuron located in the ventrolateral medulla. This neuron had a basal cardiovascular discharge and could be antidromically activated by stimulation in the IML. (From Bauer et al. 1992.)

field (LTF) in cardiovascular regulation during muscle contraction. Neurons in this region respond to muscle contraction. However, synaptic inhibition of this area with cobalt chloride does not affect the blood pressure responses to muscle contraction or caudal hypothalamic stimulation. Despite the suppression of sympathetic rhythms induced by inhibition of this region, these nuclei are apparently not necessary for expression of pressor responses induced by muscle reflexes or simulated central command. Thus, this nucleus may be involved in coordinating sympathetic outflow or provide excitatory drive to rostral ventrolateral medullary neurons excited by muscular contraction.

Electrolytic lesions in the nucleus reticularis gigantocellularis (NRG) have little effect on cardiorespiratory responses to muscle contraction. In contrast, lesions in this area potentiate heart rate and respiratory responses to hypothalamic locomotor region stimulation (Richard et al. 1989). These results are consistent with the finding that stimulation of neurons in this region of the brain inhibits cardiovascular and respiratory functioning. Apparently, the NRG is not only more likely involved in regulat-

ing respiration during exercise, but central command may depend more upon an intact NRG than on muscle reflexes. Furthermore, this is a potential brain stem site that may integrate respiratory drive initiated by central command and muscle reflexes during exercise.

The above findings provide results that elucidate some of the central circuitry responsible for mediation of the cardiovascular responses evoked by muscular contraction. Input from contracting muscles traveling in groups III and IV afferent fibers synapses on dorsal horn cells in the lumbar spinal cord. This information then ascends to the medulla and likely synapses in the nucleus of the tractus solitarius. The exact projection of this exercise-related activity from the NTS is unknown. However, ultimately, an EAA synapse in the ventrolateral medulla is important in the mediation of the arterial pressure response evoked by muscular contraction. The efferent limb of the reflex involves an EAA synapse of VLM neurons upon IML cells in the spinal cord. Other medullary sites, including the raphe nuclei (Iwamoto et al. 1991), the nucleus reticularis gigantocellularis (Richard et al. 1989), and the lateral tegmental fields (Iwamoto and Waldrop 1996) may modulate this feedback reflex. However, these sites are not likely to be crucial components in the circuitry.

Control of Cardiorespiratory Activity by Brain Sites in Awake Animals: Utilization of C-Fos Mapping

Most of the studies described above have been performed in anesthetized cats or rats in which exercise responses are evoked rather than voluntary in nature. Thus, the possibility exists that the brain sites identified in these anesthetized preparations are not the same areas that regulate the cardiorespiratory systems during voluntary exercise. A technique has been developed in the last few years that can address this potential problem. Expression of the immediate early gene *c-fos* and its protein product Fos occurs in response to a variety of stimuli, including neuronal depolarization and growth factors (Hunt, Pini, and Evan 1987; Sagar, Sharp, and Curran 1988). As a result, the technique has been used in a variety of studies to identify brain sites involved in various reflexes and responses.

Iwamoto et al. (1996) have recently used Fos immunocytochemistry to determine which areas of the brain have an altered neuronal activity during treadmill running in awake rats. Labeling increased over control conditions during exercise in many of the same areas implicated in exercise regulation in the prior experiments involving anesthetized animals. Increased labeling during exercise was observed in the hypothalamic locomotor region, the medial portion of the nucleus tractus solitarius, and both the rostral and caudal ventrolateral medulla. Oladehin et al. (1994) found similar brain Fos immunoreactive labeling in running rats. Even though dorsal column nuclei were labeled, cells in thalamic areas associated with somatosensory function were conspicuously without labeling. These results obtained in the awake rat support the findings obtained in previous studies in anesthetized animals. In addition, the labeling observed in these studies is consistent with labeling found in metabolic mapping studies using 2-deoxyglucose following induced muscle contraction in cats (Iwamoto et al. 1984).

Integration of Central Command and Feedback From Contracting Muscles

Assessing the importance of central command relative to feedback from contracting muscles is diffi-

cult. Numerous studies have attempted to study one of these mechanisms in the absence of the other. However, these types of studies have methodological problems that complicate interpretation of results. Researchers have argued that "You may have sufficient mechanisms; each of which in a given, isolated circumstance explains the whole phenomenon. When they act simultaneously, they mask each other" (Yamamoto 1977). Many mechanisms are probably involved in the control of cardiovascular and respiratory function during exercise. Each of these likely has an important, and perhaps incompletely understood, role to play in exercise regulation.

A few studies have examined the integration that occurs between simulated central command and the feedback reflex originating in contracting muscles (Rybicki et al. 1989; Waldrop, Mullins, and Millhorn 1986). One study compared the individual and combined responses to activation of simulated central command (HLR stimulation) and feedback from contracting muscles in anesthetized cats. Both stimuli elicited increases in arterial pressure, heart rate, and ventilation when given individually. Simulated central command caused large increases in cardiorespiratory activity when presented during muscular contraction. In contrast, muscular contraction elicited small increases in respiration and only modest changes in arterial pressure and heart rate during hypothalamic stimulation. Thus, simulated central command and muscular contraction produce different magnitude responses when activated together as compared with individual activation. These findings also suggest that activation of simulated central command has a predominant effect over the responses caused by muscular contraction.

Nolan and Waldrop (1997) examined one potential neuronal mechanism responsible for the above findings. They recorded the responses of medullary neurons during independently induced hypothalamic stimulation and contraction of hind limb muscles in anesthetized cats. They observed two major types of neuronal responses in their studies. One group of neurons was stimulated by both muscular contraction and hypothalamic stimulation. In contrast, another group of medullary neurons, which were stimulated by muscular contraction, were inhibited by hypothalamic stimulation. This latter response type could explain the above finding that muscular contraction evokes only small cardiorespiratory responses during hypothalamic stimulation. Moreover, these findings demonstrate one type of central integration that occurs in the regulation of systems during evoked exercise.

Integration of Exercise Control Mechanisms and Other Cardiorespiratory Reflexes

This section discusses integration of exercise control mechanisms and other cardiorespiratory reflexes including the baroreceptor reflex and vestibulo-autonomic reflexes.

Baroreceptor Reflex

Articles have reported that the baroreceptor reflex does not produce as much slowing of the heart rate during exercise as during rest (Cunningham et al. 1972; Hobbs and McCloskey 1986). Evidence suggests that both the central command mechanism and the muscle reflex active during exercise modulate baroreceptor mechanisms.

To study the role of the muscle reflex in baroreflex functioning, McMahon, McWilliam, and Kaye (1993) performed extracellular recordings from single NTS neurons in the anesthetized cat. In 42% of the NTS neurons tested, simultaneous muscle contraction (induced by ventral root stimulation) and electric carotid sinus nerve stimulation significantly reduced the amount of evoked firing observed in NTS neurons versus carotid sinus nerve stimulation alone. Ipsilateral carotid sinus distention determined that 69% of the neurons tested were baroreceptor sensitive. This suggested that the majority of neurons that decreased their evoked responses during muscle contraction were also neurons involved in the carotid sinus baroreflex. The authors further suggested that the decrease in evoked firing was at least partially GABA mediated.

Given the excitatory connections from NTS to NA neurons, findings from the McMahon, McWilliam, and Kaye (1993) study complemented previous findings. These demonstrated a reduction of baroreceptor-mediated vagal outflow upon muscle contraction in decerebrate cats (McMahon and McWilliam 1992). Certainly, these data provide preliminary evidence for the NTS as a possible brain area for baroreflex modulation during muscle contraction. Classic brain stem-baroreflex circuitry connects NTS neurons with rostral VLM (rVLM) neurons via an inhibitory GABAergic connection through the caudal VLM. Therefore, a decrease in the evoked responses of NTS neurons might contribute to a reflex increase in firing of rVLM neurons (Spyer 1981). However, the neuronal circuitry within these areas of the brain stem is quite complex. To suggest the latter is responsible for the increase in firing of VLM neurons would be an oversimplification. As a re-

sult, the exact contribution of NTS neurons to the activity of VLM neurons during muscle contraction requires additional investigation.

Researchers have demonstrated that both electric and chemical stimulation of the HLR region of the caudal hypothalamus blunts the baroreceptor mechanism in anesthetized cats and rats (Bauer et al. 1988; Coote, Hilton, and Perez-Gonzalez 1979; Spyer 1981). Thus, the same area of the hypothalamus that produces locomotion and cardiovascular responses upon stimulation also apparently depresses baroreflex functioning. As a result, modulation of the baroreflex that occurs with exercise may arise from a descending drive of caudal hypothalamic neurons. Moreover, both simulated central command-mediated increases in cardiovascular functioning and modulation of baroreflex regulation appear to involve GABAergic mechanisms in the caudal hypothalamus (Bauer et al. 1988).

Vestibulo-Autonomic Reflexes

Humans most often exercise in the upright versus supine body position. As a result, gravitational effects can impact cardiorespiratory functioning. Baroreceptors can help regulate functioning during orthostatic stress. However, the vestibular system is also able to mediate cardiovascular and respiratory changes to help maintain homeostasis (Yates and Miller 1998). Vestibular receptors located within the inner ear may also play an integrative role in regulating cardiorespiratory functioning during exercise (Kramer and Waldrop 1998). Stimulation of cranial nerve VIII as well as natural vestibular stimulation induced by rotating the head can induce fluctuations in sympathetic nerve and efferent abdominal and phrenic nerve activity (Rossiter et al. 1996; Yates, Jakuš and Miller 1993; Yates and Miller 1994).

Furthermore, research shows that several brain stem regions, including vestibular nuclei, the nucleus tractus solitarius, and ventrolateral medullary nuclei, are involved in mediating the cardiorespiratory responses to vestibulo-autonomic reflexes (Yates and Miller 1998). Similar nuclei have been implicated in mediating responses to muscle contraction (see the section in this chapter entitled "Brain Stem Sites Involved in the Feedback Reflex Originating in Exercising Muscle"). In addition, researchers have shown that neurons in the VLM respond to stimulation of the vestibular nerve and direct hind limb somatic nerve stimulation. Further studies have indicated that cardiorespiratory-related vestibular nuclei receive input from contracting hind limb muscles and, as a result, may be involved in inte-

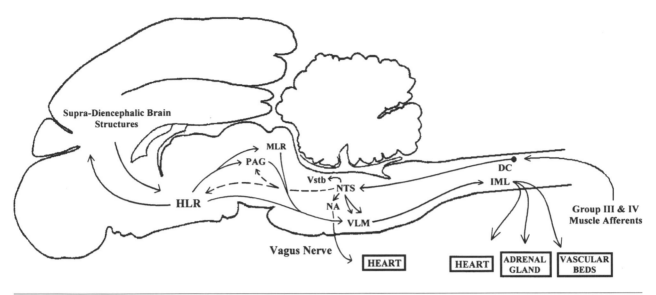

Figure 4.5   Parasagittal schematic representation of proposed neural pathways mediating cardiovascular and respiratory control during exercise. Solid lines represent established neuronal pathways; dotted lines represent hypothesized pathways. HLR = hypothalamic locomotor region; MLR = mesencephalic locomotor region; PAG = periaqueductal gray; Vstb = vestibular nuclei; NTS = nucleus tractus solitarius; NA = nucleus ambiguus; VLM = ventrolateral medulla; IML = intermediolateral cell column; DC = dorsal horn of the spinal cord.

grating these reflexes (Kramer, Beatty and Waldrop 1996). Vestibular receptors are also known to affect locomotion by coordinating outflow to locomotor and postural muscles. As a result, postural reflexes initiated by the vestibular system may quite possibly be integrated with central command drive during upright exercise. When considering the overlap of brain regions regulating the cardiorespiratory system between vestibular reflexes, muscle reflexes, and central command, further work should be devoted to examining how these drives coordinate autonomic outflow during exercise.

Summary

In the century that has passed since the first pioneering work was accomplished, we have significantly increased our understanding of how the central nervous system regulates cardiorespiratory functioning. Elucidation of the brain areas, neural pathways (see figure 4.5), and neurotransmitter mechanisms involved in both central command and the muscle feedback reflex have become and continue to be important areas of research. More importantly, how these neural control mechanisms are integrated within the central nervous system provides useful information about how the nervous system is organized. It also provides useful information about how exercise may play a role in dealing with cardiovascular disorders of a neurogenic origin. Clearly,

though, more work needs to be completed in order to understand these complex topics.

References

Bauer, R.M., G.A. Iwamoto, and T.G. Waldrop. 1989. Ventrolateral medullary neurons modulate pressor reflex to muscular contraction. *American Journal of Physiology*, 257:R1154–R1161.

Bauer, R.M., G.A. Iwamoto, and T.G. Waldrop. 1990. Discharge patterns of ventrolateral medullary neurons during muscular contraction. *American Journal of Physiology*, 259:R606–R611.

Bauer, R.M., G.A. Iwamoto, T.G. Waldrop, and M.A. Holzwarth. 1992. Properties of ventrolateral medullary neurons that respond to muscular contraction. *Brain Research Bulletin*, 28:167–178.

Bauer, R.M., P.C. Nolan, E.M. Horn, and T.G. Waldrop. 1993. An excitatory amino acid synapse in the thoracic spinal cord is involved in the pressor response to muscular contraction. *Brain Research Bulletin*, 32:673–679.

Bauer, R.M., M.B. Vela, T. Simon, and T.G. Waldrop. 1988. A GABAergic mechanism in the posterior hypothalamus modulates baroreflex bradycardia. *Brain Research Bulletin*, 20:633–641.

Bedford, T.G., P.K. Loi, and C.C. Crandall. 1992. A model of dynamic exercise—the decerebrate rat locomotor region. *Journal of Applied Physiology*, 72:121–127.

Clark, N.P., O.A. Smith, and D.W. Shearn. 1968. Topographical representation of vascular smooth muscle of limbs in primate motor cortex. *American Journal of Physiology*, 214:122–129.

Coote, J.H., S.M. Hilton, and J.F. Perez-Gonzalez. 1979. Inhibition of the baroreceptor reflex on stimulation in the brainstem defense centre. *Journal of Physiology*, 288:549–560.

Cunningham, D.J.C., E.S. Petersen, R. Teto, T.G. Pickering, and P. Sleight. 1972. Comparison of the effect of different types of exercise on the baroreflex regulation of heart rate. *Acta Physiologica Scandinavica*, 86:444–455.

Eldridge, F.L., D.E. Millhorn, J.P. Kiley, and T.G. Waldrop. 1985. Stimulation by central command of locomotion, respiration and circulation during exercise. *Respiration Physiology*, 59:313–337.

Eldridge, F.L., and T.G. Waldrop. 1991. Neural control of breathing during exercise. In *Exercise: Pulmonary physiology and pathophysiology*, eds. B. Whipp and K. Wasserman, 309–370. New York: Marcel Dekker.

Esteves, F., D. Lima, and A. Coimbra. 1993. Structural types of spinal cord marginal (lamina I) neurons projecting to the nucleus of the tractus solitarius in the rat. *Somatosensory and Motor Research*, 10:203–216.

Fink, G.R, L. Adams, J.D.G. Watson, J.A. Innes, B. Wuyam, I. Kobayashi, D.R. Corfield, K. Murphy, T. Jones, R.S.J. Frackowiak, and A. Guz. 1995. Hypernoia during and immediately after exercise in man: Evidence of motor cortical involvement. *Journal of Physiology*, 489:663–675.

Green, H.D., and E.C. Hoff. 1937. Effects of faradic stimulation of the cerebral cortex on limb and renal volumes in the cat and monkey. *American Journal of Physiology*, 1118:641–658.

Hilton, S.M., K.M. Spyer, and R.J. Timms. 1979. The origin of the hindlimb vasodilation evoked by stimulation of the motor cortex. *Journal of Physiology*, 287:545–557.

Hobbs, S.F., and D.I. McCloskey. 1986. Effect of spontaneous exercise on reflex slowing of the heart in decerebrate cats. *Journal of the Autonomic Nervous System*, 17:303–312.

Hunt, S.P., A. Pini, and G. Evan. 1987. Induction of *c-fos* like protein in spinal cord neurons following sensory stimulation. *Nature*, 328:632–634.

Iwamoto, G.A., M.E. Clement, R.D. Brtva, and R.B. McCall. 1991. Effects of muscular contraction on discharge patterns of neurons in the medullary raphe nuclei. *Brain Research*, 558:353–356.

Iwamoto, G.A., and M.P. Kaufman. 1987. Caudal ventrolateral medullary cells responsive to muscle contraction. *Journal of Applied Physiology*, 62:149–157.

Iwamoto, G.A., J.G. Parnavelas, M.P. Kaufman, B.R. Botterman, and J.H. Mitchell. 1984. Activation of caudal brainstem groups during the exercise pressor reflex in the cat as elucidated by 2-[^{14}C]deoxyglucose. *Brain Research*, 304:178–182.

Iwamoto, G.A., and T.G. Waldrop. 1996. Lateral tegmental field neurons sensitive to muscular contraction: A role in pressor reflexes? *Brain Research Bulletin*, 41:111–120.

Iwamoto, G.A., S.M. Wappel, G.M. Fox, K.A. Buetow, and T.G. Waldrop. 1996. Identification of diencephalic and brainstem cardiorespiratory areas activated during exercise. *Brain Research*, 726:109–122.

Johansson, J.E. 1893. Uber die Einwirkung der Muscdeltatigkeit auf die Atmun und die Herztatigkeit. *Skandanavian Archives of Physiology*, 5:20–66.

Kaufman, M.P., and H.B. Forster. 1996. Reflexes controlling circulatory, ventilatory and airway responses to exercise. In *Handbook of physiology, section 12, exercise: Regulation and integration of multiple systems*, ed. L.B. Rowell and J.T. Shepherd, 381–447. New York: Oxford University Press.

Kramer, J.M., J.A. Beatty, and T.G. Waldrop. 1996. Cardiorespiratory related neurons in the vestibular nuclei respond to muscle contraction in the rat. *The Physiologist*, 39:A-19.

Kramer, J.M., and T.G. Waldrop. 1998. Neural control of the cardiovascular system during exercise. An integrative role for the vestibular system. *Journal of Vestibular Research*, 8:71–80.

Krough, A., and J. Lindhard. 1913. The regulation of respiration and circulation during the initial stages of exercise. *Journal of Physiology*, 47:112–136.

McMahon, S.E., and P.N. McWilliam. 1992. Changes in R-R interval at the start of muscle contraction in the decerebrate cat. *Journal of Physiology*, 447:549–562.

McMahon, S.E., P.N. McWilliam, and C. Kaye. 1993. Hindlimb contraction inhibits evoked activity in baroreceptor-sensitive neurons in the nucleus tractus solitarii (NTS) of the anesthetized cat. *Journal of Physiology*, 467:18P.

Mense, S., and D. Craig. 1988. Spinal and supraspinal terminations of primary afferent fibers from the gastrocnemius-soleus muscle in the cat. *Neuroscience*, 26:1023–1035.

Millhorn, D.E., and F.L. Eldridge. 1986. Role of the ventrolateral medulla in regulation of respiratory and cardiovascular systems. *Journal of Applied Physiology*, 61:1249–1263.

Morrison, S.F., J. Callaway, T.A. Milner, and D.J. Reis. 1991. Rostral ventrolateral medulla: A source of glutamatergic innervation of the sympathetic intermediolateral nucleus. *Brain Research*, 562:126–135.

Morrison, S.F., and D.J. Reis. 1989. Reticulospinal vasomotor neurons in the RVL mediate the somatosympathetic reflex. *American Journal of Physiology*, 256:R1084–R1097.

Nolan, P.C., and T.G. Waldrop. 1997. Integrative role of medullary neurons of the cat during exercise. *Experimental Physiology*, 82: 547–558.

Oladehin, A., R.S. Water, K. Redmon, and B. Bell. 1994. Exercise induced *c-fos* expression in the rat brain. *Medicine and Science in Sport and Exercise*, 26:S68.

Plowey, E.D., and T.G. Waldrop. 1997. Neuronal activity in the rat mesencephalic locomotor region is altered during static contraction of the hindlimb muscles. *Society for Neuroscience Abstracts*, 23:168.

Richard, C.A., T.G. Waldrop, R.M. Bauer, J.H. Mitchell, and R.W. Stremel. 1989. The nucleus reticularis gigantocellularis modulates the cardiopulmonary responses to central and peripheral drives related to exercise. *Brain Research*, 482:49–56.

Rossiter, C.D., N.L. Hayden, S.D. Stocker, and B.J. Yates. 1996. Changes in outflow to respiratory pump muscles produced by natural vestibular stimulation. *Journal of Neurophysiology*, 76:3274–3284.

Rybicki, K.J., R.W. Stremel, G.A. Iwamoto, J.H. Mitchell, and M.P Kaufman. 1989. Occlusion of pressor responses to posterior diencephalic stimulation and static muscular contraction. *Brain Research*, 22:306–312.

Sagar, S.M., F.R. Sharp, and T. Curran. 1988. Expression of *c-fos* protein in brain: Metabolic mapping at the cellular level. *Science*, 240:1328–1331.

Shik, M.L., and G.N. Orlovski. 1976. Neurophysiology of locomotor automatism. *Physiological Reviews*, 56:465–501.

Spyer, K.M. 1981. Neural organization and control of the baroreceptor reflex. *Review of Physiology, Biochemistry and Pharmacology*, 88:123–124.

Spyer, K.M. 1994. Central nervous mechanisms contributing to cardiovascular control. *Journal of Physiology*, 474:1–10.

Waldrop, T.G., R.M. Bauer, and G.A. Iwamoto. 1988. Microinjections of GABA antagonists into the posterior hypothalamus elicits locomotor activity and a cardiorespiratory activation. *Brain Research*, 444:84–94.

Waldrop, T.G., F.L. Eldridge, G.A. Iwamoto, and J.H. Mitchell. 1996. Central neural control of respiration and circulation during exercise. In *Handbook of physiology, section 12, exercise: Regulation and integration of multiple systems*, ed. L.B. Rowell and J.T. Shepherd, 333–380. New York: Oxford University Press.

Waldrop, T.G., M.C. Henderson, G.A. Iwamoto, and J.H. Mitchell. 1986. Regional blood flow responses to stimulation of the subthalamic locomotor region. *Respiration Physiology*, 64:93–102.

Waldrop, T.G., D.C. Mullins, and D.E. Millhorn. 1986. Control of respiration by the hypothalamus and by feedback from contracting muscles in cats. *Respiration Physiology*, 64:317–328.

Waldrop, T.G., and R.W. Stremel. 1989. Muscular contraction stimulates posterior hypothalamic neurons. *American Journal of Physiology*, 256:R348–R356.

Wilson, L.B., P.T. Wall, K. Matsukawa, and J.H. Mitchell. 1992. Effect of spinal microinjections of an antagonist to substance P or somatostatin on the exercise pressor reflex. *Circulation*, 70:213–222.

Yamamoto, W.S. 1977. General discussion. In *Muscular exercise and the lung*, eds. J.A. Dempsey and C.E. Reed, 169. Madison, WI: University of Wisconsin Press.

Yates, B.J., J. Jakuš, and A.D. Miller. 1993. Vestibular effects on respiratory outflow in the decerebrate cat. *Brain Research*, 629:209–217.

Yates, B.J., and A.D. Miller. 1994. Properties of sympathetic reflexes elicited by natural vestibular stimulation: Implications for cardiovascular control. *Journal of Neurophysiology*, 71:2087–2092.

Yates, B.J., and A.D. Miller. 1998. Physiological evidence that the vestibular system participates in autonomic and respiratory control. *Journal of Vestibular Research*, 8:17–25.

Zuntz, N., and J. Geppert. 1886. Uber die Natur der normalen Atemreize und den Ort ihrer Wirkung. *Archives of Ges Physiology*, 38:337–338.

PART II

Regional Blood Flow/ Oxygen Delivery During Exercise

Chapter 5

Coronary Blood Flow and Cardiac Hemodynamics

Lennart Kaijser and Inge-Lis Kanstrup

The heart muscle is characterized by a high capacity for aerobic metabolism and a comparatively lower capacity for anaerobic metabolism than that of the skeletal musculature (Sylven, Jansson, and Olin 1983). Furthermore, 100% of its substrate for energy metabolism at any time is continuously taken up from the blood. The heart muscle contains stores of glycogen and triglycerides. However, only under extreme conditions and for short periods do these contribute significantly to the substrate supply (Kaijser 1982). Therefore, the coronary blood flow, its limits, and its regulation are of primary importance for heart function.

Coronary Blood Flow in Relation to Oxygen Demand

Myocardial oxygen demand is determined mainly by the developed pressure (or, rather, wall stress), the contractility (rate of pressure development), and the heart rate (Sarnoff et al. 1958). An increased myocardial oxygen demand is covered by a combination of increases in coronary blood flow and relative oxygen extraction. In this circumstance, coronary flow is the more important factor (see figure 5.1). Coronary blood flow increases linearly with myocardial oxygen uptake, rising almost five-fold from rest to maximal physical exercise. Simultaneously, coronary sinus oxygen saturation decreases curvilinearly from about 35% at rest to about 25% during maximal exercise (Kaijser and Berglund 1992). The linear increase in flow as a function of oxygen uptake is similar to that in skeletal muscle. The very moderate reduction in venous oxygen saturation, at first glance, seems to differ from that of skeletal muscle (Wahren 1966). However, the intact human heart is never studied under conditions of myocardial rest. If, instead, the "change" in coronary sinus oxygen saturation is compared with that in skeletal muscle, the venous saturation curves

are more similar from low submaximal to maximal exercise (Kaijser et al. 1972). The only difference between heart and skeletal muscle is that the venous oxygen saturation downstream in the myocardium never reaches as low as that in the corresponding skeletal muscle at maximal exercise.

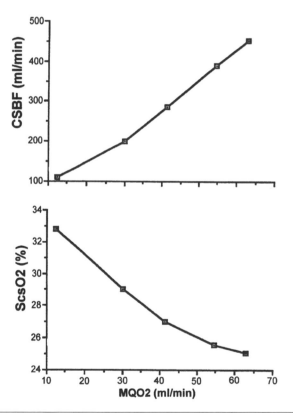

Figure 5.1 Coronary sinus blood flow and oxygen saturation in the coronary sinus as a function of myocardial oxygen uptake at rest and during exercise at loads increasing to maximal.

The *qualitative* similarity could suggest that similar mechanisms regulate myocardial and skeletal muscle blood perfusion in relation to oxygen demand. One tentative explanation for the *quantitative* difference, i.e., the higher oxygen tension in cardiac as compared with skeletal muscular venous blood, could be that in skeletal muscle, but not in the cardiac vascular bed, maximal exercise coincides with maximal local perfusion capacity. In fact, local maximal muscle perfusion capacity seems to exceed perfusion during maximal exercise with large muscle groups (Andersen and Saltin 1985; Kaijser 1970). As a consequence, when the muscle group engaged in dynamic exercise is small enough not to demand maximal cardiac output, the maximal muscle blood flow is higher and the local $(a-\bar{v})O_2$ difference is lower. This leaves muscle venous blood oxygen saturation higher as in the coronary sinus.

Alternatively, the difference in levels of venous oxygen tension may be related to a shorter capillary transit time in the heart than in the skeletal muscle. The difference may also depend on the need to maintain a higher diffusion gradient from capillary to mitochondria. In other words, the mitochondrial fraction of the cell is far greater in myocardial than in skeletal muscle fibers. This may also explain why the coronary sinus oxygen tension does not decrease when the perfusion limit is reached during hypoxia (Grubbstrom, Berglund, and Kaijser, 1993).

Exercise at a load corresponding to individual oxygen uptake with individual maximal heart rate would represent the highest oxygen demand that can be placed on the myocardium under physiological conditions. However, the maximal perfusion capacity of the myocardium seems to be greater than the flow measured under these conditions. This is obvious from at least two established findings. The absence of myocardial lactate production during maximal physical exercise indicates that the blood perfusion limit is not reached (Kaijser and Berglund 1993). More importantly, if maximal physical exercise is executed under conditions of acute hypoxemia (arterial oxygen saturation 85%) with the same maximal heart rate and arterial blood pressure as during air breathing, a 30% greater maximal coronary blood flow will be reached (Grubbstrom, Berglund, and Kaijser 1993). On the other hand, in this situation, the maximal myocardial oxygen uptake achieved falls short of that expected from the rate pressure product. By utilizing the isotope-labeled lactate technique, a significant release of lactate is detected. This indicates that the perfusion limit is reached. Thus, the human myocardium seems to have a flow reserve about 30% above the maximal flow required. This flow reserve is used when maximal myocardial oxygen demand induced by physiological means occurs under normal atmospheric conditions.

Divergent Flow/Oxygen Extraction Relationship During Provocations Increasing Myocardial Oxygen Demand

The main determinants of myocardial oxygen consumption are developed pressure, level of contractility, and heart rate (Sarnoff et al. 1958). Different modes of increasing cardiac work and thereby myocardial oxygen demand give rise to divergent relative contributions of the myocardial perfusion increase and the oxygen extraction increase. This probably occurs because different combinations of increases in pressure, heart rate, and contractility affect the mechanical myocardial tissue pressure conditions. These differentially control both local metabolic and autonomic stimulation of smooth muscle tone in the coronary vascular bed. The following will discuss some examples (see figure 5.2)

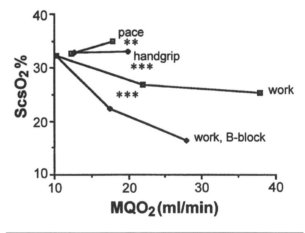

Figure 5.2   Coronary sinus oxygen saturation as a function of myocardial oxygen uptake during atrial pacing, isometric handgrip, and bicycle exercise without and with B-adrenergic blockade. ** and *** denotes $p < 0.01$ and $p < 0.001$, respectively, for the difference between the adjoining data points.

By using atrial pacing, the heart rate can be increased to about 130 beats/min in a healthy young man. Atrioventricular blocking prevents any further increase (Kaijser and Berglund 1993). This heart rate increase is accompanied by almost no change in arterial pressure, no change in cardiac output, and no change in autonomic stimulation of the heart. The myocardial oxygen consumption increases to the same level as the rate-pressure product (almost totally represented by the heart rate increase), i.e., by about 70%. This increase is covered by significantly more than a

70% increase in coronary blood flow with a simultaneous, significant decrease in oxygen extraction (arterial-coronary sinus oxygen difference) (Kaijser and Berglund 1993). Noteworthily, the myocardial oxygen consumption is increased in spite of no increase in cardiac output or arterial pressure, i.e., no increase in effective external work. This pattern emphasizes the importance of the heart rate, among other factors, for myocardial oxygen consumption, as Sarnoff et al. (1958) pointed out long ago.

An isometric handgrip, held at one-third of maximal voluntary contraction force for 2 min, is accompanied by a substantial increase in mean arterial pressure and a moderate increase in heart rate. The rate-pressure product and the myocardial oxygen uptake increase in a manner similar to that as during maximal atrial pacing. However, in the case of isometric handgrip, the increased oxygen uptake is covered by a coronary flow increase of exactly the same magnitude as the myocardial oxygen uptake increase. However, no change in oxygen extraction occurs. Thus, when comparing pacing and handgrip, a similar increase in myocardial oxygen uptake produces a greater coronary flow increase during pacing than during handgrip.

An analysis of the differences in flow increase between handgrip and pacing may contribute to elucidating the mechanisms regulating coronary blood flow. By assuming that the release of metabolically linked vasodilating substances is similar at a comparable myocardial metabolic rate during handgrip and pacing, the differences in flow may be related to mechanical factors. During pacing, the arterial pressure is unaltered. Since the stroke volume and the heart size are reduced (Bevegard et al. 1967), the intramural pressure should, according to Laplace's law, be lower. In contrast, during handgrip, the arterial pressure increases. The stroke volume and heart size are unaltered, therefore the intramural pressure increases. Thus, a similar stimulus for vasodilatation of intramyocardial resistance vessels would lead to a greater increase in myocardial blood flow in the case of pacing because of less mechanical impedance to flow. The difference in autonomic nervous stimulation of the heart may also play a role. A substantial increase in sympathetic activity accompanies isometric handgrip (Sachs, Hamberger, and Kaijser 1985) but not pacing. Sympathetic nervous activity may predominantly stimulate vasconstrictive (alpha-adrenergic) receptors attenuating metabolically linked vasodilatation during handgrip (Feigl 1983).

Heavy dynamic exercise on the cycle ergometer leads to a large increase in heart rate and substantial increases in both arterial blood pressure and cardiac contractility. Myocardial oxygen uptake may increase 5–6 fold. In this case, the increase in myocardial oxygen uptake is covered by a coronary flow increase. This increase is smaller than the increase in oxygen uptake to which a significant increase results from a higher oxygen extraction. Thus, at a given myocardial oxygen uptake, the myocardial blood flow increase is smaller during dynamic work with a large muscle mass than during the handgrip.

Sympathetic Neural Regulation of Coronary Blood Flow

When healthy subjects are exposed to the Stroop's color-word confusion test, increases in heart rate and arterial blood pressure occur. An increase in rate pressure product and myocardial oxygen consumption occurs in a manner similar to the handgrip and the atrial pacing as previously described. The exposure to stress also leads to an increase in arterial plasma concentration of noradrenaline and a greater coronary sinus-arterial difference in noradrenaline concentration. These suggest a local increase in sympathetic activity in the myocardial vascular bed. When compared with both handgrip and atrial pacing, the increased myocardial oxygen consumption during psychological stress is met by a smaller increase in coronary blood flow and, instead, by an increase in oxygen extraction. This suggests the contribution of a vasoconstrictive factor, which may be the increased local sympathetic nervous activity.

The importance of local sympathetic nervous activity in the heart for blood perfusion was also studied by comparing coronary blood flow after and before the administration of the beta-blocker propranolol. Propranolol had no significant effect on the relationship between coronary blood flow and myocardial oxygen consumption at rest. This is consistent with low sympathetic nervous activity and no measurable cardiac release of noradrenaline at rest. During physical exercise, however, at every level of myocardial oxygen consumption, the coronary blood flow was lower and the arterial-coronary sinus oxygen difference was higher after beta-blockade (Kaijser, Grubbstrom, and Berglund 1991). β-adrenergic effects probably contribute to vasodilatation. The effect of beta-blockade was probably to "unmask" the vasoconstrictive alpha-adrenergic effect (Feigl 1983).

Rapidity of Changes in Coronary Blood Flow

When extremely heavy physical exercise starts instantaneously, without a gradual warm-up phase, muscle blood flow seems to increase as rapidly as or faster than the increase in muscle oxygen uptake. This is indicated by a gradual decrease in muscular venous oxygen saturation. However, an overshoot phase with reduced oxygen saturation does not occur (Kaijser 1970). The question of whether similar conditions prevail in the myocardium was studied in healthy male volunteers in whom a catheter was placed in the coronary sinus for blood sampling. The subjects were instructed to start exercise instantaneously on the cycle ergometer at a load 10% above that corresponding to their maximal oxygen uptake. Coronary sinus oxygen saturation was followed at 10–15-s intervals. On transition from rest to heavy exercise, coronary sinus oxygen saturation decreased rapidly to reach its lowest level within 30 s. A gradual, small increase followed within the next few minutes. This resulted in a steady level similar to that prevailing during steady-state exercise at a load approximately corresponding to individual maximal oxygen uptake (see figure 5.3). Thus, the heart is characterized by an ability to increase myocardial oxygen consumption faster than factors increasing local blood flow are activated. This seems to differ from the conditions in skeletal muscle where a transient phase of extremely reduced muscle venous oxygen saturation is not seen. The factor responsible for the difference between heart and skeletal muscle may be the far greater fraction of mitochondria per unit cell mass in the heart compared with that in the skeletal muscle.

Cardiac Hemodynamics

Assessments of left ventricular function at rest as well as during exercise has been a main topic for exercise physiologists in the whole century. Traditional function parameters include heart rate (HR), cardiac output (CO), stroke volume (SV), arterial pressures, and, through the use of newer, noninvasive techniques such as echocardiography, nuclear cardiology, and MR-imaging, left ventricular end-diastolic (LVEDV) and end-systolic volumes (LVESV), ejection fraction (EF), and volume-derived variables such as peak ejection (PER) and peak filling rates (PFR).

The three determinants of ventricular function are preload, contractility, and afterload.

Preload refers to the extent of filling of the ventricles at the end of diastole. It is a function of the end-diastolic pressure and end-diastolic volume. Increasing preload leads to an increased output, known as the Starling effect. In normal subjects this relationship tends to plateau at an end-diastolic pressure of about 15 mmHg, and thereafter a declining ventricular function is seen. *Contractility* is defined as the rate of pressure rise within the ventricle during isovolumic systole. Catecholamines increase contractility and thus the response to a given preload. *Afterload* is defined as the impedance to the ejection of blood from the left ventricle and is commonly represented by the systolic arterial pressure. In the case of an acute increase in arterial pressure, this increased resistance to flow results in a decline in the rate of blood ejected from the left ventricle. Normally this is compensated by an increase in preload in order to maintain a normal stroke volume. Each ventricle has a limit of preload reserve, which can be enhanced by increasing contractility.

Cardiac Output

Cardiac output (CO) is the volume of blood pumped by the left ventricle per unit of time and is usually expressed in litres/min. The cardiac output at rest is about 4.5 – 6 ($l \times min^{-1}$). As the value is markedly dependent on body size, the cardiac index (cardiac output divided by the body surface area) is often used, the average cardiac index being about 3 ($l \times min^{-1}$) in males. With increasing oxygen uptake there is an almost rectilinear increase in cardiac output. In normal subjects with $\dot{V}O_2$max of 3.0–3.5 ($l \times min^{-1}$) cardiac output raises to 18–23 ($l \times min^{-1}$), in well trained endurance athletes even higher (a value

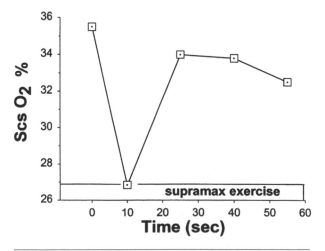

Figure 5.3   Coronary sinus oxygen saturation at rest immediately before and at 10–15-s intervals during a short bout of supramaximal exercise on the cycle ergometer.

slightly above 42 (l × min⁻¹) at a V̇O₂ of 6.2 (l × min⁻¹) has been reported; Kanstrup and Ekblom 1978).

With advancing age there is a decline in max cardiac output (Strandell 1964; Julius et al. 1967), while there is some discrepancy regarding cardiac output at submaximal work intensity with reports of both unchanged or increased cardiac output values for a given submaximal oxygen uptake. With endurance training, increased maximal values are seen in accordance with increased aerobic capacity, while at submaximal intensities most authors find unchanged cardiac output values (Rowell 1974; Clausen 1977). With detraining/inactivity, a comparable fall in the peak values is seen in accordance with a decrease in V̇O₂max. The influence of "pure ageing" is very modest in comparison with the changes that may be caused by detraining/inactivity. Females may have a slightly higher CO (increased HR, lower SV) than males at submaximal work loads, probably as a compensation for the lower arterial oxygen content (Åstrand et al. 1964). Maximal CO is reduced (SV lower, HR similar, arterio-venous oxygen difference reduced) compared to male subjects. With training, detraining, or increased age, similar adaptations occur in women as in men (Kilbom 1971; Kanstrup and Ekblom 1978).

Stroke Volume

The increase in cardiac output with exercise is mainly due to the increased heart rate (figure 5.4).

But the increase in the sympathetic stimulation also leads to a venoconstriction with an augmented venous return to the heart, which leads to an increased preload and stroke volume (SV) (Frank-Starling mechanism). This means that during upright exercise SV increases from rest to about 40% of maximal aerobic power (less pronounced during supine exercise). Thereafter it levels off or there is a small decline at maximal work intensity. At rest in the upright position, SV is about 50–70% of the peak value during exercise (Åstrand et al. 1964). The increase in SV at the low exercise level is mainly caused by an increased LVEDV due to the increased venous return (figure 5.4; Poliner et al. 1980; Adams et al. 1992). LVESV is progressively slightly reduced with increasing exercise (20 ml ~ 35% in figure 5.1) and is responsible for the maintenance of an elevated SV. An increased cardiac contractility is thought to elicit the more complete emptying during systole.

Ejection Fraction

As a result of the increased stroke volume and only slightly changed LVEDV, the ejection fraction (EF),

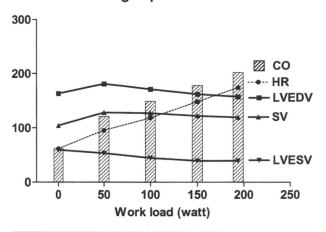

Left ventricle volumes during supine exercise

Figure 5.4 Mean values for left ventricle volumes during supine rest and exercise in ten healthy males. CO = cardiac output; HR = heart rate; LVEDV and LVESV = left ventricle end-diastolic and end-systolic volume; SV = stroke volume. (Data from Kanstrup et al. 1995.)

which is the ratio of the stroke volume to end-diastolic volume, increases. The volume thus represents the percentage of the blood in the ventricle that is ejected per beat. In normal resting subjects, values between 53 and 74% are seen. With exercise, EF increases to a plateau in parallel with SV, thus from 64% at rest to 71% (50 W), 75% (100 W), and 77% (max) in a recent study (figure 5.5).

Upright exercise elicits higher increments than supine exercise (recalculated to be 85% vs. 79% at

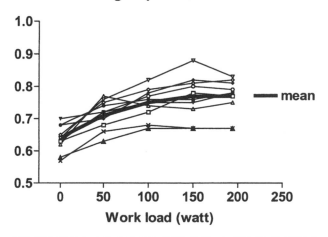

Left Ventricle Ejection Fraction during supine exercise

Figure 5.5 Individual and mean values for left ventricle ejection fractions during supine rest and exercise in ten healthy males. (Data from Kanstrup et al. 1995.)

maximum in the study by Poliner et al. 1980) due to lower LVEDV. Endurance training tends to reduce max, if anything. EF slightly secondary to the volume increases (Rubal et al. 1986).

Ejection fraction is often used as a measure for the ventricular contractility, but as seen is dependent on preload (EDV) as well as afterload. Renlund et al. (1990) describe a correlation between changes in ESV and EDV with orthostatic and exercise stresses, and an inverse correlation between changes in EF and ESV (but no direct relation between EF and EDV). This makes it difficult to discern between the determinants, and other parameters could be looked for to better evaluate the cardiac contractility and performance with exercise and training

Peak Filling Rate and Peak Ejection Rate

In patients with ischemic heart disease, symptoms of heart failure due to diastolic dysfunction have been the goal for a number of studies. But with the easier access to non-invasive techniques as echocardiography, nuclear cardiology, and MR-imaging, a growing interest in the dynamics of the left ventricular emptying and filling in normals is also seen. Left ventricular peak filling rate (PFR) is one of the most widely use parameters of LV diastolic function (figure 5.6).

Left Ventricle Volume Curve

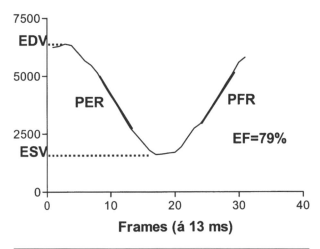

Figure 5.6 Left ventricle volume curve from an exercising subject. HR = 138 bpm; EF = 79%; peak filling rate = 5.7 EDV $\times$ s^{-1}; peak ejection rate = 5.7 EDV $\times$ s^{-1}.

Together with peak ejection rate (PER), peak filling rate is traditionally given in units of end-diastolic volume per sec. (EDV $\times$ s^{-1}). However, to be able to compare individuals of different sizes, absolute rates in ml $\times$ s^{-1} and SV $\times$ s^{-1} may also be used. The importance of this has been stressed for patients surviving acute myocardial infarction (Gadsbøll et al. 1991), but not yet evaluated in healthy subjects.

Both PFR and PER increase with increasing exercise in healthy subjects (figure 5.4), where data from ten healthy men (33–51 years old) obtained by radionuclide cardiography are shown at rest and during graded supine exercise. Both parameters increased with exercise intensity in all subjects, and mean PFR was higher than PER on all exercise loads. At maximum, a wide range, especially in PFR, was seen (861–1965 ml $\times$ s^{-1}). Other studies have confirmed an increase with exercise (Mancini et al. 1983; Kanstrup, Marving, and Høilund-Carlsen 1992; Schulman et al. 1992; Brandao et al. 1993). Physical training has been found to increase PFR significantly during exercise (Brandao et al. 1993; Gledhill, Cox, and Jannik 1994), which means that the actual state of physical training should ideally be taken into account for interpretation of the data as is the case with other cardiac volumes (LVEDV, LVESV, SV). Also an influence of age has been demonstrated with lower values for a given relative work intensity, but unchanged values for a given absolute work load (Schulman et al. 1992). The authors interpreted this as not being secondary to a decline in physical conditioning with age but rather related to a decrease in beta-adrenergic responsiveness in older individuals. In the data presented in figure 5.7, no attempt to discern between younger and older individuals, nor trained or less trained has been tried, but the highest absolute values for both PFR and PER were obtained in the three youngest and best-trained subjects.

A very close relationship between either PFR or PER and HR was found. The best correlations were for PFR (EDV $\times$ s^{-1}) with r^2 = 0.86 (figure 5.8), for PER (EDV $\times$ s^{-1}) r^2 was 0.80, and when using velocities in ml $\times$ s^{-1} the relationship grew weaker (r^2 = 0.52 and 0.48, respectively).

As mentioned PER was consistently found lower than PFR during exercise. Others have found comparable values of PFR and PER (Brandao et al. 1993). Gledhill, Cox, and Jannik (1994) also found quite similar values in an untrained group, but marked differences in trained subjects due to especially increased PFR and more modest increases in PER, while Brandao et al. (1993) could not demonstrate any effect of training on this systolic function parameter. Plasma expansion, which induces an increased ventricular filling pressure and probably simulates the circulatory changes with endurance training, was found to increase both PFR and PER

PFR (left) and PER (right) with increasing exercise

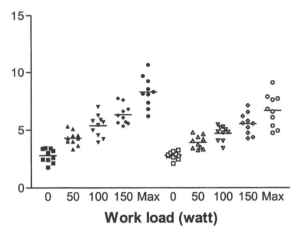

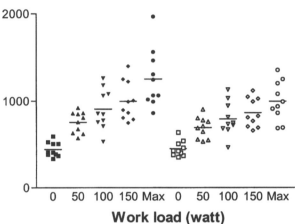

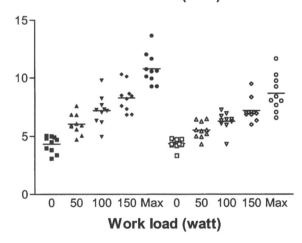

Peak Filling Rate (EDV x s⁻¹) vs HR during supine exercise

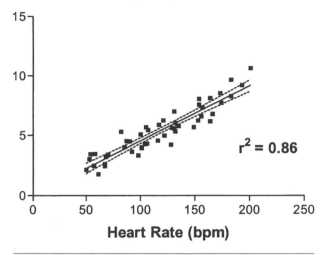

$r^2 = 0.86$

Figure 5.8 PFR vs. HR. Linear regression y = 0.046, x – 0.069 with 95% confidence intervals is shown. $r^2 = 0.86$. Values from ten healthy men performing supine exercise. (Data from Kanstrup et al. 1995.)

and Jannik's conclusions that endurance-trained athletes rely on augmentations in ventricular filling to a far greater degree than on enhancements in emptying.

Left Ventricle Work

The myocardial work/oxygen consumption is considered to be proportional to the pressure/volume area during one cardiac cycle, being the sum of the stroke work (SW) and the potential energy (PE) (Suga, Hayashi, and Shirahata 1981). In figure 5.9, mean SW for ten healthy men obtained at rest and during exercise are shown in a simplified form. SW increases initially from rest to exercise and then levels off with higher intensities.

The other component of PVA:PE, represented by the triangle V_o-ESV-sBP, is unchanged with exercise. As mentioned PVA is proportional to the myocardial oxygen consumption (Suga, Hayashi, and Shirahata 1981), but in humans quite high absolute values have been obtained based on these calculations although expected relative changes with exercise were found (Kanstrup et al. 1995).

The slope of the line V_o-sBP = E_{max} has been proposed to be a better expression of the ventricular contractility as it is independent of preload and afterload and is little influenced by heart rate (Suga et al. 1983). An increase from 2.2 (rest) to 5.3 (max exercise) mmHg $\times$ ml⁻¹ was seen in the subjects in figure 5.9; similar values have been reported by

Figure 5.7 PFR (filled symbols) and PER (open symbols) in ten healthy men during supine exercise shown in different units. Top: EDV $\times$ s⁻¹. Middle: ml $\times$ s⁻¹. Bottom: SV $\times$ s⁻¹. For each work load the mean value is marked. (Data from Kanstrup et al. 1995.)

with 19 and 14% at rest, respectively, and 33 and 15% at submaximal exercise (Kanstrup, Marving, and Høilund-Carlsen 1992), supporting Gledhill, Cox,

Left ventricle pressure-volume diagram

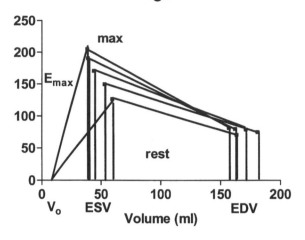

Left ventricle Pressure-Volume Area and its components during supine exercise

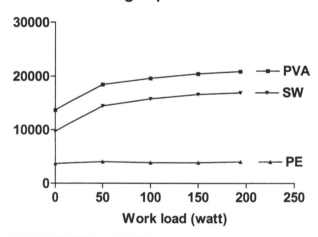

Figure 5.9 Simplified drawing of stroke work (the area within the trapezium) at rest and with increasing exercise in ten healthy subjects. At rest and during maximum, a line representing E_{max} is included; the area in the triangles represents the potential energy. V_o is the (theoretical) ventricular volume at zero pressure (upper part). The mean values for the pressure/volume area, stroke work, and potential energy are shown in the lower part. (Data from Kanstrup et al. 1995.)

others during supine exercise (Kronenberg et al. 1988). After acute plasma expansion, E_{max} was (in accordance with Suga's findings) unchanged from control at rest as well as during submaximal exercise (Kanstrup, Marving, and Høilund-Carlsen 1992).

Influence of Training

During dynamic exercise, oxygen uptake and cardiac output as mentioned are almost constant for a given work load, not influenced by the individual's level of physical fitness. But after physical training, stroke volume increases and heart rate decreases for a given work load/oxygen uptake. This has been reported in both young adults and older individuals, in healthy subjects and in patients with coronary heart disease (Ekblom 1969; Rowell 1974; Clausen 1976; Scheuer and Tipton 1977; Kanstrup and Ekblom 1978; Blomqvist and Saltin 1983). Cardiac hypertrophy (increased left ventricular end-diastolic diameter and increased wall thickness) is common in endurance-trained athletes (Maron 1986). This points to certain structural changes within the myocardium secondary to physical training, but an improved "intrinsic contractility" has also been proposed (Schaible, Penparkgul, and Scheuer 1980; Rost and Hollmann 1983; Wolfe et al. 1985). A possible changed "intrinsic rate" following training has also been suggested (Lewis et al. 1980; Nylander and Dahlström 1984).

The relative importance of these changes is unknown, as training also induces marked changes extramyocardially. Reduced afterload is found in endurance-trained subjects, probably to a minor extent caused by increased capillarisation in the trained striated muscles. Of more importance is supposedly changed arteriolar contraction (Blomqvist and Saltin 1983). The primary regulatory mechanisms have not been well defined, but the changed activity or sensitivity in the autonomic nervous system is probably closely connected with reduced activity in certain chemoreceptors (group III and IV afferent nervous fibers). These unmyelinated nerve fibers end blindly in the interstitium of the striated muscles and are sensitive to certain metabolites such as lactic acid (osmolarity), potassium, temperature changes, etc. (Mitchell and Schmidt 1983; Kaufman et al. 1984; Knochel et al. 1985). Central regulatory factors may also be of some importance (Rowell 1980; Perski, Tzankoff, and Engel 1985).

Furthermore, results pointing to increased cardial preload during exercise after a period of physical training in parallel with an increased blood volume have been reported (Blomqvist and Saltin 1983). Via a Frank-Starling mechanism, this increased blood volume could increase the end-diastolic volume and thus stroke volume. In agreement with this, an enhanced stroke volume of 12–15% has been found after acute volume expansion (blood volume and plasma volume) (Fortney et al. 1981; Kanstrup and Ekblom 1982; Convertino, Mack, and Nadel 1991; Kanstrup, Marving, and Høilund-Carlsen 1992), speaking in favor of this being the main explanation for the increases in SV and CO seen with training. It has, however, been stated that the cardiac response

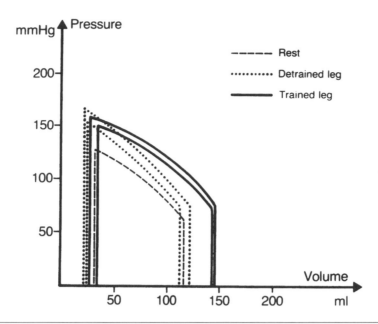

Figure 5.10 Left ventricle pressure-volume diagram in one subject at rest and during 2 submaximal exercise loads (55 W and 80 W), each performed with a trained and a detrained leg, respectively. (Reprinted from Kanstrup et al. 1991 with permission.)

to blood volume expansion depends on the myocardial state of training (Blomqvist and Saltin 1983). In support of this is the finding of an unchanged myocardial contractility, estimated as SBP/LVESV = E_{max} after acute plasma expansion, as mentioned earlier.

The probably very strong influence of extramyocardial factors on the cardiac behavior was pointed out in preliminary investigations in three subjects exercising on a fixed work load with a habitually active and an inactivated (detrained) leg, respectively (Saltin 1986). During exercise with the inactivated leg an "untrained" central response with a lower stroke volume (85 ml) and higher heart rate (158 bpm) was seen compared with the "trained" central response (high stroke volume (100 ml) and low heart rate (138 bpm)) observed during exercise with the control leg.

These results were further supported by the finding in an individual subject of a clearly different pressure-volume diagram for the left ventricle during exercise with a trained and a detrained leg, respectively (figure 5.10).

LVEDV was exactly the same on two different exercise loads with the trained leg and constantly lower when using the detrained leg or at rest. This speaks in favor of peripheral factors related to the actual state of training strongly influencing the central hemodynamic response to exercise. But it still remains to be settled which one of the two, the volume or the heart rate, is the primary determinant of the resulting cardiac output.

Also the interactions between the primary cardiac function parameters as expounded here are not completely evaluated, and it is evident that more knowledge with respect to the influence of age, physical training, inactivity, and sex is needed. In view of the complexity of the various determinants of the cardiac function it is difficult to point to *the* limiting factor to maximal performance, which in most recent papers has been linked to the cardiac pumping ability (Saltin and Strange 1992). The ventricular stroke work is a figure that increases almost asymptotic with exercise suggesting a limitation might be linked to this parameter.

References

Adams, K.F., S.M. McAllister, H. El-Ashmaway, S. Atkinson, G. Koch, and D.S. Streps. 1992. Interrelationships between left ventricular volume and output during exercise in healthy subjects. *Journal of Applied Physiology* 75:2097–2104.

Andersen, P., and B. Saltin. 1985. Maximal perfusion of skeletal muscle in man. *Journal of Physiology* 366:233–249.

Åstrand, P-O., T.E. Cuddy, B. Saltin, and J. Stenberg. 1964. Cardiac output during submaximal and maximal work. *Journal of Applied Physiology* 19:268–274.

Bevegard, S., B. Jonsson, I. Karlof, H. Lagergren, and E. Sowton. 1967. Effect of changes in ventricular rate on cardiac output and central venous pressures at rest and during exercise in patients with artifical pacemakers. *Cardiovascular Research* 1:21–33.

Blomqvist, C.G., and B. Saltin. 1983. Cardiovascular adaptations to physical training. *Annual Review of Physiology* 45:169–189.

Brandao, M.U.P., M. Wajngarten, E. Rondon, C.P. Giorgi, F. Hironaka, and C.E. Negrao. 1993. Left ventricular function during dynamic exercise in untrained and moderately trained subjects. *Journal of Applied Physiology* 75:1989–1995.

Clausen, J.P. 1976. Circulatory adjustments to dynamic exercise and effect of physical training in normal subjects and in patients with coronary heart disease. *Progress in Cardiovascular Diseases* 18(6):459–495.

Clausen, J.P. 1977. Effect of physical training on cardiovascular adjustments to exercise in man. *Physiological Reviews* 57(4):779–814.

Convertino, V.A., G.A. Mack, and E.R. Nadel. 1991. Elevated central venous pressure: A consequence of exercise training-induced hypervolemia. *American Journal of Physiology* 29:R273–R277.

Ekblom, B. 1969. Effect of physical training on oxygen transport system in man. *Acta Physiologica Scandinavica* 328(suppl):9–49.

Feigl, E.O. 1983. Coronary Physiology. *Physiological Reviews* 63:1–205.

Fortney, S.M., E.R. Nadel, C.B. Wenger, and J.R. Bove. 1981. Effect of acute alteration of blood volume on circulatory performance in humans. *Journal of Applied Physiology* 50:292–298.

Gadsbøll, N., P.F. Høilund-Carlsen, J.H. Badsberg, H. Lønborg-Jensen, P. Stage, J. Marving, and B. Hjort Jensen. 1991. Left ventricular peak filling rate, heart failure and 1-year survival in patients with acute myocardial infarction. *European Heart Journal* 12:194–202.

Gledhill, N.D. Cox, and R. Jannik. 1994. Endurance athlete's stroke volume does not plateau: Major advantage is diastolic function. *Medicine and Science in Sports and Exercise* 26:1116–1121.

Grubbstrom, J., B. Berglund, and L. Kaijser. 1993. Myocardial oxygen supply and lactate metabolism during marked arterial hypoxaemia. *Acta Physiologica Scandinavica* 149:303–310.

Julius, S., A. Amery, L.S. Whitlock, and J. Conway. 1967. Influence of age on the hemodynamic response to exercise. *Circulation* 36:222–230.

Kaijser, L. 1970. Limiting factors for physical work performance. *Acta Physiologica Scandinavica* 79(suppl. 346):1–96.

Kaijser, L. 1982. Regulatory mechanisms in the interaction of lipid and carbohydrate metabolism in cardiac and skeletal muscle in man. In metabolic risk factors in ischemic patients with

artifical pacemakers. *Cardiovascular Research* 1:21–33.

Kaijser, L., and B. Berglund. 1992. Myocardial lactate extraction and release at rest and during heavy exercise in healthy men. *Acta Physiologica Scandinavica* 144:39–45.

Kaijser, L., and B. Berglund. 1993. Coronary hemodynamics during isometric handgrip and atrial pacing in patients with angina pectoris compared to healthy men. *Cardioscience* 4:99–104.

Kaijser, L., J. Grubbstrom, and B. Berglund. 1991. Effect of β-blockade on coronary blood flow during execise under hypoxemia. *Journal of Molecular Cellular Cardiology* 23(suppl. V).

Kaijser, L., B.W. Lassers, M.I. Wahlqvist, and L.A. Carlson. 1972. Myocardial lipid and carbohydrate metabolism in fasting men during prolonged exercise. *Journal of Applied Physiology* 32:847–858.

Kaijser, L., J. Pernow, B. Berglund, and J.M. Lundberg. 1990. Neuropeptide Y is released together with noradrenaline from the human heart during hypoxia. *Clinical Physiology* 10:179–188.

Kanstrup, I-L., and B. Ekblom. 1978. Influence of age and physical activity on central hemodynamics and lung function in active adults. *Journal of Applied Physiology* 45(5):709–717.

Kanstrup, I-L., and B. Ekblom. 1982. Acute hypervolemia, cardiac performance and aerobic power during exercise. *Journal of Applied Physiology* 52:1186–1191.

Kanstrup, I-L., J. Marving, N. Gadsbøll, H. Lønborg-Jensen, and P.F. Høilund-Carlsen. 1995. Left ventricle haemodynamics and vaso-active hormones during graded supine exercise in healthy male subjects. *European Journal of Applied Physiology* 72:86–94.

Kanstrup, I-L., J. Marving, and P.F. Høilund-Carlsen. 1992. Acute plasma expansion: Left ventricular hemodynamics and endocrine function during exercise in healthy subjects. *Journal of Applied Physiology* 73:1791–1796.

Kanstrup I-L., J. Marving, P.F. Høilund-Carlsen, and B. Saltin. 1991. Left ventricular response upon exercise with trained and detrained leg muscles. *Scandinavian Journal of Medicine and Science in Sports* 1:112–118.

Kaufman, N.P., T-G. Waldrop, K.J. Rybicki, G.A. Ordway, and J.H. Mitchell. 1984. Effects of static and rhythmic twitch contractions on the discharge of group III and IV muscle afferents. *Cardiovascular Research* 18(11):663–668.

Kilbom, Å. 1971. Physical training in women. *Scandinavian Journal of Clinical and Laboratory Investigation* 28(suppl 119):7–34.

Knochel, J.P, S.D. Blachley, J.H. Johnson, and N.W. Carter. 1985. Muscle cell electrical hyperpolarization and reduced hyperkalemia in physsically conditioned dogs. *Journal of Clinical Investigation* 75:740–745.

Kronenberg, M.W., J.P. Uetrecht, W.D. Dupont, M.H. Davis, B.K. Phelan, and G.C. Friesinger. 1988. Intrinsic left ventricular contractility in normal subjects. *American Journal of Cardiology* 61:621–627.

Lewis, S.R., E. Nylander, P. Gad, and N.H. Areskog. 1980. Non-autonomic component in bradycardia of endurance trained men at rest and during exercise. *Acta Physiologica Scandinavica* 109:297–305.

Mancini, G.B.J., R.A. Slutsky, S.L. Norris, V. Bhargava, W.L. Ashburn, and C.B. Higgins. 1983. Radionuclide analysis of peak filling rate, filling fraction, and time to peak filling rate. *American Journal of Cardiology* 51:43–51.

Maron, B.M. 1986. Structural features of the athlete heart as defined by echocardiography. *Journal of American College of Cardiology* 7:190–203.

Mitchell, J.H., and R.F. Schmidt. 1983. Cardiovascular reflex control by afferent fibers from skeletal muscle receptors. In *Handbook of Physiology* edited by J.T. Shepherd and F.M. Abboud. Bethesda, MD: Waverly Press, American Physiology Society. 623–658.

Nylander, E., and U. Dahlström. 1984. Influence of long-term beta-receptor stimulation with prenalterol on intrinsic heart rate in rats. *European Journal of Applied Physiology* 53(1):48–52.

Perski, A., S.P. Tzankoff, and B.T. Engel. 1985. Central control of cardiovascular adjustments to exercise. *Journal of Applied Physiology* 58(2):431–435.

Poliner, L.R., G.J. Dehmer, S.E. Lewis, R.W. Parkey, C.G. Blomqvist, and J.T. Willerson. 1980. Left ventricular performance in normal subjects: A comparison of the responses to exercise in the upright and supine positions. *Circulation* 62:528–534.

Renlund, D.G., G. Gerstenblith, J.L. Fleg, L.C. Becker, and E.G. Lakatta. 1990. Interaction between left ventricular end-diastolic and end-systolic volumes in normal humans. *American Journal of Physiology* 258:H473–H481.

Rost, R., and W. Hollmann. 1983. Athlete's heart: A review of its historical assessment and new aspects. *International Journal of Sports Medicine* 4:147–165.

Rowell, L.B. 1974. Human cardiovascular adjustments to exercise and thermal stress. *Physiological Reviews* 54:75–159.

Rowell, L.B. 1980. What signals govern the cardiovascular responses to exercise? *Medicine and Science in Sports and Exercise* 12(5):307–315.

Rubal, B.J., J.M. Moody, S. Damore, S.R. Bunker, and N.M. Diaz. 1986. Left ventricular performance of the athletic heart during upright exercise: A heart rate-controlled study. *Medicine and Science in Sports and Exercise* 18:134–140.

Sachs, C., B. Hamberger, and L. Kaijser. 1985. Cardiovascular responses and plasma catecholamines in old age. *Clinical Physiology* 5:553–565.

Saltin, B. 1986. Physiological adaptation to physical conditioning. Old problems revisited. *Acta Medica Scandinavica* 711(suppl):11–24.

Saltin, B., and S. Strange. 1992. Maximal oxygen uptake: "Old" and "new" arguments for a cardiovascular limitation. *Medicine and Science in Sports and Exercise* 24:30–37.

Sarnoff, S.J., E. Braunwald, G.H. Welch, Jr, R.B. Case, W.N. Stainsby, and R. Marcuz. 1958. Hemodynamic determinants of oxygen consumption of heart with special reference to the tension-time index. *American Journal of Physiology* 192:148–156.

Schaible, T.F., S. Penparkgul, and J. Scheuer. 1980. Cardiac responses to exercise training in male and female rats. *Journal of Applied Physiology* 50:112–117.

Scheuer, J., and C.M. Tipton. 1977. Cardiovascular adaptations to physical training. *Annual Review of Physiology* 39:221–251.

Schulman, S.P., E.G. Lakatta, J.L. Fleg, L. Lakatta, L.C. Becker, and G. Gerstenblith. 1992. Age-related decline in left ventricular filling at rest and exercise. *American Journal of Physiology* 263:H1932–H1938.

Strandell, T. 1964. Circulatory studies on healthy old men. *Acta Medica Scandinavica* 175(suppl 414):1–44.

Suga, H., T. Hayashi, and M. Shirahata. 1981. Ventricular systolic pressure-volume area as a predictor of cardiac oxygen consumption. *American Journal of Physiology* 240:320–325.

Suga, H., R. Hisano, S. Hirata, T. Hayashi, O. Yamada, and I. Ninomiya. 1983. Heart rate-independent energetics and systolic pressure-volume area in dog heart. *American Journal of Physiology* 244:206–214.

Sylven, C., E. Jansson, and C. Olin. 1983. Human myocardial and skeletal muscle enzyme activities: Creatine kinase and its isozyme MB as related to glycolytic and oxidative enzymes and fibre types. *Clinical Physiology* 3:461–468.

Wahren, J. 1966. Quantitative aspects of blood flow and oxygen uptake in the human forearm during rhythmic exercise. *Acta Physiologica Scandinavica* 67(suppl 269):1–93.

Wolfe, L.A., R.P. Martin, D.D. Watson, R.D. Lasley, and D.E. Bruns. 1985. Chronic exercise and left ventricular structure and function in healthy human subjects. *Journal of Applied Physiology* 58(2):409–415.

Chapter 6

Pulmonary Circulation

John B. West and Odile Mathieu-Costello

There have been a number of new insights into the response of the pulmonary circulation to exercise and other challenges over the last 8 years. The central focus of our studies has been the recognition that the pulmonary capillaries face a dilemma. On the one hand, the blood-gas barrier must be extremely thin for efficient gas exchange, particularly during exercise when the demands for gas transfer are so high. On the other hand, the capillary wall needs to be immensely strong because the wall stresses become extremely high when the capillary pressure rises, particularly during heavy exercise. The capillary wall is the most vulnerable structure in the pulmonary circulation and, indeed, is the raison d'être for a separate circulation through the lungs, though this has often not been appreciated. Evolution of vertebrates to full endothermy and particularly the high aerobic performances of some mammals absolutely depend on a separate circulation to the lungs. In a similar way, ontogeny reflects phylogeny, and a crucial development in the perinatal period of the mammal is the establishment of a separate, low-pressure pulmonary circulation.

Pressures in the Pulmonary Circulation

There have been a number of recent studies of the pressures in the pulmonary circulation.

Human Pulmonary Circulation During Exercise

A common misconception is that pulmonary vascular pressures remain low during exercise. This erroneous view possibly stems from some early measurements made during cardiac catheterization. For example, in one of the first studies of

pulmonary arterial pressure during exercise, Riley and his colleagues (1948) found that mean pulmonary arterial (PA) pressure in two normal volunteers actually *decreased* in the transition from rest to exercise. In one subject (Riley himself), the mean PA pressure fell from 15 to 10 mmHg while the oxygen consumption ($\dot{V}O_2$) increased from 198 to 1246 ml/min, and in the other subject, the mean PA pressure fell from 12 to 8 mmHg in spite of an increase in $\dot{V}O_2$ from 270 to 1678 ml/min. In another pioneering study, Hickam and Cargill (1948) studied 7 patients with normal cardiovascular systems and reported that mean PA pressure rose less than 2 mmHg in the transition from rest to exercise. Dexter et al. (1951) measured both mean PA pressure and mean pulmonary arterial wedge (PW) pressure in 7 healthy individuals before and 1–2.5 min after exercise and reported that both were essentially unchanged.

More recently, Gurtner, Walser, and Füssler (1975) catheterized 19 volunteer medical students in the supine position at rest and during a workload of 160 W and found that although mean PA pressure increased by an average of 9 mmHg, the mean increase in PW pressure was only 2 mmHg. Although larger increases in PA pressure had been shown to occur in upright subjects during exercise (Bevegard, Holmgren, and Jonsson 1960), contemporary thinking in 1985 was summarized in a chapter on the pulmonary circulation in the prestigious *Handbook of Physiology*, where it was stated that exercise results in a "modest increment in the level of the mean pulmonary arterial pressure" and that the pulmonary wedge pressure is "unaffected by mild exercise but may increase slightly as the intensity of the exercise increases" (Fishman 1985).

During the last few years, measurements of the pressures in the pulmonary circulation during exercise have provided a very different picture. Wagner et al. (1986) exercised normal volunteers on a cycle ergometer at an oxygen consumption that averaged

3.7 l/min, which is about 80–90% of their maximum oxygen consumption ($\dot{V}O_2$max). Mean PA pressure measured with an indwelling Swan-Ganz catheter increased from 13.2 mmHg at rest to 37.2 mmHg during exercise. Even more remarkably, mean pulmonary arterial wedge pressure increased from 3.4 mmHg at rest to 21.1 mmHg during exercise. Thus, the increase in pulmonary arterial pressure was 24 mmHg, and this was accompanied by an increase in wedge pressure of approximately 18 mmHg.

Another study carried out shortly afterward reported similar results. Eight normal subjects exercised on a bicycle ergometer at a mean oxygen uptake of 3.4 l/min, which was 84% of their $\dot{V}O_2$max. In this instance, mean pulmonary arterial pressure averaged 33 mmHg, and pulmonary arterial wedge pressure averaged 21 mmHg (Groves et al. 1987; Reeves et al. 1990). There is close agreement between the two studies, which were done by different groups. Thus, vigorous exercise in the upright position certainly does not provoke a "modest increment" in mean pulmonary arterial pressure or cause pulmonary arterial wedge pressure to "increase slightly" as was believed a few years ago. In fact, the increases in both pressures are striking.

What is the reason for the large increases in pulmonary arterial and venous pressures? A clue comes from studies of the extremely aerobic thoroughbred racehorse. As figure 6.1 shows, mean left atrial pressures as high as 70 mmHg have been directly measured during galloping on a treadmill by using liquid-filled catheters implanted in the left atrium at thoracotomy several weeks earlier (Jones et al. 1992). Pulmonary arterial wedge pressures are consistent

with these very high left atrial pressures (Erickson, Erickson, and Coffman 1990, 1992; Manohar 1993). Thus, in these highly aerobic animals, it appears that the high pulmonary vascular pressures are a direct result of the large left ventricular filling pressures required for the high cardiac output. Presumably, this also is the case in humans during heavy exercise. The increase in pulmonary arterial pressure simply reflects the increase in venous pressure and the high pulmonary blood flow. There is no evidence of an increase in pulmonary vascular resistance. Indeed, in the study by Wagner et al. (1986), pulmonary vascular resistance fell from 1.42 to 0.67 mmHg/l^{-1}/min in the transition from rest to exercise.

What is the capillary pressure in the human lung during heavy exercise? Again, there is a common misconception that pulmonary capillary pressure is generally much closer to venous pressure than arterial pressure, as is the case in the systemic circulation. However, there is strong experimental evidence against this. Measurements by micropuncture of small pulmonary blood vessels in experimental animals have shown that capillary pressure is about halfway between arterial and venous pressure and that much of the pressure drop in the pulmonary circulation occurs in the capillary bed (Bhattacharya, Nanjo, and Staub 1982; Bhattacharya and Staub 1980). Other studies have indicated that at high pulmonary blood flows, capillary pressure is closer to arterial than to venous pressure (Younes, Bshouty, and Ali 1987). Therefore, taking capillary pressure to be an average of arterial and venous pressures is likely to be an underestimate during exercise.

If we apply this information to the data available about the human lung during exercise, the mean capillary pressure at midlung is about 29 mmHg and, allowing for the hydrostatic gradient, the capillary pressure at the bottom of the lung is about 36 mmHg. Thus, some capillaries are certainly exposed to a pressure in the mid-30s, even by conservative estimates.

As shown below, a capillary transmural pressure of this magnitude is sufficient to cause ultrastructural changes in the capillary wall in the rabbit, a condition that we call stress failure (West et al. 1991). In this animal, stress failure is consistently seen at capillary transmural pressures of 40 mmHg and occasionally occurs at pressures of 24 mmHg (Tsukimoto et al. 1991). Of course, we cannot assume that the structure of human pulmonary capillaries is the same as in the rabbit. Indeed, we know that there are species differences in the ability of capillaries to withstand high pressures. Nevertheless, these high capillary pressures emphasize the vulnerability of the capillaries during exercise.

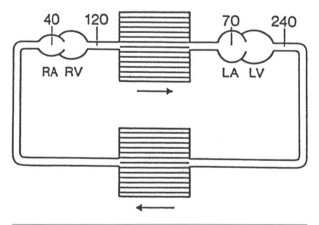

Figure 6.1   Mean vascular pressures (mmHg) in galloping thoroughbred racehorses. The right atrial (RA), right ventricular (RV), and left atrial (LA) pressures were measured with indwelling catheters. LV indicates left ventricle. (Data from Erickson, Erickson, and Coffman 1990; Jones et al. 1992; Manohar 1993.)

Phylogeny and Protection
of Pulmonary Capillaries

It is interesting to look at the evolution of vertebrate animals in the context of how they protect the pulmonary (or gill) capillaries from damage by high pressures. In fish, both teleosts and elasmobranchs, the heart consists of only two chambers in series—a single atrium and a single ventricle (see figure 6.2). Blood is pumped to the gills through the ventral aorta and then distributed to the tissues of the body via the dorsal aorta. This means that the hydrostatic pressure in the gill capillaries must exceed that in the dorsal aorta, which is responsible for delivering blood to the peripheral tissues. At first sight, this would put the gill capillaries at risk from stress failure. However, most fish are purely exothermic and, therefore, have low values of $\dot{V}O_2$max. As a consequence, their cardiac outputs are low, and the pressure in the dorsal aorta is correspondingly low.

However, in some unusually athletic fish such as the albacore tuna, *Thunnis alalunga*, which is partly endothermic, the mean dorsal aortic pressure has been measured as high as 79 mmHg (Breisch et al. 1983). The pressure in the gill capillaries must be higher because they are upstream, and therefore, the pressures greatly exceed those necessary to cause stress failure of capillaries in some mammals. The solution adopted by the fish to prevent stress failure is to maintain a relatively thick gill blood-water barrier that is therefore presumably much stronger than in the mammal. The thick barrier can be tolerated because even in a partially endothermic fish such as the skipjack tuna, $\dot{V}O_2$max is only 24 ml · min^{-1} · kg^{-1} (Stevens and Neill 1978) compared with approximately 97 ml · min^{-1} · kg^{-1} in an athletic mammal of approximately the same size such as the spring hare (Seeherman et al. 1981). Because the diffusion requirements of the barrier are so much less, the barrier can afford to be thicker and therefore stronger than in the mammal.

With the gradual evolution toward the endothermic mammal and bird, the pulmonary circulation is gradually separated from the systemic circulation so that the pulmonary capillaries are eventually exposed to lower pressures (see figure 6.2). In a modern amphibian such as a frog, the beginning of separation of the two circulations is seen. Now the atria are completely separate although there is one undivided ventricle. However, the left atrium receives oxygenated blood from the lungs, and the right atrium receives venous blood from the systemic circulation. Although the ventricle is anatomically undivided, the two streams of blood tend to remain unmixed so that oxygenated blood mainly enters the aorta and oxygen-depleted blood chiefly flows separately into the pulmonary circulation. This division of the ventricular output results in more efficient gas exchange. Thus, a functional separation has been partly achieved although the pulmonary capillaries are presumably exposed to nearly the full pressure generated by the single ventricle. However again, the blood-gas barrier can afford to be thicker than in the mammal because of the low values of $\dot{V}O_2$max and resulting lower gas diffusion requirements of the lung in these exothermic animals.

In the noncrocodilian reptile, the two atria are completely separated, and the ventricle is partially divided anatomically. This assists in the separation of the streams of oxygenated and poorly oxygenated blood with the result that there is little mixing and functionally there is a well-developed double circulation. However, the fact that the ventricles do communicate means that, again, the pulmonary capillaries are exposed to a high pressure. They can tolerate this only because, as before, the blood-barrier can afford to be thicker than in the mammal because of the low values of $\dot{V}O_2$max associated with exothermy. An interesting exception is the monitor lizard, *Varanus exanthematicus*, which has an unusually high aerobic scope and in which a ridge in the ventricle separates the two outflow tracts during systole. The result is that the pressure in the pulmonary artery is much lower than in the aorta (Burggren and Johansen 1982). One can only speculate about how the prehistoric flying reptiles such as *Pteranodon* with its 7 m wingspan (Langston 1981) solved the problem.

Only when we come to the fully endothermic vertebrates, including the mammals and birds, is it essential that the pressures in the pulmonary circulation be much lower than in the systemic circulation. The systemic circulation needs high aortic pressures for the large cardiac outputs associated with high levels of $\dot{V}O_2$max. However, the pulmonary circulation must have a low pressure because of the vulnerability of the capillary walls, which are extremely thin to allow the rapid diffusion of oxygen to satisfy the high values of $\dot{V}O_2$max. Thus, full endothermy requires complete separation of the two ventricles, and this explains why this is mandatory in mammals and birds.

Interestingly, even with complete separation of the pulmonary and systemic circulations, the vulnerability of the pulmonary capillaries may be a limiting factor in extremely aerobic animals such as the thoroughbred racehorse. As shown in figure 6.1, during galloping, mean pulmonary arterial and left atrial pressures are as high as 120 and 70 mmHg,

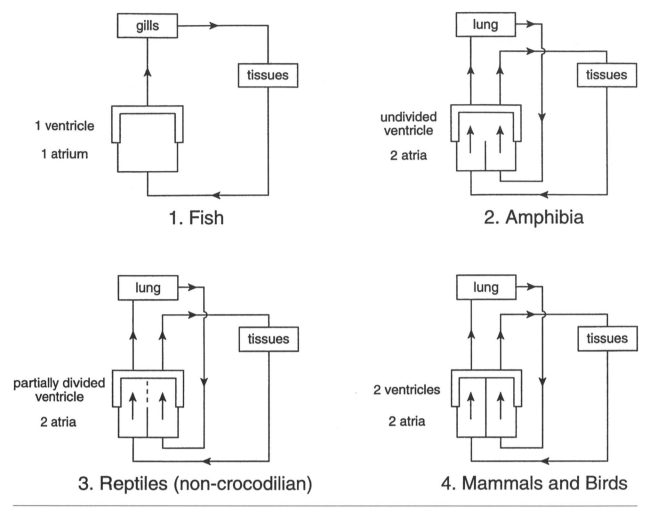

Figure 6.2 Stages in the evolution of the pulmonary circulation. In fish, the gill capillaries are exposed to the full dorsal aortic pressure. In amphibia and noncrocodilian reptiles, the pulmonary circulation is functionally partly separated for more efficient gas exchange. However, only the fully endothermic mammals and birds have achieved total separation, protecting the vulnerable lung capillaries from the high aortic pressure.

respectively. This means that the capillary transmural pressure is about 100 mmHg, and, not surprisingly, this is sufficient to cause the rupture of some capillaries. As a consequence, all thoroughbreds in training bleed into their lungs (Whitwell and Greet 1984) because of stress failure of their pulmonary capillaries (West et al. 1993).

It might be thought that this alveolar bleeding could be avoided by making the capillary walls thicker and stronger. However, these animals show clear evidence of diffusion limitation of oxygen uptake with marked arterial hypoxemia during galloping (Wagner et al. 1989). Therefore, from the point of view of increasing aerobic performance, it would be to their advantage to have a thinner blood-gas barrier to allow greater levels of $\dot{V}O_2$max. Obviously, they cannot afford to do that because the capillaries are already too weak to withstand the stresses during heavy exercise. This is apparently an insolvable dilemma unless the lungs can be

greatly increased in size to increase the area of the blood-gas interface and thus raise the diffusing capacity. Alternatively, these animals could profit from a third pump situated between the pulmonary capillaries and the left atrium!

Ontogeny and the Vulnerability of Pulmonary Capillaries

In some respects, the dramatic changes that occur in the structure and function of the circulation at the time of a mammal's birth reflect the changes seen in evolution. In the fetus, the outflow tracts of the left and right ventricles are closely connected by the large ductus arteriosus. As in the amphibians and reptiles, streaming of blood occurs in the heart with the result that the best-oxygenated blood from the placenta tends to be directed to the brain. However, the pressure in a pulmonary artery is very high, the same as that in the aorta.

How do the potentially vulnerable pulmonary capillaries tolerate this situation? First, the blood flow through the lung is very small, about 15% of the cardiac output; more is unnecessary because there is no pulmonary gas exchange. This small flow is accomplished by the high tone of the large amount of smooth muscle in the pulmonary arteries. Most of the blood from the right ventricle bypasses the lung through the patent ductus arteriosus. The result is that pulmonary capillary pressure is low. Furthermore, the potential blood-gas barrier is thicker than in the adult, but it is at risk of stress failure at birth (see figure 6.3).

At birth, dramatic changes take place rapidly. The lungs have to take over the function of gas exchange from the placenta, and pulmonary blood flow necessarily greatly increases. This is brought about by a striking reduction of resistance in the pulmonary arteries and arterioles, mostly through relief of hypoxic pulmonary vasoconstriction, but partly through an increase in lung volume. Now the capillaries are seriously at risk because they are not protected to the same extent from the pressure in the pulmonary artery. At the same time, therefore, pulmonary artery pressure must rapidly fall through constriction and ultimate closure of the ductus arteriosus. It is extraordinary that these two events, opening of the pulmonary arterial throttle and closure of the ductus arteriosus, are so well synchronized in the majority of births. If closure of the ductus does not occur, muscularization of the pulmonary arteries persists (Haworth 1978). If the vasoconstriction of the pulmonary arteries fails to reverse, the condition of persistent pulmonary hypertension of the newborn may develop (Morin and Stenmark 1995).

Effects of High Transmural Pressure on Pulmonary Capillaries

Although the micromechanics of the capillary wall during high levels of transmural pressure are not fully understood, capillary wall stress becomes extremely high. This section reviews research on the effects of high transmural pressure on capillaries and compares the responses of the rabbit, dog, and racehorse.

Ultrastructural Changes

Ultrastructural changes have been described in detail elsewhere (Costello, Mathieu-Costello, and West 1992; Fu et al. 1992; Tsukimoto et al. 1991; West et al 1991) and will be reviewed here only briefly. Most of

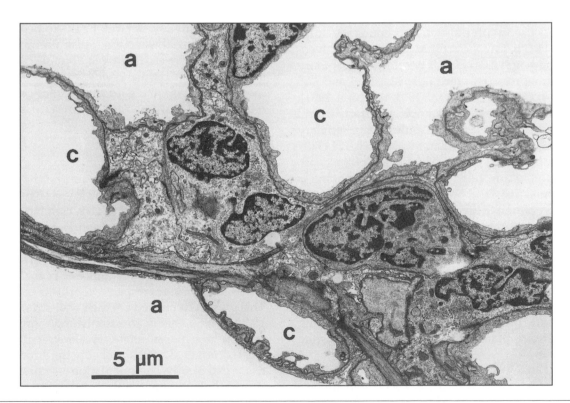

Figure 6.3   Pulmonary capillaries in the lung of a newborn rabbit. The blood-gas barrier is at risk of stress failure because the complex changes in the pulmonary circulation in the perinatal period may expose the capillaries to high transmural pressures. c: capillary; a: alveolus.

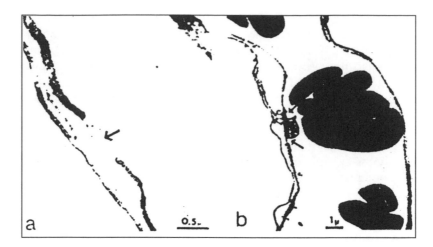

Figure 6.4 Ultrastructural changes in pulmonary capillaries exposed to high transmural pressures (rabbit). In (a), the epithelium is disrupted (arrow), but the alveolar epithelium and the two basement membranes are continuous. In (b), the alveolar epithelial layer (right) and the capillary endothelial layer (left) are disrupted. Note the platelet closely applied to the exposed endothelial basement membrane (left). (From West et al. 1991. Used by permission of the American Physiological Society.)

the data have been obtained from anesthetized rabbit preparations. Briefly, the chest was opened, cannulas were placed into the pulmonary artery and left atrium, and the lung was perfused with the rabbit's own blood. After a short time, the blood was washed out with a saline/dextran mixture, and the lungs were fixed for electron microscopy with buffered glutaraldehyde, all at the same pressure. Pulmonary arterial pressures of 20, 40, 60, and 80 cm H_2O were used with pulmonary venous pressures set 5 cm H_2O below the arterial pressure. The alveolar pressure was 5 cm H_2O. Therefore, the capillary transmural pressures were 12.5, 32.5, 52.5, and 72.5 ± 2.5 cm H_2O (Tsukimoto et al. 1991; West et al. 1991).

Figure 6.4 (a) shows an example of ultrastructural changes in the wall of the pulmonary capillary when the transmural pressure was 52.5 cm H_2O (39 mmHg). Note that there is disruption of the capillary endothelium, but its basement membrane is continuous, as is the basement membrane of the alveolar epithelial layer and the epithelial layer itself. Figure 6.4 (b) shows another example at the same capillary transmural pressure. On the right side, the alveolar epithelial layer is disrupted. On the left side, the endothelium is broken, and a platelet is closely applied to the exposed endothelial basement membrane. This membrane is electrically charged and highly reactive, and it is not surprising that it attracts platelets, red blood cells, and white blood cells.

In the rabbit preparation, stress failure of pulmonary capillaries was consistently seen at a capillary transmural pressure of 52.5 cm H_2O. No breaks occurred in preparations where the capillary transmural pressure was 12.5 cm H_2O (these were the

normal controls), but a few examples occurred at a pressure of 32.5 cm H_2O. The number of breaks further increased when the pressure was raised to 72.5 cm H_2O (Tsukimoto et al. 1991).

Micromechanics of the Capillary Wall Under High Pressure

We do not fully understand the micromechanics of the capillary wall when the transmural pressure is raised to high levels. However, a simple calculation shows that the capillary wall stress becomes extremely high. If we regard the capillary as a thin-walled cylindrical tube, the wall stress (S) is given by

$$S = \frac{P \cdot r}{t}$$

where P is the transmural pressure, r is the radius of curvature, and t is the wall thickness (Laplace relationship). In rabbit pulmonary capillaries where stress failure has been consistently seen at a transmural pressure of 52.5 cm H_2O (Tsukimoto et al. 1991), representative values for radius of curvature and wall thickness on the thin side of the capillary are 3.6 and 0.34 μm respectively (Birks et al. 1994). This gives a calculated average wall stress of 5.5 × 10^{-4} N · m^{-2}. This is an astonishingly high stress, approaching the ultimate tensile strength of collagen (West et al. 1991), probably the strongest soft tissue in the body. This calculation of wall stress is so simple that it is remarkable that the extremely high values have not been previously recognized. Probably the reason for this is that people have been misled by the small radius of the capillary that reduces wall stress, other things being equal. Indeed, the hoop or circumferential tension of the capillary wall is relatively small at about 25 mN · m⁻¹, primarily because of the small radius of curvature (West et al. 1991). What has been overlooked is the extreme thinness of the wall that, of course, follows from the gas exchange function of the blood-gas barrier.

There is good evidence that the strength of the capillary wall along the thin side can be attributed to the type IV collagen of the basement membranes. The thin side contains only three layers: capillary endothelial layer, alveolar epithelial layer, and the extracellular matrix between them which is made up of the two fused basement membranes. The evidence that the type IV collagen in the basement membrane is the main molecule responsible for the strength of the capillary wall has been presented elsewhere (West and Mathieu-Costello 1992b) and will not be repeated here.

Type IV collagen is a triple helix of three α (IV) chains, giving a molecule about 400 nm long (see figure 6.5 (a)). Two molecules join at the C-terminal end by disulfide exchange between like NC1 domains. Then, four molecules join at the N-terminal end to give an assembly somewhat like chicken wire (Timpl et al. 1981; Yurchenco and Schittny 1990) (see figure 6.5 (b)). This arrangement apparently combines great strength with porosity. The few measurements that have been made suggest that the ultimate tensile strength of basement membrane approaches that of type I collagen (Fisher and Wakely 1976; Welling and Grantham 1972; West et al. 1991).

A possible explanation of the disruptions of the capillary endothelial and alveolar epithelial layers shown in figures 6.5 (a) and (b) is that the type IV collagen assembly distorts under high tensile stress. This would be analogous to stretching a piece of chicken wire in one dimension. The result is that the mesh changes its shape and becomes elongated

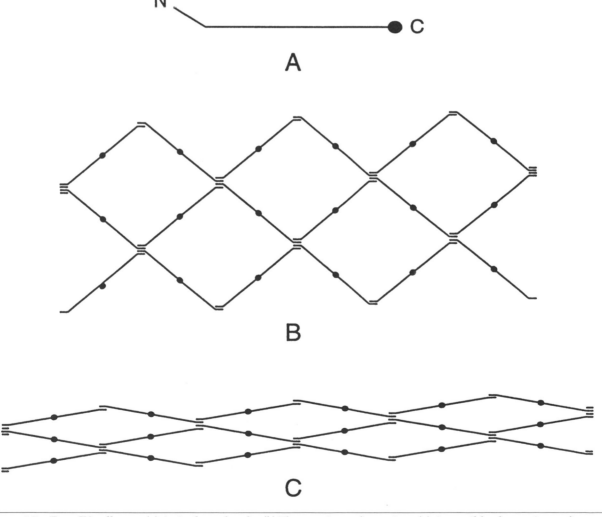

Figure 6.5 Type IV collagen. *(a)* A single molecule. *(b)* The matrix configuration. *(c)* A possible change in configuration that could explain the ultrastructural appearances of figure 6.4.

without any change of length of the individual wire pieces (see figure 6.5 (c)). This explanation of the ultrastructural pattern probably assumes that the individual endothelial and epithelial cells can rapidly move along their basement membranes. There are discrete attachment points between a cell surface and its underlying matrix through surface receptors (Albelda 1991; Buck and Horwitz 1987), and the cell could migrate along its basement membrane by disengaging from one set of receptors and reattaching to another set. The ability of cells to track along their basement membranes is well accepted in developmental biology.

Additional evidence for a mechanism such as distortion of the suprastructure of type IV collagen is the rapid reversibility of many of the endothelial and epithelial cell disruptions. We have shown that if the capillary transmural pressure is first raised to 52.5 cm H_2O and then reduced to 12.5 cm H_2O for 1 or 3 min before fixation at the low pressure, many of the disruptions are no longer visible (Fu et al. 1992). For example, the number of endothelial breaks fell by 66%, and the number of epithelial breaks fell by 70%. The disruptions that closed were those that were initially small and those associated with intact basement membranes. The rapid reduction in the number of disruptions strongly suggests that the strain associated with the increased pressure is reversible and would fit with the return of the type IV collagen assembly to its unstretched state, as shown in figure 6.5 (b).

Species Differences

Some species of animals have pulmonary capillaries that are stronger than others. Indeed, this is implicit in what has been previously stated because, if the thoroughbred racehorse had capillaries that developed stress failure at the same pressure as the rabbit, their aerobic activity would be extremely limited.

Comparisons between rabbit, dog, and racehorse show that the capillary transmural pressure required for stress failure of the capillary wall increases with the aerobic capacity of the animal (Birks et al. 1994, 1997; Mathieu-Costello et al. 1995) whereas in the rabbit lung, stress failure is consistently seen at a capillary transmural pressure of 52.5 cm H_2O (39 mmHg), in the dog the value is about 92.5 cm H_2O (46 mmHg), and in the thoroughbred racehorse about 136 cm H_2O (100 mmHg). The thickness of the blood-gas barrier and the average radius of curvature of the capillaries are consistent with the differences in pressure required for stress failure. The blood-gas barrier is thinnest in the rabbit, intermediate in the dog, and thickest in the horse whereas an average capillary

radius is largest in the rabbit, intermediate in the dog, and smallest in the horse. As a consequence, calculated capillary wall stress for a given capillary transmural pressure decreases from rabbit to dog to horse (Birks et al. 1994).

Evidence for Changes in the Permeability of the Blood-Gas Barrier During Severe Exercise

During severe exercise, there is evidence that the blood-gas barrier changes permeability. This section will first discuss the findings concerning thoroughbred racehorses and then will present the evidence dealing with humans.

Thoroughbred Racehorses

As stated previously, there is strong evidence that all thoroughbreds in training bleed into their lungs, based on the presence of hemosiderin-laden macrophages in tracheal washings (Whitwell and Greet 1984). The mechanism for the exercise-induced pulmonary hemorrhage (EIPH) was unclear for many years, but there is good evidence that it is caused by stress failure of pulmonary capillaries (West et al. 1993; West and Mathieu-Costello 1994). The reason is simply the extremely high capillary pressures that, in turn, are a consequence of the enormously high left ventricular filling pressures (see figure 6.1). Some aspects of EIPH are not fully understood, for example why the condition is principally seen in the dorsocaudal regions of the lung. Such issues have been discussed in detail elsewhere (West et al. 1993) and will not be further reviewed here.

Humans

Since stress failure of pulmonary capillaries occurs in all thoroughbreds in training, an interesting question is whether the permeability of human pulmonary capillaries is altered by severe exercise. Accumulating evidence suggests that this occurs.

McKechnie et al. (1979) described hemoptysis and hemorrhagic pulmonary edema in two athletes who took part in the extremely challenging Comrades Marathon in South Africa. This remarkable race is over a distance of 90 km, and the competitors run for 8–11 hours. Neither of the two athletes had evidence of heart or lung disease though both of them had previously developed hemorrhagic pulmonary edema in a similar race. Hemoptysis and pulmonary edema have also been described in eight swimmers taking part in a time trial, though fluid loading to prevent dehydration may have contributed (Weiler-Ravell et al. 1995). An anecdotal report describes a 35-year-old rugby player who repeatedly developed hemoptysis following extreme physical exertion during games (West et al. 1991). Extensive investigations were negative except that blood could be seen coming from peripheral regions of the lung at bronchoscopy. Another piece of evidence suggesting changes in the permeability of the blood-gas barrier during exercise is that normal volunteers who performed periodic moderate or heavy exercise over a period of 2–7 hours have been shown to have increased concentrations of protein in their bronchoalveolar lavage fluid compared with nonexercising controls (Everson 1994).

We have investigated the possibility of changes in the blood-gas barrier with intense exercise by studying six elite cyclists with a history of hemoptysis or tasting blood following exercise but who were otherwise healthy (Hopkins et al. 1997). Bronchoalveolar lavage (BAL) using normal saline was carried out with fiber-optic bronchoscopy 1 hour after a 4 km uphill sprint at maximum effort sufficient to give a mean heart rate of 177 beats/min. The controls were normal sedentary subjects who did not exercise prior to BAL. The athletes showed significantly higher concentrations of red blood cells, total protein, and leukotriene B_4 in their BAL fluid. The higher concentrations of red blood cells and protein are strong evidence of an increased permeability of the blood-gas barrier, and the higher protein level fits with the previous study by Everson (1994).

The increased leukotriene B_4 is particularly interesting. This is a potent chemotactic mediator that was previously shown to be increased in the BAL fluid of patients who had high-altitude pulmonary edema (HAPE) (Schoene et al. 1986). We believe that the mechanism may be exposure of the basement membrane of the alveolar epithelial cells. Figure 6.4 (b) shows an example of this. As pointed out earlier, the basement membrane is highly reactive and can be expected to activate alveolar macrophages that come in contact with it. Thus, this represents an inflammatory response initiated by mechanical changes. Attempts to find an inflammatory mechanism for HAPE have been fruitless to date.

Remodeling in the Pulmonary Circulation

This section deals with remodeling in the pulmonary circulation.

Pulmonary Arteries

There is a tendency to think that the structure of the adult pulmonary circulation is static. While it is well-known that there is marked involution of smooth muscle in the pulmonary arteries following birth (except at high altitude), it is sometimes assumed that once this process is complete, the morphology of the arterial system remains constant. However, striking remodeling of the pulmonary arteries can occur under some conditions, and the evolutionary advantage of this process appears to be to protect the vulnerable capillaries.

The best example of pulmonary artery remodeling occurs in alveolar hypoxia when the pulmonary artery pressure rises as a result of hypoxic pulmonary vasoconstriction. Rats made hypoxic by exposing them to air at half the normal barometric pressure show rapid morphological changes in their pulmonary arteries, including increases in the amount of vascular smooth muscle and extracellular matrix (Meyrick and Reid 1978, 1980). The increase in smooth muscle can be detected histologically within 48 hours. The remodeling persists during the period of hypoxia, and involution of both smooth muscle and extracellular matrix is rapid when the animals are returned to a normoxic environment. For example, degradation of collagen and elastin can be demonstrated within 3 days (Tozzi et al. 1991).

The rapidity of remodeling has been demonstrated by stretching excised rings of pulmonary artery in Krebs-Ringer solution to simulate an increased wall tension equivalent to a transmural pressure of 50 mmHg (Tozzi et al. 1989). Within 4 hours, there were increases in collagen synthesis (incorporation of ^{14}C-proline), elastin synthesis (incorporation of ^{14}C-valine), mRNA for pro-α1 (I) collagen, and mRNA for proto-oncogene v-sys. The last may implicate platelet-derived growth factor (PDGF) or transforming growth factor β (TGFβ) as the mediator. Interestingly, these changes were endothelium dependent because they did not occur when the endothelium was removed from the arterial rings.

What is the evolutionary advantage of this pulmonary artery remodeling? We believe that it protects the vulnerable pulmonary capillaries because in the adult lung there is meager vascular smooth muscle in the pulmonary arteries, and this is unevenly distributed. As a consequence, when hypoxic pulmonary vasoconstriction occurs, those capillaries downstream from areas of the lung lacking smooth muscle will be exposed to high pressure and will be liable to stress failure. Indeed, this is apparently the mechanism of high-altitude pulmonary edema (West and Mathieu-Costello 1992a; West et al. 1995). Remodeling of the pulmonary arteries allows the resistance of all vessels to be increased, thus protecting all the pulmonary capillaries. Note that the result is to return the structure of the pulmonary arteries to a state similar to that in the fetus where the capillaries need to be protected from the high pulmonary arterial pressure, in this case the result of the patent ductus arteriosus.

A clinical feature of high-altitude pulmonary edema (HAPE) may throw light on this process. It is known that if a person goes to high altitude, HAPE will develop within 1 week or not at all. A reasonable explanation is that the pulmonary arterial remodeling progresses far enough to completely protect the capillaries at the end of this time. However, if someone who has been at high altitude for a period goes to sea level for several days and subsequently returns to high altitude, so-called reentry HAPE may develop. This is particularly common in children. Possibly the mechanism is that the involution of vascular smooth muscle and connective tissue makes the capillaries vulnerable once again.

Capillaries

Remodeling also occurs in pulmonary capillaries when their transmural pressure is increased and again the evolutionary advantage appears to be to protect these delicate vessels. As discussed previously, the strength of the blood-gas barrier comes mainly from the type IV collagen in the basement membranes. Therefore, it is not surprising that the result of remodeling is to increase the thickness of the basement membranes. Classically, these changes are seen in patients with mitral stenosis and left ventricular failure (Haworth, Hall, and Patel 1988; Lee 1979). Figure 6.6 shows an example where obvious thickening of the basement membrane of the alveolar epithelial cells and, particularly, of the capillary endothelial cell can be seen in a patient with mitral stenosis. In these conditions where the pulmonary venous pressure is raised, the only protection available to the capillaries is to strengthen their walls. Clearly, no remodeling of the pulmonary arteries or veins will protect them.

Regulation of the structure of the capillary wall presumably occurs continuously. Otherwise, how could this enormous area of the thin side of the blood-gas barrier (over 20 m² in the human lung) be maintained extremely thin (0.2–0.4 μm in the human lung) and yet just strong enough to withstand the largest normal physiological stresses. Presumably, capillary wall stress is being continually moni-

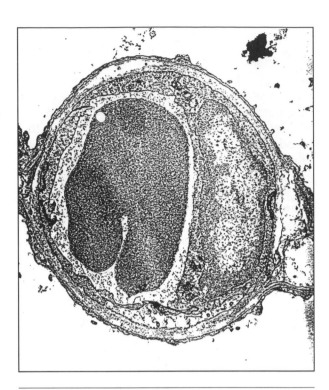

Figure 6.6   Electron micrograph of pulmonary capillary from a child aged 11 years with rheumatic mitral stenosis. Note the thickening of the basement membranes of the alveolar epithelium and, particularly, of the capillary endothelium. (Courtesy of Professor S.G. Haworth.)

tored and the structure of the wall, especially the amount of type IV collagen, continually adjusted. Surprisingly, the responsible mechanism which is central to the success of the mammalian lung has been studied little and is essentially unknown. Recently, we have been increasing capillary wall stress by ventilating one lung of anesthetized rabbits at high degrees of lung inflation. Previous work from our laboratory has shown that the incidence of stress failure is greatly increased at high states of lung inflation (Fu et al. 1992). Preliminary results show increases in mRNA for pro-αI (III) collagen, fibronectin, and TGF-β1 within 4 hours. Much further work is needed to understand this central dilemma of the mammalian lung, namely how to provide an extremely thin membrane of large area while protecting it against large mechanical stresses.

Acknowledgments

This work was supported by NIH grant R01-46910 and NHLBI Program Project grant HL-17731. We acknowledge the collaboration of Eric Birks, Michael Costello, Ann Elliott, Zhenxing Fu, James Jones,

Sadi Kurdak, Yasuo Namba, John Pascoe, Renato Prediletto, Koichi Tsukimoto, and Walter Tyler.

References

Albelda, S.M. 1991. Endothelial and epithelial cell adhesion molecules. *Am. J. Respir. Cell Mol. Biol.* 4: 195–203.

Bevegard, S., A. Holmgren, and B. Jonsson. 1960. The effect of body position on the circulation at rest and during exercise, with special reference to the influence of the stroke volume. *Acta. Physiol. Scand.* 49:279–298.

Bhattacharya, J., S. Nanjo, and N.C. Staub. 1982. Micropuncture measurement of lung microvascular pressure during 5-HT infusion. *J. Appl. Physiol.* 52:634–637.

Bhattacharya, J., and N.C. Staub. 1980. Direct measurement of microvascular pressures in the isolated perfused dog lung. *Science* 210:327–328.

Birks, E.K., O. Mathieu-Costello, Z. Fu, W.S. Tyler, and J.B. West. 1994. Comparative aspects of the strength of pulmonary capillaries in rabbit, dog and horse. *Respir. Physiol.* 97:235–246.

Birks, E.K., O. Mathieu-Costello, Z. Fu, W.S. Tyler, and J.B. West. 1997. Very high pressures are required to cause stress failure of pulmonary capillaries in thoroughbred racehorses. *J. Appl. Physiol.* 82: 1584-1592.

Breisch, E.A., F. White, H.M. Jones, and R.M. Laurs. 1983. Ultrastructural morphometry of the myocardium of *Thunnus alalunga*. *Cell. Tiss. Res.* 23(3):427–438.

Buck, C.A., and A.F. Horwitz. 1987. Cell surface receptors for extracellular matrix molecules. *Annual Review of Cell Biology* 3:179–205.

Burggren, W., and K. Johansen. 1982. Ventricular hemodynamics in the monitor lizard *Varanus exat thermaticit*: Pulmonary and systemic pressure separation. *J. Exp. Biol.* 96:343–354.

Costello, M.L., O. Mathieu-Costello, and J.B. West. 1992. Stress failure of alveolar epithelial cells studied by scanning electron microscopy. *Ann. Rev. Respir. Dis.* 145:1446–1455.

Dexter, L., J.L. Whittenberger, F.W. Haynes, W.T. Goodale, R. Gorlin, and C.G. Sawyer. 1951. Effect of exercise on circulatory dynamics of normal individuals. *J. Appl. Physiol.* 3:439–453.

Elliott, A.R., Z. Fu, K. Tsukimoto, R. Prediletto, O. Mathieu-Costello, and J.B. West. 1992. Short-term reversibility of ultrastructural changes in pulmonary capillaries caused by stress failure. *J. Appl. Physiol.* 73:1150–1158.

Erickson, B.K., H.H. Erickson, and J.R. Coffman. 1990. Pulmonary artery, aortic and esophageal pressure changes during high intensity treadmill exercise in the horse: A possible relation to exercise-induced pulmonary hemorrhage. *Equine Vet. J. Suppl.* 9:47–52.

Erickson, B.K., H.H. Erickson, and J.R. Coffman. 1992. Pulmonary artery and aortic pressure changes during high intensity treadmill exercise in the horse: Effect of furosemide and phentolamine. *Equine Veterinary Journal* 24:215–219.

Everson, R.B. 1994. Host determinants of cellular and biochemical constituents of bronchoalveolar lavage fluids. Implications for design of epidemiologic studies. *AJRCCM* 149:899–904.

Fisher, R.F., and J. Wakely. 1976. The elastic constants and ultrastructural organization of a basement membrane (lens capsule). *Proc. Roy. Soc. Lotid. Ser. B.* 193:335–358.

Fishman, A.P. 1985. Pulmonary circulation. In *The Respiratory system, volume 1: Circulation and nonrespiratory functions*, ed. A.P. Fishman. Bethesda, MD: American Physiological Society.

Fu, Z., M.L. Costello, K. Tsukimoto, R. Prediletto, A.R. Elliott, O. Mathieu-Costello, and J.B. West. 1992. High lung volume increases stress failure in pulmonary capillaries. *J. Appl. Physiol.* 73:123–135.

Groves, B.M., J.T. Reeves, J.R. Sutton, P.D. Wagner, A. Cymerman, M.K. Malconian, P.B. Rock, P.M. Young, and C.S. Houston. 1987. Operation Everest II: Elevated high-altitude pulmonary resistance unresponsive to oxygen. *J. Appl. Physiol.* 63:521–530.

Gurtner, H.P., P. Walser, and B. Füssler. 1975. Normal values for pulmonary hemodynamics at rest and during exercise in man. *Prog. Resp. Res.* 9:295–315.

Haworth, S.G. 1978. The pulmonary circulation in congenital heart disease. II. Pulmonary hypertension. *Herz.* 3:138–142.

Haworth, S.G., S.M. Hall, and M. Patel. 1988. Peripheral pulmonary vascular and airway abnormalities in adolescents with rheumatic mitral stenosis. *Int. J. Cardio.* 18:405–416.

Hickam, J.B., and W.H. Cargill. 1948. Effect of exercise on cardiac output and pulmonary arterial pressure in normal persons and in patients with cardiovascular disease and pulmonary emphysema. *J. Clin. Invest.* 27:10–23.

Hopkins, S.R., R.B. Schoene, T.R. Martin, W.R.J. Henderson, R.G. Spragg, and J.B. West. 1997. Intense exercise impairs the integrity of the pulmonary blood-gas barrier in elite athletes. *Am. J. Resp. Crit. Care. Med.* 155: 1090-1094.

Jones, J.H., B.L. Smith, E.K. Birks, J.R. Pascoe, and T.R. Hughes. 1992. Left atrial and pulmonary arterial pressures in exercising horses. *FASEB J.* 6:A2020.

Langston, H. Jr. 1981. Pterosaurs. *Scientific American* February, 122–136.

Lee, Y. 1979. Electron microscopic studies of the alveolar-capillary barrier in the patients of chronic pulmonary edema. *Jap. Circulation J.* 43:945–954.

Manohar, M. 1993. Pulmonary artery wedge pressure increases with high-intensity exercise in horses. *Am. J. Vet. Res.* 54:142–146.

Mathieu-Costello, O., D.C. Willford, Z. Fu, R.M. Garden, and J.B. West. 1995. Pulmonary capillaries are more resistant to stress failure in dog than in rabbit. *J. Appl. Physiol.* 79:908–917.

McKechnie, J.K., W.P. Leary, T.D. Noakes, J.C. Kallmeyer, E.T.M. MacSearraigh, and L.R. Olivier. 1979. Acute pulmonary edema in two athletes during a 90-km running race. *S. Afr. Med. J.* 56:261–265.

Meyrick, B., and L. Reid. 1978. The effect of continued hypoxia on rat pulmonary arterial circulation. An ultrastructural study. *Lab. Invest.* 38:188–200.

Meyrick, B., and L. Reid. 1980. Hypoxia-induced structural changes in the media and adventitia of the rat hilar pulmonary artery and their regression. *Am. J. Pathol.* 100:151–178.

Morin, F.C., and K.R. Stenmark. 1995. Persistent pulmonary hypertension of the newborn. *AJRCCM* 151:2010–2032.

Reeves, J.T., B.M. Groves, A. Cymerman, J.R. Sutton, P.D. Wagner, D. Turkevich, and C.S. Houston. 1990. Operation Everest II: Cardiac filling pressures during cycle exercise at sea level. *Respir. Physiol.* 80:147–154.

Riley, R.L., A. Himmelstein, H.L. Motley, H.M. Weiner, and A. Cournand. 1948. Studies of the pulmonary circulation at rest and during exercise in normal individuals and in patients with chronic pulmonary disease. *Am. J. Physiol.* 152:372–382.

Schoene, R.B., P.H. Hackett, W.R. Henderson, E.H. Sage, M. Chow, R.C. Roach, W.J. Mills, and T.R. Martin. 1986. High-altitude pulmonary edema. Characteristics of lung lavage fluid. *J. Am. Med. Assoc.* 256:63–69.

Seeherman, H.J., C.R. Taylor, G.M.O. Maloiy, and R.B. Armstrong. 1981. Design of the mammalian respiratory system. II. Measuring maximum aerobic capacity. *Respir. Physiol.* 44:11–23.

Stevens, E.D., and W.H. Neill. 1978. Body temperature relations of tunas, especially skipjack. In *Fish physiology,* ed. W.S. Hoar and D.J. Randall. New York: Academic Press.

Timpl, R., H. Wiedemann, V. van Delden, H. Furthmayr, and K. Kiihn. 1981. A network model for the organization of type IV collagen molecules in basement membranes. *Eur. J. Biochem.* 120:203–211.

Tozzi, C.A., F.J. Wilson, S.Y. Yu, R.F. Bannett, B.W. Peng, and D.J. Riley. 1991. Vascular connective tissue is rapidly degraded during early regression of pulmonary hypertension. *Chest* 99 (Suppl. 3):4IS–42S.

Tozzi, C.A., G.J. Poiani, A.M. Harangozo, C.D. Boyd, and D.J. Piley. 1989. Pressure-induced connective tissue synthesis in pulmonary artery segments is dependent on intact endothelium. *J. Clin. Invest.* 84:1005–1012.

Tsukimoto, K., O. Mathieu-Costello, R. Prediletto, A.R. Elliott, and J.B. West. 1991. Utrastructural appearances of pulmonary capillaries at high transmural pressures. *J. Appl. Physiol.* 71:573–585.

Wagner, P.D., G.E. Gale, R.E. Moon, J.R. Torre-Bueno, B.W. Stolp, and H.A. Saltzman. 1986. Pulmonary gas exchange in humans exercising at sea level and simulated altitude. *J. Appl. Physiol.* 61:260–270.

Wagner, P.D., J.R. Gillespie, G.L. Landgren, M.R. Fedde, B.W. Jones, R.M. DeBowes, R.L. Pieschl, and H.H. Erickson. 1989. Mechanism of exercise-induced hypoxemia in horses. *J. Appl. Physiol.* 66:1227–1233.

Weiler-Ravell, D., A. Shupak, L. Goldenberg, P. Halpem, O. Shoshani, G. Hirschhorn, and A. Margulis. 1995. Pulmonary edema and hemoptysis induced by strenuous swimming. *Br. Med. J.* 311:361–362.

Welling, L.W., and J.J. Grantham. 1972. Physical properties of isolated perfused renal tubules and tubular basement membranes. *J. Clin. Invest.* 51:1063–1075.

West, J.B., G.L. Colice, Y.-J. Lee, Y. Namba, S.S. Kurdak, Z. Fu, L.C. Ou, and O. Mathieu–Costello. 1995. Pathogenesis of high-altitude pulmonary edema: Direct evidence of stress failure of pulmonary capillaries. *Eur. Resp. J.* 8:523–529.

West, J.B., and O. Mathieu-Costello. 1992a. High altitude pulmonary edema is caused by stress failure of pulmonary capillaries. *Int. J. Sports. Med.* 13 (Suppl. 1):S54–S58.

West, J.B., and O. Mathieu-Costello. 1992b. Strength of the pulmonary blood-gas barrier. *Respir. Physiol.* 88:141–148.

West, J.B., and O. Mathieu-Costello. 1994. Stress failure of pulmonary capillaries as a mechanism for exercise-induced pulmonary hemorrhage: A review. *Equine Vet. J.* 26:441–447.

West, J.B., O. Mathieu-Costello, J.H. Jones, E.K. Birks, R.B. Logemann, J.R. Pascoe, and W.S. Tyler. 1993. Stress failure of pulmonary capillaries in racehorses with exercise-induced pulmonary hemorrhage. *J. Appl. Physiol.* 75:1097–1109.

West, J.B., K. Tsukimoto, O. Mathieu-Costello, and R. Prediletto. 1991. Stress failure in pulmonary capillaries. *J. Appl. Physiol.* 70:1731–1742.

Whitwell, K.E., and T.R.C. Greet. 1984. Collection and evaluation of tracheobronchial washes in the horse. *Equine Vet. J.* 16:499–508.

Younes, M., Z. Bshouty, and J. Ali. 1987. Longitudinal distribution of pulmonary vascular resistance with very high pulmonary blood flow. *J. Appl. Physiol.* 62:344–358.

Yurchenco, P.D., and J.C. Schittny. 1990. Molecular architecture of basement membranes. *FASEB J,* 4:1577–1590.

Chapter 7

Liver, Spleen, and Kidneys

Inge-Lis Kanstrup

The gastrointestinal organs and the kidneys are positioned below the diaphragm, bordering the thoracic organs, heart and lungs, and above the large muscles of the legs—the organs of main interest to exercise physiologists. At rest, the abdominal organs receive at least 25% of the cardiac output and extract only 15–20% of the available oxygen. Furthermore, 20–30% of the total blood volume is "stored" in the abdominal veins (Rowell and Johnson 1984). This implies that the region is regarded as a major site for blood redistribution during exercise. In 1912, August Krogh stated,

> In order to explain such an increase in flow (i.e. cardiac output during exercise) it is necessary to find a source of supply which is independent of the dilatation in the working muscles and sufficient to raise the pressure in the central veins to such a height that the heart can be filled completely during each diastole, even when the rate is doubled or trebled. Such a source of supply is the portal system as a consequence of the double set of resistances which it possesses (Krogh 1912).

Researchers later confirmed that the abdominal organs are exposed to a blood flow reduction proportional to exercise intensity in order to assist the need of the exercising muscles for extra blood supply.

As long as the organ function remains intact and does not influence exercise performance or the well-being of the subject, however, interest in further studying the abdominal organs during exercise in the healthy subject has been limited. However, does the reduction in organ blood flow mean that these organs are simply put on standby during exercise, waiting to take over their function when blood flow is reestablished after cessation of exercise? Conversely, does exercise influence organ pathology? These questions must be addressed separately for the different abdominal organs. The answers must include an evaluation of each organ's importance for exercise and the impact of physical training on long-term function.

This chapter reviews the role of the abdominal organs in the cardiovascular response to exercise. It points out some poorly elucidated themes in order to stimulate an interest in further evaluation.

Splanchnicus

Splanchnology is the study of the internal organs. *Splanchnicus* means the abdominal organs innervated by the splanchnic nerves. In the following, however, the term will be used solely for the digestive organs that drain to the liver. This means that splanchnic blood flow (SBF) represents the flow to the intestines, pancreas, spleen, and liver.

SBF is traditionally measured by the use of Fick's principle as modified by Bradley et al. (1945). The arterial-mixed hepatic venous difference of a substance exclusively eliminated from the blood by the liver is determined during constant infusion. The method demands catheterization of a hepatic vein (Hultmann 1966; Kanstrup and Winkler 1987) and thus has limited application. By using this method, the mean hepatic blood flow is estimated to be about 1500 ml/min in healthy adults at rest. About one-third of this flow originates from the hepatic artery, and the rest comes from the portal vein. The arterial hepatic circulation supplies about 50% of the oxygen consumed by the liver (Hultmann 1966). During exercise, SBF and the splanchnic blood volume (SBV) are reduced by 250–450 ml/min and 300–700 ml, respectively, in both the supine and the sitting positions (Bishop et al. 1957; Flamm et al. 1990; Wade et al. 1956).

Rowell and colleagues (Rowell 1974; Rowell, Blackmon, and Bruce 1964) and Clausen et al. (1973) performed a more systematic evaluation in a series of experiments. A reduction in SBF was consistently found with increasing exercise intensity. During heavy upright exercise, SBF was reduced to 20–30%

93

Blood Volume in Abdominal Organs

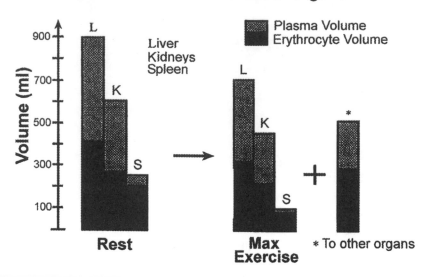

Figure 7.1 Blood volume in the abdominal organs at rest and during maximal dynamic exercise. The light shaded bars represent the plasma volume and dark shaded bars indicate the erythrocyte volume for L (liver), K (kidney), and S (spleen).

of the resting value (minimum measured value 390 ml/min). The flow reduction was strongly correlated to the relative rather than to the absolute work intensity. SBF was estimated from the clearance of ICG after a single dye injection. This may be questioned because blood flow influences dye removal (Kanstrup and Winkler 1987). Therefore, the absolute values must be taken with special precaution. In two subjects, the researchers found a hepatic (a-$\overline{v}$)O$_2$ difference of 120 ml/L during exercise. This demanded 2.20 L O$_2$ / min and a reduction in hepatic blood flow to 43% of the resting value (Bishop et al. 1957). The arterial oxygen capacity is about 200 ml/L of the volume. The contribution of blood from the portal vein during exercise and the remaining oxygen content in this is unknown. However, with a flow reduction to 20% of the resting value at maximal exercise and with the above-mentioned hepatic (a-$\overline{v}$)O$_2$ difference during moderate exercise, splanchnic ischemia is a risk (Nielsen et al. 1995).

During exercise in hot environments, even increased demands to flow reduction in the splanchnic area do exist. Thus, compared with normothermia, short-term exercise at 43 °C leads to a 20% reduction in SBF, while 60 min of exercise at 48 °C at about 50% of $\dot{V}O_2$max reduces SBF by almost 40%. At exhaustion, the hepatic venous O$_2$ content is reduced to 0.5 ml/100 ml, and hepatic lactate production suggests hepatic ischemia (Rowell 1974).

This SBF reduction, proportional to heart rate and plasma norepinephrine concentration, indicates a major role for the sympathetic nervous system in the control of the splanchnic vascular resistance. The stimulation is baroreceptor mediated. In humans, the arterial baroreflexes exert greater effects on the splanchnic region than do cardiopulmonary baroreflexes (Johnson et al. 1974). However, metabolic or vasoactive hormones may influence this response (Kjaer et al. 1993). Circulating epinephrine affects SBF and SBV differently. The beta-adrenergic effect on the splanchnic arterioles is dilation while causing splanchnic venoconstriction (Rowell and Johnson 1984). A myriad of gastroenteropancreatic and other vasoactive hormones increase during exercise (Brouns 1991; Kanstrup et al. 1995). The specific roles of these hormones still need to be elucidated. Of special interest are perhaps vasoactive intestinal peptide, atrial natriuretic peptide, and glucagon, which increase gut blood flow, and also vasopressin and angiotensin, which reduce it. Researchers have proposed that circulating epinephrine during exercise in hypoxia is responsible for SBF being unchanged in spite of marked increases in heart rate and plasma norepinephrine (Rowell et al. 1984).

An increased mixed venous oxygen content during exercise is a frequent finding with aging (Julius et al. 1967). One explanation could be an impaired blood redistribution to exercising muscles. This would direct a greater fraction of cardiac output to other organs including the splanchnic region, kidneys, and nonactive muscles with a low oxygen extraction. Thus, reduced sympathetic vasoconstriction of the

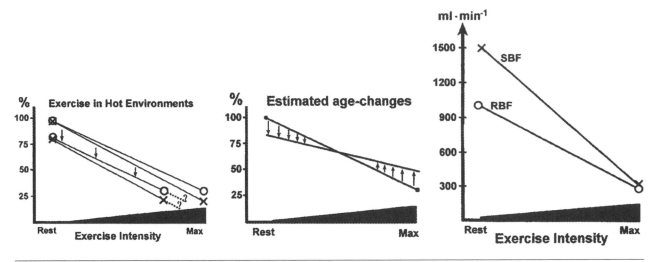

Figure 7.2 (*a*) Splanchnic (SBF) and renal (RBF) blood flow from rest through maximal dynamic exercise, and the effect of (*b*) age and (*c*) hot environments on these parameters.

vessels to tissues other than the exercising muscles could at least partly be the explanation of the observed changes in the arterial-venous oxygen difference with age. Indices of reduced neuro-humoral activity and cardiovascular responsiveness with age have been found (Gerstenblith, Lakatta, and Weisfeldt 1976). Furthermore, the maximal rise in plasma norepinephrine concentration is reduced at advanced age (Jensen et al. 1994). Kenney and Ho (1995) compared younger and older (~64-year-old) subjects with the same $\dot{V}O_2$max at two identical workloads, demanding 45% and 60% of the peak oxygen uptake, respectively. They found a less pronounced reduction in SBF with exercise in the elderly and also a minor reduction in renal blood flow (even if the absolute exercise values were lower due to a lower resting renal blood flow). This supports the view that blood redistribution is impaired with increasing age, although the researchers did not evaluate the influence of training. Thus, the older subjects may be assumed to be more fit than the younger ones because of the matching for identical $\dot{V}O_2$max.

The influence of exercise on the function of the various splanchnic organs is outside the scope of this chapter. However, the liver has an especially important role in glycogenolysis and gluconeogenesis. (For details of the neuroendocrine regulation of metabolic changes in exercise, see Kjaer and Mohr (1994).) Digestion is reduced. However, during prolonged exercise, water and carbohydrates must be supplied to continue performance. The probably reduced oxygen supply (promoted by hypohydration and possible hyperthemia) could evidently lead to a reduction in the energy-demanding absorption of substances such as glucose, amino acids, and electrolytes. Ischemia may even lead to lesions

of the mucosa, bleeding, and loss of fluids (Brouns 1991). He cites Derek Clayton, the world record holder of the marathon in 1979,

> Two hours later the elation had worn off. I was urinating large dots of blood and I was vomiting black mucus and had a lot of black diarrhea. I don't think too many people can understand what I went through for the next 48 hours.

He also mentions that 30–50% of the participants in endurance exercise complain of upper and lower gastrointestinal symptoms. The changes in gastrointestinal hormones, their release and metabolism, as well as the nervous stimulation probably are of importance. However, their specific role remains to be elucidated. Researchers also do not know to what extent training may influence the function of the splanchnic organs during exercise, i.e., improve their absorptive or glycolytic capacity. However, Kjaer and Mohr (1994) have shown that the exercise-induced rise in hepatic glucose production precedes and exceeds the peripheral glucose uptake, which is especially pronounced during intense exercise in trained subjects with presumably high liver glycogen stores. This indicates the existence of a *sports liver*.

In conclusion, SBF and SBV are reduced during exercise proportional to increased sympathetic nervous activation. However, vasoactive and gastroenteropancreatic hormones probably modulate them. The response seems to be less pronounced with aging. Splanchnic organ function during especially long-term strenuous exercise should be looked further upon, especially since ischemia may result in severe gastrointestinal lesions. Adaptations with endurance training are poorly studied.

Spleen

The spleen is the largest lymphoid organ in the body and weighs about 150–250 g. The main functions are filtration, phagocytosis of lymphocytes and monocytes, and destruction of damaged or dead erythrocytes and platelets, microorganisms, and other debris from the blood. Macrophages remove iron from hemoglobin of old red blood cells and return it to the circulation for use by the bone marrow in producing new red blood cells. The spleen also assists in lymphocyte production and antibody formation. In fetal life, it produces red blood cells. Later in life, it stores red blood cells and platelets. In this respect, the spleen serves as a blood reservoir. This section will focus on that function. The spleen's important role in the immunologic reaction to exercise is outside the scope of this paper. Researchers now accept that the erythrocyte content of the spleen alters depending on sympathetic nervous discharge or pressure in the splenic veins and the intra-abdominal pressure. The structure of the sinusoids in the vascular bed of the spleen allows erythrocytes to penetrate the vascular wall and to be packed in the parenchyma of the spleen. In this way, the spleen acts as an erythrocyte depot to be emptied into the portal circulation.

In 1927, Barcroft and Poole found an increased hemoglobin and hematocrit concentration in the splenic pulp of the dog and cat in a ratio of 3:2 to that of the general circulation. They also noticed a splenic contraction during exercise (Barcroft and Stephens 1927). Later studies have confirmed that the spleen in various animals (dog, cat, sheep, and horse) plays a role as reservoir for the erythrocytes. The spleen may empty upon stimulation from exercise, cold, hypoxia, bleeding, hypotension, and circulating catecholamines. Thus, Kramer and Luft (1951) found a considerable reduction in spleen weight (about 100–150 g) in nembutalized dogs and a concomitant increase in hemoglobin concentration of 10.5% during acute severe hypoxia. Similar hemoglobin increases were seen after administering epinephrine intravenously. The diving seal heavily depends on any extra oxygen supply during its occasionally more than 60 min dives. In these animals, researchers found that arterial hemoglobin concentration increased from 15.1 to 25.4% and hematocrit increased from 38 to 59% after dives of 17–27 min duration (Qvist et al. 1986). The authors estimate a reduction in splenic volume from 24 L to 4 L that augments the circulating red blood cell volume from 13 L to 33 L. This would add at least 8 min of diving oxygen supply.

Even though this splenic autoinfusion capability is functional for diving animals, seeing its appropriateness in humans, all exercising aerobically, is difficult. Early studies about this phenomenon failed to demonstrate the large increases in hemoglobin and hematocrit found in cats and dogs after exercise or epinephrine infusion. By using the blood volume methods then available, an increase in the circulating blood volume could be measured in humans after stimuli such as exercise, heat, and epinephrine, but only small changes occurred in the hematocrit values. Since this increase in volume occurred in splenectomized as well as in healthy subjects, researchers concluded that other organs such as the liver and skin vessels must serve as reservoirs for both erythrocytes and plasma (Ebert and Stead 1941).

Ebert and Stead (1941) found an increase in hematocrit from 44 to 49.6% and in hemoglobin from 14.6 to 16.4% (9.1 to 10.2 mmol/L) in six normal male subjects after 3–5 min of exhaustive exercise. They also measured increases from 40.2 to 43.5% and 12.7 to 13.5%, respectively, in two splenectomized subjects who were exercising slightly less. Comparable changes were found after intravenous epinephrine. They concluded that the results were similar, the small differences occurred due to the harder work performed by the healthy subjects. Ebert and Stead therefore could not differentiate between healthy and splenectomized subjects, in contrast to animal studies. The hemoconcentration was due to extravasation of plasma and not addition of noncirculating erythrocytes. Exercise physiologists ever since have accepted the view that the human spleen does not serve as an important reservoir of red cells. This belief is also based on studies demonstrating a constant body/venous hematocrit ratio over a wide hematocrit range (Chaplin, Mollison, and Vetter 1953). However, in patients with splenomegalia, Fudenberg et al. (1961) found a significantly increased ratio directly proportional to the degree of splenic enlargement. This indicates an increased amount of erythrocytes released from an enlarged spleen.

With the noninvasive blood pool imaging methods (^{99m}Tc-labeling of the erythrocytes) linked to gamma camera registration, a renewed interest in the visceral blood volume during exercise occurred. Sandler et al. (1984) noted that splenic blood radioactivity decreased by 49% after supine exercise in 10 patients. Froelich et al. (1988) performed graded upright cycling in 10 healthy adult volunteers and continuously recorded the activity in the spleen, liver, kidneys, and right lung. They found that the largest reduction in blood volume occurred in the spleen—39%. In the kidney and liver, activity decreased by 9% and 14%, respectively (these two results are not significant).

In a similar experiment, spleen, liver, and kidney blood volumes decreased to 54%, 82%, and 76% of baseline, respectively, at peak exercise. The bowel blood volume, however, remained unchanged (Flamm et al. 1990). They also demonstrated that hematocrit levels increased from 44.0 to 48.3%. The reductions in the splanchnic organs count rates were graded with increasing exercising loads. The authors suggested that autotransfusion from the spleen could be an alternative mechanism that explains the hemoconcentration observed during exercise. This would also agree with the more profound reduction in splenic volume than in hepatic volume. They explained the unchanged bowel blood volume because the subjects were exposed to a minimal baseline flow in the fasting state.

Laub et al. (1993) put forward a quantitative calculation of the importance of this suggested autotransfusion. They performed graded bicycle exercise to maximum in five normal adults. By using a slightly different calculation procedure, they found a reduction in splenic count rate to a mean of 34% (range 26–44%) of the initial count rate at supine rest and a concomitant increase in hematocrit from 44.6% to 48%. They assumed that the blood content of the human spleen under normal conditions does not exceed 250–300 ml. Two hundred ml of this stagnant blood with a hematocrit of 80%, when released into a circulatory blood volume of 6000 ml with a hematocrit of 45%, would increase the hematocrit to only 46%. Thus the spleen contraction could be responsible for only about one-fourth of the rise in hematocrit during exhaustive exercise. Other possible explanations for the hemoconcentration could be water transport across the capillary wall due to increased interstitial osmotic pressure in the exercising muscles, an extravasation of plasma water caused by a rise in hydrostatic pressure, and increased vascular permeability (Convertino et al. 1981; Kanstrup and Ekblom 1984; Senay, Rogers, and Jooste 1980).

The mechanism for the decrease in splenic volume is uncertain because smooth muscular tissue is very sparse in the human spleen. In animals, in contrast, a muscular splenic capsule exists that may physically squeeze the splenic contents. However, Pinkus et al. (1986), by using highly specific antibodies to smooth muscle myosin, found an extensive network of smooth muscle elements in the human spleen. Within the red pulp, these were arranged in short, regular, repetitive bands aligned parallel to the long axis of the sinus and extending between contiguous ring fibers. The authors state that this may be the basis for cyclic volume changes. Besides, contractile proteins in the sinusoid-lining

cells may also serve to alter the cell permeability of the sinus wall.

Concluding that the human spleen plays a certain, although not decisive, role for the hemoconcentration during exercise seems reasonable. This is also supported by the normal exercise capacity found in splenectomized, but otherwise healthy, human subjects. However, no systematic studies have been performed. The question arises about whether a *sports spleen* exists. The enlarged spleens reported are all related to hematological illnesses. At present, no evidence indicates the existence of a sports spleen regarding the spleen's size and ability to release erythrocytes in endurance-trained athletes.

In diving humans, the release of oxygenated erythrocytes would be of further importance. Studies in Korean amas have demonstrated a 20% decrease in splenic volumes (measured using an ultrasonograph) and a 10% increase in hemoglobin concentration and hematocrit after three hours of repetitive diving (Hurford et al. 1990). No changes were observed in Japanese male divers who did not normally practice breath-hold diving, pointing to a certain influence of training on contraction ability. Espersen et al. (1992) readdressed this subject by measuring splenic contraction in control subjects and in submarine rugby players and divers—although the last mentioned were not elite. They did not find any difference in the resting size of the spleen or the contractility during the diving stimulus, which elicited about 30% contraction of the spleen in both groups at maximum breath-hold time.

In conclusion, recent evidence in humans confirms the results found in animal studies. The spleen has a certain role as reservoir for erythrocytes to be released into the general circulation during exercise, cold, anxiety, and diving. All of these activities are connected to increased sympathetic nervous activity. In addition to the redistribution from the splanchnic region, this assists in increasing the oxygen transport capacity of the organism. A sports spleen in human divers would be important and also potentially influence endurance-trained subjects. However, this has not yet been documented. Instead, it might be a functional adjustment in athletes with high aerobic performance.

Kidneys

The kidneys could influence the circulatory regulation during exercise as part of the blood redistribution described for the splanchnic system. However, also through the production of hormones, they might play a role in cardiovascular regulation. Finally, a change in the renal excretion of salt and water during exercise is of utmost importance but will be discussed here only briefly. Renal function during exercise was studied as early as 1936 by Covian and Rehberg in the zoophysiologic laboratory in Copenhagen. In two subjects, they found an increasing reduction in creatinine clearance with moderate-to-severe exercise and a concomitant reduction in diuresis. They observed proteinuria in both subjects during the most demanding exercises.

White and Rolf (1948) measured renal plasma flow (RPF) as PAH clearance and inulin clearance after running for 11 to 20 min. During moderate exercise, PAH and inulin clearance fell to half or less of their resting values, while brief maximum exercise lowered the clearances to 20% or less of their resting values. Proteinuria was again observed after the most strenuous exercises. This led the authors to conclude, "These findings thus support, although they do not prove, the view that under great stress many glomeruli may cease functioning". Also in 1948, Chapman et al. measured the effect of exercise on RPF in nine healthy male subjects who walked on a treadmill in 16-min periods at various speeds/ inclination. During the heaviest work, they found a reduction in RPF to an average of 65% of the basal RPF (613 ml/min). They stated that the decline in RPF is directly related to the severity of the exercise within certain, as yet incompletely defined, limits. Radigan and Robinson (1949) put extra stress on their five subjects, having them exercise on a treadmill both in normal and hot (50 °C) temperatures. The first produced a reduction in RPF to 49% (422 ml/min) of the resting level, while the heat reduced plasma flow even further to a mean of 293 ml/min. They also found that the renal plasma clearance (of mannitol) was less affected. Thus, they found no reduction during exercise in the cold and a reduction of 30% in the hot environments.

Castenfors (1967) investigated this topic more thoroughly. He confirmed a marked decrease in RPF during exercise, related to the intensity, and a smaller reduction in GFR, leading to an increased filtration rate. Furthermore, he found a decrease in urine production and urinary sodium excretion and variable changes in potassium excretion. He also noted proteinuria connected to exercise, which was more marked during than after exercise. The proteins consisted mainly of albumin and lower-molecular-weight proteins. He stated that the main factors causing proteinuria were related to increased glomerular permeability and to renal vasoconstriction. However, he also said that individual differences in the glomerular membrane might be of importance.

The RBF reduction is elicited mainly via a sympathetic nervous stimulation as is the case for the splanchnic circulation. The kidneys are abundantly innervated by sympathetic nerves concerning both the afferent and the efferent arterioles to the glomerular capsules. Renal vasoconstriction is elicited upon falling blood pressure or pulse pressure (reduced stimulation of arterial baroreceptors) or pronounced arterial hypoxemia (via glomerular chemoreceptors) (Honig 1989). However, the initial increase in renal sympathetic nervous activity during exercise seems to be caused by descending input from higher brain centers (central command) parallel with heart rate stimulation (Matsukawa et al. 1991). The vasoconstriction occurs after electric stimulation of the renal nerves or injection of alphaadrenergic substances. Stimulation of beta-adrenergic receptors are thought to promote the release of renin. The kidneys possess a pronounced vascular autoregulation, mediated primarily by resistance changes in the afferent arterioles. Efferent arteriolar resistance, though, changes little until perfusion pressure falls below about 80 mmHg. The autoregulation is seemingly mediated primarily by an intrinsic myogenic mechanism at the level of the afferent arteriole (Stein 1990).

Various vasoactive substances released during exercise may, however, modulate the renal vascular tone. The increased sympathetic nervous activity and reduced perfusion lead to primary stimulation of the renin activity, promoting the synthesis of angiotensin (ANG) II (Mannix et al. 1990). ANG II is a potent vasoconstrictor in the kidneys, leading to both an afferent and an efferent arteriolar vasoconstriction (Stein 1990). This probably assists the overall perfusion reduction during exercise. Atrial natriuretic peptide (ANP) is released mainly from the right atrium in response to atrial distension and increased heart rate. It stimulates peripheral vasodilation, diuresis, and natriuresis (thus also an increase in RPF). ANP increases during exercise (Freund et al. 1991; Kanstrup et al. 1995; Mannix et al. 1990). However, under baseline circumstances, this probably does not influence the vasoconstriction during exercise. In acute hypoxia, however, it might be one of the possible mechanisms behind the renal response where a similar flow reduction was found in normoxia and hypoxia at the same relative workload in spite of a marked increased heart rate and norepinephrine concentration in hypoxia (Olsen et al. 1992). This could parallel the splanchnic blood flow measurements during hypoxic exercise (Rowell et al. 1984).

Prostaglandins are thought to act as local regulators of the renal vascular tone in situations of renal stress as in renal ischemia, hypotension, dehydration, and heart failure (Stein 1990). Thus, locally synthesized prostaglandins are thought to counterbalance the vasoconstricting hormones and preserve RPF and GFR. Indomethacin administration (a prostaglandin inhibitor) results in increased renal vascular resistance and depressed RPF for more than 2 h after 30 min strenuous exercise (Walker et al. 1994). This points to a certain role for prostaglandins in maintaining renal perfusion during exercise. Other vasoactive substances such as vasopressin, adenosine, kinins, and endothelin among others are thought to moderate regulation of the renal circulation (Stein 1990). However, the final role remains to be elucidated.

Another hormone produced by the kidneys, erythropoietin, might be of long-term importance for the hemodynamics during exercise. Erythropoietin stimulates erythropoiesis, counterbalancing the continuous loss of aged blood cells. One could speculate that this plays a role in the increased erythrocyte volume seen in endurance-trained subjects (Convertino 1991). Tissue hypoxia is the primary stimulus for erythropoietin production (Jelkmann 1992). Exercise-induced increases in circulating blood volume have an important role. Against this background is the finding that erythropoietin remains almost unchanged after exercise. In addition, a similar concentration of erythropoietin exists in males and females (Jelkmann 1992). Finally, apparently increased erythrocyte volumes in endurance-trained female subjects are normalized when related to lean body mass (Suetta, Kanstrup, and Fogh-Andersen 1996).

As mentioned, exercise-induced proteinuria is a frequent finding. It has raised the question of whether exercise exerts a potential damaging effect on the renal glomeruli and, subsequently, total renal function. This has led to certain precautions when prescribing exercise to people with impaired renal function. This aspect was investigated in rats that had an experimentally reduced renal function. Heifets et al. (1987) found that exercise improved the renal function, while the control nonexercising group had a constant fall in renal function. In patients with a moderately to severely reduced renal function, confirming an improved effect of exercise on the reduced renal function was not possible. On the other hand, the study did not show an extra loss of glomerular function in the trained patients (Eidemak et al. 1997). Furthermore, training is believed to have a positive effect on the impaired insulin-glucose tolerance found in the patients with reduced renal function as compared with healthy controls. Whether a *sports kidney* exists, i.e., a kidney able to excrete salt and water at reduced perfusion for longer periods

or adapt to or influence strenuous exercise in other ways, remains to be clarified.

In conclusion, RBF is reduced during exercise related to increased sympathetic nervous activity. However, perfusion may be modulated by various vasoactive substances as the renin-angiotensin system, ANP, and prostaglandins. Normoxic exercise apparently does not stimulate erythropoietin production. Exercise has not—in contrast with earlier expectations—been found to promote renal glomerular injury. The literature still sparsely describes renal function during especially long-term exercise and renal adaptations with training.

References

Barcroft, J., and L.T. Poole. 1927. The blood in the spleen pulp. *Journal of Physiology* 64:23–29.

Barcroft, J., and J.G. Stephens. 1927. Observations upon the size of the spleen. *Journal of Physiology* 64:1-22.

Bishop, J.M., K.W. Donald, S.H. Taylor, and P.N. Wormald. 1957. Changes in arterial-hepatic venous oxygen content difference during and after supine leg exercise. *Journal of Physiology* 137:309–317.

Bradley, S.E., F.J. Ingelfinger, G.P. Bradley, and J.J. Curry. 1945. The estimation of hepatic blood flow in man. *Journal of Clinical Investigation* 24:890–897.

Brouns, F. 1991. Etiology of gastrointestinal disturbances during endurance events. *Scandinavian Journal of Medicine and Science in Sports* 1:66–77.

Castenfors, J. 1967. Renal function during exercise. *Acta Physiologica Scandinavia* 70: suppl. 293:1–44.

Chaplin, H., P.L. Mollison, and H. Vetter. 1953. The body/venous hematocrit ratio: Its constancy over a wide hematocrit range. *Journal of Clinical Investigation* 32:1309–1316.

Chapman, C.B., A. Henschel, J. Minckler, A. Forsgren, and A. Keys. 1948. The effect of exercise on renal plasma flow in normal male subjects. *Journal of Clinical Investigation* 27:639–644.

Clausen, J.P., K. Klausen, B. Rasmussen, and J. Trap-Jensen. 1973. Central and peripheral circulatory changes after training of the arms or legs. *American Journal of Physiology* 225:675–682.

Convertino, V.A. 1991. Blood volume: Its adaptation to endurance training. *Medicine and Science in Sports and Exercise* 23:1338–1348.

Convertino, V.A., L.C. Keil, E.M. Bernauer, and J.E. Greenleaf. 1981. Plasma volume, osmolality, vasopressin and renin activity during graded exercise in man. *Journal of Applied Physiology* 50:123–128.

Covian, F.G., and P.B. Rehberg. 1936. Über die Nierenfunktion während schwerer Muskelarbeit. *Scandinavian Archives of Physiology* 75:21–37.

Ebert, R.V., and E.A. Stead. 1941. Demonstration that in normal man no reserves of blood are mobilized by exercise, epinephrine, and hemorrhage. *American Journal of Medicine and Science* 201:655–664.

Eidemak, I., A-B. Haaber, B. Feldt-Rasmussen, I-L. Kanstrup, and S. Strandgaard. 1997. Exercise training and the progression of chronic renal failure. *Nephron* 75:36-40.

Espersen, K., H. Frandsen, T. Lorentzen, I-L. Kanstrup, and N.J. Christensen. 1992. Hemodynamic importance of the human spleen in simulated diving. *Acta Physiologica Scandinavia* 146:74.

Flamm, S.D., J. Taki, R. Moore, S.F. Lewis, F. Keech, F. Maltais, M. Ahmad, R. Callahan, S. Dragotakes, N. Alpert, and H.W. Strauss. 1990. Redistribution of regional and organ blood volume and effect on cardiac function in relation to upright exercise intensity in healthy human subjects. *Circulation* 81:1550–1559.

Freund, B.J., E.M. Shizuru, G.M. Hashiro, and J.R. Claybaugh. 1991. Hormonal, electrolyte, and renal responses to exercise are intensity dependent. *Journal of Applied Physiology* 70:900–906.

Froelich, J.W., H.W. Strauss, R.H. Moore, and K.A. McKusick. 1988. Redistribution of visceral blood volume in upright exercise in healthy volunteers. *Journal of Nuclear Medicine* 29:1714–1718.

Fudenberg, H., M. Baldini, J.P. Mahoney, and W. Dameshek. 1961. The body hematocrit/venous hematocrit ratio and the "splenic reservoir." *Blood* 17:71–82.

Gerstenblith, G., E.G. Lakatta, and M.L. Weisfeldt. 1976. Age changes in myocardial function and exercise response. *Progress in Cardiovascular Diseases* 19:1–21.

Heifets, M., T.A. Davis, E. Tegtmeyer, and S. Klahr. 1987. Exercise training ameliorates progressive renal disease in rats with subtotal nephrectomy. *Kidney International* 32:815–820.

Honig, A. 1989. Peripheral arterial chemoreceptors and reflex control of sodium and water homeostasis. *American Journal of Physiology* 257:R1282–R1302.

Hultmann, E. 1966. Blood circulation in the liver under physiological and pathological conditions. *Scandinavian Journal of Clinical and Laboratory Investigation* 18: suppl. 92:27–41.

Hurford, W.E., S.K. Hong, Y.S. Park, D.W. Ahn, K.

Shiraki, M. Mohri, and W.M. Zapol. 1990. Splenic contraction during breath-hold diving in the Korean ama. *Journal of Applied Physiology* 69:932–936.

Jelkmann, W. 1992. Erythropoietin: Structure, control of production, and function. *Physiological Reviews* 72:449–489.

Jensen, E.W., K. Espersen, I-L. Kanstrup, and N.J. Christensen. 1994. Exercise-induced changes in plasma catecholamines and neuropeptide Y: Relation to age and sampling times. *Journal of Applied Physiology* 76:1269–1273.

Johnson, J.M., L.B. Rowell, M. Niederberger, and M.M. Eisman. 1974. Human splanchnic and forearm vasoconstrictor responses to reductions of right atrial and aortic pressure. *Circulation Research* 34:515–524.

Julius, S., A. Amery, L.S. Whitlock, and J. Conway. 1967. Influence of age on the hemodynamic response to exercise. *Circulation* 36:222–230.

Kanstrup, I-L., and B. Ekblom. 1984. Blood volume and hemoglobin concentration as determinants of maximal aerobic power. *Medicine and Science in Sports and Exercise* 16:256–262.

Kanstrup, I-L., J. Marving, N. Gadsbøll, H. Lønborg-Jensen, and P.F. Høilund-Carlsen. 1995. Left ventricular hemodynamics and vasoactive hormones during graded supine exercise in healthy male subjects. *European Journal of Applied Physiology and Occupational Physiology* 72:86–94.

Kanstrup, I-L., and K. Winkler. 1987. Indocyanine green plasma clearance as a measure of changes in hepatic blood flow. *Clinical Physiology* 7:51–54.

Kenney, W.L., and C-W. Ho. 1995. Age alters regional distribution of blood flow during moderate-intensity exercise. *Journal of Applied Physiology* 79:1112–1119.

Kjaer, M., K. Engfred, A. Fernandes, N.H. Secher, and H. Galbo. 1993. Regulation of hepatic glucose production during exercise in humans: Role of sympathoadrenergic activity. *American Journal of Physiology.* 265:E275–E283.

Kjaer, M., and T. Mohr. 1994. Substrate mobilization, delivery, and utilization in physical exercise: Regulation by hormones in healthy and diseased humans. *Critical Reviews in Physical and Rehabilitation Medicine* 6:317–336.

Kramer, K., and U.C. Luft. 1951. Mobilization of red cells and oxygen from the spleen in severe hypoxia. *American Journal of Physiology* 165:215–228.

Krogh, A. 1912. The regulation of the supply of blood to the right heart. *Scandinavian Archive of Physiology* 27:227–248.

Laub, M., K. Hvid-Jacobsen, P. Hovind, I-L. Kanstrup, N.J. Christensen, and S.L. Nielsen. 1993. Spleen emptying and venous hematocrit in humans during exercise. *Journal of Applied Physiology* 74:1024–1026.

Mannix, E.T., P. Palange, G.R. Aronoff, F. Manfredi, and M.O. Farber. 1990. Atrial natriuretic peptide and the renin-aldosterone axis during exercise in man. *Medicine and Science in Sports and Exercise* 22:785–789.

Matsukawa, K., J.H. Mitchell, P.T. Wall, and L.N. Wilson. 1991. The effect of static exercise on renal sympathetic nerve activity in conscious cats. *Journal of Physiology* 434:453–467.

Nielsen, H.B., L.B. Svendsen, T.H. Jensen, and N.H. Secher. 1995. Exercise-induced gastric mucosal acidosis. *Medicine and Science in Sports and Exercise* 27:1003–1006.

Olsen, N.V., I-L. Kanstrup, J-P. Richalet, J.M. Hansen, G. Plazen, and F-X. Galen. 1992. Effects of acute hypoxia on renal and endocrine function at rest and during graded exercise in hydrated subjects. *Journal of Applied Physiology* 73:2036–2043.

Pinkus, G.S., M.J. Warhol, E.M. O'Connor, C.L. Etheridge, and K. Fujiwara. 1986. Immunohistochemical localization of smooth muscle myosin in human spleen, lymph node, and other lymphoid tissues: Unique staining patterns in splenic white pulp and sinuses, lymphoid follicles, and certain vasculature, with ultrastructural correlations. *American Journal of Pathology* 123:440–453.

Qvist, J., R.D. Hill, R.C. Schneider, K.J. Falke, G.C. Liggins, M. Guppy, R.L. Elliot, P.W. Hochachka, and W.M. Zapol. 1986. Hemoglobin concentrations and blood gas tensions of free-diving Weddell seals. *Journal of Applied Physiology* 61:1560–1569.

Radigan, L.R., and S. Robinson. 1949. Effects of environmental heat stress and exercise on renal blood flow and filtration rate. *Journal of Applied Physiology* 2:185–191.

Rowell, L.B. 1974. Human cardiovascular adjustments to exercise and thermal stress. *Physiological Reviews* 54:75–159.

Rowell, L.B., J.R. Blackmon, and R.A. Bruce. 1964. Indocyanine green clearance and estimated hepatic blood flow during mild to maximal exercise in upright man. *Journal of Clinical Investigation* 43:1677–1690.

Rowell, L.B., J.R. Blackmon, M.A. Kenny, and P. Escourrou. 1984. Splanchnic vasomotor and metabolic adjustments to hypoxia and exercise in humans. *American Journal of Physiology* 247:H251–H258.

Rowell, R.V., and J.M. Johnson. 1984. Role of the splanchnic circulation in reflex control of the cardiovascular system. In *Physiology of the intestinal circulation*, eds. A.P. Shepherd and D.N. Granger, 153–163. New York: Raven Press.

Sandler, M.P., M.W. Kronenberg, M.B. Forman, O.H. Wolfe, J.A. Clanton, and C.L. Partain. 1984. Dynamic fluctuations in blood and spleen radioactivity: Splenic contraction and relation to clinical radionuclide volume calculations. *Journal of American College of Cardiology* 3:1205–1211.

Senay, L.C., Jr., C. Rogers, and P. Jooste. 1980. Changes in blood plasma during progressive treadmill and cycle exercise. *Journal of Applied Physiology* 49:59–65.

Stein, J.H. 1990. Regulation of the renal circulation. *Kidney International* 38:571–576.

Suetta, C., I-L. Kanstrup, and N. Fogh-Andersen. 1996. Hematological status in elite long-distance runners, influence of body composition. *Clinical Physiology* 16:563-574.

Wade, O.L., B. Combes, A.W. Childs, H.O. Wheeler, A. Cournand, and S.E. Bradley. 1956. The effect of exercise on the splanchnic blood flow and splanchnic blood volume in normal man. *Clinical Science* 15:457–463.

Walker, R.J., J.P. Fawcett, E.M. Flannery, and D.F. Gerrard. 1994. Indomethacin potentiates exercise-induced reduction in renal hemodynamics in athletes. *Medicine and Science in Sports and Exercise* 26:1302–1306.

White, H.L., and D. Rolf. 1948. Effects of exercise and of some other influences on the renal circulation in man. *American Journal of Physiology* 152:505–516.

Chapter 8

Neural Control of Skin Circulation

Susanne F. Vissing

The traditional view of the sympathetic nervous system as a slow-reacting system that activates en masse to produce a uniform outflow to different parts of the body is no longer valid. For example, the control of sympathetic discharge to skin versus skeletal muscle differs both during exercise and during orthostatic stress (see figures 8.1 and 8.2) (Vissing, Scherrer, and Victor 1991, 1994). Also, the control may be differentiated within one type of nerve containing fibers to different effectors. In skin nerves, this occurs during body cooling, when vaso-

constrictor impulses are preferentially activated, and during body heating, preferentially activating sudomotor and probably vasodilator impulses (see figure 8.3) (Bini et al. 1980b; Vissing, Scherrer, and Victor 1991).

Regional differences also occur in thermoregulatory vasomotor and sudomotor tone. For example, in the skin areas innervated by the median and peroneal nerves, vasoconstrictor fibers execute reflex thermoregulatory functions to a major extent. Whereas in the skin areas innervated by the poste-

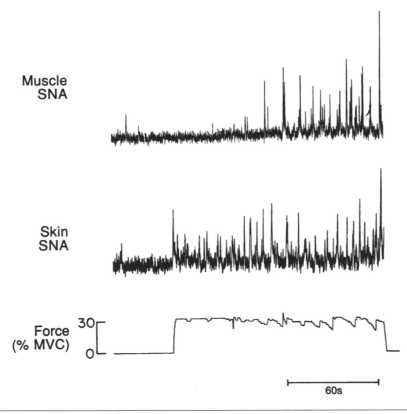

Figure 8.1 Illustrative record from one subject showing mean voltage neurograms of muscle and skin sympathetic nerve activity (SNA) recorded simultaneously from left and right peroneal nerves. Static handgrip at 30% maximal voluntary contraction (MVC) markedly increased both skin and muscle SNA. Skin SNA increased rapidly at the onset of handgrip. However, muscle SNA increased much more slowly with a latency of almost 1 min from the onset of handgrip to the onset of sympathetic activation. (From Vissing, Scherrer, and Victor 1991 with permission.)

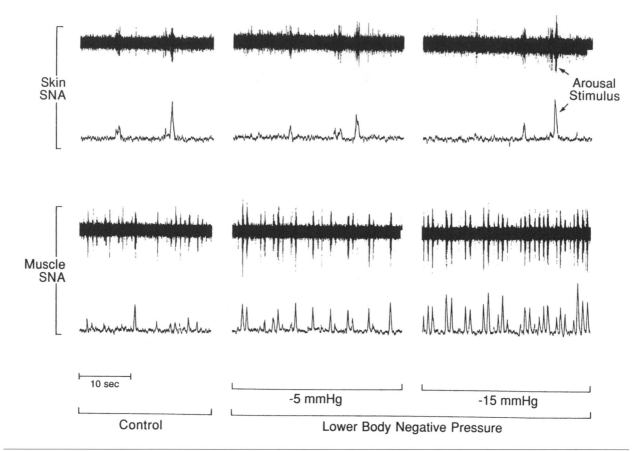

Figure 8.2 Differential responses of skin and muscle sympathetic discharge to low levels of lower body negative pressure. Segments of an original record from one subject showing microneurographic recordings of both skin and muscle sympathetic nerve activity (SNA) from the right peroneal nerve. Note that LBNP at both –5 and –15 mmHg had no effects on skin SNA even though these levels evoked large and graded increases in muscle SNA. The arousal stimulus verifies the adequacy of the skin recording. (From Vissing, Scherrer, and Victor 1994 with permission.)

rior antebrachial and superficial radial nerves, sudomotor fibers execute the thermoregulatory functions to a large extent (Bini et al. 1980a). In contrast, remarkable parallelism exists between neurograms from the left and right peroneal muscle nerve both at rest and during various stimuli, such as lower-body negative pressure (Vissing, Scherrer, and Victor 1989) and static exercise (Wallin, Victor, and Mark 1989). Also, recordings from muscle nerves of the arm (radial nerve) and leg (peroneal nerve) show remarkable parallelism at rest and during lower-body negative pressure (Rea and Wallin 1989) and static handgrip exercise (Wallin, Victor, and Mark 1989). Differences between sympathetic drives to the radial and peroneal muscle nerves is described during mental arithmetic (Anderson, Wallin, and Mark 1987) and posthandgrip muscle ischemia (Wallin, Victor, and Mark 1989). In addition, differences have been shown in sympathetic drive to contracting versus resting muscles (Wallin, Burke, and Gandevia 1992).

The strength of the different effectors' responses to sympathetic discharge depend not only on the characteristics of the effector organ but also on the pattern of frequency of the sympathetic discharge. Thus, experiments in blood vessels of the rat, cat, and human sweat glands have shown that these effectors respond more vigorously to irregular than to monotone impulse discharge (Andersson, Bloom, and Järhult 1983; Nilsson et al. 1985). These findings indicate that the irregularity of baseline discharge seen in recordings from both skin and muscle neurons in humans have physiological importance.

Thermoregulatory Control

Thermoregulatory reflexes play a major role in regulating the cutaneous circulation. At neutral temperature conditions, both vasomotor and electrodermal responses accompany volleys of skin sympathetic discharge (Bini et al. 1980b). Cooling

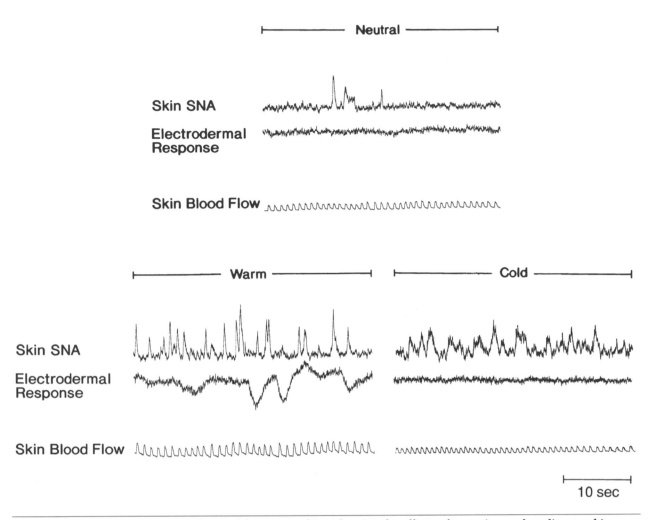

Figure 8.3   Segments of an original record from one subject showing the effects of warming and cooling on skin sympathetic nerve activity (SNA), electrodermal activity, and laser Doppler skin blood flow. Warming the subject produced a large increase in presumable skin sympathetic sudomotor and perhaps vasodilator activity. This was accompanied by increased skin blood flow and increased electrodermal activity. Cooling the subject produced a large increase in presumably skin sympathetic vasoconstrictor activity. This was accompanied by decreased skin blood flow with no signs of sudomotor activation. Note that sudomotor (and vasodilatory) bursts seem to be narrowly based while vasoconstrictor bursts seem to be more broadly based.

causes cutaneous vasoconstriction via a generalized increase of sympathetic vasoconstrictor activity (see figure 8.3) (Bini et al. 1980b; Vallbo et al. 1979; Vissing, Scherrer, and Victor 1991). Moderate heating reduces skin sympathetic discharge, probably by withdrawal of skin vasoconstrictor outflow. This suggestion is based on experiments with five subjects where decreases in skin SNA during moderate heating was associated with increased skin blood flow and absence of sudomotor activity (Vissing and Powell, unpublished observations). More intense heating increases skin SNA, an effect attributed to an enhancement of sudomotor activity (see figure 8.3) (Normell and Wallin 1974; Vissing, Scherrer, and Victor 1991).

Three different types of nerve fibers may execute increases in skin blood flow during warm conditions:

1. Vasoconstrictor
2. Vasodilator
3. Sudomotor neurons

Vasoconstrictor Nerve Fibers

Withdrawal of sympathetic vasoconstrictor activity may increase skin blood flow. This is suggested by the finding that withdrawal of skin sympathetic discharge caused by epidural or axillary blockade increased skin blood flow (Lundin et al. 1990; Vissing,

Secher, and Victor 1996). In addition, exposure to moderately warm conditions reduced skin SNA. At the point of almost complete withdrawal, i.e., virtually neural silence, the increase in skin blood flow was maximal without any signs of sudomotor activity (Vissing and Powell, unpublished observations).

Vasodilator Nerve Fibers

No direct evidence for the existence of vasodilator neurons is available. However, indirect evidence has been provided by Edholm, Fox, and MacPherson (1957) and Shepherd (1963) who concluded that vasodilator neurons must have contributed to the vasodilated state before a nerve section or block. A recent hemodynamic study in humans using atropine to block muscarinic receptors and botulinum toxin to produce presynaptic cholinergic nerve blockade indicated that a cotransmitter released from cholinergic nerves mediated heat-induced cutaneous vasodilation (Kellogg et al. 1995). In cats, Bell and colleagues (1985) have provided direct evidence for the existence of skin vasodilator neurons. Whether the release of substances from sweat can induce vasodilation is controversial (Rowell 1981). However, evidence suggests that active vasodilation is linked to sweat gland activity. First, patients with congenital absence of sweat glands cannot actively vasodilate the skin (Brengelmann et al. 1981). These patients have a normal local response to heating and presumably normal vasculature and autonomic function. Second, patients with pathological absence of sweat glands cannot actively vasodilate the skin (Freund et al. 1981).

Interaction of Thermoregulatory Reflexes

The local and reflex drives to the skin interact, in some cases additively, so that a higher skin blood flow occurs for a given core temperature when local skin temperature is elevated (Johnson, Brengeimann, and Rowell 1976).

The thermal state of the body is mainly controlled by three factors:

1. Core temperature
2. Mean skin temperature
3. Local skin temperature

The relative importance of the core temperature versus skin temperature have been studied extensively despite the great difficulty in separately controlling these two variables. Researchers have conducted studies in conscious baboons with a chronically implanted heat exchanger in a femoral arterial-venous shunt (Proppe 1981; Proppe, Brengelmann, and Rowell 1976) and in humans (Wyss et al. 1974, 1975). They found a 10 to 20 times greater influence of core temperature, as compared with skin temperature, on skin blood flow and sweating. Furthermore, in hot environments, even the highest skin temperature (36–38°C) is not sufficient to raise skin blood flow to more than one-half to two-thirds of the highest values induced by vasodilation produced by rising core temperature (Barcroft and Edholm 1943).

Two different mechanisms may execute thermoregulatory adjustments to changes in local skin temperature. First, a change in blood temperature may stimulate central thermoreceptors (Gibbon and Landis 1932; Pickering 1933; Snell 1954; Stijrub et al. 1935). Second, application of radiant heat to the skin produces reflex vasodilation by stimulation of thermosensitive nerve endings whose fibers run with the sympathetic nerves (Kersiake and Cooper 1950). However, in very warm subjects, increased local hand temperature led to an increase in skin SNA in the ipsilateral median nerve with simultaneous reduction in finger blood flow. This suggests increased cutaneous vasoconstrictor activity (Nagasaka et al. 1990). Thus, the danger of hyperthermia is counteracted by decreased skin blood flow to reduce further heat uptake through the skin.

That the thermoregulatory state influences the degree and direction of various reflex responses in the cutaneous circulation is well documented. For example, intraneural stimulation, mental stress, arousal stimuli, and deep breaths are accompanied by vasoconstriction in glabrous skin of warm subjects. In cold subjects, this was accompanied by vasodilation (Blumberg and Wallin 1987; Elam and Wallin 1987; Oberle et al. 1988). Static handgrip exercise produced vasodilation in hairy skin in warm subjects and vasoconstriction in cold subjects (Vissing, Scherrer, and Victor 1991).

Regulation During Exercise

Regulation of sympathetic discharge during exercise has traditionally been attributed to three different reflex mechanisms:

1. Central command
2. Exercise pressor reflexes
3. Baroreflexes

The central motor command signal emanates from the rostral brain and radiates to autonomic circuits in the brain stem, causing parallel activation of motor and sympathetic neurons (Johansson 1895; Krogh and Lindhard 1913). In conscious humans, central motor command refers to neural signals associated with voluntary motor effort. The exercise pressor reflexes originate in chemically and mechanically sensitive muscle afferents of the contracting muscle (Alam and Smirk 1937; Kaufman et al. 1983; McCloskey and Mitchell 1972; Victor et al. 1989; Vissing et al. 1991; Volkmann 1841). The baroreceptor reflexes originate in mechanically sensitive afferents in the wall of the carotid sinus and aortic arch (Mancia and Mark 1983). They also originate in walls of vessels situated in the cardiopulmonary region, in the right atria, as well as in the ventricular walls (Bishop, Malliani, and Thorén 1983; Hainsworth 1991; Thorén 1979).

In humans, the pattern of sympathetic activation of skin produced by static exercise suggest that central command drives the sympathetic outflow (Vissing and Hjortsø 1996; Vissing, Scherrer, and Victor 1991). The increase in skin SNA precedes the onset of tension development and increases throughout the exercise period (see figure 8.1). The finding that cutaneous sympathetic activation precedes the onset of muscle tension can be explained by a central neural rather than a peripheral reflex mechanism. The progressive increase in skin sympathetic activity during the exercise period can be explained by augmentation of central command. This pattern of sympathetic activation is reproduced by attempted handgrip exercise during neuromuscular blockade with decreased or abolished tension (i.e., decreased stimulation of metaboreceptor and mechanoreceptor afferents) (Vissing and Hjortso 1996). Increases in skin sympathetic discharge also are intensity dependent and related to perceived effort (Vissing, Scherrer, and Victor 1991).

Metaboreceptor afferents are unlikely to cause the exercise-induced sympathetic activation in skin. When stimulated by static exercise in animals, metaboreceptor afferents show a slow and progressive increase in activity that corresponds to the progressive accumulation of intramuscular metabolites within the vicinity of these afferent endings (Kaufman et al. 1983; Mitchell and Schmidt 1983; Rotto, Stebbins and Kaufman 1989). The efferent sympathetic outflow to muscle evoked by the static handgrip reflects this slow pattern of activation (see figure 8.1) (Mark et al. 1985; Vissing, Scherrer, and Victor 1991; Wallin, Victor, and Mark 1989).

Stimulation of metaboreceptor afferents with eliminated input from central command and mechanoreceptor afferents produced by posthandgrip muscle ischemia do not maintain the exercise-induced increases in skin sympathetic activity. However, muscle sympathetic activity remains elevated (Vissing, Scherrer, and Victor 1991).

Mechanoreceptor afferents also do not likely explain the exercise-induced increases in skin SNA. In experimental animals, static contractions evoke increases in SNA that show an initial burst of activity beginning approximately 1 s after the onset of tension development. Rapid adaptation follows (Kaufman et al. 1983; Victor et al. 1989). In skin, increases in sympathetic discharge are seen before, rather than after, the onset of muscle tension. Then the sympathetic discharge continues to increase during sustained contraction.

The exercise-induced activation of skin sympathetic fibers probably consists of a mixture of sudomotor and vasoconstrictor fibers. This suggestion is based on the finding that at normothermia, exercise produced an initial increase in vascular resistance followed by return of vascular resistance to the baseline value (Vissing, Scherrer, and Victor 1991). The gradual decline of vasomotor tone despite increased skin SNA and electrodermal activity may be explained by increased activity in vasoconstrictor fibers obscured by progressive neurohumoral vasodilation. This results from the gradual release of vasoactive substances from sweat glands.

This suggestion is further emphasized by the additional finding that during attenuated sudomotor activation (cooling), increases in estimated skin vascular resistance were maintained throughout the exercise period (Vissing, Scherrer, and Victor 1991). These studies suggest that thermoregulatory reflexes do not produce the increase in sudomotor and active vasodilator outflow during exercise as previously suggested (Rowell 1986). Instead, increases in electrodermal activity and active vasodilation were seen during levels of exercise that did not increase core and skin temperatures.

Regulation During Upright Posture

Activation of the sympathetic nervous system plays a pivotal role in producing circulatory adjustments while assuming an upright posture. This includes vasoconstriction in the renal, splanchnic, and skeletal muscle vascular beds (Rowell 1993). Abundant evidence shows that these are reflex responses evoked largely by unloading of both cardiopulmonary and arterial receptors (Jacobsen et al. 1993; Mark and Mancia 1983; Wieling and Wesseling 1993). Persuasive evidence also shows that the cutaneous circulation is an important site of vasoconstriction during upright posture (Beiser et al. 1970; Johnson et al. 1973; Mosley 1969; Rowell 1986, 1993; Rowell,

Wyss, and Brengeimann 1973). However, the mechanisms responsible for vasoconstriction in skin during upright posture have been disputed.

Researchers well know that baroreceptor afferents regulate muscle vasoconstrictor outflow. Studies in skin with simultaneous measurements of sympathetic discharge and blood flow have demonstrated that unloading both cardiopulmonary and sinoaortic afferents with nonhypotensive and hypotensive levels of lower-body negative pressure (LBNP) had no effect on skin sympathetic outflow or skin vascular resistance at normothermia (see figure 8.1) (Vissing, Scherrer, and Victor 1994; Vissing, Secher, and Victor 1996). In contrast, cutaneous vasoconstriction during upright posture was explained by a local neurogenic, presumably venous-arteriolar, reflex (Vissing, Secher, and Victor 1996).

Regulation of the cutaneous circulation in warm conditions becomes important with respect to blood pressure regulation because heating causes a greater proportion of blood to be situated in the cutaneous circulation. Thus, while baroreceptor regulation of the cutaneous circulation seems to be of minor importance at normothermia, this is not the case in hyperthermia. In hyperthermia, unloading cardiopulmonary afferents with low levels of lower-body negative pressure decreased skin sympathetic outflow, decreased electrodermal (i.e., sudomotor) outflow, and increased skin vascular resistance. This effect of the LBNP was ascribed to a decrease in skin temperature evoked by a cooling effect of air leaking into the LBNP chamber during performance of LBNP. However, the possibility remains that baroreceptor unloading with LBNP, in addition to the cooling effect, might have also induced withdrawal of skin sympathetic discharge to sweat glands and thus indirectly decreased skin blood flow. In that respect, research has shown that baroreceptor unloading can decrease skin blood flow by withdrawal of vasodilator outflow (Kellogg, Johnson, and Kosiba 1990). That conclusion was based on the finding that LBNP at hyperthermia increased cutaneous vascular resistance comparably in areas of skin with and without selective blockade of sympathetic nerve fibers produced by iontophoresis with bretylium tosylate.

Taken together, at neutral temperature conditions with little or no sudomotor activity, a local reflex mechanism drives skin vasoconstriction to a major extent. At warm temperatures, however, baroreceptor-induced withdrawal of sudomotor activity also participates in producing skin vasoconstriction during upright posture.

References

Alam, M., and F.H. Smirk. 1937. Observations in man upon a blood pressure raising reflex arising from the voluntary muscles. *Journal of Physiology* 89:372–383.

Anderson, E.A., B.G. Wallin, and A.L. Mark. 1987. Dissociation of sympathetic nerve activity in arm and leg muscle during mental stress. *Hypertension* 9:114–119.

Andersson, P.O., S.R. Bloom, and J. Järhult. 1983. Colonic motor and vascular responses to pelvic nerve stimulation and their relation to local peptide release in the cat. *Journal of Physiology* 334:293–307.

Barcroft, H., and O.G. Edholm. 1943. The effect of temperature on blood flow and deep temperature in the human forearm. *Journal of Physiology* 102:5–20.

Beiser, G.D., R. Zelis, S.E. Epstein, D.T. Mason, and E. Braunwald. 1970. The role of skin and muscle resistance vessels in reflexes mediated by the baroreceptor system. *The Journal of Clinical Investigation* 49:225–231.

Bell, C., W. Jänig, H. Kümmel, and H. Xu. 1985. Differentiation of vasodilator and sudomotor responses in the cat paw pad to preganglionic sympathetic stimulation. *Journal of Physiology* 364:93–104.

Bini, G., K.E. Hagbarth, P. Hynninen, and B.G. Wallin. 1980a. Regional similarities and differences in thermoregulatory vaso- and sudomotor tone. *Journal of Physiology* 306:553–565.

Bini, G., K.E. Hagbarth, P. Hynninen, and B.G. Wallin. 1980b. Thermoregulatory and rhythm-generating mechanisms governing the sudomotor and vasoconstrictor outflow in human cutaneous nerves. *Journal of Physiology* 306:537–552.

Bishop, V.S., A. Malliani, and P. Thorén. 1983. Cardiac mechanoreceptors. In *Handbook of physiology. The cardiovascular system. Peripheral circulation and organ blood flow*, Vol. III, Sect. 2, Part 2, eds. J.T. Shepherd, F.M. Abboud, and S.R. Geiger, 497–555. Bethesda, MD: American Physiological Society.

Blumberg, H., and B.G. Wallin. 1987. Direct evidence of neurally mediated vasodilatation in hairy skin of the human foot. *Journal of Physiology* 382:105–121.

Brengelmann, G.L., P.R. Freund, L.B. Rowell, J.E. Olerud, and K.K. Kraning. 1981. Absence of active cutaneous vasodilation associated with congenital absence of sweat glands in humans. *American Journal of Physiology* 240:H571–H575.

Edholm, O.G., R.H. Fox, and R.K. MacPherson. 1957. Vasomotor control of the cutaneous blood vessels in the human forearm. *Journal of Physiology* 139:455–465.

Elam, M., and B.G. Wallin. 1987. Skin blood flow responses to mental stress in man depend on body temperature. *Acta Physiologica Scandinavica* 129:429–431.

Freund, P.R., G.L. Brengelmann, L.B. Rowell, L. Engrav, and D.M. Heimbach. 1981. Vasomotor control in healed grafted skin in humans. *Journal of Applied Physiology* 51:168–171.

Gibbon, J.H., and E.M. Landis. 1932. Vasodilatation in the lower extremities in response to immersing the forearms in warm water. *Journal of Clinical Investigation* 11:1019–1036.

Hainsworth, R. 1991. Reflexes from the heart. *Physiological Reviews* 71:617–658.

Jacobsen, T.N., B.J. Morgan, U. Scherrer, S.F. Vissing, R.A. Lange, N. Johnson, W.S. Ring, P.S. Rahko, P. Hanson, and R.G. Victor. 1993. Relative contributions of cardiopulmonary and sinoaortic baroreflexes in causing sympathetic activation in the human skeletal muscle circulation during orthostatic stress. *Circulation Research* 73:367–378.

Johansson, J.E. 1895. Über der Einwirkung der Muskelhätigkeit auf die Athmung und die Herzthatigkeit. *Skandinavisches Archives für Physiologi* 5:20–66.

Johnson, J.M., G.L. Brengelmann, and L.B. Rowell. 1976. Interactions between local and reflex influences on human forearm skin blood flow. *Journal of Applied Physiology* 41:826–831.

Johnson, J.M., M. Niederberger, L.B. Rowell, M.M. Eisman, and G.L. Brengelmann. 1973. Competition between cutaneous vasodilator and vasoconstrictor reflexes in man. *Journal of Applied Physiology* 35:798–803.

Kaufman, M.P., J.C. Longhurst, K.J. Rybicki, J.H. Wallach, and J.H. Mitchell. 1983. Effects of static muscular contraction on impulse activity of groups III and IV afferents in cats. *Journal of Applied Physiology* 55:105–112.

Kellogg, D.L., J.M. Johnson, and W.A. Kosiba. 1990. Baroreflex control of the cutaneous active vasodilator system in humans. *Circulation Research* 66:1420–1426.

Kellogg, D.L., P.E. Pérgola, K.L. Pieast, W.A. Kosiba, C.G. Crandall, M. Grossmann, and J.M. Johnson. 1995. Cutaneous active vasodilation in humans is mediated by cholinergic nerve cotransmission. *Circulation Research* 77:122–128.

Kerslake, D., and K.E. Cooper. 1950. Vasodilatation in the hand in response to heating the skin elsewhere. *Clinical Science* 9:31–46.

Krogh, A., and J. Lindhard. 1913. The regulation of respiration and circulation during the initial stages of muscular work. *Journal of Physiology* 47:112–136.

Lundin, S., K. Kirnö, B.G. Wallin, and M. Elam. 1990. Effects of epidural anesthesia on sympathetic nerve discharge to the skin. *Acta Anaesthesiologica Scandinavica* 34:492–497.

Mancia, G., and A.L. Mark. 1983. Arterial baroreflexes in humans. In *Handbook of physiology, section 2: The cardiovascular system, peripheral circulation and organ blood flow*, Vol. III, Part 2, eds. J.T. Shepherd and F.M. Abboud, F.M., 755–785. Bethesda, MD: American Physiological Society.

Mark, A.L., and G. Mancia. 1983. Cardiopulmonary baroreflexes in humans. In *Handbook of physiology, section 2: The cardiovascular system, peripheral circulation and organ blood flow*, Vol. III, Part 2, eds. J.T. Shepherd and F.M. Abboud, 795–813. Bethesda, MD: American Physiological Society.

Mark, A.L., R.G. Victor, C. Nerhed, and B.G. Wallin. 1985. Microneurographic studies of the mechanisms of sympathetic nerve responses to static exercise in humans. *Circulation Research* 57:461–469.

McCloskey, D.I., and J.H. Mitchell. 1972. Reflex cardiovascular and respiratory responses originating in exercising muscle. *Journal of Physiology* 224:173–186.

Mitchell, J.H., and R.F. Schmidt. 1983. Cardiovascular reflex control by afferent fibers from skeletal muscle receptors. In *Handbook of physiology, section 2: The cardiovascular system, peripheral circulation and organ blood flow*, Part 2, Vol. III, eds. J.T. Shepherd and F.M. Abboud, 623–658. Bethesda, MD: American Physiological Society.

Mosley, J.G. 1969. A reduction in some vasodilator responses in free-standing man. *Cardiovascular Research* 3:14–21.

Nagasaka, T., K. Hirata, T. Mano, S. Iwase, and Y. Rosseti. 1990. Heat induced finger vasoconstriction controlled by skin sympathetic nerve activity. *Journal of Applied Physiology* 68:71–75.

Nilsson, H., B. Ljung, N. Sjöblom, and B.G. Wallin. 1985. The influence of the sympathetic impulse pattern on contractile responses of rat mesenteric arteries and veins. *Acta Physiologica Scandinavica* 123:303–309.

Normell, L.A., and B.G. Wallin. 1974. Sympathetic skin nerve activity and skin temperature changes in man. *Acta Physiologica Scandinavica* 91:417–426.

Oberle, J., M. Elam, T. Karlsson, and B.G. Wallin. 1988. Temperature dependent interaction between vasoconstrictor and vasodilator mechanisms in human skin. *Acta Physiologica Scandinavica* 132:459–469.

Pickering, G.W. 1933. The vasomotor regulation of heat loss from the human skin in relation to external temperature. *Heart* 16:115–135.

Proppe, D.W. 1981. Influence of skin temperature on central thermoregulatory control of leg blood flow. *Journal of Applied Physiology* 50:974–978.

Proppe, D.W., G.L. Brengelmann, and L.B. Rowell. 1976. Control of baboon limb blood flow and heart rate-role of skin vs. core temperature. *American Journal of Physiology* 231:1457–1465.

Rea, R.F., and B.G. Wallin. 1989. Sympathetic nerve activity in arm and leg muscles during lower body negative pressure in humans. *Journal of Applied Physiology* 66:2778–2781.

Rotto, D.M., C.L. Stebbins, and M.P. Kaufman. 1989. Reflex cardiovascular and ventilatory responses to increasing hydrogen ion activity in cat hindlimb muscle. *Journal of Applied Physiology* 67:256–263.

Rowell, L.B. 1981. Active neurogenic vasodilatation in man. In *Vasodilatation,* eds. P.M. Vanhoutte and L. Leusen, 1–17. New York: Raven Press.

Rowell, L.B. 1986. Nonthermoregulatory reflex control of skin blood flow. *Human circulation: Regulation during physical stress.* 178–183. New York: Oxford University Press.

Rowell, L.B. 1993. Reflex control during orthostasis. *Human cardiovascular control.* 51–56. New York: Oxford University Press.

Rowell, L.B., C.R. Wyss, and G.L. Brengelmann. 1973. Sustained human skin and muscle vasoconstriction with reduced baroreceptor activity. *Journal of Applied Physiology* 34:639–643.

Shepherd, J.T. 1963. Nervous control of the blood vessels in the skin. *Physiology of the circulation in human limbs in health and disease.* 9–41. Philadelphia: W.B. Saunders.

Snell, E.S. 1954. The relationship between the vasomotor response in the hand and heat changes in the body induced by intravenous infusions of hot or cold saline. *Journal of Physiology* 125:361–372.

Stürup, G.B., B. Bolton, D.J. Williams, and E.A. Carmochael. 1935. Vasomotor responses in hemiplegic patients. *Brain* 58:456–469.

Thorén, P. 1979. Role of cardiac vagal C-fibers in cardiovascular control. *Reviews of Physiology, Biochemistry and Experimental Pharmacology* 86:1–94.

Vallbo, A.B., K.-E. Hagbarth, H.E. Torebjork, and B.G. Wallin. 1979. Somatosensory, proprioceptive, and sympathetic activity in human peripheral nerves. *Physiological Reviews* 59:919–957.

Victor, R.G., D.M. Rotto, S.L. Pryor, and M.P. Kaufman. 1989. Stimulation of renal sympathetic activity by static contraction: Evidence for mechanoreceptor induced reflexes from skeletal muscle. *Circulation Research* 64:592–599.

Vissing, S.F., and E.M. Hjortsø. 1996. Central motor command activates sympathetic outflow to the cutaneous circulation in humans. *Journal of Physiology* 492:931–939.

Vissing, S.F., U. Scherrer, and R.G. Victor. 1989. Relation between sympathetic outflow and vascular resistance in the calf during perturbations in central venous pressure: Evidence for cardiopulmonary afferent regulation of calf vascular resistance in humans. *Circulation Research* 65:1710–1717.

Vissing, S.F., U. Scherrer, and R.G. Victor. 1991. Stimulation of skin sympathetic nerve discharge by central command: Differential control of sympathetic outflow to skin and skeletal muscle during static exercise. *Circulation Research* 69:228–238.

Vissing, S.F., U. Scherrer, and R.G. Victor. 1994. Increase of sympathetic discharge to skeletal muscle but not to skin during mild lower body negative pressure in humans. *Journal of Physiology* 481:233–241.

Vissing, S.F., N.H. Secher, and R.G. Victor. 1997. Mechanisms of cutaneous vasoconstriction during upright posture. *Acta Physiologica Scandinavica* 159:131–138.

Vissing, J., L.B. Wilson, J.H. Mitchell, and R.G. Victor. 1991. Static muscle contraction reflexly increases adrenal sympathetic nerve activity in rats. *American Journal of Physiology* 261:R1307–R1312.

Volkmann, A.W. 1841. Die Bewegungen des Athmens und Schluckens, mit besonderer Berücksichtigung neurologischer Streitragen. In *Archiv für Anatomie, Physiologie und Wissenschaftliche Medizin,* 332–360. Berlin, Germany: G. Eichler.

Wallin, B.G., D. Burke, and S.C. Gandevia. 1992. Coherence between the sympathetic drives to relaxed and contracting muscles of different limbs of human subjects. *Journal of Physiology* 455:219–233.

Wallin, B.G., R.G. Victor, and A.L. Mark. 1989. Sympathetic outflow to resting muscles during static handgrip and postcontraction muscle ischemia. *American Journal of Physiology* 256:H105–H110.

Wieling, W., and K.H. Wesseling. 1993. Importance of reflexes in the circulatory adjustments to postural changes. In *Cardiovascular reflex control in health and disease,* eds. R. Hainsworth and A.L. Mark. 35–43.

Wyss, C.R., G.L. Brengelmann, J.M. Johnson, L.B. Rowell, and M. Niederberger. 1974. Control of skin blood flow, sweating, and heart rate: Role of skin vs. core temperature. *Journal of Applied Physiology* 36:726–733.

Wyss, C.R., G.L. Brengelmann, J.M. Johnson, L.B. Rowell, and D. Silverstein. 1975. Altered control of skin blood flow at high skin and core temperatures. *Journal of Applied Physiology* 38:839–845.

Chapter 9

Cerebral Blood Flow and Metabolism

Lisbeth G. Jørgensen, Markus Nowak, Kojiro Ide, and Niels H. Secher

More than with any other organ, the results of an evaluation of the brain's responses to exercise are dependent on the evaluation method chosen. Organ blood flow is derived as the arterial input function, as the washout of, for example, a radioactive substance, or as an evaluation of the venous outflow. For the brain, the pioneering work was by assessment of the venous outflow (Kety-Schmidt method), which has demonstrated no increase in flow or oxygen and glucose uptakes (Madsen et al. 1993). In contrast, an evaluation of regional blood flow advanced to a tomographic representation demonstrates an increase in flow and glucose uptake corresponding to the cortical representation of the involved muscle groups (Lassen, Ingvar, and Skinhøj 1978). Also, the arterial evaluation of cerebral blood flow increases during exercise. The flow in the internal carotid artery increases, as does the transcranial Doppler-derived flow velocity in basal cerebral arteries. With the later method it is possible to demonstrate the regional distribution of flow—for example, an increase in the flow velocity of the left (but not in the right) middle cerebral artery during rhythmic handgrip of the right arm. In accordance with the regional blood flow, this increase in flow velocity is eliminated by regional anesthesia of the working arm, suggesting that it is dependent on integration of nerve signals from the arm (Jørgensen, Perko, Payne, et al. 1993). One problem with evaluation of cerebral blood flow by its venous outflow is that reported values represent only light to moderate exercise, where the increase in arterial flow (velocity) is moderate. Of larger significance may be that, in most subjects, the jugular vein of the two sides of the neck represents outflow from either the cortex or the basal area of the brain, and it is not known to what extent the values reported during exercise represent one or the other. To obtain a coherent description of the cerebral responses to exercise, an evaluation of the venous outflow needs to define what jugular vein is investigated, include intense work rates, and probably also include more metabolites than glucose.

The integrity of the body depends to such an extent on neural control that the importance of cerebral blood flow (CBF) is obvious not only for the nutrition of neurons but also for control of all body portions. CBF is recognized as being so crucial for the maintenance of an adequate supply of oxygen and glucose to brain tissue that the control of circulation is thought to be designed, ultimately, to protect CBF. This is the case both at rest and in a variety of circumstances, including exercise, where CBF has to compete with the perfusion of other capillary beds. In contrast to muscle cells, neurons are very sensitive to a lack of oxygen, and, possessing only a limited amount of glycogen, they critically depend on an adequate CBF. As a result, unconsciousness is manifest only a few seconds after the arterial supply to the brain is compromised.

A cornerstone in the physiology of CBF was the discovery by Mogens Fog as developed by Lassen (1964) that CBF is subject to "autoregulation"—that is, that CBF stays close to constant (~50 ml $\cdot$ 100 g^{-1} min^{-1}) over a wide range of arterial blood pressures. Only at a mean arterial pressure of more than approximately 150 mmHg is there a proportional increase in CBF (Paulson, Strandgaard, and Edvisson 1990). Conversely, CBF is reduced when the mean arterial pressure becomes lower than 60 to 80 mmHg. CBF autoregulation protects the brain against hyperperfusion, and in patients with arterial hypertension CBF is maintained at a normal level. In situations where autoregulation is lost, cerebral edema and, in turn, ischemia, develops, with imminent death (Larsen et al. 1995).

Another important feature of CBF is its susceptibility to changes in the arterial carbon dioxide tension (its $PaCO_2$ reactivity). CBF increases 3–4% per mmHg elevation in $PaCO_2$ (Lassen 1959). Besides the "global" influences of blood pressure and $PaCO_2$ on CBF, "local" metabolic regulation of regional perfusion (rCBF) is exhibited as flow is coupled specifically to discrete regions of the brain activated during, for example, motor tasks or visual stimulation (Lassen, Ingvar, Skinhøj 1978).

Although these three regulatory mechanisms (CBF autoregulation, the CO_2 response, and the local metabolic regulation) form the basis for control of CBF, they do not fully account for CBF under all circumstances. For example, it takes time for CBF autoregulation to become manifest, and, therefore, changes of blood pressure may be too rapid to allow the cerebral vessels to adapt. It is also likely that the basis for regulation of CBF depends on the premise that there is a sufficient cardiac output available to provide the brain with blood. In extreme situations such as hypovolemic shock and fainting, CBF becomes reduced (Jørgensen, Perko, Perko, et al. 1993), resulting in a decrease in cerebral oxygenation (Madsen et al. 1998), while the perfusion to skeletal muscles is enhanced (Barcroft et al. 1944) with an associated increase in their oxygenation (Madsen, Lyck, et al. 1995). Thus, in contrast to the common belief that the brain has regulatory priority for flow, during hypovolemic shock the circulation provides skeletal muscle with a blood flow at the expense of perfusion of the brain!

In association with motor activation, neuronal activity in the brain is considered an important regulatory mechanism for the cardiovascular response to exercise. Attention is now focused on the localization and regulatory function of "central command" by more direct methods designed to assess the activation, metabolism, and blood flow regulation of regions of the brain important for autonomic function during exercise.

Cerebral Blood Flow During Exercise

The original method for evaluation of CBF in humans was developed by Kety and Schmidt (1948). Kety and Schmidt made the subject inhale nitrous oxide (N_2O) and determined the arterial to jugular venous difference for N_2O [(a-v) diff] while the brain is loaded with N_2O. More recently, the method has been modified as N_2O is replaced by ^{133}Xe and the clearance of ^{133}Xe from the brain is determined. With this modification, the faster decrease in the arterial concentration than in the venous concentration after inhalation of ^{133}Xe is discontinued allows for the calculation of blood flow (Madsen et al. 1993; Madsen, Hasselbach, et al. 1995).

Despite the nonspecific description of the exercise protocols used for determination of CBF in humans, the general finding from both the modified and the original Kety-Schmidt methods is that CBF does not change significantly during exercise (Jørgensen 1995). Whether this finding accurately describes the overall flow response to the brain is

uncertain, as there are three reports on carotid artery blood flow during exercise using a duplex-scanning Doppler method in which a rather uniform 20% increase from rest is observed (Huang et al. 1991; Huang et al. 1992; Hellström, Fischer-Colbrie, et al. 1996). A physiological basis for these deviating blood flow patterns during exercise may in part be explained by the tight control of CBF by $PaCO_2$. During low-intensity exercise, $PaCO_2$ initially increases. As exercise continues and work rate increases, $PaCO_2$ decreases to resting levels and often continues to decrease below resting values as ventilation progressively increases during exhaustive exercise. With such changes in $PaCO_2$, it becomes difficult to evaluate if CBF is influenced by $PaCO_2$ alone or if there is an additional effect on blood flow by the neuronal activation associated with exercise. The separation of these two effects on CBF is difficult especially because it is not known if the resting CO_2 reactivity is applicable to the exercise condition.

An important methodological concern with the Kety-Schmidt technique for evaluation of CBF is whether the jugular venous blood is a representative mixture of blood from the whole brain or if it represents specific regions. In humans, the two jugular veins are very different, in that one jugular vein is much larger than the other and the two veins drain different parts of the brain. The larger (most often the right) jugular vein drains blood from the central sinus and thereby from the cortical structures on both sides of the brain, while the smaller (most often the left) jugular vein drains blood from deeper brain structures (figure 9.1). From the two parts of the brain, the jugular venous blood is not the same. As an example, in hypertensive patients, "norepinephrine spillover" is elevated corresponding to subcortical brain regions (Ferrier et al. 1993) and equally related to (muscle) sympathetic nerve activity in normal humans (Lambert et al. 1997). Thus, the Kety-Schmidt method is influenced by which jugular vein is cannulated, and it requires ideally an independent mapping of the brain structures from which blood is drained into the two veins. However, such data are not available for the evaluation of CBF during exercise.

Regional Cerebral Blood Flow

In contrast to techniques designed to measure flow to the brain as a whole, methods have been developed to evaluate changes in flow to the area associated with the neuronal integration of specific tasks.

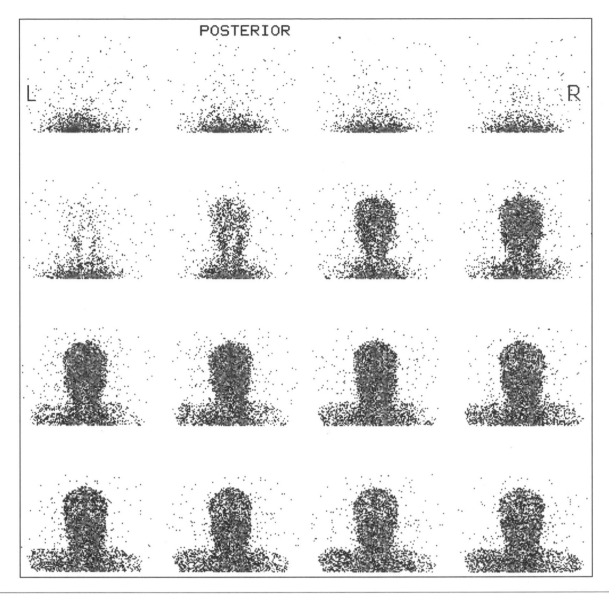

Figure 9.1   Posterior view of the head during the transient of a bolus of technetium labeled human albumin. The predominant flow from the central sinus is to the right jugular vein.

In fact, in humans, most mapping of neuronal activation is based on changes in rCBF. Several methods are available for a determination of rCBF during cerebral activation and, in general, they convey the same information. The original work investigated the clearance of Kr or ^{133}Xe (Lassen, Ingvar, and Skinhøj 1978), measured after intra-arterial injection. More recent methods include SPECT (single-photon emission computerized tomography), PET (positron emission tomography), near-infrared spectroscopy (NIRS), and functional MRI ([nuclear] magnetic resonance imaging; fMRI).

With the involvement of skeletal muscles, rCBF increases corresponding to the sensorimotor cortex and when the contractions include movement also corresponding to the supplementary motor area (figure 9.2). In consequence, the supplementary motor area is assumed to represent planning of more complicated movement (Orogozo and Larsen 1979). During exercise there is also activation corresponding to the cerebellum, where the first neurophysiological evidence of motor activity is recorded (Evarts 1973). Research has also been directed to define cerebral activation related to the changes in circulatory variables during exercise. In this respect, the insula is described as a "motor cortex" for autonomic function (Cechetto and Saper 1990), and evidence is accumulating that rCBF increases corresponding to the activity in the insula when the motor task becomes difficult—that is, with a contribution from the "central command" (Williamson et al. 1997) (see also figure 9.3).

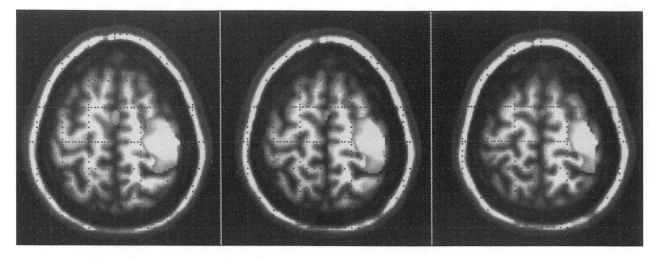

Figure 9.2   PET-determined changes in regional cerebral blood flow during rhythmic handgrip with the left arm. Values are superimposed on an MR representation of a standard brain (Talairach and Tournoux 1988) corresponding to 6 cm above the AC-PC (anterior commissure-posterior commissure) line. Activation is noted corresponding to the right sensorimotor cortex and to the supplementary motor area.

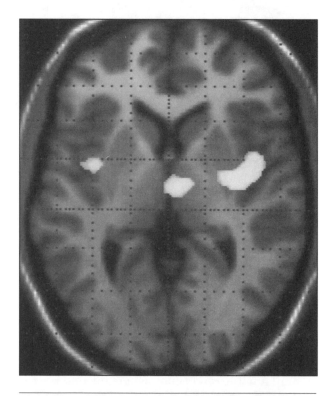

Figure 9.3   fMRI-determined regional changes in oxygenation (BOLD effect) during rhythmic handgrip with the left arm. Activation is noted corresponding to the right sensorimotor cortex and to the supplementary motor area. The recorded activation is lateral to the recording made by PET (figure 9.2).

Of the methods available for evaluation of rCBF during exercise, fMRI provides information of special interest. This technique evaluates the different magnetic properties of hemoglobin and oxyhemo-

globin (BOLD effect), and it is consistently found that oxygenation increases corresponding to the activated brain areas (Ogawa et al. 1993) (figure 9.4). From figure 9.4 it is seen that with fMRI the detected activation is obtained from a position on the outer margin of the brain, which is a very different region than where the activation is seen with other techniques such as PET (figure 9.2). This difference is due to the fact that fMRI is more sensitive to the signal changes in draining veins. The conclusion is that within the brain, the increase in flow is larger than that of oxygen utilization, and this has been confirmed by NIRS (Villringer et al. 1993; Obrig et al. 1996). Using PET, it is also found that in the activated areas of the brain, the uptake of oxygen is small compared to the increase in flow (Fox and Raichle 1986).

Transcranial Doppler

While most methods for evaluation of CBF require extensive apparatus and often involve the use of radioactive substances, transcranial Doppler (TCD) can be applied in almost any circumstance, including vigorous exercise such as rowing (Pott, Knudsen, et al. 1997). Along with the fact that TCD assessment of cerebral perfusion may be performed during almost any human endeavor, it is an advantage of TCD that it has a high temporal resolution (100 Hz), allowing monitoring of blood velocity within one heart beat. The evaluation of cerebral perfusion by TCD is based on a determination of flow velocity in basal cerebral arteries, where it is possible to follow

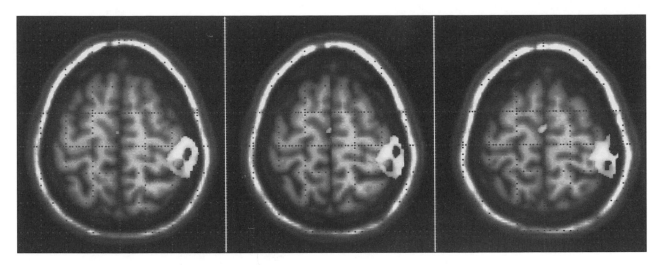

Figure 9.4 PET-determined changes in regional cerebral blood flow during rhythmic handgrip with the left arm (same experiment as figure 9.2). Values are superimposed on an MR representation of a standard brain (Talairach and Tournoux 1988) corresponding to 2 mm above the AC-PC (anterior commissure-posterior commissure) line. Activation is noted corresponding to the insula on both sides of the brain.

flow velocity responses to posture or the sudden change in blood pressure at the onset and cessation of exercise.

A limitation of a TCD evaluation of cerebral perfusion is that flow *velocity* rather than flow *volume* is recorded since the diameter of the insonated vessel is not known. The extent to which changes in the mean flow velocity (V_{mean}) of the middle cerebral artery (MCA) reflect flow can therefore be assessed only indirectly, taking into account contributions from blood pressure and $PaCO_2$ and by comparison to methods providing for an estimate of CBF.

The increase in MCA V_{mean} during exercise has been validated with CBF derived by external detection of ^{133}Xe clearance from the brain (Jørgensen, Perko, and Secher 1992; Jørgensen, Perko, Hanel, et al. 1992) and with flow in the carotid artery (Hellström, Fischer-Colbrie, et al. 1996). During exercise, the increase in MCA V_{mean} is of the same order of magnitude as that recorded in the internal carotid artery serving the brain, but it is unrelated to flow in the external carotid artery providing blood to other tissues of the head. Also, with dynamic exercise, the increase in MCA V_{mean} corresponds to the magnitude of brain blood flow to gray and white matter combined as determined by the [133]Xe "initial slope index" method. When flow specifically to the cortex is considered with the use of the "F1" index of CBF, the increase with exercise is about two times as large (see Jørgensen, Perko, and Secher 1992; Jørgensen, Perko, Hanel, et al. 1992).

During exercise, an influence of blood pressure on V_{mean} can be estimated by comparison of the response to static and dynamic exercise and to

postexercise muscle ischemia (Jørgensen, Perko, Hanel, et al. 1992). With a similar elevation in blood pressure, an increase in V_{mean} takes place only during dynamic exercise associated with the increase in CBF (Thomas et al. 1989; Rogers et al. 1990) and also the largest increase in rCBF (Orogozo and Larsen 1979; Friedman et al. 1991; Friedman et al. 1992). Another attempt to evaluate the validity of V_{mean} for cerebral perfusion involved estimating the sympathetic nerve activity that might constrict the artery and thereby increase V_{mean} at a given volume flow. It was found that MCA V_{mean} stays constant despite the increase in (muscle) sympathetic nerve activity associated with postexercise muscle ischemia and consequent peripheral vasoconstriction (Pott, Ray, et al. 1997). In addition, a moderate elevation of plasma catecholamines during exercise appears to be of no consequence on MCA V_{mean}. In support of this conclusion, during handgrip exercise, the increase in MCA V_{mean} takes place with no elevation in plasma catecholamines (Pott et al. 1996). Only during maximal exercise in elite bicyclists with an elevation in plasma epinephrine and norepinephrine to 4 and 13 nmol 1^{-1}, respectively, is there an increase in MCA V_{mean} of about 50% rather than the normal increase of 25–30% (Pott et al. 1996).

With the changes in $PaCO_2$ during exercise, it is critical whether the CO_2 reactivity is maintained. MCA V_{mean} tends to decrease during continued exercise as ventilation increases (Jørgensen, Perko, Hanel, et al. 1992), and a "correction" by approximately 3% mm Hg^{-1} makes it stable (Linkis et al. 1995). It is therefore likely that the resting value can be applied also during exercise, but as the intersubject variability is large, it is preferable to apply an individually determined value. V_{mean} is calculated from the maximal Doppler shift, and it is assumed to record flow velocity in the center of the vessel. It would be preferable to measure the integrated Doppler shift as the average flow velocity for the (whole) vessel. In situations where the head can remain motionless, as is the case during release of an occluding cuff around a leg, the changes in the two expressions of flow velocity follow each other (Aaslid et al. 1989). However, it has been reported that during cycling the increase in the integrated signal is smaller (4%) than that in V_{mean} (14%) (Poulin, Syed, and Robbins 1999). With the head movement so prevalent during cycling, it may well be that, from rest to exercise, part of the vessel is "lost," for insonation or the angle of insonation changes so that the derived change in velocity becomes an artifact. Newer duplex scanning methods allow visualization of intracranial vessels and may therefore help to maintain visualization of the whole artery under investigation during exercise, as is possible for evaluation of flow in peripheral vessels.

From the recording of MCA V_{mean}, it is known that it takes 2–3 seconds for cerebral autoregulation to be established (Aaslid et al. 1989). During exercise, the delayed response of the cerebral vessels to a sudden change in blood pressure has implications because blood pressure often fluctuates rapidly. Large and rapid changes in blood pressure are noted during weight lifting and rowing where there is an increase in blood pressure at the onset of each muscle contraction (Clifford, Hanel, and Secher 1992). During rowing with a stroke rate of 35 min⁻¹, MCA V_{mean} fluctuates almost in parallel with blood pressure (figure 9.5; Pott, Knudsen, et al. 1997).

The regional distribution of flow is limited by the artery that supplies blood to a given part of the cortex, and TCD reflects rCBF to some extent. For example, the increase in flow velocity is confined to MCA V_{mean} during hand movement but to the anterior cerebral artery during movement of one foot (Linkis et al. 1995). Also, with the use of the right hand, MCA V_{mean} increases exclusively on the left side of the brain and vice versa for the use of the left hand (Jørgensen, Perko, Perko, et al. 1993).

With TCD it is possible to obtain some insight into the central neuronal integration of skeletal muscle control, and the conclusions obtained by TCD agree with those derived from a determination of CBF by intra-arterial injection of ¹³³Xe (Orogozo and Larsen 1979) and SPECT during hand contractions (Friedman et al. 1991, 1992). During normal hand contractions, MCA V_{mean} increases; but this is not the case during attempted handgrip after regional anesthesia of the arm, when not only the forearm muscles

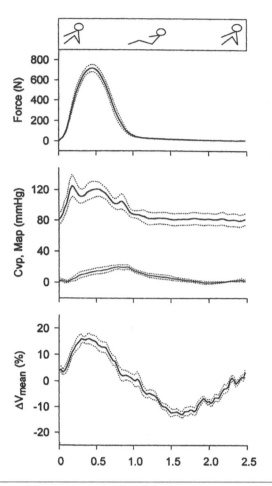

Figure 9.5   Force on the "oar" during ergometer rowing (top panel) together with mean arterial pressure (MAP) and central venous pressure (CVP) (middle panel), and changes in mean flow velocity (V_{mean}) in the middle cerebral artery (lower panel). Values are mean and standard error for more than 1000 strokes in seven subjects. (Pott, Knudsen, et al. 1997. Used by permission.)

are weak but also the afferent input from the forearm is attenuated or eliminated by the block of the nerves at the level of the shoulder (Jørgensen, Perko, Payne, et al. 1993). In other words, during attempted handgrip, in the sensorimotor cortex the neuronal activation is too small to increase rCBF and, in turn, MCA V_{mean}. The dominant neuronal activation would seem to be related to the central integration of afferent input from the arm rather than to the "command" signal(s). Only with the use of PET is it possible to detect what could be "command"-related activation in the motor cortex—that is, an increase in rCBF during attempted handgrip with regional anesthesia of the arm (Nowak et al. 1999).

To what extent such afferent input to the brain comes from the muscles, the tendons, or the joints is not known. Each of these components may contribute to the increase in rCBF. There is only a small increase in cerebral perfusion when tendons and joints are stimulated with "no load" exercise (Jørgensen, Perko, and Secher 1992), and selective stimulation of the muscle afferents during postexercise muscle ischemia is of little consequence for cerebral perfusion (Jørgensen, Perko, and Secher 1992; Williamson et al. 1996). Also, static exercise is of no consequence for CBF (Rogers et al. 1990; Jørgensen, Perko, Hanel, et al. 1992). However, the presumably combined influences of muscle, tendons, and joints during normal cycling are associated with an increase in cerebral perfusion that is linked to the exercise intensity (Jørgensen, Perko, Hanel, et al. 1992).

An interesting, but unsettled, question is whether CBF is described fully by the balance between arterial pressure, $PaCO_2$, and changes in (regional) metabolism associated with cerebral activation. With TCD, there is evidence for an additional effect of cardiac output in situations where there may be a competition between the perfusion of different organs. An effect of cardiac output on MCA V_{mean} seems to explain observations in patients with orthostatic intolerance who demonstrate a normal CO_2 reactivity when supine that apparently is lost during standing where their cardiac output is significantly reduced (Harms et al. 1998). The first report of an influence of cardiac output during exercise was on the recording of MCA V_{mean} in patients with chronic heart failure (Hellström, Magnusson, et al. 1996). These patients demonstrated no significant increase in MCA V_{mean} during exercise with one leg, and MCA V_{mean} decreased when exercise was performed with both legs. This response was associated with a lower cardiac output compared to control subjects. Similar observations are available for patients with atrial fibrillation in whom cardiac output fails to increase normally during exercise (figure 9.6).

A possible effect of cardiac output on cerebral perfusion during exercise can be evaluated experimentally. After the administration of a β-1 selective blocking agent during cycling, the increase in cardiac output is reduced, as is muscle blood flow (Pawelczyk et al. 1992) and the increase in MCA V_{mean} (from 22% to 12%; Ide et al. 1998). In contrast, when the need for an increase in cardiac output is small, as during handgrip exercise, the increase in V_{mean} is unaffected by beta blockade. In these studies, however, the direct influence of cardiac output is difficult to ascertain, although the attenuation of the MCA V_{mean} response to exercise was established with no change in $PaCO_2$. Differentiation of the independent influences of cardiac output and $PaCO_2$ on CBF during exercise may require more sophisticated methods for CBF quantification applied in conjunction with studies controlling the level of cardiac output.

Near-Infrared Spectroscopy (NIRS)

With the use of near-infrared (NIR) light, it is possible to penetrate the scalp to a depth sufficient to assess the content of hemoglobin and oxyhemoglobin in the brain (Madsen et al. 1998). To verify that the signal depends critically on perfusion of the brain as opposed to extracranial tissue, it is demonstrated that the cerebral oxygenation determined by NIRS increases as $PaCO_2$ increases, and conversely that it decreases when subjects hyperventilate (Madsen et al. 1998). With a distance between the optodes of approximately 5 cm, in the adult it is possible to detect only regional changes in presumably cortical oxygenation during cerebral activation such as visual stimulation or motor tasks. During exercise, MCA V_{mean} may, as an example, increase from 60 to 68 cm s^{-1}, and an increase in CBF is supported by an elevated cerebral oxyhemoglobin concentration of 25 μmol l^{-1}.

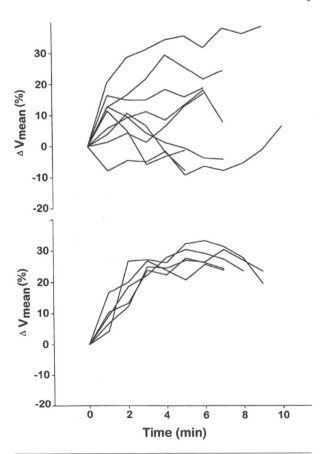

Figure 9.6 Middle cerebral artery mean blood velocity (ΔV_{mean}) during cycling. Upper panel, patients with atrial fibrillation; lower panel, control subjects. (From Ide et al., 1999. Used by permission.)

An interesting observation made by NIRS is on the cerebral response to rowing (Nielsen et al. 1999). During rowing, the arterial oxygen tension and saturation often decrease markedly (e.g., to 81 mm Hg and 93%, respectively), and the arterial desaturation becomes manifest as a reduction in the NIRS-determined cerebral oxygenation (figure 9.7). This reduction in cerebral oxygenation is so prominent that it corresponds to the reduction seen during fainting (Madsen, Lyck, et al. 1995, Madsen et al. 1998). As during rowing, the arterial deoxygenation is eliminated following the administration of 30% oxygen, the cerebral oxygenation remains stable, and the subject is able to increase work capacity (over 6 minutes) by 2.5%. This occurs even though the similarly NIRS-determined muscle oxygenation remains unaffected by the administration of oxygen. The effect of oxygen on work capacity may accordingly be attributed to prevention of "central fatigue."

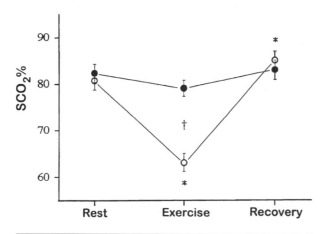

Figure 9.7 Cerebral oxygenation as determined by near-infrared spectroscopy during rowing in nine subjects breathing atmospheric air and air with an oxygen fraction of 30%. Values are mean ± SE. (Nielsen et al. 1999.)

Cerebral Metabolism

The traditional view is that not only CBF but also the energy turnover rate of the brain remains stable during exercise. The finding by Madsen and colleagues (1993) was that both CBF and the cerebral oxygen uptake were unaffected by exercise. However, recent data indicate that the cerebral metabolic rate of oxygen alone is inadequate for an evaluation of the metabolic response of the brain to activation. During visual stimulation, Fox and colleagues (1988) found that the regional ratio between oxygen and glucose uptake decreases. This ratio would be ex-

pected to be around 6.0 as glucose is oxidized to CO_2 and water, but from the calculations by Fox and colleagues (1988) the ratio was only 4.0 at rest. More importantly, during cerebral activation the ratio decreased to a value as low as 0.1. For the brain as a whole, Madsen, Lyck, and colleagues (1995) have provided data more similar to the expected values based on the (a-v) diff. At rest the ratio was 5.9; it decreased to 5.4 during visual stimulation and, importantly, stayed low even during the recovery.

During exercise, a similar phenomenon of a decrease in the molar ratio between oxygen and carbohydrate uptake by the brain may be observed, and with a marked elevation in blood lactate, the (a-v) diff for lactate becomes important. There is little increase in cerebral glucose uptake during low-intensity exercise (figure 9.8). During maximal exercise, the (a-v) diff for glucose increases, and it becomes largest in the first few minutes of recovery

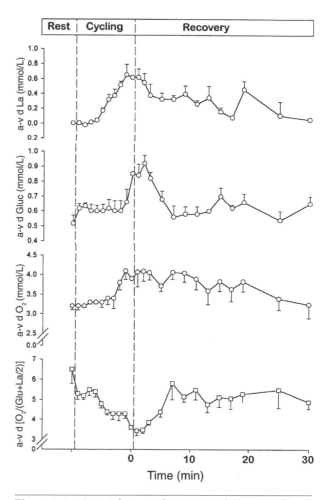

Figure 9.8 Arterial to jugular venous differences [(a-v) diff] for lactate (La), glucose, and oxygen (O_2) at rest, during incremental dynamic exercise, and recovery. Also shown is the molar ratio between the brain uptake of O_2 and carbohydrate. Values are mean ± SE for 6 subjects. (Ide et al., 1998. Used with permission.)

from exercise. There appears to be no increase in the uptake of lactate during low-intensity exercise; but as blood lactate levels increase and pH decreases, lactate extraction by the brain is elevated and becomes maximal during recovery. With respect to oxygen, the pattern is similar to that of glucose. Although there is an increase in the brain oxygen uptake during maximal exercise and in the recovery, it is not large enough to correspond to the uptake of carbohydrate (glucose + lactate/2). The ratio of oxygen to carbohydrate decreases from a resting value close to 6.0 to about 4.0 during exercise, with a further decrease to about 3.0 during early recovery.

A decrease in the ratio between oxygen and carbohydrate uptake by the brain might be thought to reflect an anaerobic metabolic capacity. Before such a conclusion is made, it should be remembered that the metabolic ratio would also decrease in response to the likely assumption that metabolic intermediates are accumulated during neuronal activation and enhanced cell metabolism. Yet, such a buildup of intermediates would be broken down as soon as the neuronal activation decreases or stops. Then the ratio between oxygen and carbohydrate uptake would increase; though in fact the ratio decreases even more during early recovery from exercise (figure 9.8). It is tempting to base an interpretation on two rather unorthodox hypotheses. The first one is that the brain does possess an anaerobic potential; the second is that the brain uses glycogen during neuronal activation. This would be similar to the well-known physiological principle in skeletal muscle. If glycogen is replenished immediately after exercise, it would explain that the ratio decreases to a very low level during recovery. For the brain, a glycogen level of about 4 mmol · kg^{-1} has been reported (Norberg, Quistorff, and Sjesji 1975), compared with 300 mmol · kg^{-1} for skeletal muscle.

Conclusion

Neuronal activation associated with exercise is accompanied by an increase in rCBF corresponding to the sensorimotor cortex, and it is likely to be dominated by integration of signals from the working limbs rather than by the "will" to perform the contractions. Also, activation in the supplementary motor area is present when complex movements are planned or performed, and activation in the cerebellum may be detected, reflecting initiation of motor function. Activation in the insula is established when exercise is intense; such activation may represent the so-called central command influence on the autonomic nervous system. Whether or not an increase in rCBF is large enough to manifest itself as an increase in flow for the whole brain is controversial. This may be due to the difficulty of accounting for the often-marked changes in PaCO$_2$ during exercise and the uncertainty about which jugular vein the blood should be obtained from. Also, the independent influence of the level of cardiac output on CBF is unclear in cases where there is only a limited capacity to increase cardiac output. Newer aspects of brain function indicate that rCBF increases out of proportion to the metabolic demand ("focal uncoupling"), and it is followed by a regional increase in the concentration of oxyhemoglobin. Although there appears to be oxygen available in the activated areas of the brain, it also seems that cerebral activation is associated with a larger uptake of carbohydrate (including lactate) than of oxygen, and the comparatively small increase in oxygen metabolism appears to be prevalent especially in the immediate recovery from exercise. These observations support the hypotheses that although the brain has but a small anaerobic potential, the neurons use glycogen during their activation. However, more direct assessments of these hypotheses are needed before the data can be interpreted in detail.

References

Aaslid, R., K.-E. Lindegaard, W. Sorteberg, and H. Nornes. 1989. Cerebral autoregulation dynamics in humans. *Stroke* 20:45–52.

Barcroft, H., O.G. Edholm, J. McMichael, and E.P. Sharpey-Schafer. 1944. Posthaemorrhage fainting. Study by cardiac output and forearm flow. *Lancet* 1:489–91.

Cechetto, D.F., and C.B. Saper. 1990. Role of the cerebral cortex in autonomic function. In *Central regulation of autonomic functions*, ed. A.D. Loewy and K.M. Spyer, 208–23. London: Oxford University Press.

Clifford, P.S., B. Hanel, and N.H. Secher. 1992. Arterial blood pressure response to rowing. *Medicine & Science in Sports & Exercise* 26:715–19.

Evarts, E.V. 1973. Brain mechanisms in movement. *Scientific American* 229:96–103.

Ferrier, C., G.L. Jennings, G. Eisenhofer, G. Lambert, H.S. Cox, V. Kalff, M. Kelly, and M.D. Esler. 1993. Evidence for increased noradrenaline release from subcortical brain regions in essential hypertension. *Journal of Hypertension* 11:1217–27.

Fox, P.T., and M.E. Raichle. 1986. Focal physiological uncoupling of cerebral blood flow and oxidative metabolism during somatosensory stimulation in human subjects. *Proceedings of the National Academy of Science* 83:1140–44.

Fox, P.T., M.E. Raichle, M.A. Mintun, and C. Dence. 1988. Nonoxidative glucose consumption during focal neural activity. *Science* 241:462–64.

Friedman, D.B., L. Friberg, J.H. Mitchell, and N.H. Secher. 1991. Effect of axillary blockade on regional cerebral blood flow during static handgrip. *Journal of Applied Physiology* 71:651–56.

Friedman, D.B., L. Friberg, G. Payne, J.H. Mitchell, and N.H. Secher. 1992. Effects of axillary blockade on regional cerebral blood flow during dynamic hand contractions. *Journal of Applied Physiology* 73:2120–25.

Harms, M., J.W.M. Lenders, W. Wieling, N.H. Secher, and J.J. van Lieshout. 1998. Arterial pressure level modifies cerebral blood velocity response to step changes in end-tidal CO_2 in patients with sympathetic failure. (abstract). *Nordiaska CBF Mötet* (Höör, Sweden) 13:e.

Hellström, G., W. Fischer-Colbrie, N.G. Wahlgren, and T. Jorgenstrand. 1996. Carotid artery flow and middle cerebral artery blood flow velocity during physical exercise. *Journal of Applied Physiology* 81:413–18.

Hellström, G., B. Magnusson, N.G. Wahlgren, A. Gordon, A. Sylvén, and B. Saltin. 1996. Physical exercise may impair cerebral perfusion in patients with chronic heart failure. *Cardiology in the Elderly* 4:191–94.

Huang, S.Y., S. Sun, T. Droma, J. Zhuang, J.X. Tao, R.G. McCullough, R.E. McCullough, A.J. Micco, J.T. Reeves, and L.G. Moore. 1992. Internal carotid arterial flow velocity during exercise in Tibetan and residents of Lhasa (3658 m). *Journal of Applied Physiology* 73:2638–42.

Huang, S.Y., K.W. Tawney, P.R. Bender, B.M. Groves, R.E. McCullough, R.G. McCullough, A.J. Micco, M. Manco-Johnson, A. Cymerman, E.R. Greene, and J.T. Reeves. 1991. Internal carotid flow velocity with exercise before and after acclimatization to 4300 m. *Journal of Applied Physiology* 71:1469–76.

Ide, K., F. Pott, J.J. van Lieshout, and N.H. Secher. 1998. Middle cerebral artery blood velocity depends on cardiac output during exercise with a large muscle mass. *Acta Physiologica Scandinavica* 162:13–20.

Ide, K., A.L. Gulløv, F. Pott, J.J. van Lieshout, and B.G. Koefoed, P. Pedersen, and N.H. Secher. 1999. Middle cerebral artery blood velocity during exercise in patients with atrial fibrillation. *Clin. Physiol* 19 (In Press).

Jørgensen, L.G. 1995. Transcranial Doppler ultra-

sound for cerebral perfusion. *Acta Physiolgica Scandinavica* 154: suppl. 154.

Jørgensen, L.G., G. Perko, G. Payne, and N.H. Secher. 1993. Effect of limb anesthesia on middle cerebral artery response to handgrip. *American Journal of Physiology* 264:H553–59.

Jørgensen, L.G., G. Perko, and N.H. Secher. 1992. Regional cerebral artery mean flow velocity and blood flow during dynamic exercise in humans. *Journal of Applied Physiology* 73:1825–30.

Jørgensen, L.G., M. Perko, B. Hanel, T.V. Schroeder, and N.H. Secher. 1992. Middle cerebral artery flow velocity and blood flow during dynamic exercise and muscle ischemia in humans. *Journal of Applied Physiology* 72:1123–32.

Jørgensen, L.G., M. Perko, G. Perko, and N.H. Secher. 1993. Middle cerebral artery velocity during head-up tilt induced hypovolaemic shock in humans. *Clinical Physiology* 13:323–36.

Kety, S.S., and C.F. Schmidt. 1948. The nitrous oxide method for the quantificative determination of cerebral blood flow in man: Theory, procedure and normal values. *Journal of Clinical Investigation* 27:476–83.

Lambert, G.W., J.M. Thompson, A.G. Turner, H.S. Cox, D. Wilkinson, M. Vaz, V. Kalff, M.J. Kelly, G.L. Jennings, and M.D. Esler. 1997. Cerebral noradrenaline spillover and its relation to muscle sympathetic nervous activity in healthy human subjects. *Journal of the Autonomic Nervous System* 64:57–64.

Larsen, F.S., F. Pott, B.A. Hansen, E. Ejlersen, G.M. Knudsen, J.D. Clemmesen, and N.H. Secher. 1995. Transcranial Doppler sonography may predict brain death in patients with fulminant hepatic failure. *Transplantation Proceedings* 27:3510–11.

Lassen, N.A. 1959. Cerebral blood flow and oxygen consumption in man. *Physiological Reviews* 39:183–238.

Lassen, N.A. 1964. Autoregulation of cerebral blood flow. *Circulation Research* 15 (suppl 1):201–4.

Lassen, N.A., D.H. Ingvar, and E. Skinhøj. 1978. Brain function and blood flow. *Scientific American* 239:62–71.

Linkis, P., L.G. Jørgensen, H.L. Olesen, P.L. Madsen, N.A. Lassen, and N.H. Secher. 1995. Dynamic exercise enhances regional cerebral artery mean flow velocity. *Journal of Applied Physiology* 78:709–716.

Madsen, P., F. Lyck, M. Pedersen, H.L. Olesen, H.B. Nielsen, and N.H. Secher. 1995. Brain and muscle oxygenation saturation during head-up tilt-induced central hypovolaemia in humans. *Clinical Physiology* 15:523–33.

Madsen, P., F. Pott, S.B. Olsen, H.B. Nielsen, I. Burcev, and N.H. Secher. 1998. Near-infrared spectrophotometry determined brain oxygenation during fainting. *Acta Physiologica Scandinavica* 162:501–7.

Madsen, P.L., S.G. Hasselbach, L.P. Hagemann, K.S. Olsen, J. Bülow, S. Holm, G. Wildschiødtz, O.B. Poulson, and N.A. Lassen. 1995. Persistent resetting of the cerebral oxygen/glucose uptake ratio by brain activation: Evidence obtained with the Kety-Schmidt technique. *Journal of Cerebral Blood Flow and Metabolism* 15:485–91.

Madsen, P.L., B.K. Sperling, T. Warming, J.F. Schmidt, N.H. Secher, G. Wildschiødtz, S. Holm, and N.A. Lassen. 1993. Middle cerebral artery blood velocity and cerebral blood flow and O_2 uptake during dynamic exercise. *Journal of Applied Physiology* 74:245–50.

Nielsen, H.B., R. Boushel, P. Madsen, and N.H. Secher. 1999. Cerebral desaturation during exercise reversed by O_2 supplementation. *Am. J. Physiol.* (In Press).

Norberg, K., B. Quistorff, and B.K. Sjesji. 1975. Effects of hypoxia of 10–45 seconds on energy metabolism in the cerebral cortex of unanaesthetized and anaesthetized rats. *Acta Physiologica Scandinavica* 95:301–10.

Nowak, M., K.S. Olsen, I. Law, S. Holm, O.B. Paulson, and N.H. Secher. 1999. Command-related distribution of regional cerebral blood flow during attempted handgrip. *Journal of Applied Physiology* 86(3):819–824.

Obrig, H., C. Hirth, J.G. Junge-Hülsing, C. Döge, T. Wolf, U. Dirnagl, and A. Villringer. 1996. Cerebral oxygenation changes in response to motor stimulation. *Journal of Applied Physiology* 81:1174–83.

Ogawa, S., R. S. Menon, D. W. Tank, S. G. Kim, H. Merkle, J. M. Ellermann, and K. Ugurbil. 1993. Functional brain mapping by blood oxygenation level-dependent contrast magnetic resonance imaging. A comparison of signal characteristics with a biophysical model. *Biophysical Journal* 64:803–12.

Orogozo, J.M., and B. Larsen. 1979. Activation of the supplementary motor area during voluntary movement in man suggests it works as a supramotor area. *Science* 206:847–50.

Paulson, O.B., S. Strandgaard, and L. Edvisson. 1990. Cerebral autoregulation. *Cerebrovascular Brain Metabolism Reviews* 2:161–92.

Pawelczyk, J.A., B. Hanel, R.A. Pawelczyk, J. Warberg, and N.H. Secher. 1992. Leg vasoconstriction with reduced cardiac output. *Journal of Applied Physiology* 73:1838–46.

Pott, F., K. Jensen, H. Hansen, N.J. Christensen, N.A. Lassen, and N.H. Secher. 1996. Middle cerebral artery blood velocity and plasma catecholamines during exercise. *Acta Physiologica Scandinavica* 158:349–56.

Pott, F., L. Knudsen, M. Nowak, H.B. Nielsen, B. Hanel, and N.H. Secher. 1997. Middle cerebral artery blood velocity during rowing. *Acta Physiologica Scandinavica* 160:251–55.

Pott, F., C.A. Ray, H.L. Olesen, K. Ide, and N.H. Secher. 1997. Middle cerebral artery blood velocity, arterial diameter and muscle sympathetic nerve activity during post-exercise muscle ischemia. *Acta Physiologica Scandinavica* 160:43–47.

Poulin, M.J., R.J. Syed, and P.A. Robbins. 1999. Assessments of flow by transcranial Doppler ultrasound in the middle cerebral artery during exercise in humans. *Journal of Applied Physiology.* 86:1632–1637.

Rogers, S.N., J. Schroeder, N.H. Secher, and J.H. Mitchell. 1990. Cerebral blood flow during static exercise in man. *Journal of Applied Physiology* 68:2358–61.

Talairach, J., and P. Tournoux. 1988. *Co-planar stereotaxic atlas of the human brain.* Stuttgart: Georg Thieme Verlag.

Thomas, S.N., J. Schroeder, N.H. Secher, and J.H. Mitchell. 1989. Cerebral blood flow during submaximal and maximal dynamic exercise in man. *Journal of Applied Physiology* 67:744–48.

Villringer, A., J. Planck, C. Hock, L. Schleinkofer, and U. Dirnagl. 1993. Near-infrared spectroscopy (NIRS): A new tool to study hemodynamic changes during activation of brain function in human adults. *Neuroscience Letters* 154:101–4.

Williamson, J.W., D.B. Friedman, J.H. Mitchell, N.H. Secher, and L. Friberg. 1996. Mechanisms regulating regional cerebral activation during dynamic handgrip in humans. *Journal of Applied Physiology* 81:1884–90.

Williamson, J.W., A.C.L. Nobrega, R. McColl, D. Mathews, P. Winchester, L. Friberg, and J.H. Mitchell. 1997. Activation of the insular cortex during dynamic exercise in humans. *Journal of Physiology* 503:277–83.

Chapter 10

Muscle Blood Flow and Its Regulation

Bengt Saltin, Göran Rådegran, Maria Koskolou,
Robert C. Roach, and Janice M. Marshall

The basic mechanisms for regulating oxygen supply to muscle and its utilization have been unraveled by the use of isolated muscle preparations (Barclay 1988; Cain 1977; Renkin and Rosell 1962; Stainsby and Andrew 1988). However, the extent to which these findings relate to the in vivo situation in humans is not always clear (Rowell 1988). The most apparent example is peak blood flow of skeletal muscle. In the various muscle preparations used, peak perfusion may reach 50–60 ml/100 g/min, and peak oxygen uptake may plateau at 80–100 ml/100 g/min. Similar low values have been reported for contracting human skeletal muscle using pleth-ysmography or the ^{133}Xe-washout technique (for references, see Mellander and Johansson 1968). From recent work on intact animals (microspheres) including humans (dilution techniques), researchers have well demonstrated that true peak values may be threefold to fivefold higher than ever observed in isolated muscle preparations (Andersen and Saltin 1985; Richardson et al. 1993; for further references, see Laughlin et al. 1996). Thus, the study of muscle blood flow regulation in humans warrants using an in vivo exercise model that biomechanically and physiologically resembles exercise in normal life. In this chapter, such studies will be preferentially used to give an account of the present knowledge about exercise-induced muscle hyperemia in humans.

Skeletal Muscle Blood Flow

The following discusses skeletal muscle blood flow at various times. It describes the blood flow at the onset of exercise, during the steady-state phase, and at peak effort.

Onset of Exercise

The introduction of ultrasound Doppler has allowed determination of blood velocity in vessels that are the main supplier of blood flow to a specified region (Gill 1979, 1985; Wesche 1986). To be able to estimate the blood flow from blood velocity, one has to know the vessel diameter. This can be accomplished with an acceptable accuracy, at least for larger vessels such as the femoral artery (Rådegran 1997). The diameter of this vessel just distal to the inguinal ligament is unaltered from rest to short-term exercise. This allows researchers to make very precise determinations of the vessel diameter at rest and to apply it also in the exercising condition. Not only the femoral but also the brachial arteries have been used to study the inflow of blood to the contracting muscles. These studies have used the forearm musculature (Tschakovsky, Shoemaker, and Hughson 1995), the whole leg (Shoemaker, Hodge, and Hughson 1994), or the knee extensor muscles (Walloe and Wesche 1988).

These studies collectively demonstrate that at onset of exercise, a very rapid elevation in the blood velocity and thus the blood flow occurs. The fluctuations are large and synchronized with the contractions. Indeed, although the force developed in each contraction may be only a fraction of the peak force, the arterial inflow occurs almost exclusively between contractions. Since the quality of the ultrasound Doppler signal may deteriorate with motion where the insonation occurs, researchers have often employed intermittent static contractions (Eriksen et al. 1990). However, performing the measurements during regular dynamic work and near to the exhaustive level with the knee extensors is possible (Rådegran 1997). Data from such studies confirm

This chapter was modified from a recent review article by B. Saltin, G. Rådegran, N.D. Koskolou, and R.C. Roach. 1998. Skeletal muscle blood flow in humans and its regulation during exercise. *Acta Physiologica Scandinavica* 162:421–436.

the earlier observation of a fast increase in blood flow at the start of exercise and that the variation in blood velocity is a function of the contraction cycle (see figure 10.1). By using these continuous measurements of the arterial inflow, researchers can calculate the time constants for the elevation in blood flow at various work levels (Rådegran and Saltin 1998). At very light workloads, the time to reach half-peak value ($t_{1/2}$) is less than 5 s. It increases to <10 s at higher workloads including close-to-peak exercise intensifies for the knee extensor muscle (see figure 10.2). This means that the absolute increase in muscle perfusion is from ~0.3 at rest and increases up to 10 l/min at peak exercise, a thirtyfold or more increase in ≤10 s.

The blood flow velocity and thus the blood flow varies during contraction-relaxation phases of a movement. This occurs not only during more intense exercises but quite apparently at the lightest workloads (Rådegran and Saltin 1998). Another factor, however, causes fluctuations in blood flow. The pulse pressure varies independently from rate of contraction. Thus, depending upon where systole and diastole occur during a cycle of contraction and relaxation, pulse pressure affects the inflow between contractions (see figure 10.1). Blood velocity is the highest when peak pulse pressure occurs in between contractions. The systolic pressure also has a small effect during the contraction when they coincide (see figure 10.1).

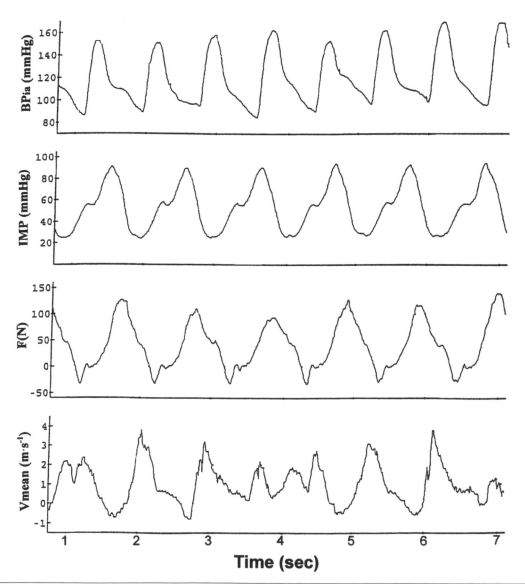

Figure 10.1   The variation in femoral arterial inflow (V_{mean} m · s^{-1}; ultrasound Doppler) (Rådegran 1997) during dynamic knee extensor exercise. Included in the graph is also knee extensor force (F), intramuscular pressure (IMP), and intra-arterial blood pressure (BPia). Note the close coupling between mechanical hindrance to flow and femoral artery blood velocity (Rådegran and Saltin 1998). The top graph shows a sample of duty cycles where the interaction of blood pressure (BPia) on arterial inflow can be observed (Rådegran and Saltin 1998).

The Steady-State Phase

Depending on the work intensity, blood flow level stabilizes within 30–90 s. A minor further elevation may occur during very intense exercise (see figure 10.2). If the work can be continued for hours, the limb blood flow and oxygen uptake remain quite stable (see figure 10.3) (Savard, Kiens, and Saltin 1987b). In this steady-state phase of the exercise, methods other than the ultrasound Doppler, e.g., the reverse dye (Wahren 1966) or the thermodilution method (Andersen and Saltin 1985), can also be used to determine the blood flow precisely (Kim et al. 1995). A word of caution may be warranted, however. The ultrasound Doppler and the dye dilution methods measure arterial inflow, whereas the thermodilution method is preferentially applied to the femoral vein. Although the methods give the same value for the blood flow, these may not represent identical blood flows. Of course, the major portion is identical but probably not 100%. By infusing green dye into the femoral artery at the level of the inguinal ligament and determining the recovery

in the femoral vein, researchers can account for a maximum of ~90% of the dye (Kim et al. 1995). Placing the tip of the arterial catheter proximal to the inguinal ligament will reduce the recovery of dye in the femoral vein. Since the observed blood flow is the same when measured as arterial inflow or venous drainage, apparently ~10% of the inflow by the femoral artery is drained by veins other than the femoral vein. Thus, an inflow drained by the femoral vein during exercise occurs in addition to the femoral artery to the knee extensors.

Also of note is that the blood flow measured in these large vessels perfuse tissues other than the muscle. At rest, the relative fraction perfusing the skin and other tissues is substantial. However, during exercise, the distribution is reversed. In neutral temperature (-20–25 °C), the contribution from the saphenous vein to the femoral venous blood flow is up to ~200 ml/min, or less than 10% of the total blood flow at the lower exercise intensities, and < 3% at peak exercise intensities (Gonzales-Alonso unpublished data; Savard et al. 1988). Since the distribution of the arterial inflow to other tissues is

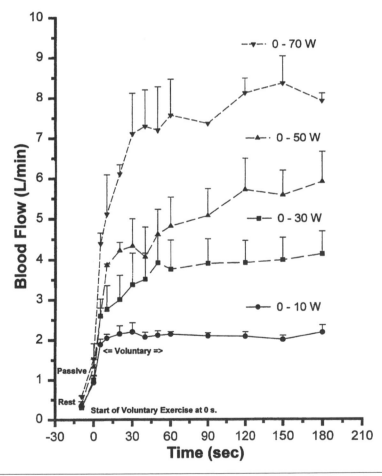

Figure 10.2   Blood flow measured with ultrasound Doppler (Rådegran 1997) in the femoral artery at rest, during passive movement of the knee extensor muscle group, and continuously when exercising at different power outputs with the knee extensors of one leg (Rådegran and Saltin 1998).

One-legged knee-extension

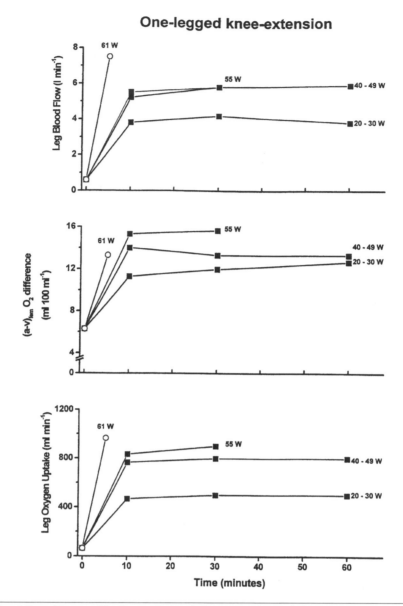

Figure 10.3   Limb blood flow (thermodilution technique) during prolonged dynamic knee extensor exercise. This demonstrates very stable levels after 6–10 min of exercise with negligible drift even when the exercise lasts for a long time and is quite intense. The (a-v)O$_2$difference also remains unaltered during the exercise. This results in stable limb (muscle) oxygen uptake (data redrawn from Savard, Kiens, and Saltin 1987b.)

even less than to the skin, it appears safe to assume that ~90–98% of the measured blood flow either to or from the leg in dynamic knee extensor exercise perfuses the muscle. This assumption most likely also applies to ordinary bicycle work.

When relating the limb (muscle) blood flow at steady state in knee extensor exercise to power output, the coupling is very tight over the entire range of exercise intensities (see figure 10.4) (Andersen and Saltin 1985; Richardson et al. 1993). Note that regardless of the individual's exercise capacity and training status at a given submaximal exercise intensity, power output primarily sets limb blood flow. Thus, a coupling exists between blood

flow and mechanical work or oxygen uptake at the muscle level as it does for cardiac output and oxygen uptake at the systemic level (5 l blood/1 O$_2$ uptake/min). In the knee-extensor model, 7 l/min in blood flow gives an oxygen consumption of 1 l/ min (Andersen and Saltin 1985). Corresponding blood flow in ordinary bicycle exercise is ~6 l/min (Savard et al. 1987b; Wahren et al. 1974). This coupling between work and limb blood flow remains linear up to peak exercise levels (see figure 10.4). Thus, peak blood flows vary closely with peak power output. In knee extensor exercise, peak blood flow values from 4–10 l/min have been reported in subjects reaching 40–100 W at exhaustion.

To convert the limb blood flow values to muscle perfusion, the muscle mass engaged in the exercise needs to be known. Limb volume measurements have been used in the past. However, with today's imaging techniques, very precise muscle volume/mass measurements are possible. The knee extensors (quadriceps femoris m.) of young males and females is usually in the range of 1.5–3.0 kg (Blomstrand, Rådegran, and Saltin 1997). Relating muscle blood flow per unit muscle to power output or oxygen uptake results in the same close linear relationship as when using total blood flow to the limb (Saltin 1985a, 1985b). However, since limb blood flow is a function of the performed work regardless of muscle size, blood flow per unit muscle,

although linearly related to power output, varies markedly between individuals. Thus, at a given submaximal exercise intensity, mean muscle perfusion will be inversely related to knee extensor muscle mass. What is not known is how heterogeneous the blood flow is in contracting skeletal muscle of humans. Data from studies about other species indicate that the muscle perfusion is far from homogenous (Piiper and Haab 1991).

The above-given values for muscle blood flow apply to normal hemoglobin and arterial oxygen content (C_aO_2) levels. With hypoxia, limb blood flow is elevated to compensate for the lower C_aO_2 that fully occurs at submaximal work intensities but hardly occurs at maximal effort even when the hypoxia is severe (Koskolou, Calbet, et al. 1997; Rowell

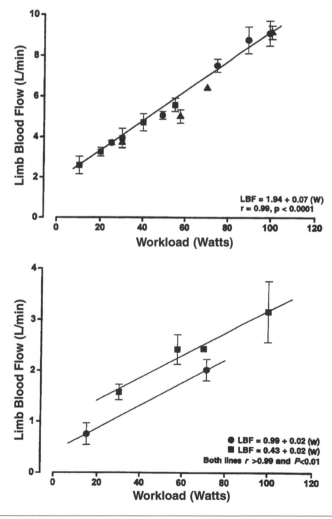

Figure 10.4   Upper panel: Limb blood flow expressed in L/min (thermodilution technique) at different power outputs when performing one-legged dynamic knee extensor work. (Data from Andersen and Saltin 1985; Blomstrand, Rådegran, and Saltin 1997; Richardson et al. 1993). Lower panel: Limb blood flow (thermodilution technique) normalized for knee extensor muscle mass measured with CT scan (Blomstrand, Rådegran, and Saltin 1997) or anthropometric measures (Koskolou, Calbet, et al. 1997; Koskolou, Roach, et al. 1997). The downward parallel displacement of the latter regression line is due to an overestimation of the muscle mass with the anthropometric method.

et al. 1986). In hyperoxia, blood flow is reduced (Welch 1982). Thus, limb blood flow is adjusted at a given power output to be above or below the normal blood flow level in relation to the deviation in C_aO_2 due to varying arterial oxygen pressures (P_aO_2). Interestingly, a variation in hemoglobin concentration ([Hb]) of the blood affects the blood flow similarly (Koskolou, Roach, et al. 1997). An increase in blood flow compensates for the lowering of the [Hb]. With elevated [Hb], blood flow is reduced (Saltin et al. 1986). The match is well tuned, i.e., the oxygen delivery is maintained although [Hb] varies from 10–17 g/100 ml. At maximal effort, although the muscle engaged in the exercise is limited (3–7 kg of muscles; one or two knee extensor muscle groups exercising), oxygen delivery is not maintained. Oxygen consumption, and thus power output, is reduced accordingly (Koskolou, Calbet, et al. 1997; Koskolou, Roach, et al. 1997). The degree of compensation observed at submaximal exercises when varying the C_aO_2 either by O_2 tensions, [Hb], or a combination of low [Hb] and hypoxia appears to be similar. Thus, relating oxygen delivery to power output including all of the above-mentioned conditions produces a relationship between these two variables approaching 1.0 (see figure 10.5). This suggests that the variable controlled is oxygen delivery and muscle blood flow is the means by which it can be adjusted.

Peak Muscle Perfusion

In the first studies using the knee extensor muscle model, researchers observed peak muscle perfusion of 200–220 ml/100 g/min (Andersen and Saltin 1985). Soon, however, others showed in trained subjects or when arterial hypoxemia was induced that 300 ml/100 g/min could be reached (Rowell et al. 1986). This does not appear to be the upper limit, though. In more recent studies using exact estimations of muscle size, researchers have reported more than 380 ml/100 g/min in perfusion in well-trained bicyclists (Blomstrand, Rådegran, and Saltin 1997), confirming Richardson et al.'s findings (1993). These values are high and considerably higher (fourfold to sixfold higher) than ever observed with the plethysmographic or ^{133}Xe-washout techniques (the reason for the discrepancy has been discussed elsewhere; see Kim et al. 1995). However, they are in the same range as that observed using the microsphere technique in skeletal muscles of other species, such as the rat, but higher than what is found in pigs and ponies. The perspectives for this very high muscle perfusion for systemic cardiovascular regulation in humans are discussed below.

Peak perfusion of a limb has another aspect. When Rowell et al. (1986) used the one-legged knee extensor model and hypoxia to evaluate the extent to which limb blood flow could be raised, they

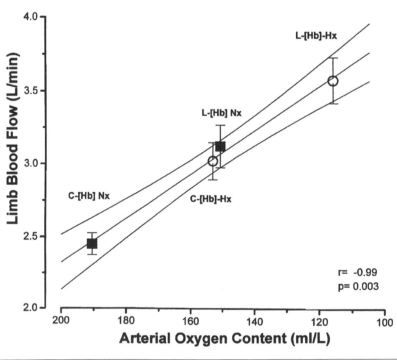

Figure 10.5   Limb blood flow (thermodilution technique) at different arterial oxygen contents (C_aO_2). Acute isovolemic anemia resulted in lower hemoglobin concentration in the two L-[Hb] conditions. The C_aO_2 was lowered in the C-[Hb]-Hx condition by having volunteers breathe hypoxic gas ($F_IO_2 = 0.11$).

summarized that it apparently has no limit. It is true that in their study, there was at intense exercise a further increase in the peak limb blood flow with hypoxia. However, this is at least partly due to their subjects not reaching true normoxic peak blood flow levels. Later studies have confirmed that peak limb blood flow is attained when exercising intensely with one or two knee extensor groups under control conditions. Low [Hb], hypoxia, or a combination do not further elevate the highest attained blood flow. This is surprising since the limb blood flow amounts to only a fraction of the attainable maximal cardiac output. Blood pressure is not further enhanced either. Moreover, there is some indication of a maintained perfusion of the splanchnic region in spite of the exhaustive effort and heart rates in the range of up to 150–160 bpm (Koskolou, Calbet, et al. 1997; Koskolou, Roach, et al. 1997).

The question is then what sets the upper limit? The explanation may have to be sought for in the limb. One of the following central factors could have contributed to an even higher limb blood flow: elevation of the cardiac output, elevation of the blood pressure, or further reduction in noncontracting tissue blood flow. However, none of these adjustments occurred. Before explaining possible peripheral limitations, the authors should emphasize that all of the previously described adjustments can occur in humans when exercising. The exercise must not be confined to a small muscle group, i.e., like ordinary bicycle exercise. In submaximal ordinary bicycle exercise, cardiac output is elevated in several conditions such as hypoxia (Stenberg et al. 1966), anemia (Sproule, Mitchell, and Miller 1960), and muscle metabolic disorders when blood-borne substrates are critical for the muscle energy turnover (Haller et al. 1983). Elevation in blood pressure is observed in arm exercise (Astrand et al. 1965).

In several of these conditions and in patients with a low cardiac reserve, splanchnic blood flow is also further reduced (Rowell 1974). The likely explanation for none of these responses occurring in small muscle group exercise with the leg must be that the proper stimulus is lacking. In other words, the sympathetic nervous system activity is not markedly enhanced (Secher and Saltin 1997). One could argue that the central command does not drive the cardiovascular nuclei in the hypothalamus because the power output is small and only a small fraction of the muscle mass is engaged in the exercise. Although this is true, one should remember that all three of the above-mentioned central adjustments do occur. They take place even though several muscle groups are active, the total muscle mass engaged in the exercise may be small, and the power output may be small.

The same argument could be made about a lack of adjustments by central command for the peripheral receptors including the muscle reflex. Are they muscle-mass or muscle-group dependent as well? In contrast with static contraction, the muscle group and mass involvement in dynamic exercise are major factors elevating the sympathetic nervous activity (Secher and Saltin 1997). Indeed, static contraction with the fingers or the forearm has to be compared with contraction of both legs to obtain significant differences in blood pressure and heart rate responses (Schibye et al. 1981). In dynamic exercise, however, the former would barely affect the central hemodynamics. During this type of exercise, the latter elicits a heart rate and a redistribution of the blood flow that approach the maximal response (Lewis et al. 1983; Rowell 1974). Note, however, that in addition to the importance of the size of the muscle mass in the dynamic exercise, the demand for blood flow is apparently sensed in relation to the capacity of the heart and integrated in the signaling to and from nuclei of the sympathetic nervous system. This conclusion is based on studies where an elevation in the cardiac output is somewhat or severely limited (Magnusson et al. 1997; Pawelczyk et al. 1992).

To return to a possible limitation of limb blood flow on the peripheral level, vessel size may be critical. Thus, in the limb, the size of conduit vessels as well as the capillaries could play a role. Researchers are accumulating knowledge about quite marked adaptations with physical activity level of not only the capillary net but also of the larger vessels (Huonker, Halle, and Keul 1996). The latter are wider in the physically active limbs. Even if both femoral arteries are enlarged as in bicyclists, the aorta has the dimension expected from body size. Thus, there appears to be a need for an increased limb vessel diameter to allow for very high peripheral blood flows. By estimating the Reynolds number for the femoral artery with a diameter of ~10 mm and a blood flow of up to 9–10 l/min, it is still in the range for laminar flow. However, the knee extensors are primarily supplied with blood flow from a. femoralis profundus, which has a diameter in the range of 5–6 mm. If it is assumed that this vessel provides two-thirds to three-fourths of the knee extensor blood flow, turbulence may well occur. Vessel size may, in a sense, become limiting. However, when relating femoral artery size with peak blood flow or oxygen uptake, no correlation exists that implies that the feeding artery is limiting (Rådegran, Blomstrand, and Saltin in press).

On another note, it can be questioned whether an elevated blood flow would have any functional significance in elevating muscle oxygen uptake. MTT is already low (~500 ms), and O_2 extraction is compromised with peak perfusion (Saltin 1985b). Any further lowering of the passage time for the red cells may only elevate femoral venous CO_2. In whole-body exercise, however, in all studies where muscle oxygen delivery is elevated, the oxygen uptake is too (Ekblom, Wilson, and Astrand 1976), whereas it is more uncertain in one-legged exercise (Pedersen et al., personal communication).

Regulation of Muscle Blood Flow

In the transition from rest to exercise, a very rapid initial elevation occurs in blood flow. This is followed by one or two more phases with less-pronounced acceleration in the increase of the arterial inflow to the contracting muscle group. Figure 10.6

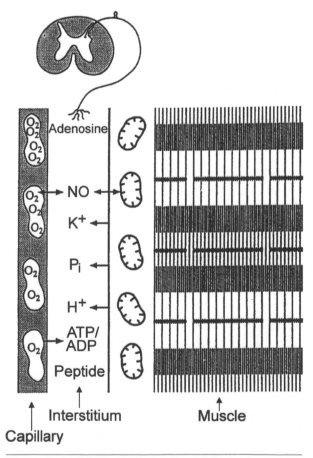

Figure 10.6   Schematic illustration of the various compounds in skeletal muscle that may cause vasodilatation. The graph also illustrates that several of the substances may activate group III and IV sensory nerve endings and thereby elicit a muscle reflex activating the sympathetic nervous system.

outlines the various compounds discussed to explain the exercise-induced hyperemia.

Transitional Phase

Gaskell (1877) first described and discussed the very fast and early increase in muscle blood flow with exercise. In light of the very quick response, Gaskell proposed it to be of neurogenic origin. This has been a recurrent hypothesis. Many species have a cholinergic vasodilator system in skeletal muscle as an integrated part of the defense reaction (Folkow and Neil 1971). In primates, however, the cholinergic vasodilator branch does not appear to exist (Bolme et al. 1970; Buelbring and Bum 1936; Dietz et al. 1997). Honig and colleagues (Honig 1979; Honig and Frierson 1976) have also argued for a nervous component to explain this first marked and very rapid elevation in muscle blood flow. Their proposal was an axon reflex linked with the motor nerve activation of the muscle. However, no such nerves have been demonstrated. More lately, acetylcholine (Ach) has been suggested to be the link between the motor nerve activation of the muscle and the hyperemia (Kurjiaka and Segal 1995). An overflow of Ach from the end-plate region has been suggested to be the source of the Ach. More recently, evidence has been presented for endothelial cells to produce Ach (Panavelas and Bumstock 1985).

In spite of the various proposals for a nervous or Ach component in the early blood flow response, the question is if muscle mechanical factors do not explain the instantaneous elevation in blood flow after the very first contraction (Laughlin 1987; Sheriff, Rowell, and Scher 1993; Tschakovsky, Shoemaker, and Hughson 1996). This possibility has been both experimentally and theoretically penetrated and offers the best and most satisfactory explanation. The observations in humans seen by using ultrasound Doppler to follow the changes in blood velocity definitely support this notion (Rådegran and Saltin 1998). As figure 10.1 depicts, the very first contraction clearly blocks the inflow. However, with the release of the contraction and concomitant drop in intramuscular pressure and thereby mechanical hindrance to flow, an immediate marked elevation occurs in the blood velocity. This peaks just before the next contraction at a level some 60% higher than at rest. In the femoral artery with a diameter of ~10 mm, this means a blood volume of ~50 ml. This amount should be compared with the volume of blood in the capillary bed of the knee extensor muscle, which ranges from 30–90 ml depending upon degree of capillarity (Saltin 1985b). Ample evidence from the fluctuations in the temperature of the femoral vein blood also show that such an amount of blood is squeezed out of the muscle with each contraction (Andersen and Saltin 1985). It is true that femoral venous blood flow may not be zero. However, it reaches a nadir that approaches zero between contractions and peaks during a contraction. Thus, the pattern of flow in and out of a contracting muscle is phase shifted. When considering the moment just after the first contraction at the start of an exercise period, the difference compared with the resting state is that the capillary bed and the collecting venules are empty of blood and pressure is zero or less to the point of the first valves. The elevated pressure gradient between mean arterial pressure and the venous end of the capillary is most likely sufficient to explain the very early (1–3 s) elevation in blood velocity and blood flow without postulating any concomitant vasodilatation.

Secondary Phase

For the next phase of an increase in blood flow, active vasodilatation is a prerequisite since arterial blood pressure is not yet elevated and local pressure in the muscle is unaltered (see figure 10.1). One hypothesis is that the sheer stress elicited by the first elevation in blood flow induces a release of a vasodilator substance. The most commonly proposed vasoactive agent is NO (nitric oxide) (Furchgott and Zawadzki 1980; Pohl et al. 1991; Wilson and Kapoor 1993). Interventions making NO more available in the muscle vasculature, as when giving sodium nitroprusside, elevate muscle blood flow both at rest and during exercise (Rådegran and Saltin 1999). The time response is somewhat slow but possibly within the range of the blood flow increase observed over the first 3–12 s of exercise. However, the absolute magnitude of the flow increase is limited in part due to reduced blood pressure. Even more important is that an NO synthesis blockade with L-NMMA, which causes a distinct reduction in blood flow (by 50 %) at rest and during recovery, has no effect on the blood flow during voluntary or passive exercise (see figure 10.7). This relates to the early elevation as well as the steady state or the peak level attained. These findings in humans during knee extensor exercise have support from similar findings in humans as well as

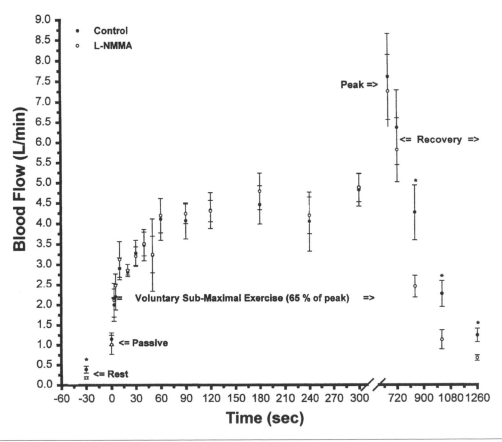

Figure 10.7 The effect of NOS inhibition by L-NMMA on femoral arterial inflow (ultrasound Doppler) (Rådegran 1997) at rest, during exercise, and in recovery from exercise (Rådegran and Saltin 1999).

in other species. However, some reports favor NO playing a role in muscle vasodilatation also during exercise (Delp and Laughlin 1998).

As indicated above, endothelial cells can produce Ach. This is an attractive alternative to NO since Ach has a vasodilator effect at its site of release, causing ascending vasodilatation (Kurjiaka and Segal 1995). An alternative to NO and to Ach could be adenosine. Several indications are at hand for it also having a role in skeletal muscle (Skinner and Marshall 1996). In dogs adenosine is elevated in venous blood, draining the contracting skeletal muscle. Injections of adenosine into a vascular bed cause vasodilatation (Sollevi 1986). In more systematic attempts to evaluate a possible role for adenosine, Rådegran and Calbet (1999) have evaluated the dose and time response for injected/infused adenosine needed to cause vasodilatation as well as the magnitude of elevation in flow and its persistence (see figure 10.8). Minute amounts of adenosine injected into the femoral artery induce an increase in flow that occurs within seconds and lasts as long as the same dose of adenosine is infused. Upon termination of infusion of adenosine, the time for return of blood flow to resting levels is as quick as the onset of vasodilatation, i.e., $t_{1/2}$ is 10–15 s (see figure 10.9; compare also with data in figure 10.2).

The elevation in blood flow is a direct function of the adenosine dose. Levels of up to 8–10 1/min are achieved with an infusion rate of 1–2 mg/min/1 thigh volume (a cuff below the knee is inflated to ~220 mmHg). Also, during exercise with the knee extensors and the hamstring causing an increase in the blood flow to 5–6 1/min, adenosine similarly enhances the blood flow further by 2–3 1/min (Rådegran and Calbet 1999). Thus, adenosine is quite apparently both a rapid and a powerful dilator substance in skeletal muscle of humans. However, whether it has a role in human voluntary exercise cannot as yet be concluded.

The major limitation in evaluating a role for adenosine is that there is not yet a specific A_2 receptor blocker. Theophylline and caffeine, which are nonspecific adenosine blockers, have been used. The blood flow at a given submaximal workload may be reduced with 0.5–1 1/min at high workloads with caffeine. Theophylline, which is both a nonspecific blocker of adenosine and at certain doses at rest and during light exercise, a vasodilator, reduces the blood flow during intense exercise to a similar extent as caffeine (Rådegran and Calbet 1999). If adenosine is given during voluntary contraction to elevate the blood flow, most of the adenosine-induced vasodilatation can be eliminated with theo-

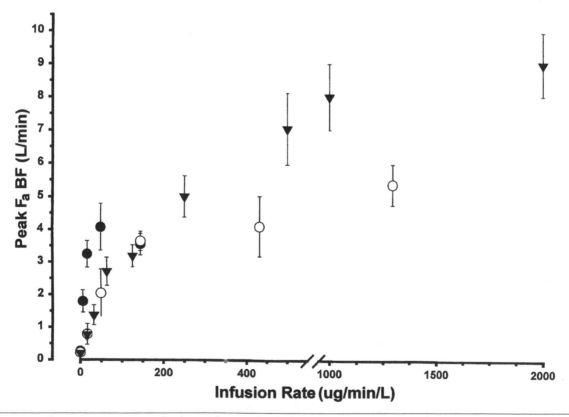

Figure 10.8   Dose-response curve for femoral arterial inflow for sodium nitroprusside (SNP ●), acetylcholine (○), and adenosine (ultrasound Doppler ▼) (Rådegran and Calbet 1999; Rådegran and Saltin 1999.)

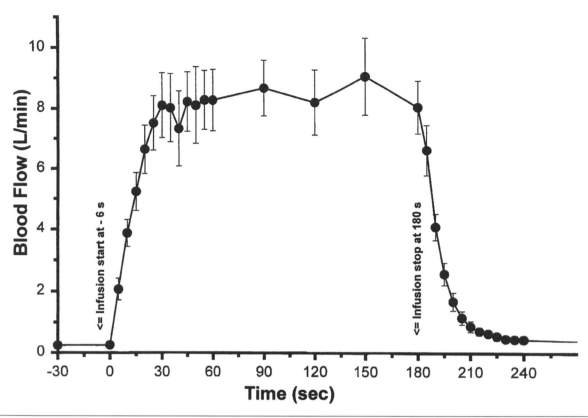

Figure 10.9   The time course for the change in femoral arterial inflow (ultrasound Doppler) when adenosine is infused at a rate of 2 mg/min/1 thigh volume (Rådegran and Calbet 1999.)

phylline. Although this set of results is inconclusive, it does not rule out a role for adenosine.

An important finding shows that adenosine is produced in skeletal muscle of humans. By using the microdialysis technique in skeletal muscle during dynamic knee extensor exercise, adenosine is found in the dialysate in increasing concentrations related to the work level (Hellsten et al. 1998). At rest, the estimated level is low or around 0–50 μg/ 1, increasing to 500–800 μg/1 at peak effort. These levels of adenosine appear high enough to have physiological significance. In the experiments where adenosine was infused into the femoral artery, we have, however, failed to detect any adenosine in the interstitial space. The explanation cannot be that the red cells take up all the adenosine since some systemic effects, elevated pulmonary ventilation, heart rate, and sympathetic activity occur (MacLean et al. 1997). The possibility exists that adenosine does not reach the interstitium due to the endothelial cells functioning as a barrier. We are then left with the fact that adenosine given in an artery quite precisely mimics the blood flow response observed during voluntary exercise, but it acts without entering the interstitial space of the muscle. On the other hand, during voluntary contractions, adenosine is produced and accumulated in the interstitium prob-

ably in amounts sufficient to be of physiological significance, i.e., causing vasodilatation (Hellsten et al. 1998).

Further evidence for adenosine to be found in skeletal muscle with contractions comes from primary cell culture studies. Spontaneously contracting muscle fibers in culture produce adenosine. A further increase is observed with electric activation of the fibers (Hellsten and Frandsen 1997). Endothelial cells do not produce adenosine, but when in contact with the muscle fibers, the adenosine production is enhanced. The substrate for this adenosine production is AMP (adenosine monophosphate). Evidence suggests that the 5′-nucleotidase located on the outer membrane of the sarcolemma catalyzes the production since the formed adenosine is found in the medium but not in the muscle fiber (Hellsten and Frandsen 1997; Marshall 1995). From the above, adenosine clearly fulfills critical criteria for being a vasodilator substance in voluntary contraction. It is produced in the interstitium in what appears to be functionally significant amounts in relation to exercise intensity. Thus, it could be the link between muscle contraction, metabolic rate, and vasodilatation. What remains to be shown is whether the timing of its production is fast enough and, most importantly, to establish the causal link.

Another possible regulation has been newly proposed. In the kidney, P_{450} enzyme linked with the cytochromes is PO_2 sensitive. The lower the PO_2, the higher the enzyme activity (PO_2 range of 50–80 mmHg). This mechanism also appears to be operative in skeletal muscle. The P_{450} enzyme catalyzes the formation of vasoactive metabolites from arachnoid acid. The attractiveness of this mechanism is that the PO_2 sensor is identified, and it is at a critical location.

Vasodilator Candidates

Other aspects are involved in the mediation and regulation of the muscle blood flow during exercise. In studies varying arterial oxygen content either by hyperoxia/hypoxia or high/low [Hb], the close coupling between C_aO_2 and the steady-state level muscle perfusion call upon a role for not only P_aO_2 but also for C_aO_2 or the Hb molecule for the regulation of muscle blood flow. Marshall (1995) has proposed that muscle tissue PO_2 affects the production of adenosine. This, in turn, should open an ATP-sensitive K^+ channel with interstitial $[K^+]$ modulating the degree of vasodilatation. During hypoxia in animals, K^+ is released from muscle via ATP-sensitive K^+ channels on the skeletal muscle fibers since the increase in venous $[K^+]$ can be blocked by glibenclamide, which inhibits these channels (Marshall 1995). However, the increase in venous $[K^+]$ is also blocked by blockade of adenosine receptors. Moreover, infusion of adenosine into muscle induces vasodilatation and an increase in venous $[K^+]$ that are inhibited by adenosine receptor blockade. Further, the vasodilator effects of adenosine receptor stimulation can be blocked by glibenclamide. Thus, researchers have concluded that adenosine can induce vasodilatation by stimulating adenosine receptors on the skeletal muscle fibers that are coupled to K^+-ATP channels, thereby causing release of K^+, which acts as a vasodilator. This is an attractive hypothesis. However, the observed levels of venous $[K^+]$ in normoxic and hypoxic exercise in humans hardly support this notion (Koskolou, Calbet, et al. 1997). Moreover, if adenosine is a vasodilator, there is barely a need for its function to be primarily via an enhanced K^+ release.

The dilatation induced by intra-arterial infusion of adenosine is also greatly reduced by L-NAME. Studies performed using intravital microscopy in animals have shown that when adenosine is applied to the extraluminal surface of the blood vessels, then the resulting dilatation is not affected by L-NAME. Thus, these results suggest that a major part of the adenosine-mediated component of the hypoxia-induced dilatation is caused by adenosine acting on adenosine receptors on the endothelium and releasing NO. They also suggest that a large part of the adenosine must be released from the intraluminal surface of the blood vessels rather than from the skeletal muscle fibers. This is consistent with evidence that adenosine itself is released from vascular endothelium during hypoxia. It is also consistent with evidence that hypoxia also causes the release of ATP from the endothelium since ATP can be degraded to AMP and then broken down enzymatically to adenosine by 5'-nucleotidase.

Novel proposals for C_aO_2 and [Hb] to have regulatory functions have been put forward by Ellsworth and coworkers (1995), Jia and colleagues (1996), and Stamler et al. (1997). The former suggest that a release of ATP from the erythrocytes occurs when they pass through peripheral vascular beds that is related to the number of occupied O_2 binding sites on the Hb molecules (Ellsworth et al. 1995). The idea that nucleotides play a role for vasodilatation in the muscle is not new (Forrester and Lind 1969). This suggestion is attractive, but the support for its existence is limited. This is also true for how ATP may induce vasodilatation. Ellsworth et al. (1995) propose that an ATP receptor triggers formation of nitric oxide (NO) or a prostaglandin (PGI_2), causing a relaxation. An alternative could be that adenosine causes the vasodilatation. The ATP serving as a substrate for its formation could possibly be catalyzed enzymatically by 5'-nucleotidase. The other possibility as proposed by Jia et al. (1996) and Stamler et al. (1997) is that the hemoglobin molecule functions as a scavenger of NO in the peripheral vascular bed with low [Hb], causing more NO to be available to induce vasodilatation. An attractive feature with these two proposals is that they take into consideration a role for the variable level of [Hb], a key to explaining a high blood flow in anemia.

Conclusion

In conclusion, although the mechanisms for exercise-induced hyperemia still are far from elucidated, several very attractive and likely possibilities exist. Adenosine and the P_{450} system combined with a [Hb]-related mechanism would well serve not only to cause vasodilatation but also to direct the blood at the microcirculatory level to the active fibers. This being the case, it is of note that a surprising amount of O_2 remains in the veins draining contracting skeletal muscle. The explanation for this is unresolved despite major efforts devoted to bring this problem to a close (Piiper and Haab 1991; also see the chapter "Oxygen Transport in Blood and to Mitochondria"). The very high perfusion capacity of skeletal muscles of humans (well above 200 ml/100 g/min) constitutes a regulatory challenge on the systemic level. The pumping capacity of the heart cannot supply all muscles of the human body with their peak flow capacity (see the chapter "Integration of Muscle Blood Flow and Cardiac Output").

Acknowledgments

The original studies were performed at the Copenhagen Muscle Research Centre, which is funded by the Danish National Research Foundation (grant #504-14).

References

Andersen, P., and B. Saltin. 1985. Maximal perfusion of skeletal muscle in man. *J. Physiol.* 366:233–249.

Åstrand, P.-O., B. Ekblom, R. Messin, B. Saltin, and J. Stenberg. 1965. Intra-arterial blood pressure during exercise with different muscle groups. *J. Appl. Physiol.* 20:253–256.

Barclay, J.K. 1988. Physiological determinants of Qmax in contracting canine skeletal muscle in situ. *Med. Sci. Sports Exerc.* 20:113–118.

Blomstrand, E., G. Rådegran, and B. Saltin. 1997. Maximum rate of oxygen uptake by human skeletal muscle in relation to maximal activities of enzymes in the Krebs cycle. *J. Physiol.* 501:455–460.

Bolme, P., J. Novotna, B. Uvnds, and P.G. Wright. 1970. Species distribution of sympathetic cholinergic vasodilator nerves in skeletal muscle. *Acta. Physiol. Scand.* 78:60–64.

Buelbring, E., and J.H. Bum. 1936. Sympathetic vasodilator fibres in the hare and the monkey compared with other species. *J. Physiol.* 88:341–360.

Cain, S.M. 1977. Oxygen delivery and uptake in dogs during anemic and hypoxic hypoxia. *J. Appl. Physiol.* 42:228–234.

Delp, M.D., and M.H. Laughlin. 1998. Regulation of skeletal muscle perfusion during exercise. *Acta. Physiol. Scand.* 162:411-419.

Dietz, N.M., K.A. Engelke, T.T. Samuel, R.T. Fix, and M.J. Joyner. 1997. Evidence for nitric oxide-mediated sympathetic forearm vasodilation in humans. *J. Physiol.* 498:334–340.

Ekblom, B., G. Wilson, and P.-O. Astrand. 1976. Central circulation during exercise after ven-esection and reinfusion of red blood cells. *J. Appl. Physiol.* 40:379–383.

Ellsworth, M.L., T. Forrester, C.G. Ellis, and H.H. Dietrich. 1995. The erythrocyte as a regulator of vascular tone. *Am. J. Physiol.* 269:H2155–H2161.

Eriksen, M., B.A. Waaler, L. Walloe, and J. Wesche. 1990. Dynamics and dimensions of cardiac output changes in humans at the onset and at the end of moderate rhythmic exercise. *J. Physiol.* 426:423–437.

Folkow, B., and E. Neil. 1971. *Circulation.* New York: Oxford University Press.

Forrester, T., and A.R. Lind. 1969. Identification of adenosine triphosphate in human plasma and the concentration in the venous effluent of forearm muscles before, during and after sustained contractions. *J. Physiol.* 204:347–364.

Furchgott, R.F., and J.V. Zawadzki. 1980. The obligatory role of endothelial cells in relaxation of arterial smooth muscle by acetylcholine. *Nature* 288:373–376.

Gaskell, W.H. 1877. On the changes of the blood stream in muscle through stimulation of their nerves. *J. Anat.* 11:380–402.

Gill, R.W. 1979. Pulsed Doppler with b-mode imaging for quantitative blood flow measurement. *Ultrasound Med. Biol.* 5:223–235.

Gill, R.W. 1985. Measurement of blood flow by ultrasound: Accuracy and sources of error. *Ultrasound Med. Biol.* 11:625–641.

Haller, R.G., S.F. Lewis, J.D. Cook, and C.G. Blomqvist. 1983. Hyperkinetic circulation during exercise in neuromuscular disease. *Neurology* 33:1283–1287.

Hellsten, Y., and U. Frandsen. 1997. Adenosine formation in contracting primary rat skeletal muscle cells and endothelial cells in culture. *J. Physiol* 504:695-704.

Hellsten, Y., D. MacLean, G. Rådegran, B. Saltin, and J. Bangsbo. 1998. Adenosine concentrations in the interstitium of resting and contracting human skeletal muscle. *Circulation* 98:6-8.

Honig, C.R. 1979. Contributions of nerves and metabolites to exercise vasodilation: A unifying hypothesis. *Am. J. Physiol.* 236:H705–H719.

Honig, C.R., and J.L. Frierson. 1976. Neurons intrinsic to arterioles initiate post-contraction vasodilation. *Am. J. Physiol.* 230:493–507.

Huonker, M., M. Halle, and J. Keul. 1996. Structural and functional adaptations of the cardiovascular system by training. *Int. J. Sports Med.* 17:164–172.

Jia, L., C. Bonaventura, J. Bonaventura, and J.S. Stamler. 1996. S–nitrosohaemoglobin: A dynamic activity of blood involved in vascular control. *Nature* 380:221–226.

Kim, C.K., S. Strange, J. Bangsbo, and B. Saltin. 1995. Skeletal muscle perfusions in electrically induced dynamic exercise in humans. *Acta Physiol. Scand.* 153:279–287.

Koskolou, M.D., J.A.L. Calbet, G. Rådegran, and R.C. Roach. 1997. Hypoxia and the cardiovascular response to dynamic knee extensor exercise. *Am. J. Physiol.* 272:H2655—H2663.

Koskolou, M.D., R.C. Roach, J.A.L. Calbet, G. Rådegran, and B. Saltin. 1997. Cardiovascular responses to dynamic exercise with acute anemia in humans. *Am. J. Physiol* 273:H1787-H1793.

Kurjiaka, D.T., and S.S. Segal. 1995. Conducted vasodilation elevates flow in arteriole networks of hamster striated muscle. *Am. J. Physiol.* 269:H1723–H1728.

Laughlin, M.H. 1987. Skeletal muscle blood flow capacity: Role of muscle pump in exercise hyperemia. *Am. J. Physiol.* 22:H993–H1004.

Laughlin, M.H., R.J. Korthuis, D.J. Duncker, and R.J. Bache. 1996. Control of blood flow to cardiac and skeletal muscle during exercise. In *Handbook of physiology, XII: Exercise: Regulation and integration of multiple systems*, eds. L.B. Rowell and J.T. Shepherd, 705–769. New York: Oxford University Press.

Lewis, S.F., W.F. Taylor, R.M. Graham, W.A. Pettinger, J.E. Shutte, and C.G. Blomqvist. 1983. Cardiovascular responses to exercise as functions of absolute and relative work load. *J. Appl. Physiol.* 54:1314–1323.

MacLean, D.A., B. Saltin, G. Rådegran, and L. Sinoway. 1997. Femoral arterial injection of adenosine in humans elevates MSNA via central but not peripheral mechanisms. *J. Appl. Physiol.* 83:1045-1053.

Magnusson, G., L. Kaijser, C. Sylvdn, K.-E. Karlberg, B. Isberg, and B. Saltin. 1997. Peak skeletal muscle perfusion is maintained in patients with chronic heart failure when only a small muscle mass is exercised. *Cardiovascular Res.* 33:297–306.

Marshall, J.M. 1995. Skeletal muscle vasculature and systemic hypoxia. *NIPS* 10:274–280.

Mellander, S., and B. Johansson. 1968. Control of resistance, exchange and capacitance vessels in the peripheral circulation. *Pharmacol. Rev.* 20:117–196.

Panavelas, J.G., and G. Bumstock. 1985. Ultrastructural localization of choline acetyltransferase in vascular endothelial cells in rat brain. *Nature* 316:724–725.

Pawelczyk, J.A., B. Hanel, R.A. Pawelczyk, J. Warberg, and N.H. Secher. 1992. Leg vasoconstriction during dynamic exercise with reduced cardiac output. *J. Appl. Physiol.* 73:1838–1846.

Piiper, J., and P. Haab. 1991. Oxygen supply and uptake in tissue models with unequal distribution of blood flow and shunt. *Respir. Physiol.* 84:261–271.

Pohl, U., K. Herlan, A. Huang, and E. Bassenge. 1991. EDRF-mediated shear-induced dilation opposes myogenic vasoconstriction in small rabbit arteries. *Am. J. Physiol.* 30:H2016–H2033.

Rådegran, G. 1997. Ultrasound Doppler estimates of femoral artery blood flow during dynamic knee extensor exercise in man. *J. Appl. Physiol.* 83:1383–1388.

Rådegran, G., and B. Saltin. 1999. Human femoral arterial diameter in relation to knee extensor muscle mass, peak blood flow, and oxygen uptake. *Am. J. Physiol.* (in press).

Rådegran, G., and J.A.L. Calbet. 1999. Skeletal muscle vascular response to adenosine in humans during rest and exercise. *Am. J. Physiol.* (in press).

Rådegran, G., and B. Saltin. 1998. Muscle blood flow at onset of dynamic exercise in man. *Am. J. Physiol.* 274:H314–H322.

Rådegran, G., and B. Saltin. Nitric oxide in the regulation of vasomotor tone in human skeletal muscle. 1999. *Am. J. Physiol.* 276:H1951–1957.

Renkin, R.M., and S. Rosell. 1962. Effects of different types of vasodilator mechanisms on vascular tonus and on transcapillary exchange of diffusible material in skeletal muscle. *Acta Physiol. Scand.* 54:241–251.

Richardson, R.S., D.C. Poole, D.R. Knight, S.S. Kurdak, M.C. Hogan, B. Grassi, E.C. Johnson, K. Kendrick, B.K. Erickson, and P.D. Wagner. 1993. High muscle blood flow in man: Is maximal O_2 extraction compromised? *J. Appl. Physiol.* 75:1911–1916.

Rowell, L.B. 1974. Human cardiovascular adjustments to exercise and thermal stress. *Physiol. Rev.* 54:75–159.

Rowell, L.B. 1988. Muscle blood flow in humans: How high can it go? *Med. Sci. Sports Exerc.* 203:S97–S103.

Rowell, L.B., B. Saltin, B. Kiens, and N.J. Christensen. 1986. Is peak quadriceps blood flow in humans even higher during exercise with hypoximia? *Am. J. Physiol.* 251:H1038–H1044.

Saltin, B. 1985a. Hemodynamic adaptations to exercise. *Am. J. Cardiol.* 55:42D–47D.

Saltin, B. 1985b. Malleability of the system in overcoming limitations: Functional elements. *J. Exp. Biol.* 115:345–354.

Saltin, B., B. Kiens, G. Savard, and P.K. Pedersen. 1986. Role of hemoglobin and capillarization for oxygen delivery and extraction in muscular exercise. *Acta Physiol. Scand.* 128:21–32.

Savard, G., B. Kiens, and B. Saltin. 1987a. Central cardiovascular factors as limits to endurance: With a note on the distinction between maximal oxygen uptake and endurance fitness. In *Exercise: Benefits, Limits and Adaptations*, eds. D. Macleod, R. Maughan, M. Nimmo, T. Reilly, and C. Williams, 162–180. London: E. & F.N. Spon.

Savard, G.K., B. Kiens, and B. Saltin. 1987b. Limb blood flow in prolonged exercise: Magnitude and implication for cardiovascular control during muscular work in man. *Can. J. Appl. Sports Sci.* 12:S89–S101.

Savard, G.K., B. Nielsen, L. Laszczynzka, B. Elmann–Larsen, and B. Saltin. 1988. Muscle blood flow is not reduced in man during moderate exercise and heat stress. *J Appl Physiol* 64:649–657.

Savard, G.K., E.A. Richter, S. Strange, B. Kiens, N.J. Christensen, and B. Saltin. 1989. Norepinephrine spillover from skeletal muscle during dynamic exercise in man: Role of muscle mass. *Am. J. Physiol.* 257:H1812–H1818.

Schibye, B., J.H. Mitchell, F.C. Payne, and B. Saltin. 1981. Blood pressure and heart rate response to static exercise in relation to electro-myographic activity and force development. *Acta Physiol. Scand.* 113:61–66.

Secher, N.H., and B. Saltin. 1997. Blood flow regulation during exercise in man. In *The physiology and pathophysiology of exercise tolerance*, eds. S.A. Ward and J. Steinacker, 97–102. London: Plenum Press.

Sheriff, D.D., L.B. Rowell, and A.M. Scher. 1993. Is rapid rise in vascular conductance at onset of dynamic exercise due to muscle pump? *Am. J. Physiol.* 265:H1227–H1234.

Shoemaker, J.K., L. Hodge, and R.L. Hughson. 1994. Cardiorespiratory kinetics and femoral artery blood velocity during dynamic knee extension exercise. *J. Appl. Physiol.* 77:2625–2632.

Skinner, M.R., and J.M. Marshall. 1996. Studies on the roles of ATP, adenosine and nitric oxide in mediating muscle vasodilatation induced in the rat by acute systemic hypoxia. *J. Physiol.* 495:553–560.

Sollevi, A. 1986. Cardiovascular effects of adenosine in man: Possible clinical implications. *Progr. Neurobiol.* 27:319–349.

Sproule, B.J., J.H. Mitchell, and W.F. Miller. 1960.

Cardiopulmonary physiological responses to heavy exercise in patients with anemia. *J. Clin. Invest.* 39:378–388.

Stainsby, W.N., and G.M. Andrew. 1988. Maximal blood flow and power output of dog muscle in situ. *Med. Sci. Sports Exerc.* 20:109–112.

Stamler, J.S., L. Jia, J.P. Eu, T.J. McMahon, I.T. Demchenko, J. Bonaventura, K. Gemert, and C.A. Piantadosi. 1997. Blood flow regulation by s-nitrosohemoglobin in the physiological oxygen gradient. *Science* 276:2034–2037.

Stenberg, I., P.-O. Astrand, B. Ekblom, J. Royce, and B. Saltin. 1966. Hemodynamic response to work with different muscle groups, sitting and supine. *J. Appl. Physiol.* 22:61–70.

Tschakovsky, M.E., J.K. Shoemaker, and R.L. Hughson. 1995. Beat-by-beat forearm blood flow with Doppler Ultrasound and strain-gauge plethysmography. *J. Appl. Physiol.* 79:713–719.

Tschakovsky, M.E., J.K. Shoemaker, and R.L. Hughson. 1996. Vasodilation and muscle pump contribution to immediate exercise hyperemia. *Am. J. Physiol.* 271:H1697–H1701.

Wahren, J. 1966. Quantitative aspects of blood flow and oxygen uptake in the human forearm during rhythmic exercise. *Acta Physiol. Scand.* 67:5–93.

Wahren, J., B. Saltin, L. Jorfeldt, and B. Pernow. 1974. Influence of age on the local circulatory adaptation to leg exercise. *Scand. J. Clin. Lab. Invest.* 33:79–86.

Walloe, L., and J. Wesche. 1988. Time course and magnitude of blood flow changes in human quadriceps muscle during and following rhythmic exercise. *J. Physiol.* 405:257–273.

Welch, H.G. 1982. Hyperoxia and human performance: A brief review. *Med. Sci. Sports Exerc.* 149:253–262.

Wesche, J. 1986. The time course and magnitude of blood flow changes in the human quadriceps muscles following isometric contraction. *J. Physiol.* 377:445–462.

Wilson, J.R., and S. Kapoor. 1993. Contribution of endothelium-derived relaxing factor to exercise-induced vasodilation in humans. *J. Appl. Physiol.* 75:2740–2744.

Chapter 11

Dynamics of Microvascular Control in Skeletal Muscle

Steven S. Segal

Blood flow control to skeletal muscle represents the physiology of aerobic performance. The correspondence between energy expenditure, oxygen consumption, and muscle blood flow (Andersen and Saltin 1985; Laughlin and Armstrong 1982; Mackie and Terjung 1983; Ordway et al. 1984; Saltin and Gollnick 1983; Secher et al. 1977) imply a coupling between muscle fiber activity and capillary perfusion through a range that exceeds resting values by as much as 50-fold (Andersen and Saltin 1985; Laughlin and Armstrong 1982). Vasoactive stimuli include chemical, electrical, and physical events manifested between muscle fibers and microvessels. In turn, these events reflect muscle fiber recruitment, the transmission of signals to and among vascular cells, and rhythmic movement. Such dynamic interactions ultimately govern the contractile activity of vascular smooth muscle cells and thereby control vascular resistance. The goal of this chapter is to explore the anatomic organization and physiological mechanisms that underlie these interactions and, where information is lacking, to identify fruitful avenues for future research efforts.

Organization of the Resistance Network

The vascular resistance network begins with the feed arteries (diameter, 100–500 μm) external to the muscle and encompasses the arteriolar network (diameter, 10–100 μm) embedded among the muscle fibers (Folkow, Sonnenschein, and Wright 1971; Granger, Goodman, and Granger 1976; Segal 1994; Williams and Segal 1993). This anatomic arrangement results in loci of flow control that are either removed from or in direct contact with the myocytes (see figure 11.1). Moreover, the anatomic distinction between intra- and extramuscular sites of flow control implies that the respective components of the network must cooperate during exercise; such reasoning is central to this chapter. Nevertheless, the

mechanisms that underlie the changes in vessel diameter (and thereby control vascular resistance) can vary with location in the network (Folkow, Sonnenschein, and Wright 1966; Kuo, Davis, and Chilian 1992; Segal 1994).

The wall of resistance vessels typically consists of smooth muscle cells wrapped around the vessel lumen, which is lined by a continuous layer of endothelial cells in physical contact with the blood (Rhodin 1967). Vasoactive substances are produced by vascular cells (Furchgott and Zawadzki 1980) as well as muscle fibers (Haddy and Scott 1975; Shepherd 1983; Sparks 1980) and by autonomic (sympathetic) nerves, which are distributed as a perivascular plexus throughout the resistance network (see figure 11.1) (Fleming et al. 1989; Marshall 1982; Welsh and Segal 1996). The vasculature is continually exposed to such physical forces as transmural pressure and luminal shear stress, which give rise to myogenic (Folkow 1964; Meininger and Davis 1992) and flow-induced (Davies 1995) mechanisms of vasomotor control. The interaction between these opposing forces is viewed as maintaining the basal contractile state of the smooth muscle cells (Kuo, Chilian, and Davis 1991; Pohl et al. 1991), which can be modified by stimuli arising from muscle fiber activity.

Ascending Vasodilation and Locus of Flow Control

When skeletal muscle is at rest, the resistance vasculature maintains a high level of tone, and the removal of oxygen from the blood is only a few volumes percent (Andersen and Saltin 1985; Ordway et al. 1984; Saltin 1985). In response to exercise, most of the available oxygen is removed from the blood as vasodilation ensues. Moreover, the effect of exercise on oxygen extraction and muscle blood flow follow a characteristic pattern that reflect the dilation of distal and proximal branches of the resistance

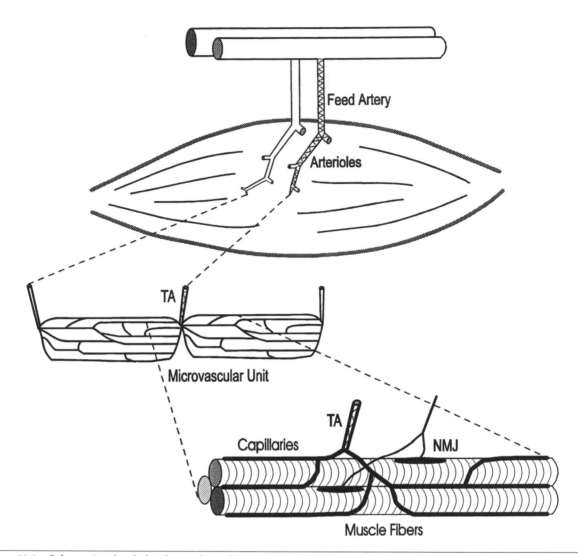

Figure 11.1 Schematic of a skeletal muscle and its vascular supply with an area enlarged to illustrate a microvascular unit. A central portion of the microvascular unit (MVU) is enlarged to illustrate the neuromuscular junction (NMJ) and its proximity to capillaries and terminal arteriole (TA). Perivascular (sympathetic) innervation, indicated by the diagonal lines, encompass feed arteries through terminal arterioles.

network, respectively. As exercise begins, the distal arterioles dilate first (Honig, Odoroff, and Frierson 1980; Segal 1991; Welsh and Segal 1997). These branches control capillary perfusion and blood flow distribution within the tissue (Emerson and Segal 1997; Sweeney and Sarelius 1989). The increase in capillary perfusion coupled with a fall in muscle fiber PO_2 thereby explains the large increase in oxygen extraction at exercise onset (Andersen and Saltin 1985; Folkow, Sonnenschein, and Wright 1971; Honig, Odoroff, and Frierson 1980). As metabolic demand increases, vasodilation ascends into the proximal arterioles and feed arteries, which govern the total volume of flow into the muscle (Folkow, Sonnenschein, and Wright 1971; Granger, Goodman, and Granger 1976; Honig, Odoroff, and Frierson 1980; Segal 1994; Segal and Duling 1986a; Welsh and

Segal 1997; Williams and Segal 1993). Such interaction within and among vascular segments can be explained by the transmission of signals along the vessel wall (Hilton 1959; Segal 1991; Segal and Duling 1986b; Welsh and Segal 1998), as this chapter later describes.

The respective contributions of distal arterioles and proximal feed arteries to the control of blood flow can be appreciated by a simple model of resistance elements connected in series with each other (Segal and Duling 1986a; Williams and Segal 1993). With substantial vasomotor tone in the upstream feed arteries, dilation of downstream arterioles would have a limited effect on increasing muscle blood flow because of the high proximal resistance. In contrast, when feed arteries dilate in concert with arterioles, a profound increase in muscle perfusion occurs (Segal and Duling 1986a; Welsh and Segal 1997; Williams and Segal 1993). Thus, dilation of upstream (i.e., feed artery) segments concomitant with downstream segments (i.e., intramuscular arterioles) is necessary for maximal perfusion of muscle fibers.

Functional Units of Blood Flow Control

The following discusses the functional units of blood flow control. This information is based on both motor units and microvascular units.

Motor Units

Muscle blood flow and oxygen uptake increase in proportion to the intensity of aerobic exercise (Andersen and Saltin 1985; Ordway et al. 1984). These responses coincide with the progressive recruitment of muscle fibers (Burke 1981). This indicates that the utilization and supply of oxygen are closely coupled and that much of this coupling occurs within the muscle (Shepherd 1983; Shepherd et al. 1973). The functional unit of skeletal muscle is the motor unit, which describes the group of muscle fibers activated by the firing of a single motor neuron in the spinal cord (Burke 1981). Muscle fibers within a motor unit may span the entire length of the muscle, which is often many centimeters. The fibers within a motor unit are similar in metabolic properties and are normally dispersed throughout a muscle rather than lying side by side (Burke 1981; Saltin and Gollnick 1983).

Microvascular Units

An alternative definition for the functional unit is based upon the smallest volume of tissue to which blood flow can be actively controlled (Bloch and Iberall 1982; Lund, Damon, and Duling 1987). In skeletal muscle, a group of capillaries fed by a common terminal arteriole is recognized as a capillary unit. Each capillary unit consists of 12–20 capillaries that run parallel to muscle fibers for a distance of ≤ 1 mm to a collecting venule (Bloch and Iberall 1982; Delashaw and Duling 1988; Lund, Damon, and Duling 1987); blood flow through these capillaries is concurrent (Lund, Damon, and Duling 1987). One terminal arteriole supplies blood to two adjacent capillary units (one on each side of the arteriole), collectively known as a unit pair (Delashaw and Duling 1988). Because this is the

smallest element of control for capillary perfusion (Bloch and Iberall 1982; Emerson and Segal 1997; Fuglevand and Segal 1997), it is referred to hereafter as a microvascular unit (MVU) (see figure 11.1). Due to the differences in length between capillaries (< 1 mm) and muscle fibers (often several centimeters), many MVUs are required to span the distance of each muscle fiber (see figure 11.1) (Emerson and Segal 1997). This organization of MVUs appears common to skeletal muscle across many species, from amphibians to primates (Eriksson and Myrhage 1972; Honig, Odoroff, and Frierson 1980; Lund, Damon, and Duling 1987; Plyley, Sutherland, and Groom 1976; Skalak and Schmid-Schonbein 1986; Weibel 1984). The perfusion of MVUs upon dilation of terminal arterioles is essential to the control of oxygen transport to muscle fibers. Nevertheless, researchers do not understand how MVU perfusion is controlled in accord with muscle fiber recruitment. Resolving this interaction is central to ongoing research in our laboratory.

Localization of Metabolic Demand

The following describes the localization of metabolic demand and the interaction of motor units with microvascular units. It also describes microvascular unit alignment.

Motor Unit and Microvascular Unit Interaction

The volume of muscle within a MVU (approximately 0.1 mm^3 (Delashaw and Duling 1988; Emerson and Segal 1997)) is comprised of segments of ~30 muscle fibers (Emerson and Segal 1997). Each of these fibers may derive from a distinct motor unit (Burke 1981). A particular capillary can lie between as many as three or four muscle fibers (Saltin and Gollnick 1983). Therefore, a capillary may have none, one, or all of the adjacent muscle fibers drawing oxygen from it at any given moment (Fuglevand and Segal 1997). These anatomic relationships imply that the metabolic demand within each MVU varies both spatially and temporally. When a muscle fiber contracts, metabolism is stimulated throughout its entire length. This will require blood flow through consecutive MVUs that supply corresponding segments of the fiber. Given the proportionality between muscle work and blood flow (Andersen and Saltin 1985; Ordway et al. 1984), the control of motor units must somehow interact with that of the MVUs that perfuse the muscle fibers.

Microvascular Unit Alignment

One should recognize that an increase in blood flow to an active muscle fiber without overperfusion of inactive fibers would require MVUs to be no wider than a single fiber (which is certainly not the case, as described above) and to be perfectly aligned along the active fiber. In a recent study (Emerson and Segal 1997), we utilized the parallel-fibered hamster retractor muscle (Nakao and Segal 1995) to investigate whether adjacent MVUs encompass the same group of fibers along the muscle axis. If this occurred, it would be indicated by complete alignment of adjacent MVUs along muscle fibers. In practice, we found that neither the size nor the alignment of MVUs is conducive to selectively increasing blood flow to a single muscle fiber nor to a particular fiber bundle (Emerson and Segal 1997). Instead, capillary perfusion apparently must increase through relatively large regions along the muscle to accommodate those capillaries associated with each active fiber. This hypothesis requires additional testing using intravital microscopy to observe patterns of red blood cell perfusion in adjacent MVUs.

Coactivation of Muscle Fibers and Microvessels

The next part of the chapter explains coactivation of muscle fibers and microvessels. It delves into four areas: spatial relationships, metabolic vasodilation, functional implications, and conducted vasodilation.

Spatial Relationships

The imperfect register of adjacent MVUs along muscle fibers (Emerson and Segal 1997) implies that for capillary perfusion to increase along the length of a muscle, a mechanism is required whereby many MVUs can function together (Fuglevand and Segal 1997; Sarelius 1993; Segal 1991). Furthermore, with hemodynamic resistance distributed throughout arteriolar networks, dilation of arterioles as well as feed arteries must correspond with local (i.e., muscle fiber) demands for blood flow. Because feed arteries are external to the muscle, the vasodilator signal must ascend from the arterioles to achieve peak perfusion of the tissue (Segal and Duling 1986a; Welsh and Segal 1997; Williams and Segal 1993). These requirements point to a mechanism that can be triggered by muscle fiber activity and that pro-

motes the correspondence between metabolic demand and oxygen delivery to the working myocytes.

Metabolic Vasodilation

The microcirculation is separated from muscle fibers cells only by the interstitial space. The traditional explanation of functional hyperemia centers on the metabolic theory of local flow control. This states that vasodilator substances (e.g., adenosine, carbon dioxide, and potassium), released from active fibers in proportion to energy expenditure, diffuse through the interstitium to relax the smooth muscle cells of arterioles (Gorczynski, Klitzman, and Duling 1978; Haddy and Scott 1975; Sparks 1980). To complement those substances that act directly on smooth muscle cells (Sparks 1980), several stimuli elicit smooth muscle relaxation by stimulating endothelial cells (Furchgott and Zawadzki 1980; Koller and Kaley 1991; Pohl et al. 1986; Segal 1994; Welsh and Segal 1997). Additionally, vasodilator metabolites carried in venous blood may, via countercurrent exchange, diffuse from the venules to adjacent arterioles and induce vasodilation in proximal branches (Hester 1990). This explanation of flow control implies that the action of vasodilator substances is confined to the vicinity of production and diffusion or to regions influenced by metabolites carried in (and diffusing from) the venous blood. Metabolic vasodilation would thereby increase blood flow in proportion to the local demand of skeletal muscle fibers.

Limitations to Metabolic Control of Muscle Blood Flow

Studies investigating the action of specific vasodilator metabolites have revealed that while a variety of substances released by muscle fibers can contribute to functional hyperemia, no one stimulus can explain the integrated response of blood flow to muscle contraction (Haddy and Scott 1975; Sparks 1980). Metabolic vasodilation requires at least 5–6 s for substances to be produced by skeletal muscle fibers and diffuse to arteriolar smooth muscle cells in sufficient concentration to elicit vasodilation (Gorczynski, Klitzman, and Duling 1978). In practice, these processes are too slow to account for the rapid (1–2 s) onset of functional hyperemia (Honig, Odoroff, and Frierson 1980; Shepherd 1983; Segal 1991). Discrete application of a variety of substances onto arteriole segments using micropipettes has revealed distinctive vasodilator responses. The first type involves those localized to the site of release, as seen in the dilation of short arteriole segments adjacent to stimulated muscle fibers (Gorczynski, Klitzman, and Duling 1978). The second type of response involves those that spread rapidly along the vessel wall (see "Conducted Vasodilation" below) over several millimeters and encompass multiple branches (Delashaw and Duling 1991; Duling and Berne 1970; Hilton 1959; Kurjiaka and Segal 1995a; Segal and Duling 1986b; Welsh and Segal 1997, 1998). Examples of vasodilators that produce local effects are adenosine and sodium nitroprusside (Delashaw and Duling 1991; Kurjiaka and Segal 1995a). In contrast, acetylcholine (ACh) rapidly (in 1–2 s) triggers the conduction of vasodilation throughout arteriolar networks (Duling and Berne 1970; Segal 1991; Welsh and Segal 1997).

Functional Implications

When vasodilation was localized to the vicinity of a particular muscle fiber, blood flow did not increase above rest. In contrast, the conduction of vasodilation (triggered with ACh) rapidly increased flow into arteriolar networks (Kurjiaka and Segal 1995a; Welsh and Segal 1997). These observations illustrate that ascending vasodilation into proximal branches increases network perfusion. We therefore suggest that the conduction of vasodilation contributes to the initial rapid elevation in arteriolar perfusion with the onset of hyperemia (Kurjiaka and Segal 1995a; Segal 1991). Conducted responses triggered on distal branches can be summed and integrated in parent vessels (Segal, Damon, and Duling 1989; Welsh and Segal 1997) and thereby produce a graded increase in network perfusion. Thus, conducted vasodilation could explain the rapid increase in capillary perfusion (Fuglevand and Segal 1997; Honig, Odoroff, and Frierson 1980; Segal 1991) with metabolic vasodilation governing the steady-state response (Sparks 1980). While such relationships are apparent from intravital studies (Gorczynski, Klitzman, and Duling 1978; Kurjiaka and Segal 1995a; Segal 1991) and integrative models (Fuglevand and Segal 1997), they remain to be confirmed in exercising humans.

Conducted Vasodilation

The conduction of vasomotor responses refers to direct coupling (as enabled by gap junctions) between the endothelial cells and/or smooth muscle cells that comprise the resistance vasculature (Little, Xia, and Duling 1995; Rhodin 1967; Segal and Bény 1992; Welsh and Segal 1998). Given the preceding

scenario, the difference between the effects of nonconducting vasodilators (e.g., adenosine and nitroprusside) and those that trigger conduction (e.g., ACh) on muscle blood flow can be explained by their respective mechanisms of action (Kurjiaka and Segal 1995a). Upon release (e.g., from a muscle fiber or micropipette onto an arteriole), the stimulus may activate guanylate cyclase in smooth muscle cells. This thereby elicits relaxation and vasodilation in the vicinity of the stimulus (Kurjiaka and Segal 1995a) via pharmacomechanical coupling (Somlyo and Somlyo 1994).

Acetylcholine dilates arterioles by activating muscarinic receptors on endothelial cells (Rivers and Duling 1992; Segal 1991). With ACh, not only can endothelial cells release nitric oxide (Kaley et al. 1992) to activate guanylate cyclase, but they also hyperpolarize (Chen and Cheung 1992; Segal and Bény 1992). This change in membrane potential can spread (i.e., is conducted) from cell to cell along the wall of arterioles via gap junctions (Hirst and Neild 1978; Little, Beyer, and Duling 1995; Segal and Bény 1992; Segal and Duling 1986b; Welsh and Segal 1998). The conducted vasomotor response can thereby give rise to smooth muscle cell relaxation at remote locations via electromechanical coupling (Segal, Damon, and Duling 1989; Somlyo and Somlyo 1994; Welsh and Segal 1998; Xia and Duling 1995). This explanation implies that the generation of an electric signal that is rapidly transmitted from its site of initiation underlies the ability of the resistance network to respond in a coordinated manner. It also gives rise to a fundamental question: What is the physiological stimulus for triggering conduction in vivo?

Neuromuscular Junctions

The neuromuscular junction (NMJ) of skeletal muscle is the site of synthesis, storage, and release of ACh as well as the apparent physiological source of ACh in this tissue (Katz and Miledi 1973; Pierzga and Segal 1994; Welsh and Segal 1997). The relaxing effect of ACh on arterioles and the interplay among microvessels during blood flow regulation (Sarelius 1993; Segal 1991; Segal, Damon, and Duling 1989; Sweeney and Sarelius 1989; Welsh and Segal 1997) suggests a mechanism by which the recruitment of muscle fibers may be coupled to the local control of blood flow (Honig 1979; Pierzga and Segal 1994). We reasoned that such a relationship may be reflected in the pattern of motor innervation with respect to arteriolar network topology. In other words, the distances between NMJs and microvessels may reflect the likelihood of ACh released at NMJs acts as a paracrine substance on the microvasculature (Pierzga and Segal 1994). In hamster striated muscle, we found that capillaries, which may conduct vasomotor signals into terminal arterioles (Dietrich and Tyml 1992; Segal 1991; Song and Tyml 1993), were typically located within a distance of 10–15 μm of NMJs irrespective of location within the muscle. Thus, the time required for ACh to diffuse from a NMJ to capillary endothelial cells would be ~0.1 s, which is consistent with the half-life of ACh in the synaptic cleft (Katz and Miledi 1973).

A Working Hypothesis

Of all the arterioles sampled, those from which MVUs arise were consistently closest to NMJs and are therefore the most likely resistance vessels to be reached by stimuli derived from NMJs (Pierzga and Segal 1994). We therefore hypothesized that the physiological recruitment (i.e., via motor nerves) of muscle fibers correspondingly triggers the perfusion of MVUs, resulting in a rapid increase in capillary perfusion. One should recognize that this hypothesis contrasts with the localized metabolic vasodilation seen when fibers are activated electrically (Gorczynski, Klitzman, and Duling 1978), i.e., without ACh release. A potential constraint on this relationship is that each fiber typically has only one NMJ (Burke 1981), whereas many MVUs are required to encompass the length of a single muscle fiber (see figure 11.1) (Bloch and Iberall 1982; Emerson and Segal 1997).

Nevertheless, given the ability of ACh to trigger responses conducted through multiple branches (Kurjiaka and Segal 1995a; Segal 1991; Segal, Damon, and Duling 1989), the activation of NMJs could facilitate the distribution of blood flow among terminal arterioles. This could perfuse MVUs in accord with the recruitment of particular muscle fibers (Emerson and Segal 1997). Furthermore, ascending vasodilation into proximal arterioles would progressively increase the volume of capillary flow as exercise intensity (i.e., motor unit activation) increases (Welsh and Segal 1997).

Synthesis

Each MVU encompasses many muscle fibers (see figure 11.1) (Bloch and Iberall 1982; Emerson and Segal 1997; Fuglevand and Segal 1997). Therefore, we propose that an increase in blood flow to an active fiber or motor unit requires increased perfusion to many fibers surrounding those actually re-

cruited during performance (Emerson and Segal 1997; Fuglevand and Segal 1997). The release of ACh at NMJs could facilitate this concerted response by triggering the conduction of vasodilation among distal branches of the arteriolar network. Indeed, such an arrangement for MVU perfusion may constitute a feed-forward mechanism whereby promoting the availability of oxygen in capillaries may actually precede the activation of many of the muscle fibers. Hence, upon the firing of additional motor units, this augmented capillary perfusion would facilitate oxygen consumption (via increased extraction) prior to significant increases in muscle blood flow (Fuglevand and Segal 1997; Honig, Odoroff, and Frierson 1980). Moreover, this explanation is consistent with the findings from exercising humans (Andersen and Saltin 1985; Saltin 1985) discussed earlier. As exercise continues and intensity increases, substances released from contracting fibers can effectively govern steady-state flow in accord with metabolic demand. The progressive ascending vasodilation of proximal branches is reflected as a corresponding increase in muscle perfusion. This occurs with a magnitude ultimately determined by vasodilation encompassing the feed arteries (Segal and Duling 1986a; Welsh and Segal 1997; Williams and Segal 1993). Furthermore, this progression implies that once the distal branches of the network dilate, capillary blood flow can increase only by dilation of proximal branches.

Physical Forces and Blood Flow Control

The following information describes physical forces and blood flow control. This section discusses the muscle pump and active vasomotor responses to changing muscle length.

The Muscle Pump

During exercise, skeletal muscle undergoes rhythmic changes in length and tension, producing mechanical perturbations within the tissue that intermittently reduce and augment muscle blood flow (Anrep and Von Saalfeld 1935; Barcroft and Millen 1939; Supinski et al. 1986; Wisnes and Kirkebo 1976). As seen in the myocardium (Feigl 1983), peak flow occurs during the relaxation between contractions, with impediment of flow during contractions. The intermittent compression and relaxation of intramuscular veins correspondingly lowers postcapillary pressure (Folkow et al. 1971; Laughlin 1987). In the presence of a constant arterial pressure, these mechanical effects during rhythmic exercise will increase the driving force for capillary perfusion (Sheriff, Rowell, and Scher 1993) as well as impart substantial energy to the movement of blood though the muscle (Folkow et al. 1971; Laughlin 1987). Because these physical forces are exerted upon vessels located within the muscle (or between muscle fascicles (Gray, Carlsson, and Staub 1967; Gray and Staub 1967), muscular contraction will impede perfusion even during maximal vasodilation.

The original studies of the mechanical forces exerted by muscle on its vascular supply (Anrep and Von Saalfeld 1935; Barcroft and Millen 1939) were based upon measurements of total muscle blood flow. The vasculature was considered to be passive and assumed to be uniformly affected. In marked contrast, we have recently found that the effect of changing muscle length on vascular geometry highly depended upon the orientation of vessels to the muscle fibers (Nakao and Segal 1995). Muscle shortening resulted in axial compression and kinking of vessels aligned with muscle fibers, whereas vessels perpendicular to fibers were simultaneously extended and straightened. Reciprocal changes occurred upon muscle lengthening. Although total muscle blood flow indeed reflects the sum of all mechanical (as well as other) effects on its vascular supply, researchers and physicians should now recognize that substantive differences exist in the mechanical effects of muscle fibers on individual vessels (Nakao and Segal 1995). Moreover, such differences may contribute to the pumping action of skeletal muscle during rhythmic exercise in ways that remain to be defined.

Active Vasomotor Responses to Changing Muscle Length

We recently investigated whether resistance vessels were actually sensitive to changes in muscle length (Welsh and Segal 1997). A novel preparation of the hamster retractor muscle (Nakao and Segal 1995) enabled precise control of muscle and sarcomere length through a range defined by the classic length-tension relationship (Lieber, Loren, and Friden 1994; Rack and Westbury 1969) while concomitantly evaluating microvascular hemodynamics. At rest, increasing muscle length produced a progressive constriction of arterioles and a reduction in blood flow. Passive shortening resulted in arteriolar dilation and elevated flow. These active vasomotor responses were eliminated by application of a vasodilator (which eliminates smooth muscle cell reactivity), by blockade of action potentials with tetradotoxin, by pharmacological inhibition of alpha-adrenoceptors on the vasculature, and by depleting perivascular sympathetic nerve terminals of neurotransmitter (Welsh and Segal 1997). These findings lead to the conclusion that lengthening-induced vasoconstriction results from the generation of action potentials in periarteriolar sympathetic nerves (see figure 11.1) (Fleming et al. 1989; Kurjiaka and Segal 1995b; Marshall 1982; Welsh and Segal 1997). This releases norepinephrine to produce contraction of vascular smooth muscle. Furthermore, these responses initiated on arterioles within the muscle were propagated antidromically into the feed arteries. These findings from resting muscle demonstrate that resistance vessels can *actively respond* to changes in muscle length independent of changes in tissue metabolism.

At the cellular level, these findings reveal a novel mechanotransduction sequence in which changing muscle fiber length per se actively regulates the caliber of resistance vessels by directing perivascular sympathetic nerve activity. Although this mechanism contrasts the cell-to-cell conduction described earlier, these responses have two effects. First, they reinforce the concept that events triggered on arterioles within the muscle can effect feed artery resistance external to the muscle. Second, they exemplify the coordinated interaction among distal and proximal branches of the resistance network (Segal 1994). One outcome of feed arteries responding in concert with arterioles is the maintenance of microvascular perfusion pressure (Segal and Duling 1986a; Welsh and Segal 1997). Moreover, these active vasomotor responses to changing muscle length indicate that the traditional explanation of the muscle pump (which assumes a passive vasculature) may require revision.

Sympathetic Innervation and Local Flow Control

As exercise intensifies, the proportion of cardiac output directed to skeletal muscle increases (Rowell 1986). This systemic redistribution of blood flow is mediated primarily by sympathetic constriction of the splanchnic and renal circulations (Rowell 1986). Blood flow to skeletal muscle can also be limited by sympathetic vasoconstriction, particularly when another large mass of muscle is active simultaneously (Rowell 1986; Saltin 1985; Seals 1989; Secher et al. 1977). Indeed, when the majority of cardiac output is directed to skeletal muscle, the restriction of muscle blood flow becomes a highly effective means for controlling arterial pressure (O'Leary,

Rowell, and Scher 1991). Thus, during whole-body exercise, each muscle group receives only a fraction of the blood flow it would otherwise get if it were the only active group in the body (Andersen and Saltin 1985; Saltin 1985; Secher et al. 1977). In practice, the sympathetic nervous system can thereby restrict oxygen consumption during exercise (Costin and Skinner 1971; Thompson and Mohrman 1983). However, muscle contraction also results in functional sympatholysis, whereby sympathetic vasoconstriction is overcome by metabolic demand of the muscle fibers (Costin and Skinner 1971; Kjellmer 1965; Remensnyder, Mitchell, and Sarnoff 1962; Thomas and Victor 1998). For example, substances released by muscle fibers (e.g., nitric oxide) and motor nerves (e.g., acetylcholine) can inhibit norepinephrine (NE) release and thereby promote vasodilation (McGillivray-Anderson and Faber 1990; Thomas and Victor 1998; Vanhoutte 1974; Vanhoutte, Verbeuren, and Webb 1981). The NE released during exercise may also stimulate endothelial cells to release factors that oppose smooth muscle cell contraction (Cocks and Angus 1983).

Sympathetic Vasoconstriction and Conducted Vasodilation

Based upon relationships considered in the preceding discussion, we recently investigated whether vasoconstriction induced by sympathetic nerves would interact with the conduction of vasodilation in arterioles (Kurjiaka and Segal 1995b). We found that when triggered with ACh, the conduction of vasodilation significantly attenuated sympathetic vasoconstriction. This suggests a novel explanation for functional sympatholysis. Furthermore, the activation of sympathetic nerves significantly depressed conducted vasodilation (Kurjiaka and Segal 1995b). This indicates that sympathetic nerve activity can influence cell-to-cell communication in arterioles and thereby affect the coordination of local flow control. Note that the suppression of NE release by tissue metabolites and nitric oxide release (Thomas and Victor 1998) would indirectly promote conduction and thereby facilitate ascending vasodilation.

Locus of Flow Control Revisited

The competition between sympathetic vasoconstriction and the vasodilation arising from muscle fiber activity leads to the fundamental question of where in the resistance network the volume of blood flow is actually controlled during exercise (see figure 11.1).

Capillary Perfusion

The extraction of oxygen by active muscle increases when the muscle's blood flow is restricted (Granger, Goodman, and Granger 1976; Secher et al. 1977; Shepherd et al. 1973; Stainsby and Otis 1964). This response appropriately facilitates aerobic energy production despite any limitation in oxygen delivery. It also indicates that the terminal arterioles that control capillary (i.e., MVU) perfusion remain dilated. Since microvessels are embedded within the muscle, vasodilator stimuli induced by muscle fiber contraction apparently maintain the arterioles in a dilated state. The lowering of venous PO_2 (Andersen and Saltin 1985; Saltin 1985; Secher et al. 1977) further indicates that active muscle cells are more effective in removing oxygen as blood flow becomes limiting, which can be explained by a corresponding fall in intracellular PO_2.

Feed Artery Control

The location of feed arteries external to the muscle physically removes them from the direct influence of the vasoactive products of muscular activity. Indeed, evidence suggests that at this level, sympathetic vasoconstriction effectively limits blood flow during exercise (Folkow, Sonnenschein, and Wright 1971; Lind and Williams 1979). Thus, feed arteries may integrate autonomic vasoconstrictor signals descending along sympathetic nerves with vasodilator signals ascending from within the muscle. The diameter of feed arteries would thereby reflect the sum of these influences at any particular moment. Under such conditions, the volume of blood flow to active muscle (and thereby its oxygen consumption) could be effectively limited external to the muscle by sympathetic nerve activity. Simultaneously, extraction could be maximized by vasodilation of arterioles within the muscle.

Regional Distribution of Flow

Blood flow to exercising muscle varies in proportion to differences in oxidative capacity. Thus, one determinant of blood flow distribution appears to be the proportion of oxidative fibers recruited during exercise (Laughlin and Armstrong 1982; Mackie and Terjung 1983). However, the anatomic and physiological basis of such variation has been ambiguous. Recent work has shown differences between slow-twitch (soleus) and fast-twitch (extensor digitorum longus) muscles of the rat with respect to both microvascular architecture (Williams and Segal 1992) and feed artery reactivity (Williams and Segal 1993).

We therefore suggest that the organization and control of resistance vessels may vary in accord with the physiological properties of the muscles involved in the exercise task. For example, the dilation of particular feed arteries supplying respective muscles could be explained by ascending vasodilation originating in the vicinity of active muscle fibers. This would facilitate the distribution of cardiac output in accord with muscle recruitment patterns (Burke 1981). In turn, differential responses within the arteriolar network could explain the preferential increments in flow to specific regions of a muscle, e.g., red versus white gastrocnemius (Laughlin and Armstrong 1982; Mackie and Terjung 1983). The metabolic activity of respective muscle fiber types may also exert differential effects on the resistance vasculature, though this remains to be determined.

Summary and Conclusions

This chapter is concerned with the dynamic processes that give rise to the control of muscle blood flow during exercise. At the cellular level, we have considered three factors. The first is how the activity of muscle fibers gives rise to vasodilator responses that encompass progressively greater portions of the resistance network. The second involves how signaling between smooth muscle and endothelial cells contributes to the appropriate distribution and magnitude of muscle blood flow. The third is concerned with how motor and autonomic innervation may contribute to the coupling between muscle work and microvascular perfusion. A theme central to each of these relationships is that blood flow control is exerted not only within the muscle but through external sites as well (see figure 11.1). Moreover, respective mechanisms of vasomotor control do not appear to be uniformly effective at each location in the resistance network.

Within the muscle, we propose that conducted and metabolic mechanisms of vasodilation work with the muscle pump against a background of myogenic smooth muscle cell tone (maintained by transmural pressure) that may be modulated by endothelial cell production of vasodilators (determined by shear stress on endothelial cells). Whereas sympathetic nerves can induce arteriolar constriction and suppress conduction, this effect can be effectively overridden in response to the activity of muscle fibers. Not only does conduction appear to be important in coordinating MVU perfusion and blood flow distribution within the muscle, it can also give rise to ascending vasodilation and thereby influence blood flow magnitude.

Externally, feed arteries govern the volume of flow entering the muscle. In the presence of resting vasomotor tone, ascending vasodilation will increase total flow. However, sympathetic nerve activity will be a primary controller of vessel diameter because the feed arteries are physically removed from the products of muscle fiber contraction. This role in governing total flow may be particularly important with respect to maintaining arterial pressure during intense aerobic activity.

Resolving how the perfusion of MVUs and corresponding regions of the arteriolar network are adjusted to the pattern of motor unit activity will provide fundamentally new insight into the coupling between the musculoskeletal and cardiovascular systems. We anticipate that this information will contribute to minimizing the pathophysiological consequences of inactivity that are associated with aging or injury as well as to suggesting novel strategies for enhancing physical performance. Our long-term goal is to provide this insight and thereby enhance the quality of life through promoting the benefits of physical activity.

Acknowledgments

Geoffrey G. Emerson, Donald G. Welsh, Andrew J. Fuglevand, and Colin J. Carati contributed valuable discussion and critique of this chapter. Dr. Segal's research is supported by the National Institutes of Health (grants HL56786 and HL41026) and the American Heart Association. This chapter was prepared during the tenure of an Established Investigatorship award from the American Heart Association and Genentech, Inc.

References

Andersen, P., and B. Saltin. 1985. Maximal perfusion of skeletal muscle in man. *Journal of Physiology-London* 366:233–249.

Anrep, G.V., and E. Von Saalfeld. 1935. The blood flow through the skeletal muscle in relation to its contraction. *Journal of Physiology-London* 85:375–399.

Barcroft, H., and J.L.E. Millen. 1939. The blood flow through muscle during sustained contraction. *Journal of Physiology-London* 97:17–31.

Bloch, E.H., and A.S. Iberall. 1982. Toward a concept of the functional unit of mammalian skeletal muscle. *American Journal of Physiology* 242:R411–R420.

Burke, R. 1981. Motor units: Anatomy, physiology, and functional organization. In *Handbook of Physi-*

ology, ed. V.B. Brooks, 345–422. Bethesda, MD: American Physiological Society.

Chen, G., and D. Cheung. 1992. Characterization of acetylcholine induced membrane hyperpolarization in endothelial cells. *Circulation Research* 70:257–263.

Cocks, T.M., and J.A. Angus. 1983. Endothelium-dependent relaxation of coronary arteries by noradrenaline and serotonin. *Nature* 305:627–630.

Costin, J.C., and N.S. Skinner, Jr. 1971. Competition between vasoconstrictor and vasodilator mechanisms in skeletal muscle. *American Journal of Physiology* 220:462–466.

Davies, P.F. 1995. Flow-mediated endothelial mechanotransduction. *Physiological Reviews* 75:519–560.

Delashaw, J.B., and B.R. Duling. 1988. A study of the functional elements regulating capillary perfusion in striated muscle. *Microvascular Research* 36:162–171.

Delashaw, J.B., and B.R. Duling. 1991. Heterogeneity in conducted arteriolar vasomotor response is agonist dependent. *American Journal of Physiology* 260:H1276–H1282.

Dietrich, H.H., and K. Tyml. 1992. Capillary as a communicating medium in the microvasculature. *Microvascular Research* 43:87–99.

Duling, B.R., and R.M. Berne. 1970. Propagated vasodilation in the microcirculation of the hamster cheek pouch. *Circulation Research* 26:163–170.

Emerson, G.G., and S.S. Segal. 1997. Alignment of microvascular units along skeletal muscle fibers of hamster retractor. *Journal of Applied Physiology* 82:42–48.

Eriksson, E., and R. Myrhage. 1972. Microvascular dimensions and blood flow in skeletal muscle. *Acta Physiologica Scandinavica* 86:211–222.

Feigl, E.O. 1983. Coronary physiology. *Physiological Reviews* 63:1–205.

Fleming, B.P., I.L. Gibbins, J.L. Morris, and B.J. Gannon. 1989. Noradrenergic and peptidergic innervation of the extrinsic vessels and microcirculation of the rat cremaster muscle. *Microvascular Research* 38:255–268.

Folkow, B. 1964. Description of the myogenic response. *Circulation Research* 15:279–287.

Folkow, B., U. Haglund, M. Jodal, and O. Lundgren. 1971. Blood flow in the calf muscle of man during heavy rhythmic exercise. *Acta Physiologica Scandinavica* 81:157–163.

Folkow, B., R.R. Sonnenschein, and D.L. Wright. 1966. Differential influences of nervous and local humoral factors on large and small precapillary vessels of skeletal muscle. In *Circulation in*

Skeletal Muscle, ed. O. Hudlicka, 165–169. New York: Pergamon Press.

Folkow, B., R.R. Sonnenschein, and D.L. Wright. 1971. Loci of neurogenic and metabolic effects on precapillary vessels of skeletal muscle. *Acta Physiologica Scandinavica* 81:459–471.

Fuglevand, A.J., and S.S. Segal. 1997. Simulation of motor unit recruitment and microvascular unit perfusion: Spatial considerations. *Journal of Applied Physiology* 83:1223–1234.

Furchgott, R.F., and J.V. Zawadzki. 1980. The obligatory role of endothelial cells in the relaxation of arterial smooth muscle by acetylcholine. *Nature* 288:373–376.

Gorczynski, R.J., B. Klitzman, and B.R. Duling. 1978. Interrelations between contracting striated muscle and precapillary microvessels. *American Journal of Physiology* 235:H449–H504.

Granger, H.J., A.H. Goodman, and D.N. Granger. 1976. Role of resistance and exchange vessels in local microvascular control of skeletal muscle oxygenation in the dog. *Circulation Research* 38:379–385.

Gray, S.D., E. Carlsson, and N.C. Staub. 1967. Site of increased vascular resistance during isometric muscle contraction. *American Journal of Physiology* 213:683–689.

Gray, S.D., and N.C. Staub. 1967. Resistance to blood flow in leg muscles of dog during tetanic isometric contraction. *American Journal of Physiology* 213:677–682.

Haddy, F.J., and J.B. Scott. 1975. Metabolic factors in peripheral circulatory regulation. *Federation Proceedings* 34:2006–2011.

Hester, R.L. 1990. Venular-arteriolar diffusion of adenosine in hamster cremaster microcirculation. *American Journal of Physiology* 258:H1918–H1924.

Hilton, S.M. 1959. A peripheral arterial conducting mechanism underlying dilatation of the femoral artery and concerned in functional vasodilatation in skeletal muscle. *Journal of Physiology-London* 149:93-111.

Hirst, G.D., and T.O. Neild. 1978. An analysis of excitatory junctional potentials recorded from arterioles. *Journal of Physiology.* London 80:87–104.

Honig, C.R. 1979. Contributions of nerves and metabolites to exercise vasodilation: A unifying hypothesis. *American Journal of Physiology* 236:H705–H719.

Honig, C.R., C.L. Odoroff, and J.L. Frierson. 1980. Capillary recruitment in exercise: Rate, extent, uniformity, and relation to blood flow. *American Journal of Physiology* 238:H31–H42.

Kaley, G., A. Koller, J.M. Rodenburg, E.J. Messina, and M.S. Wolin. 1992. Regulation of arteriolar tone and responses via L-arginine pathway in skeletal muscle. *American Journal of Physiology* 262:H987–H992.

Katz, B., and R. Miledi. 1973. The binding of acetylcholine to receptors and its removal from the synaptic cleft. *Journal of Physiology-London* 231:549–574.

Kjellmer, I. 1965. On the competition between metabolic vasodilatation and neurogenic vasoconstriction in skeletal muscle. *Acta Physiologica Scandinavica* 63:450–459.

Koller, A., and G. Kaley. 1991. Endothelial regulation of wall shear stress and blood flow in skeletal muscle microcirculation. *American Journal of Physiology* 260:H862–H868.

Kuo, L., W.M. Chilian, and M.J. Davis. 1991. Interaction of pressure- and flow-induced responses in porcine coronary resistance vessels. *American Journal of Physiology* 261:H1706—H1715.

Kuo, L., M.J. Davis, and W.M. Chilian. 1992. Endothelial modulation of arteriolar tone. *News in Physiological Sciences* 7:5–9.

Kurjiaka, D.T., and S.S. Segal. 1995a. Hemodynamic responses to conducted vasodilation in arteriolar networks. *American Journal of Physiology* 269:H1723–H1728.

Kurjiaka, D.T., and S.S. Segal. 1995b. Interaction between conducted vasodilation and sympathetic nerve activation in arterioles of hamster striated muscle. *Circulation Research* 76:885–891.

Laughlin, M.H. 1987. Skeletal muscle blood flow capacity: Role of muscle pump in exercise hyperemia. *American Journal of Physiology* 253:H993–H1004.

Laughlin, M.H., and R.B. Armstrong. 1982. Muscular blood flow distribution patterns as a function of running speed in rats. *American Journal of Physiology* 243:H296–H306.

Lieber, R.L., G.J. Loren, and J. Friden. 1994. In vivo measurement of human wrist extensor muscle sarcomere length changes. *Journal of Neurophysiology* 71:874–881.

Lind, A.R., and C.A. Williams. 1979. The control of blood flow through human forearm muscles following brief isometric contractions. *Journal of Physiology-London* 288:529–547.

Little, T.L., E.C. Beyer, and B.R. Duling. 1995. Connexin 43 and connexin 40 gap junctional proteins are present in arteriolar smooth muscle and endothelium in vivo. *American Journal of Physiology* 268:H729–H739.

Little, T.L., J. Xia, and B.R. Duling. 1995. Dye tracers define differential endothelial and smooth muscle coupling patterns within the arteriolar wall. *Circulation Research* 76:498–504.

Lund, N., D.N. Damon, and B.R. Duling. 1987. Capillary grouping in hamster tibialis anterior muscles: Flow patterns, and physiological significance. *International Journal of Microcirculation:Clinical and Experimental* 5:359–372.

Mackie, B.G., and R.L. Terjung. 1983. Blood flow to different skeletal muscle fiber types during contraction. *American Journal of Physiology* 245:H265–H275.

Marshall, J.M. 1982. The influence of the sympathetic nervous system on individual vessels of the microcirculation of skeletal muscle of the rat. *Journal of Physiology-London* 332:169–186.

McGillivray-Anderson, K.M., and J.E. Faber. 1990. Effect of acidosis on contraction of microvascular smooth muscle by α_1- and α_2-adrenoreceptors Implications for neural and metabolic regulation. *Circulation Research* 66:1643–1657.

Meininger, G.A., and M.J. Davis. 1992. Cellular mechanisms involved in the vascular myogenic response. *American Journal of Physiology* 263:H647–H659.

Nakao, M., and S.S. Segal. 1995. Muscle length alters geometry of arterioles and venules in hamster retractor. *American Journal of Physiology* 268:H336–H344.

O'Leary, D.S., L.B. Rowell, and A.M. Scher. 1991. Baroreflex-induced vasoconstriction in active skeletal muscle of conscious dogs. *American Journal of Physiology* 260:H37–H41.

Ordway, G.A., D.L. Floyd, J.C. Longhurst, and J.H. Mitchell. 1984. Oxygen consumption and hemodynamic responses during graded treadmill exercise in the dog. *Journal of Applied Physiology* 57:601–607.

Pierzga, J.M., and S.S. Segal. 1994. Spatial relationships between neuromuscular junctions and microvessels in hamster cremaster muscle. *Microvascular Research* 48:50–67.

Plyley, M.J., G.J. Sutherland, and A.C. Groom. 1976. Geometry of the capillary network in skeletal muscle. *Microvascular Research* 11:161–173.

Pohl, U., K. Herlan, A. Huang, and E. Bassenge. 1991. EDRF-mediated shear-induced dilation opposes myogenic vasoconstriction in small rabbit arteries. *American Journal of Physiology* 261:H2016–H2023.

Pohl, U., J. Holtz, R. Busse, and E. Bassenge. 1986. Crucial role of endothelium in the vasodilator response to increased flow in vivo. *Hypertension* 8:37–44.

Rack, P.M.H., and D.R. Westbury. 1969. The effects of length and stimulus rate on tension in the isometric cat soleus muscle. *Journal of Physiology-London* 204:443–460.

Remensnyder, J.P., J.H. Mitchell, and S.J. Sarnoff. 1962. Functional sympatholysis during muscular activity. *Circulation Research* 11:370–380.

Rhodin, J.A.G. 1967. The ultrastructure of mammalian arterioles and precapillary sphincters. *Journal of Ultrastructural Research* 18:181–223.

Rivers, R.J., and B.R. Duling. 1992. Arteriolar endothelial cell barrier separates two populations of muscarinic receptors. *American Journal of Physiology* 262:H1311–H1315.

Rowell, L.B. 1986. Circulatory adjustments to dynamic exercise. In *Human circulation: Regulation during physical stress*, ed. L.B. Rowell, 213–256. New York: Oxford University Press.

Saltin, B. 1985. Hemodynamic adaptations to exercise. *American Journal of Cardiology* 55:42D–47D.

Saltin, B., and P.D. Gollnick. 1983. Skeletal muscle adaptability: Significance for metabolism and performance. In *Handbook of physiology*, ed. L.D. Peachey, R.H. Adrian, and S.R. Geiger, 555–631. Bethesda, MD: American Physiological Society.

Sarelius, I.H. 1993. Cell and oxygen flow in arterioles controlling capillary perfusion. *American Journal of Physiology* 265:H1682–H1687.

Seals, D.R. 1989. Influence of muscle mass on sympathetic neural activation during isometric exercise. *Journal of Applied Physiology* 67:1801–1806.

Secher, N., J.P. Clausen, I. Noer, and J. Trap-Jensen. 1977. Central and regional circulatory effects of adding arm exercise to leg exercise. *Acta Physiologica Scandinavica* 100:288–297.

Segal, S.S. 1991. Microvascular recruitment in hamster striated muscle: Role for conducted vasodilation. *American Journal of Physiology* 261:H181–H189.

Segal, S.S. 1994. Cell-to-cell communication coordinates blood flow control. *Hypertension Dallas* 23 (part 2):1113–1120.

Segal, S.S., and J.L. Bény. 1992. Intracellular recording and dye transfer in arterioles during blood flow control. *American Journal of Physiology* 263:H1–H7.

Segal, S.S., D.N. Damon, and B.R. Duling. 1989. Propagation of vasomotor responses coordinates arteriolar resistances. *American Journal of Physiology* 256:H832–H837.

Segal, S.S., and B.R. Duling. 1986a. Communication between feed arteries and microvessels in hamster striated muscle: Segmental vascular responses are functionally coordinated. *Circulation Research* 59:283–290.

Segal, S.S., and B.R. Duling. 1986b. Flow control among microvessels coordinated by intercellular conduction. *Science* 234:868–870.

Shepherd, A.P., H.J. Granger, E.E. Smith, and A.C. Guyton. 1973. Local control of tissue oxygen delivery and its contribution to the regulation of cardiac output. *American Journal of Physiology* 225:747–755.

Shepherd, J.T. 1983. Circulation to skeletal muscle. In *Handbook of physiology. The Cardiovascular system. Peripheral circulation and organ blood flow, part 1*, ed. J.T. Shepherd, and F.M. Abboud, 319–370. Bethesda, MD: American Physiological Society.

Sheriff, D.D., L.B. Rowell, and A.M. Scher. 1993. Is rapid rise in vascular conductance at onset of dynamic exercise due to muscle pump? *American Journal of Physiology* 265:H1227–H1234.

Skalak, T.C., and G.W. Schmid-Schonbein. 1986. The microvasculature in skeletal muscle IV. A model of the capillary network. *Microvascular Research* 32:333–347.

Somlyo, A.P., and A.V. Somlyo. 1994. Signal transduction and regulation in smooth muscle. *Nature* 372:231–236.

Song, H., and K. Tyml. 1993. Evidence for sensing and integration of biological signals by the capillary network. *American Journal of Physiology* 265:H1235–H1242.

Sparks, H.V. 1980. Effect of local metabolic factors on vascular smooth muscle. In *Handbook of physiology. The Cardiovascular system. Vascular smooth muscle*, ed. D.F. Bohr, A.P. Somlyo, and H.V. Sparks, 475–513. Bethesda, MD: American Physiological Society.

Stainsby, W.N., and A.B. Otis. 1964. Blood flow, blood oxygen tension, oxygen uptake, and oxygen transport in skeletal muscle. *American Journal of Physiology* 206:858–866.

Supinski, G.S., H. Bark, A. Guanciale, and S.G. Kelsen. 1986. Effect of alterations in muscle fiber length on diaphragm blood flow. *Journal of Applied Physiology* 60:1789–1796.

Sweeney, T.E., and I.H. Sarelius. 1989. Arteriolar control of capillary cell flow in striated muscle. *Circulation Research* 64:112–120.

Thomas, G.D., and R.G. Victor. 1998. Nitric oxide mediates contraction-induced attenuation of sympathetic vasoconstriction in rat skeletal muscle. *Journal of Physiology-London* 506:817–826.

Thompson, L.P., and D.E. Mohrman. 1983. Blood flow and oxygen consumption in skeletal muscle during sympathetic stimulation. *American Journal of Physiology* 245:H66–H71.

Vanhoutte, P.M. 1974. Inhibition by acetylcholine of adrenergic neurotransmission in vascular smooth muscle. *Circulation Research* 34:317–326.

Vanhoutte, P.M., T.J. Verbeuren, and R.C. Webb. 1981. Local modulation of adrenergic neuroeffector interaction in the blood vessel well. *Physiological Reviews* 61:151–247.

Weibel, E.R. 1984. Delivering oxygen to the cells. Structure and function in the mammalian respiratory system. In *The Pathway for Oxygen*, 175–210. Cambridge, MA: Harvard University Press.

Welsh, D.G., and S.S. Segal. 1996. Muscle length directs sympathetic nerve activity and vasomotor tone in resistance vessels of hamster retractor. *Circulation Research* 79:551–559.

Welsh, D.G., and S.S. Segal. 1997. Coactivation of resistance vessels and muscle fibers with acetylcholine release from motor nerves. *American Journal of Physiology* 273:H156–H163.

Welsh, D.G., and S.S. Segal. 1998. Endothelial and smooth muscle cell conduction in arterioles controlling blood flow. *American Journal of Physiology* 274:H178–H186.

Williams, D.A., and S.S. Segal. 1992. Microvascular architecture in rat soleus and extensor digitorum longus muscles. *Microvascular Research* 43:192–204 [erratum corrected in *Microvascular Research* 43:358].

Williams, D.A., and S.S. Segal. 1993. Feed artery role in blood flow control to rat hindlimb skeletal muscles. *Journal of Physiology-London* 463:631–646.

Wisnes, A., and A. Kirkebø. 1976. Regional distribution of blood flow in calf muscles of rat during passive stretch and sustained contraction. *Acta Physiologica Scandinavica* 96:256–266.

Xia, J., and B.R. Duling. 1995. Electromechanical coupling and the conducted vasomotor response. *American Journal of Physiology* 269:H2022–H2030.

Chapter 12

Integration of Muscle Blood Flow and Cardiac Output

Niels H. Secher, Atsuko Kagaya, and Bengt Saltin

Almost a century ago, Athanasiu and Carvallo (1898) demonstrated a 25% reduction in skeletal muscle volume during contraction due to the squeezing of blood toward the heart by the muscle pump (Beecher, Field, and Krogh 1936). The enhancement of central blood volume facilitates the increase in cardiac output necessary for muscle perfusion. However, activation of the muscle pump alone is insufficient to maintain blood pressure as illustrated in patients with autonomic dysfunction (see chapter 24). With the orthostatic challenge of upright posture in humans, blood pressure is tightly regulated to optimize the perfusion of the tissues most in need of oxygen, including the brain. The recent finding that maximal cardiac output in humans is insufficient to supply the peak muscle blood flow capacity of all the muscles simultaneously engaged in exercise underscores the importance of the sympathetic nervous system for blood pressure regulation. It also shows the importance of selectivity in the effectiveness of sympathetically mediated vasoconstrictor activity.

At the regional level, blood flow regulation may be viewed as a balance between local vasodilation and sympathetic vasoconstriction whereby the available cardiac output is directed toward muscles most in demand of oxygen. Whether oxygen delivery or blood pressure is the regulated variable is a subject of debate. One could argue that blood pressure is preserved at the expense of blood flow during whole-body exercise. However, when two small muscle groups are contracting simultaneously, sympathetic vasoconstriction reduces blood flow in the muscle contracting at a lighter workload in order to raise perfusion pressure in the muscle group contracting at a more intense load. This latter pattern occurs despite a presumable reserve of cardiac output and without the threat of a fall in blood pressure. In this situation, sympathetic activity, and therefore blood pressure, may subserve the more primary signals for oxygen delivery.

Ohm's Law and Poisseule's Equation

Evaluation of cardiovascular control during exercise is complicated by the fact that both blood pressure and muscle blood flow depend on cardiac output and peripheral resistance. However, the relationship between blood pressure and resistance is opposite to that of flow and resistance. Application of Ohm's law illustrates the relationships between cardiac output, blood pressure, and peripheral resistance:

$$\text{blood pressure} = \text{cardiac output} \times \text{vascular resistance}$$

and Poisseule's equation (simplified) defines tissue blood flow:

$$\text{blood flow} = \text{perfusion pressure} \times \text{vessel area}$$

While blood pressure is supported by peripheral resistance, tissue blood flow depends on a lowering of vascular resistance. Within a given range of cardiac output, blood pressure subserves muscle blood flow by establishing an adequate perfusion pressure. However, the vasodilator capacity of muscle can sufficiently lower vascular resistance to threaten blood pressure. Thus, in humans, regulation of perfusion pressure is a fundamental regulatory challenge.

From the above equations, one can deduce that cardiac output is also inversely related to vascular resistance. This concept has been demonstrated experimentally. Especially notable is the close relationship between local vascular adaptations with training and the observed increase in cardiac output and $\dot{V}O_2max$ without any change in blood pressure (Clausen 1976) (see chapter 18).

One approach for insight into the control of the circulation specific to exercise is to compare the exercise response with other situations associated with a significant increase in cardiac output but with little sympathetic activation. The transient decrease in blood pressure when moving from the supine to upright posture is also a well-known clinical problem. This especially occurs in the young because the pressure drop may last long enough to affect cerebral blood flow and oxygenation. This could then cause dizziness and eventual fainting (Madsen et al. 1995; Sprangers et al. 1991). In this context, it may be a problem that the influence on the cardiovascular system from central command is directed toward the heart while the delayed influence from the muscle controls peripheral resistance (Williamson et al. 1996). Thus, in situations where peripheral resistance is uncontrolled, the fall in blood pressure affects muscle perfusion and oxygen delivery.

Exercise

At the onset of dynamic exercise, peripheral resistance decreases rapidly and regularly leads to a drop in blood pressure (Holmgren 1956; Sprangers et al. 1991). This especially happens when the intensity of exercise is low (see figure 12.1). In fact, central command may support blood pressure during the rapid decrease in vascular resistance at the onset of exercise (Secher, Kjaer, and Galbo 1988). A need exists to evaluate blood pressure at the onset of exercise in view of the time taken for translocation of blood. Although the increase in cardiac output may occur parallel with that of leg blood flow (Eriksen et al. 1990), it is regularly delayed with a time constant of approximately 1 min versus < 0.5 min for leg blood flow (Rådegran and Saltin 1997; Shoemaker, Hodge, and Hughson 1994).

During continued exercise exertion, the situation differs greatly. The central blood volume is elevated by about 50% (Flamm et al. 1990), and cardiac output is enhanced proportionally. Also, heart rate and blood pressure increase in proportion to exercise intensity. In other words, the level of sympathetic activity is adjusted to allow peripheral resistance to decrease only to a level somewhat smaller than the percentage increase in cardiac output. As an example, one-legged training increases $\dot{V}O_2$max and cardiac output only when exercise is performed with the trained leg (Gleser 1973). This emphasizes the important role of local adaptation (Saltin et al. 1976). Evaluating the relative role of modulation of flow to passive tissues as well as that directed toward the muscles engaged in exercise is important. Also of interest is the regulatory priority for blood flow to different muscle groups contracting simultaneously.

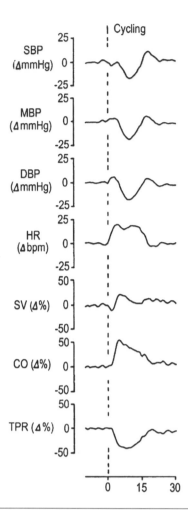

Figure 12.1 Cardiovascular responses following a 3-sec bout of cycling exercise at 50 watts. At time = 0, the exercise began, and after 3 sec the subjects were again motionless while cardiovascular responses were monitored up to 30 sec. SBP = systolic blood pressure; MBP = mean arterial pressure; DBP = diastolic blood pressure; HR = heart rate; SV = stroke volume; CO = cardiac output; TPR = total peripheral resistance. Note the fall in MBP and TPR and the increase in CO shortly after the onset of exercise. (Modified from Sprangers et al. 1991.)

Regional Blood Flow

The following section describes regional blood flow. It discusses two areas of this topic: passive tissues and active muscle.

Passive Tissues

Many tissues have a weak internal control mechanism (autoregulation). In other words, the surplus of cardiac output significantly influences flow. These tissues function as a reservoir when mobilization of blood to the central circulation is required. Krogh (1912) argued that in response to clamping the aorta above the mesenteric vessels, the increase in blood pressure reflects mobilization of blood from the splanchnic region. During exercise, liver blood flow decreases in proportion to the increase in sympathetic nervous activity, as indicated by heart rate (Clausen 1976; Rowell 1973), through a baroreceptor mechanism also affecting kidney and skin blood flow (Hales and Ludbrook 1988). Thus, the splanchnic circulation has been termed the blood giver of the circulation (Katz and Robard 1939). In addition, in many animals and notably in the dog and thoroughbred racing horse, exercise causes constriction of the spleen and increases hematocrit to ~60%. In humans, constriction of the spleen mobilizes only about 50% of its blood during exercise (Flamm et al. 1990), but because splenic hematoccrit is similar to that in the systemic circulation, the increase in hematocrit by 3% primarily reflects muscle edema (Rasmussen et al. 1992). When taken together, the splanchnic region mobilizes about 20% of its blood when exercising at $\dot{V}O_2$max (Flamm et al. 1990). Conversely, when no sympathetic tone exists, as in the legs of paraplegics, the volume of inactive regions, such as the calf muscles, changes little (Hopman, Verheijen, and Binkhorst 1993).

An internal autoregulatory mechanism influences renal blood flow (Holstein-Rathlou and Marsh 1994). However, during exercise, it is markedly reduced, and its blood volume is reduced by 24% (Flamm et al. 1990). Similarly, blood flow to the resting forearm is reduced in response to exercise, and its blood volume is reduced (Athanasiu and Carvallo 1898). This reflects, in part, that blood flow to the skin is lowered when exercise is performed at a certain intensity and related to activation of central command and venoarterial reflexes (Vissing 1997; Vissing et al. 1997). Yet, skin blood flow increases again when thermoregulation becomes a priority (see chapter 17). Conversely, flow to the skin of the head increases during exercise (Cotzias and Marshall 1992).

Flow to the resting forearm (but not the calf) muscle has been reported to increase for about 0.5 min through a β-adrenergic mechanism followed by α-adrenergic vasoconstriction (Eklund and Kaijser 1976). Alternatively, vasodilatation of the resting forearm vasculature may be related to a cholinergic mechanism (Sanders, Mark, and Ferguson 1989). On the other hand, when the display of the electromyographic activity establishes resting muscle activity, vasoconstriction prevails (Cotzias and Marshall 1993). Vasodilatation may then represent an anticipatory response (Armstrong, Vandenakker and Laughlin 1985).

Active Muscle

One may question to what extent muscle blood flow is a regulated variable during exercise. Obviously, the mechanical forces accompanying muscle contraction hinder flow as demonstrated during both static and dynamic exercise (Bonde-Petersen, Mork, and Nielsen 1975; Rådegran and Saltin 1998.) This concept has been evaluated by comparing flow during exercise with that established during reperfusion after ischemia. As determined by plethysmography, the maximal forearm and calf blood flow during exercise is 22–45 ml/100 ml/min (Black 1959; Byström and Kilbom 1990; Kagaya 1994; Sinoway et al. 1986). The maximal vasodilatation for these regions elicits a flow in the range of 40–60 ml/100 ml/min (Snell et al. 1987; Victor and Seals 1989). However, muscle contraction promotes venous return and, during muscle relaxation, the venous system may induce suction of blood through the capillaries, providing for rapid filling of the vascular bed. Thus, muscle flow is pulsatile as demonstrated by the Doppler technique (Eriksen et al. 1990; Kagaya and Ogita 1992a; Rådegran 1997) and constant thermodilution (Andersen and Saltin 1985). This confirms the original observation by Barcroft (1953). During recovery from exercise, blood flow velocity diminishes in the first seconds after exercise ceases and the muscle pump is inactivated (Shoemaker, Hodge, and Hughson 1994). This is also reflected in a transient decrease in cardiac output (Eriksen et al. 1990).

In favor of muscle vasodilatation being a regulated variable is its almost perfect relationship with oxygen delivery (see the chapter Muscle Blood Flow and its Regulation). Furthermore, the hyperdynamic circulation of patients with muscle metabolic deficits (see the chapter "Circulatory Regulation in Muscle Disease") argues for regulation of muscle flow beyond the mechanical influence of contraction. With local regulation of muscle blood flow during exercise, researchers have argued that sympathetic control is lost (Remensnyder, Mitchell, and Sarnoff 1962; Donald, Rowlands, and Ferguson 1970). This phenomenon has been called "functional sympatholysis." This may be the case only when the muscle mass is small and the vasodilator signals are strong.

Strandell and Shepherd (1967) demonstrated an increase in muscle blood flow (^{133}Xe clearance) with exercise intensity during rhythmic handgrip. They also evaluated the influence of lower-body negative pressure. At low exercise intensities, lower-body negative pressure reduced muscle blood flow. It had no effect during intense handgrip. Similarly,

Hansen et al. (1996) showed a reduction in muscle O_2 saturation at light handgrip intensities during increases in MSNA produced by lower-body negative pressure (figure 12.2). Increased MSNA had no effect on muscle O_2 saturation when the handgrip load was intense. Thus, this finding was interpreted as functional sympatholysis.

Different results have been found when researchers undertook a somewhat parallel approach with a larger muscle mass. Sinoway et al. (1988) evaluated the influence of light and severe leg exercise followed by postexercise muscle ischemia on the peak forearm vascular conductance as determined during reperfusion after 10 min of ischemia. Both intense leg exercise and postexercise muscle ischemia opposed metabolic vasodilatation (34% and 52%, respectively). Lower-body negative pressure at −40 mmHg was of no consequence. One could argue that for both forearm and leg exercise, lower-body negative pressure is not a strong enough sympathetic stimulus to reduce blood flow since activation of the muscle metaboreflex apparently opposes vasodilation during both forearm and leg exercise. Joyner (1992) showed that when the muscle metaboreflex was activated in the forearm by positive pressure that reduced forearm perfusion, the muscle metaboreflex reduced blood flow even further. Sympathetic blockade abolished this response. The study of Strange et al. (1990) suggests a similar curtailment of muscle blood flow by the metaboreflex. They applied neck suction to activate the carotid baroreceptors during

cycling exercise. At low exercise intensities, neck suction increased leg vascular conductance. However, at a high cycling intensity, no effect occurred to leg vascular conductance, suggesting sympathetic vasoconstriction induced by the metaboreflex.

Similar observations have been obtained in studies on the limitations to $\dot{V}O_2$max. Addition of extra muscle groups to ongoing exercise is well-known to have only a small effect, if any, on $\dot{V}O_2$max (see figure 12.3) (Åstrand and Saltin 1961; Secher et al. 1974; Secher and Oddershede 1975; Stenberg et al. 1967; Taylor, Buskirk, and Henschel 1955) and cardiac output (Kilbom and Persson 1981). At the same time, researchers argue that muscle blood flow would exceed the maximal cardiac output if all muscles in the body were to receive the flow determined when they are exercising in isolation (300–400 ml/100 g/min) (Andersen and Saltin 1985; Richardson et al. 1993; Rowell et al. 1986). This argument requires that all muscles, in fact, are that well perfused. However, the values reported for the forearm and the calf are, as mentioned, lower. To some extent these discrepancies reflect the content of bone and skin in the sample volume, the difficulties in determination of xenon clearance, or the muscle mass supplied by the femoral artery and vein is not known.

With peak muscle perfusion being 300–400 ml/100 g/min, it is quite apparent that a limit may be set by the heart for how large an amount of active muscle mass can be optimally perfused. A key find-

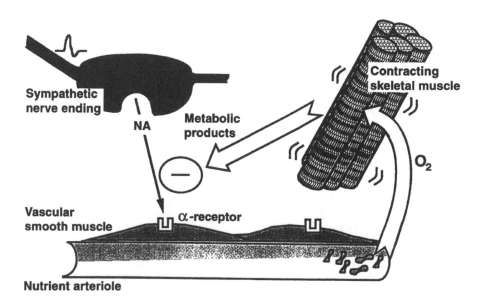

Figure 12.2   A schematic illustration of possible mechanisms for a functional sympatholysis to occur. Norepinephrine (NE) is released in relation to an elevated sympathetic nerve activity. However, its effect is reduced or completely inhibited by some compound related to skeletal muscle contraction. This thereby secures ample supply of blood flow and oxygen to the muscles. (Hansen et al. 1996.)

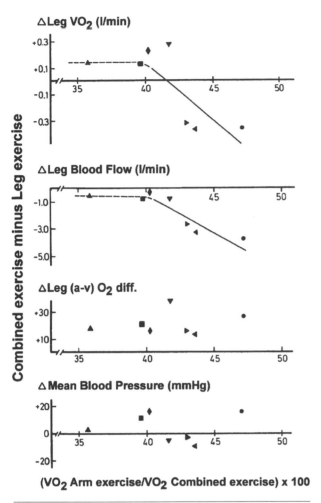

Figure 12.3 The effect of adding arm exercise to ongoing leg exercise. Individual changes in leg oxygen uptake (Leg V̇O₂), leg blood flow, leg aterial-venous oxygen difference (Leg (a-v̄) O₂ diff.), and mean arterial blood pressure. (From Secher et al. 1977 with permission from *Acta Physiologica Scandinavica*.)

ing in ordinary exercise either performed with the legs or with the legs and arms is that blood pressure is indeed not only maintained but also becomes elevated in relation to the exercise intensity (Åstrand et al. 1965). Thus, in addition to the elevation in cardiac output occurring with more muscle mass involvement in the exercise and higher power outputs, vasoconstriction of various vascular beds of vital organs is brought about (Rowell 1973).

The critical question is whether this also occurs in vessels feeding contracting skeletal muscles during intense exercise with a large muscle mass as suggested by Secher et al. (1977). Two lines of reasoning suggest it may occur. It is true that adding arm to exhaustive leg exercise usually causes an increase in oxygen uptake. However, the magnitude of the elevation is small and in the order of 0–10%. The relative contribution of arms and legs to produce the power may be critical for the magnitude of the observed enhancement of the oxygen uptake (Bergh, Kanstrup, and Ekblom 1976; Secher et al. 1974). Of note is that this increase in maximal oxygen uptake is primarily an effect of the widening of the (a-v) O₂ difference with no or very minor contribution of the cardiac output (Clausen et al. 1973; Klausen et al. 1982). A ceiling appears to exist for the pump capacity of the heart for a given individual at a given training level (Saltin et al. 1968). This upper limit is already taxed with exercises engaging both legs (Savard et al. 1987). As a consequence, when adding the arms to perform exhaustive exercise, the blood flow to the legs has to be reduced in order to allow some perfusion of musculature in the upper body, as demonstrated by Secher et al. (1977). Vasoconstriction, by further lowering the blood flow to vital organs, would possibly jeopardize their function, and the blood flow provided would be insufficient. Thus, if the cardiac output cannot be elevated and noncontracting tissue blood flow is already as low as it can be, the blood flow to the exercising limbs has to be redistributed for the arm exercise to be performed. As pointed out previously, Secher et al. (1977) in their classic study also found a lowering of leg blood flow when the arms were added to ongoing leg exercise. This occurred at a relative oxygen uptake of ~80% of the subjects' maximal oxygen uptake. Thus, a concordance exists between the more theoretical and the experimental lines of arguments.

A problem has arisen, however. No one has clearly been able to confirm the observations made by Secher et al. (1977). Several have tried with similar protocols and exercise levels (Richter et al. 1992; Savard et al. 1989). In some of these studies, several of the subjects reduced their leg blood flow when the arms were added, but far from all subjects did. For a whole group of subjects, no study has been able to demonstrate a significant drop. However, in all studies adding arm to leg exercise, a lowered leg conductance occurred due to an elevation in the blood pressure response. This indicates vasoconstriction to further inhibit increases in leg blood flow.

In one attempt to confirm Secher et al.'s (1977) results, mild hypoxia was used during the exercise in healthy young subjects (Bangsbo et al. 1997). Again, a reduction in leg conductance was apparent when adding the arm to the ongoing leg exercise, but the legs received an unchanged blood flow. By reducing heart pump capacity by giving a β-blocker to healthy subjects, leg conductance was lowered primarily because of a reduced leg blood flow

(Pawelczyk et al. 1992). Also, in patients with a low cardiac reserve, increasing the muscle mass engagement in the exercise slightly reduced peak muscle perfusion significantly. Thus, in chronic heart failure patients, the engagement of only 4 kg of muscle will lower limb blood flow (Magnusson et al. 1997).

Several explanations could show why demonstrating a reduced limb blood flow at intense exercise engaging a very large muscle mass is difficult. One possibility is the selection of the total work load and its split between arms and legs, respectively. On one side, the power output cannot be too low. However, Secher et al. (1977) demonstrated the phenomenon at a relative exercise intensity as low as ~80% of the maximal oxygen uptake. It cannot be too high either since the exercise time will be too short for performing the adequate measurements. The problem might be to find the balance between the arm and leg muscle power and, at the same time, truly increase the muscle mass engaged in the exercise. To this should be added that in healthy subjects who have not especially trained their upper body, the capacity for perfusion of arm/shoulder muscles is low. In Ahlborg and Jensen-Urstad's study (1991) measuring arm blood flow, the increase observed during exercise appears to be less pronounced than at equal power output by the legs. This occurs in spite of blood pressure being much more elevated with arm than with leg exercise at a given workload. This resistance to flow is hardly due to a low capillarization since at least the upper arm and shoulder muscles have an equal number of capillaries as do the muscles of the legs (Saltin 1990). An indication of poor perfusion of the arms could be the very pronounced lactate release with upper-body exercise (Ahlborg and Jensen-Urstad 1991; Klausen et al. 1982; Richter et al. 1992). The sympathetic activity also appears to be larger than in two-legged exercise (Ray et al. 1993). Moreover, in combined arm and leg exercise, peak plasma norepinephrine levels are more than twice as high as in leg exercise alone (Secher and Saltin 1997). Skeletal muscles of the upper arms and shoulders may be underperfused in upper-body exercise and the more so in combined intense exercise with legs. However, we still lack an explanation as to why leg blood flow is not reduced when the cardiac output has reached its maximal level and the arms are added to continuing leg exercise.

The observed reduced leg conductance in arm and leg exercise due to an elevation in blood pressure could result from a myogenic autoregulatory response. What may speak against autoregulation is that it does not last long (Bacchus et al. 1981). After an initial quite rapid myogenic reflex due to the elevated pressure, more metabolic vasodilator substances accumulate. These override the myogenic response, at least temporarily. Thus, if the elevated conductance was due to the myogenic reflex, oscillatory alterations in blood flow and conductance would have been anticipated. Instead, a permanent change with stable blood flow and blood pressure is observed. This speaks in favor of a sympathetically mediated vasoconstriction (see figure 12.4).

Sympathetic nerve stimulation causes vasoconstriction (Thompson and Mohrman 1983), but its effect on blood flow may be minor (Strandell and Shephard 1967). Ample evidence shows that the sympathetic activity to muscle, directly recorded by microneurography, gradually increases with increasing workloads (Ray et al. 1993; Seals and Victor 1991). This sympathetic activity is directed to active as well as to inactive muscle (Hansen et al. 1996). Furthermore, researchers have, in fact, demonstrated higher hind limb blood flow during exercise after sympathectomy in the rat (Peterson, Armstrong and Laughlin 1988). In addition, forearm blood flow during intense handgrip is higher after sympathetic blockade (Joyner 1992). It is true that dynamic work is less studied than static exercise and that very intense dynamic work cannot be studied with this technique. However, norepinephrine (NE) in plasma reflects the overall sympathetic activity. Thus, when NE spillover is measured, it gradually becomes elevated over a dynamically exercising limb (Savard et al. 1989). This elevation is quite marked and accounts for over 90% of NE found in plasma. In one of the few studies where NE spillover was determined simultaneously with limb blood flow while adding more and more muscles to the exercise to produce a larger power output and utilizing a larger fraction of the maximal oxygen uptake, no effect of the markedly higher muscle sympathetic activity (approximate leg NE spillover) on leg blood flow was encountered (Savard et al. 1989). This is functional sympatholysis (Remensnyder, Mitchell, and Sarnoff 1962) (see figures 12.2 and 12.5)

The possibility also exists that local vasodilators are produced in the muscle with contractions inducing an hyperemia. However, the actual blood flow may be the net result of vasodilator substances and some vasoconstrictor activity mediated by the sympathetic nervous system. The studies by Strange et al. (1993) provide support for this alternative. They studied electrically induced muscle contractions combined with epidural anesthesia. These researchers found that blocking the sympathetic outflow to the legs results in a higher limb blood flow than in the control situation (Strange et al. 1993). Moreover, observations by Saito et al. (1992), Kagaya (1993), and Kagaya et al. (1994) can best be explained by the

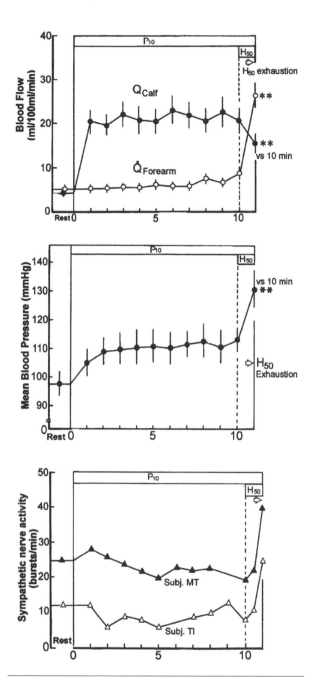

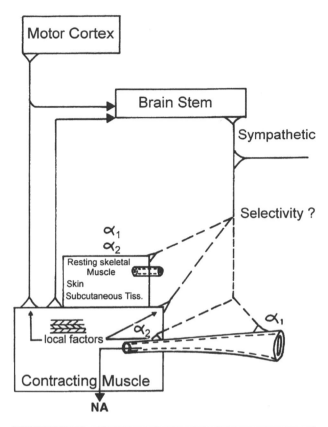

Figure 12.5   A schematic illustration of possible interaction between locally elicited muscle hyperemia and an overriding sympathetically mediated vasoconstriction. The sympathetic nerve activity to the vessels in the muscles comes on gradually but causes no or only a small reduction in blood flow (functional sympatholysis) (Hansen et al. 1996; Savard et al. 1989). This continues until it becomes critical to reduce limb blood flow to maintain blood pressure. In the graph, only α_1 receptors are depicted since they may be the main site for an action being located on the resistance vessels. In contrast, the α_2 receptors are more in the microcirculatory region, playing a role for the distribution of the blood flow within the capillary net.

Figure 12.4   Reduced calf blood flow associated augmented mean arterial pressure and muscle sympathetic nerve activity after adding handgrip exercise to plantar flexion exercise. (Reprinted with permission from A. Kagaya, M. Saito, F. Ogita, and M. Shinohara. 1994.) Exhausting handgrip exercise reduces the blood flow in the active calf muscle exercising at low intensity. (*European Journal of Applied Physiology* 68:252–257.)

actual blood flow to a muscle during exercise being the net occurrence of locally induced vasodilatation and sympathetically mediated vasoconstrictor activity.

One example is Saito et al.'s study (1992) (figure 12.6). They measured calf blood flow (plethysmog-

raphy), skin blood flow, and MSNA (opposite leg) during plantar flexion exercise. Adding static handgrip exercise caused MSNA to become elevated, but calf blood flow was unaltered. However, when ischemia was applied to the statically contracting forearm, MSNA was elevated further and muscle blood flow of the lower leg became reduced. These data suggest that the vasodilatation during muscle contraction can be counteracted by the sympathetic activity. However, the vasoconstrictor effect may mainly occur in muscle performing a low level of exercise.

Japanese investigators have made a more detailed analysis by combining exercises with different intensities (Kagaya 1992, 1993; Kagaya and Ogita

1992b; Kagaya et al. 1994). These studies involved only the small muscle mass of the forearm and the calf. This makes it unlikely that the capacity of the heart was of any consequence for a limitation of regional blood flow. Exercise with the calf muscles was kept at a constant intensity (1 Hz contractions at 10% maximal voluntary contraction [MVC]). However, the intensity of the hand contractions was increased from 30% to 70% MVC and maintained at the same frequency (see figure 12.6). When following this procedure, forearm flow increased with exercise intensity. The calf muscle flow was uninfluenced by hand contractions at 30% MVC. However, with increasing forearm work, calf blood flow became reduced and significantly so at 70% MVC. Even with hand contractions at 50% MVC, calf blood flow became reduced when exercise was carried out to exhaustion. With the involvement of the somewhat larger muscle mass of the elbow continued to near exhaustion, the level of exercise that attenuated calf blood flow was reduced to 50% MVC (Kagaya, Ogita, and Koyama 1996). Interestingly, during the first two minutes after discontinuation of elbow flexion, calf vascular conductance remained low.

The reverse protocol has also been evaluated, i.e., increasing levels of calf muscle exercise was added to ongoing arm exercise. Researchers confirmed

that exhaustive rhythmic handgrip at 30% MVC reduces calf blood flow when the intensity of calf muscle exercise was low but not when it is intense. However, in comparison with the sum of $\dot{V}O_2$ during exercise with a single muscle group, $\dot{V}O_2$ is reduced during combined exercise only when intense arm exercise is added to ongoing mild leg exercise (Ogita and Kagaya 1996).

The same work rate that elicits a decrease in muscle blood flow during combined exercise also causes a marked increase in blood pressure and sympathetic nerve activity to muscle (Kagaya et al. 1994; Saito et al. 1992). This points again to the importance of the sympathetic nervous system. The fact that relative vasoconstriction persists in the first minutes after exercise with the added muscle group further indicates that the response is elicited by way of metabosensitive nerve fibers rather than through central command or activation of mechanoreceptors (Kagaya, Ogita, and Koyama 1996).

At one end of the spectrum, therefore, we have a situation with small muscle mass exercise in humans where local vasodilator factors are at play. These factors have the dominating influence since sympathetic activity is small although the work is quite intense. At the other end of the spectrum, during very intense whole-body work, local factors inducing hyperemia are very pronounced. However, their effect is attenuated and caused by an escalating sympathetically mediated vasoconstrictor activity.

Conclusion

The present discussion may be considered an evaluation of whether blood pressure, flow, or peripheral resistance is the primarily regulated variable during exercise. The discussion is complicated by the fact that all three variables change markedly. In addition, they are linked by the modification of the Ohm's equation for the circulation and therefore cannot be viewed independently. The results of studies heretofore conform to the concept that blood pressure is a regulated variable. However, one could equally argue that this regulation is coupled to the primary signals of oxygen demand. The fact that blood pressure increases during exercise argues for a resetting of the baroreceptor control mechanism (see the chapter "Arterial Baroreceptors"). This argument holds even though the blood pressure increases do not happen in the hyperdynamic phase of reperfusion of the previously ischemic organ (or after a meal). Flow control may be directed to provide for the largest value possible, as dictated by

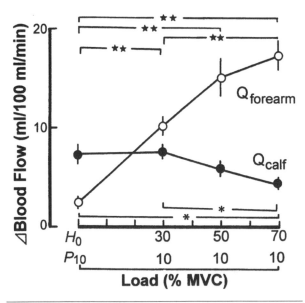

Figure 12.6 Reduced exercise hyperemia of calf muscles (Q_{calf}) in response to increasing levels of handgrip contractions performed together with lower-leg exercise. (Reprinted with permission from A. Kagaya 1993.) Relative contraction force producing a reduction in calf blood flow by superimposition of forearm exercise on lower leg exercise. (*European Journal of Applied Physiology* 66:309–314.)

muscle metabolites, that does not affect blood pressure at the set level. Such control is illustrated by the larger leg blood flow in the supine rather than in the upright position.

In this scheme, a primary adaptation to endurance training is to allow the subject to exercise at a lower level of sympathetic activation. This may occur with respect to central command as already suggested by Johansson (1895). It may also happen for the reflex activation of the sympathetic nervous system from the muscles. Many aspects of endurance training act to down-regulate sympathetic activation at a given exercise intensity with consequence for regional blood flow.

With training, the movement pattern becomes atomized and can presumably be carried out with less involvement of central command. Also, perfection of muscle control may involve a movement pattern that induces less mechanical hindrance to flow. Furthermore, the enhanced mitochondrial function of trained muscle allows for metabolism to be covered with less metabolite accumulation and, therefore, less sympathetic activation. Also, the tendency of endurance training to enhance blood volume and, presumably, also the central blood volume combine to enhance the pumping capacity of the heart (see chapter 18, Cardiovascular Regulation with Endurance Training).

References

Aars, H., P. Brodin, and E. Andersen. 1993. A study of cholinergic and 8-adrenergic components in the regulation of blood flow in the tooth pulp and gingiva in man. *Acta Physiologica Scandinavica* 148:441–447.

Ahlborg, G., and M. Jensen-Urstad. 1991. Arm blood flow at rest and during arm exercise. *Journal of Applied Physiology* 70(2):928–933.

Andersen, P., and B. Saltin. 1985. Maximal perfusion of skeletal muscle in man. *Journal of Physiology London* 366:233–249.

Armstrong, R.B., C.B. Vandenakker, and M.H. Laughlin. 1985. Muscle blood flow patterns during exercise in partially curarized rats. *Journal of Applied Physiology* 59:698–701.

Åstrand, P.O., B. Ekblom, R. Messin, B. Saltin, and J. Stenberg. 1965. Intra-arterial blood pressure during exercise with different muscle groups. *Journal of Applied Physiology* 20:253–256.

Åstrand, P.O., and B. Saltin. 1961. Maximal oxygen uptake and heart rate in various types of muscular activity. *Journal of Applied Physiology* 16:977–981.

Athanasiu, J., and J. Carvallo. 1898. Le travail musculaire et le rythme du coeur. *Archives de Physiologie* 30:347–362.

Bacchus, A., G. Gamble, D. Anderson, and J. Scott. 1981. Role of the myogenic response in exercise hyperemia. *Microvascular Research* 21:91–102.

Bangsbo, J., C. Juel, Y. Hellsten, and B. Saltin. 1997. Dissociation between lactate and proton exchange in muscle during intense exercise in man. *Journal of Physiology* 504:489–499.

Barcroft, H. 1953. *Sympathetic Control of Human Blood Vessels.* London: Arnold.

Beecher, H.K., M.E. Field, and A. Krogh. 1936. The effect of walking on the venous pressure at the ankle. *Skandinaviesches Archiv fiir Physiologie* 73:133–141.

Bergh, U., I-L. Kanstrup, and B. Ekblom 1976. Maximal oxygen uptake during exercise with various combinations of arm and leg work. *Journal of Applied Physiology* 41:191–196.

Black, J.E. 1959. Blood flow requirements of the human calf after walking and running. *Clinical Science* 18:89–93.

Bonde-Petersen, F., A. Mork, and E. Nielsen. 1975. Local muscle blood flow and sustained contractions of human arm and back muscles. *European Journal of Applied Physiology* 34:43–50.

Bonde-Petersen, F., L.B. Rowell, R.G. Murray, G.C. Blomqvist, R.W. White, J. Karlsson, E.W. Chambell, and J.H. Mitchell. 1978. Role of cardiac output in the pressor responses to graded muscle ischemia in man. *Journal of Applied Physiology* 45:574–580.

Boushel, R. 1998. *The Muscle Metaboreflex During Exercise in Humans.* Doctor of Science Thesis. Bell & Howell Co: UMI Dissertation Services, Ann Arbor, MI, USA.

Bystrøm, S.E., and A. Kilblom. 1990. Physiological response in the forearm during and after isometric intermittent handgrip. *European Journal of Applied Physiology* 60:457–466.

Clausen, J.P. 1976. Circulatory adjustments to dynamic exercise and effects of physical training in normal subjects and in patients with coronary artery disease. *Progress in Cardiovascular Disease* 18:459–495.

Clausen, J.P., K. Klausen, B. Rasmussen and J. Trap-Jensen. 1973. Central and peripheral circulatory changes after training of the arms or legs. *American Journal of Physiology* 225:675–682.

Cotzias, C., and J.M. Marshall. 1992. Differential effects of isometric exercise on the cutaneous circulation of different regions. *Clinical Autonomic Research* 2:235–241.

Cotzias, C., and J.M. Marshall. 1993. Vascular and electromyographic responses evoked in forearm muscle by isometric contraction of the contralateral forearm. *Clinical Autonomic Research* 3:21–30.

Dodd-O, J.M., and P.A. Gwirtz. 1996. Coronary alpha-adrenergic constrictor tone varies with intensity of exercise. *Medicine and Science in Sports and Exercise* 28(1):62–71.

Donald, D.E., D.J. Rowlands, and D.A. Ferguson. 1970. Similarity of blood flow in the normal and the sympathectomiced dog limb during graded exercise. *Clinical Research* 26:185–199.

Ejlersen, E., C. Skak, K. Moller, F. Pott, T. Mogensen, and N.H. Secher. 1995. Volumenterari storet af venos iltsaturation og elektrisk impedans under levertransplantation (in Danish). *Dra'ben* 10:46–59.

Ejlersen, E., C. Skak, K. Moller, F. Pott, and N.H. Secher. 1995. Central cardiovascular responses to surgical incision at a maximal mixed venous oxygen saturation. *Transplantation Proceedings* 27:3500.

Eklund, B., and L. Kaijser. 1976. Effect of regional a- and i3-adrenergic blockade on blood flow in the resting forearm during contralateral isometric handgrip. *Journal of Physiology London* 262:39–50.

Eriksen, M., B.A. Waaler, L. Walloe, and J. Wesche. 1990. Dynamics and dimensions of cardiac output changes in humans at the onset and at the end of moderate rhythmic exercise. *Journal of Physiology London* 426:423–437.

Flamm, S.D., J. Taki, R. Moore, S.F. Lewis, F. Keech, F. Maltais, M. Ahmad, R. Callahan, S. Dragotakes, N. Alpert, and H.W. Strauss. 1990. Redistribution of regional and organ blood volume and effect on cardiac function in relation to upright exercise intensity in healthy human subjects. *Circulation* 81:1550–1559.

Friedman, D.B., J. Brennum, F. Sztuk, O.B. Hansen, P.S. Clifford, F.W. Bach, L. Arendt-Nielsen, J.H. Mitchell, and N.H. Secher. 1993. The effect of epidural anaesthesia with 1 % lidocaine on the pressor response to dynamic exercise in man. *Journal of Physiology London* 470:681–691.

Friedman, D.B., J.M. Johnson, J.H. Mitchell, and N.H. Secher. 1991. Neural control of the forearm cutaneous vasoconstrictor response to dynamic exercise. *Journal of Applied Physiology* 71:1892–1896.

Gleser, M.A. 1973. Effects of hypoxia and physical training on hemodynamic adjustments to one-legged training. *Journal of Applied Physiology* 34:655–659.

Hales, J.R.S., and J. Ludbrook. 1988. Baroreflex participation in redistribution of cardiac output at onset of exercise. *Journal of Applied Physiology* 64:627–634.

Hansen, J., G.D. Thomas, T.A. Jacobsen, and R.G. Victor. 1996. Muscle metaboreflex parallel sympathetic activation in exercising and resting human skeletal muscle. *American Journal of Physiology* 266:H2508–H2514.

Hellstrøm, G., G. Magnusson, N.G. Wahlgren, A. Gordon, C. Sylven and B. Saltin. 1996. Physical exercise may impair cerebral perfusion in patients with chronic heart failure. *Cardiology in the Elderly* 4:191–194.

Holmgren, A. 1956. Circulatory changes during muscular work in man. *Scandinavian Journal of Clinical and Laboratory Investigation* 8(suppl):24.

Holstein-Rathlou, N-H., and D.J. Marsh. 1994. Renal blood flow autoregulation: A case study in nonlinear dynamics. *Physiological Reviews* 74:637–681.

Hopman, M.T.E., P.H.E. Verheijen, and R.A. Binkhorst. 1993. Volume changes in the legs of paraplegic subjects during arm exercise. *Journal of Applied Physiology* 75:2079–2083.

Ide, K., A.L. Gollov, F. Pott, J.J. van Lieshout, B.K. Kofoed, P. Petersen, and N.H. Secher. 1999. Middle cerebral artery blood velocity during exercise in patients with atrial fibrillation. *Clinical Physiology* (in press).

Johansson, J.E. 1895. Ueber die Einwirkung der Muskelthdtigkeit auf die Athmung und die Hertzhiitigkeit. *Skandinaviesches Archiv fiir Physiologie* 5:20–66.

Jørgensen, L.G. 1995. Transcranial Doppler ultrasound for cerebral perfusion. *Acta Physiologica Scandinavica* 625:1–44.

Joyner, M.J. 1992. Sympathetic modulation of blood flow and O_2 uptake in rhythmically contracting human forearm muscles. *American Journal of Physiology* 263:H1078–H1083.

Kagaya, A. 1992. Reduced exercise hyperaemia in calf muscles working at high contraction frequencies. *European Journal of Applied Physiology* 64:298–303.

Kagaya, A. 1993. Relative contraction force producing reduction in calf blood flow by superimposing forearm exercise on lower leg exercise. *European Journal of Applied Physiology* 66:309–314.

Kagaya, A. 1994. Maximal exercise-induced vasodilatation and pulmonary oxygen uptake during dynamic forearm and calf exercise. *Journal of Exercise Science* 4:55–62.

Kagaya, A., and F. Ogita. 1992a. Blood flow during muscle contraction and relaxation in rhythmic exercise at different intensifies. *The Annals of Physiological Anthropology* 11:251–256.

Kagaya, A., and F. Ogita. 1992b. Peripheral circulatory readjustment to superimposition of rhythmic handgrip exercise to plantar flexion of different duration. *Journal of Exercise Science* 2:512.

Kagaya, A., F. Ogita, and A. Koyama. 1996. Vasoconstriction in active calf persists after discontinuation of combined exercise with high-intensity elbow flexion. *Acta Physiologica Scandinavica* 157:85–92.

Kagaya, A., M. Saito, F. Ogita, and M. Shinohara. 1994. Exhaustive handgrip exercise reduces the blood flow in the active calf muscle exercising at low intensity. *European Journal of Applied Physiology* 68:252–257.

Katz, L.N., and S. Robard. 1939. The integration of the vasomotor responses in the liver with those in other systemic vessels. *Journal of Pharmacology and Experimental Therapeutics* 67:407–422.

Kilbom, A., and J. Persson. 1981. Cardiovascular responses to combined dynamic and static exercise. *Circulation Research* 48 (suppl I):93–97.

Klausen, K., N.H. Secher, J.P. Clausen, O. Hartling, and J. Trap-Jensen. 1982. Central and regional circulatory adaptations to one-leg training. *Journal of Applied Physiology* 52(4):976–983.

Krogh, A. 1912. The regulation of the supply of blood to the right heart. *Skandinavishe Archiv fiir Physiologie* 27:229–248.

Madsen, P., F. Lyck, M. Pedersen, H.L. Olesen, H.B. Nielsen, and N.H. Secher. 1995. Brain and muscle oxygen saturation during head-up-tilt-induced central hypovolaemia in humans. *Clinical Physiology* 15:523–533.

Magnusson, G., L. Kaijser, C. Sylven, K-E. Karlberg, B. Isberg, and B. Saltin. 1997. Peak skeletal muscle perfusion is maintained in patients with chronic heart failure when only a small muscle mass is exercised. *Cardiovascular Research* 33:297–306.

Ogita, F., and A. Kagaya. 1996. Differential cardiorespiratory response to combined exercise with different combinations of forearm and calf exercise. *European Journal of Applied Physiology.* 73:511–513.

Pawelczyk, J.A., B. Hanel, R.A. Pawelczyk, J. Warberg, and N.H. Secher. 1992. Leg vasoconstriction during dynamic exercise with reduced cardiac output. *Journal of Applied Physiology* 73:1838–1846.

Pawelczyk, J.A., R.A. Pawelczyk, J. Warberg, J.H. Mitchell, and N.H. Secher. 1996. Cardiovascular and catecholamine responses to static exercise in partially curarized man. *Acta Physiologica Scandinavica* 160:23–28.

Perko, M., P. Madsen, G. Perko, T.V. Schroeder, and N.H. Secher. 1997. Mesenteric artery response to head-up-tilt-induced central hypovolemia and hypotension. *Clinical Physiology* 17:487–496.

Peterson, D.F., R.B. Armstrong, and M.H. Laughlin. 1988. Sympathetic neural influence on muscle blood flow in rats during submaximal exercise. *Journal of Applied Physiology* 65:434–440.

Rådegran, G. 1997. Ultrasound Doppler estimates of femoral artery blood flow during dynamic knee extensor exercise in man. *Journal of Applied Physiology* 83:1383–1388.

Rådegran, G., and B. Saltin. 1998. Muscle blood flow at onset of dynamic exercise in man. *American Journal of Physiology* 274:H314–H322.

Rasmussen J., B. Hanel, K. Saunamaki, and N.H. Secher. 1992. Recovery of pulmonary diffusing capacity after maximal exercise. *Journal of Sports Sciences* 10:525–531.

Ray, C., R. Rea, M. Clary, and A.L. Mark. 1993. Muscle sympathetic nerve responses to dynamic one-legged exercise: Effect of body posture. *American Journal of Physiology* 264:H1–H7.

Remensnyder, J.P., J.H. Mitchell, and S.J. Sarnoff. 1962. Functional sympatholysis during muscular activity: Observations on influence of carotid sinus on oxygen uptake. *Circulation Research* 11:370–380.

Richardson, R.S., D.C. Poole, D.R. Knight, S.S. Kurdak, M.C. Hogan, B. Grassi, E.C. Johnson, K.F. Kendrick, B.K. Erickson, and P.D. Wagner. 1993. High muscle blood flow in man: Is maximal O_2 extraction compromised? *Journal of Applied Physiology* 75:1911–1916.

Richter, E.A., B. Kiens, M. Hargreaves, and M. Kjaer. 1992. Effects of armcranking on leg blood flow and noradrenaline spillover during leg exercise in man. *Acta Physiologica Scandinavica* 144:9–14.

Rowell, L.B. 1973. Regulation of splanchnic blood flow in man. *The Physiologist* 16:127–142.

Rowell, L.B., B. Saltin, B. Kiens, and N.J. Christensen. 1986. Is peak quadriceps blood flow in humans even higher during exercise with hypoxemia? *American Journal of Physiology* 251:H1038–H1044.

Saito, M., A. Kagaya, F. Ogita, and M. Shinohara. 1992. Changes in muscle sympathetic nerve activity and calf blood flow during combined leg and forearm exercise. *Acta Physiologica Scandinavica* 146:449–456.

Saltin, B. 1990. Maximal oxygen uptake: Limitation and malleability. In K. Nazar, R.L. Terjung, H. Kabiuba-Uscilko, and L. Budohoski (eds.) *International Perspectives in Exercise Physiology,* pp 26–40. Champaign, IL: Human Kinetics.

Saltin, B., G. Blomqvist, J.H. Mitchell, R.L. Johnson,

K. Wildenthal, and C.B. Chapman. 1968. Responses to exercise after bedrest and after training. *Circulation* 38 (suppl VII):1–72.

Saltin, B., and P.D. Gollnick. 1983. Skeletal muscle adaptability: Significance for metabolism and performance. In L.D. Peachey, R.H. Adrian, and S.R. Geiger (eds.) *Handbook of Physiology: Skeletal Muscle*, 555–631. Bethesda, MD: American Physiological Society.

Saltin, B., K. Nazar, D.L. Costill, E. Stein, E. Jansson, B. Essen, and P.D. Gollnick. 1976. The nature of the training response: Peripheral and central adaptations to one-legged training. *Acta Physiologica Scandinavica* 96:289–305.

Sanders, J.S., A.L. Mark, and D.W. Ferguson. 1989. Evidence for cholinergically mediated vasodilatation at the beginning of isometric exercise in humans. *Circulation* 79:815–824.

Savard, G.K., E.A. Richter, S. Strange, B. Kiens, N.J. Christensen, and B. Saltin. 1989. Norepinephrine spillover from skeletal muscle during exercise in humans: Role of muscle mass. *American Journal of Physiology* 257:H1812–H1818.

Savard, G.K., S. Strange, B. Kiens, E.A. Richter, N.J. Christensen, and B. Saltin. 1987. Noradrenaline spillover in active versus resting skeletal muscle in man. *Acta Physiologica Scandinavica* 131:507–515.

Schmidt, T.A., H. Bundgaard, H.L. Olesen, N.H. Secher, and K. Kjeldsen. 1995. Digoxin affects potassium homeostasis during exercise in patients with heart failure. *Cardiovascular Research* 29:506–511.

Seals, D.R., and R.G. Victor. 1991. Regulation of muscle sympathetic nerve activity during exercise in humans. *Exercise and Sport Science Reviews* 19:313–349.

Secher N.H. 1992. Central nervous influence on fatigue. In R.J. Shephard and P.O. Åstrand (eds.) *The Olympic Book of Endurance Sports*, 96–107. London: Blackwell Scientific.

Secher, N.H., J.P. Clausen, K. Klausen, I. Noer, and J. Trap-Jensen. 1977. Central and regional circulatory effects of adding arm exercise to leg exercise. *Acta Physiologica Scandinavica* 100:288–297.

Secher, N.H., M. Kjaer, and H. Galbo. 1988. Arterial blood pressure at the onset of dynamic exercise in partially curarized man. *Acta Physiologica Scandinavica* 133(2):233–237.

Secher, N.H., and I. Oddershede. 1975. Maximal oxygen uptake during swimming and bicycling. In L. Levilleie, and J.P. Clarys (eds.) *Swimming II*, 137–142. Baltimore: University Park Press.

Secher, N.H., N. Ruberg-Larsen, R.A. Binkhorst, and F. Bonde-Petersen. 1974. Maximal oxygen uptake during arm and combined arm plus leg exercise. *Journal of Applied Physiology* 36:315–318.

Secher, N.H., and B. Saltin. 1997. Blood flow regulation during exercise in man. In S.A. Ward and J. Steinacker (eds). *The Physiology and Pathophysiology of Exercise Tolerance*, pp 97–102. London: Plenum Press.

Shoemaker, J.K., L. Hodge, and R.L. Hughson. 1994. Cardiorespiratory kinetics and femoral artery blood velocity during dynamic knee extension exercise. *Journal of Applied Physiology* 77:2625–2632.

Sinoway, L.I., T.I. Musch, J.R. Minotti, and R. Zelis. 1986. Enhanced maximal metabolic vasodilatation in the dominant forearms of tennis players. *Journal of Applied Physiology* 61:673–678.

Sinoway, L., and S. Prophet. 1990. Skeletal muscle metaboreceptor stimulation opposes peak metabolic vasodilation in humans. *Circulation Research* 66:1576–1586.

Sinoway, L., J.S. Wilson, R. Zelis, J. Shenberger, D.P. McLaughlin, D.L. Morris, and F.P. Day. 1988. Sympathetic tone affects human limb vascular resistance during a maximal metabolic stimulus. *American Journal of Physiology* 255:H937–H946.

Snell, P.G., W.H. Martin, J.C. Burckey, and C.G. Blomqvist. 1987. Maximal vascular leg conductance in trained and untrained men. *Journal of Applied Physiology* 62:606–610.

Sprangers, R.L.H., K.H. Wesseling, A.L.T. Imholz, B.P.M. Imholz, and W. Wieling. 1991. Initial blood pressure fall on stand up and exercise explained by changes in total peripheral resistance. *Journal of Applied Physiology* 70:523–530.

Stenberg, J., P.O. Åstrand, B. Ekblom, J. Royce, and B. Saltin. 1967. Hemodynamic response to work with different muscle groups, sitting and supine. *Journal of Applied Physiology* 22:61–70.

Strandell, T., and J.T. Shepherd. 1967. The effect in humans of increased sympathetic activity on the blood flow to active muscles. *Acta Medica Scandinavica* 472 (suppl):146–167.

Strange, S., L.B. Rowell, N.J. Christensen, and B. Saltin. 1990. Cardiovascular responses to carotid sinus baroreceptor stimulation during moderate to severe exercise in man. *Acta Physiologica Scandinavica* 138:145–153.

Strange, S., N. Secher, J.A. Pawelczyk, J. Karpakka, N.J. Christensen, J.H. Mitchell, and B. Saltin. 1993. Neural control of cardiovascular responses and of ventilation during dynamic exercise in man. *Journal of Physiology* 470:693–704.

Taylor, H.L., E. Buskirk, and A. Henchel. 1955. Maximal oxygen intake as an objective measure of cardiorespiratory performance. *Journal of Applied Physiology* 8:73–80.

Thompson, L.P., and D.E. Mohrman. 1983. Blood flow and oxygen consumption in skeletal muscle during sympathetic stimulation. *American Journal of Physiology* 245:H66–H71.

Victor, R.G., and D.R. Seals. 1989. Reflex stimulation of sympathetic outflow during rhythmic exercise in humans. *American Journal of Physiology* 257:H2017–H2024.

Vissing, S.F. 1997. Differential activation of sympa-

thetic discharge to skin and skeletal muscle in humans. *Acta Physiologica Scandinavia* (suppl) 639:1–32.

Vissing, S.F., N.H. Secher, and R.G. Victor. 1997. Mechanisms of cutaneous vasoconstriction during upright posture. *Acta Physiologica Scandinavia* 159:131–138.

Williamson, J.W., H.L. Olesen, F. Pott, and N.H. Secher. 1996. Central command increases cardiac output during static exercise in humans. *Acta Physiologica Scandinavica* 156(4):429–434.

Chapter 13

Oxygen Transport in Blood and to Mitochondria

John W. Severinghaus

The underlying mechanistic limit to work is closely associated with O_2 supply to the working muscle (Groebe and Thews 1988; Hogan et al. 1988; Honig et al. 1984). Several factors are known to limit maximum exercise. These include cardiac output, ventilation, pulmonary function, blood hemoglobin concentration, blood volume, and, especially, diffusing capacity (Gutierrez et al. 1990; Stainsby, Snyder, and Welch 1988). In each case, these affect maximum exercise by limiting oxygen delivery to muscle capillaries. This chapter focuses on six other factors involved in O_2 transport from capillary red cells to mitochondrial cytochrome oxidase:

1. Limits of cytochrome oxidase and the theory of $\dot{V}O_2$max at zero PO_2
2. Effects on muscle redox of oxygen supply limitations
3. Effects of redox on pyruvate and lactate concentrations
4. Lactic acid excretion and the Bohr effect of pH on capillary PO_2
5. Diffusing capacity of the tissue path from red cells to cytochrome
6. Location of myoglobin and its role

This chapter presents a model whose predictions of myoglobin saturation conflict with new human exercise MRI measurements of myoglobin saturation.

Hypothesis: Zero Cytochrome Oxidase PO_2 Determines $\dot{V}O_2$max

In very dilute solutions where O_2 gradients may be neglected, the $\dot{V}O_2$ of mitochondria remains constant as PO_2 falls to about 0.1 mm Hg in the medium. At this point, cytochrome is > 99% reduced (Li et al. 1987; Naqui, Chance, and Cadenas 1986). This final 0.1 mm Hg of PO_2 is sufficient to maintain an unmeasurable microdiffusion gradient from medium to the reaction sites. Takahashi and Doi (1995) recently demonstrated oxygen tension gradients between myoglobin and

cytochrome at a medium PO_2 of 1.3 mmHg even in quiescent isolated cardiac myocytes by using three-wavelength microspectrophotometry. Wagner (1992), in reviewing this work, proposed that the inherent limitation of exercise was the delivery of O_2 to cytochrome such that at $\dot{V}O_2$max, the oxygen tension at cytochrome must essentially be zero. In order to model oxygen transport, one must assume that at $\dot{V}O_2$max, all the O_2 reaching the oxidase metal's reaction radius was reduced, none remaining to diffuse away, a condition defined herein as $P_xO_2 = 0$ (Severinghaus 1994). This is comparable to the function of a polarographic O_2 electrode in which a negative potential, typically –0.6 V on the metal, reacts with every O_2 molecule that diffuses within striking distance. Hence, none remains nonreacted *at the surface*, i.e., $PO_2 = 0$. In an electrode and in muscle, the diffusion gradient, and thus oxygen consumption, is maximum when oxidase PO_2 is zero.

Effects of Oxygen Supply on Muscle Redox Potential

Cytochrome redox state varies from fully oxidized at $PO_2 > 20$ mmHg to essentially fully reduced at PO_2 (medium) = 0.1 mmHg (Li et al. 1987; Naqui, Chance, and Cadenas 1986). Full reduction means that the metal has a maximum negative potential, comparable to the –0.6 V of the polarographic cathode needed to reduce all O_2 diffusing to its surface. As energy use increases in exercise, the flux of metabolites and electrons within mitochondria increases. The increased O_2 consumption causes P_xO_2 to fall, and this increases the rate of O_2 diffusion from the capillary.

Effects of Redox on Pyruvate and Lactate Concentrations

O_2 reduction and increased work cause lactate and pyruvate concentrations to rise. As P_xO_2 falls, cytochrome aa_3 becomes progressively reduced and electrons accumulate, i.e., a more negative electric potential forms. At each step in the citric acid cycle, substrate flux determines the potential gradient. At maximum flux, potential and concentration gradients are maximal. Therefore, as cytochrome becomes reduced, concentrations of each intermediary, including pyruvate, rise long before O_2 consumption fails to keep pace with electron flux. NADH and [H+] rise, causing an increase of the lactate/pyruvate ratio. These two processes lead to lactate diffusion or spill to blood. This *anaerobic threshold* or lactic acid threshold (Wasserman 1994) becomes evident at 50–70% of $\dot{V}O_2$max. Since ATP is generated (1 mole/mole of lactate), anaerobic metabolism contributes (slightly) to total energy production.

Pyruvate also rises with increasing work. Its entry into the mitochondrial citric acid cycle is an enzymatically facilitated process that requires a concentration gradient proportional to the rate of metabolism or flux, called the *concentration effect* (Roth and Brooks 1990; Stanley et al. 1985). Lactate and HLa (lactic acid) diffuse passively from muscle to blood. In steady high work, as blood lactate back pressure rises, diffusion out from muscle is reduced, as is the anaerobic ATP production. This shifts metabolism aerobically (Bangsbo et al. 1992; Jorfeldt, Juhlin-Dannfelt, and Karlsson 1978), but this does not impair maximum power (Bogdanis, Nevill, and Lakony 1994).

Muscle Lactic Acid and the Bohr Effect

As [HLa] rises, the acid raises capillary PO_2 by the Bohr effect. This causes leg venous PO_2 to remain nearly constant as work increases from 60% to 100% of $\dot{V}O_2$max (Wasserman 1994). [H+] enters the capillary from muscle both as [HLa] and as CO_2. CO_2 diffusion occurs faster, so the initial effect of a rise of myocyte [HLa] is titration of tissue HCO_3 to CO_2.

Diffusing Capacity of Capillary Wall and Muscle

At $\dot{V}O_2$max, in order to sustain a maximum O_2 flux from capillary to cytochrome, maximum gradients of PO_2 occur across the plasma film in the blood, to the capillary endothelium and basement membrane, into the interstitial space, from the muscle into the myoglobin, from myoglobin to the mitochondria,

and into the cytochrome. In human muscle at $\dot{V}O_2$max, the venous SO_2 can fall to about 10% and sometimes less. PO_2 is usually reported to reach a minimum of about 12–15 mmHg, although some reports suggest values as low as 8 mmHg (Roca et al. 1989, 1992; Wasserman 1994). Hypoxemia or anemia limits exercise (Gutierrez et al. 1990; Hogan et al. 1988; Schaffartzik et al. 1993). By varying F_IO_2, or blood Hb concentration, and driving work to $\dot{V}O_2$max, the relationship of either leg venous or computed mean capillary PO_2 to $\dot{V}O_2$max is linear. The intercept occurs near zero as illustrated in figure 13.1 (Roca et al. 1989; Wagner et al. 1990). The slope of this relationship is the muscle-diffusing capacity.

Roca et al. (1992) studied femoral venous blood composition at $\dot{V}O_2$max in 12 normal sedentary subjects at three levels of inspired O_2 before and after training. P_vO_2 at $\dot{V}O_2$max was a function of F_IO_2, falling from about 20 mmHg in breathing air to 14 mm Hg with 12% O_2 in direct proportion to the fall of leg $\dot{V}O_2$max. The relationship of endpoint P_vO_2 to leg $\dot{V}O_2$ at $\dot{V}O_2$max was approximately linear, lying on lines extrapolating to zero. Training permits work to proceed to a lower muscle venous PO_2, implying an increased diffusing capacity (Klausen et al. 1982). Downstream PO_2 at the cytochrome thus either approaches zero at $\dot{V}O_2$max (Gayeski, Federspiel, and Honig 1988; Hogan et al. 1988; Honig et al. 1984; Wagner 1988) or represents maldistribution of blood flow or other inhomogeneity (Piiper 1990). Recently, Knight et al. (1993) found that dur-

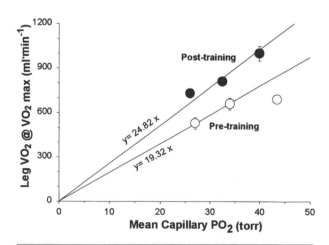

Figure 13.1 The effect of hypoxemia and training on $\dot{V}O_2$max with $F_IO_2 = 21\%$ and 12%. Hypoxia sufficient to reduce (estimated) mean capillary PO_2 also reduced $\dot{V}O_2$max. Training increased the $\dot{V}O_2$max at each F_IO_2, indicating improved diffusion capacity of the tissue. However, the relationship between $\dot{V}O_2$max and mean capillary PO_2 remained approximately proportional. (From Roca et al. 1992.)

ing hyperoxia, $\dot{V}O_2$max rose less than expected from the rise of both P_vO_2 and calculated mean capillary PO_2, assuming constant diffusing capacity of muscle. This suggested to them that another undefined factor became limiting, perhaps because subjects were not trained to work in hyperoxia.

Myoglobin Location and Function

Gayeski, Federspiel, and Honig (1988) showed that myoglobin PO_2 during maximum electrically stimulated exercise in dog gracilis muscle fell as low as 1 mmHg but was usually limited to 3–5 mmHg and remarkably uniform. This suggests that much of the

diffusion gradient for O_2 occurred between red cells and myocytes. Myoglobin is 50% saturated at about 3.2 mmHg at 37 °C and has no Bohr effect. Its dissociation curve is a rectangular hyperbola. O_2 diffuses so rapidly within myoglobin that as a first approximation, one may assume myoglobin to be at a uniform PO_2 (Gayeski and Honig 1991). Mancini et al. (1994) reported that maximum exercise (plantar flexion) led to significant but unquantified falls of $S_{Mb}O_2$ (myoglobin saturation) as detected by 1 H magnetic resonance spectroscopy (MRS). The tissue hemoglobin, measured by infrared light scattering, became 71% desaturated. These studies suggest that in heavy exercise, $S_{Mb}O_2$ remains significantly higher than venous and capillary SO_2.

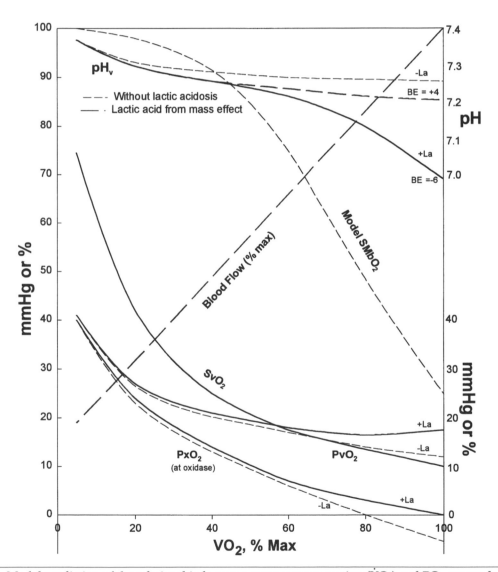

Figure 13.2 Model prediction of the relationship between oxygen consumption ($\dot{V}O_2$) and PO_2 at cytochrome oxidase. This is based on the assumption that this PO_2 reaches zero at $\dot{V}O_2$max. PO_2 is nearly constant or may rise from 60% to 100% of VO_2 due to the Bohr effect. Dotted lines indicate the pH at P_vO_2 and PO_2 without the Bohr effect. By assuming a constant diffusing capacity, myoglobin O_2 saturation was predicted to fall steeply late in exercise. (From Severinghaus 1994.)

Modeling O_2 Transport at $\dot{V}O_2$max

Figure 13.2 shows an example of the model constructed from these six factors (Severinghaus 1994). The independent variable is percentage of $\dot{V}O_2$max. pH was assumed to be an exponential function of $\dot{V}O_2$, reaching 7.0 at $\dot{V}O_2$max. $\dot{V}O_2$ was assumed to be exactly proportional to work rate. Blood flow (% of max) was assumed to be a linear function of $\dot{V}O_2$, venous SO_2 falling from 75% at rest to 10% at $\dot{V}O_2$max. The researchers assumed that 90% of the gradient between capillary, red cells, and cytochrome lies before myoglobin and that 10% lies after. At this level, the model myoglobin saturation fell to about 25% at $\dot{V}O_2$max with a capillary PO_2 of 18 mmHg. The Bohr effect contributed 20% of $\dot{V}O_2$max, illustrated by the dotted line labeled P_xO_2

(at oxidase). Without the Bohr effect, this reaches zero at 80% of $\dot{V}O_2$max. The acid injection is also responsible for the terminal small rise in P_vO_2.

Discrepancy Between Model Theory and New Evidence of Constant $S_{Mb}O_2$

In recent experiments, Richardson et al. (1995) used 1 H MRS quantification of the quadriceps myoglobin signal with one-leg exercise in human volunteers. The results indicated that $S_{Mb}O_2$ fell with mild exercise (50% of $\dot{V}O_2$max) to about 50% but remained at that level at higher work levels (approaching $\dot{V}O_2$max). Inhalational hypoxia ($F_IO_2 = 12\%$) reduced $S_{Mb}O_2$ from 49% to 40% and calculated $P_{Mb}O_2$ from 3.1 ± 0.3 mmHg to 2.1 ± 0.2 mmHg. The

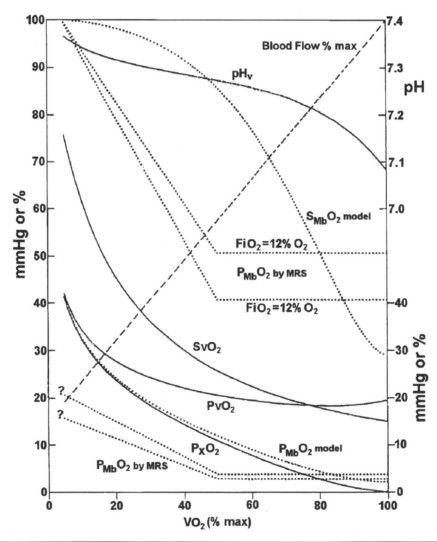

Figure 13.3   Illustration of the difference between myoglobin saturation and oxygen pressure as modeled (for air breathing only) and as observed during human exercise in both normoxia and hypoxia by Richardson et al. (1995). At rest, experimental values of $S_{Mb}O_2$ and PO_2 were not measurable but are assumed. The difference appears to be due to increasing diffusion capacity or conductance of myoglobin as it desaturates.

independence of $S_{Mb}O_2$ on work rate (above 50% of $\dot{V}O_2$max) varied with the prediction of the model (see figure 13.3).

Muscle P_vO_2 at rest and in work is a function of the blood flow profile. Flow may become intermittent or heterogeneous at rest such that muscle tissue PO_2 may be much lower than P_vO_2 at rest. However, by using various assumptions of capillary recruitment, i.e., heterogeneous flow, rationalizing these new data has not been possible. The high initial S_vO_2 and its fall as work increases could not be accounted for with either a recruitment or a two-compartment model in which muscle venous blood as collected might consist of a mix from nonworking and working muscle with the proportion from nonworking (or shunt) areas falling with increased work.

The model treated myoglobin as if it were located at a single diffusional distance. In reality, it probably fills most of the intracellular diffusion path where it facilitates O_2 diffusion, leaving most of the O_2 gradient to occur within the blood capillary wall and extracellular fluid $\dot{V}O_2$. The model assumed a constant diffusing capacity of the capillary to myoglobin path, whereas myoglobin facilitation of oxygen diffusion is greatest at lowest PO_2 (Honig and Gayeski 1993). Richardson et al. (1995) calculated muscle red cell to myoglobin O_2 conductance from mean capillary PO_2 and $P_{Mb}O_2$. It increased linearly with work and was greater during hypoxemia. They concluded that as work increases, increased diffusing capacity between red cells and myoglobin is the most probable explanation for their observations. Therefore, the improvement of diffusing capacity with training reported by Roca et al. (1992) probably hinged on either increased or relocated myoglobin.

Other Candidates for Limitation of Work

Bangsbo and colleagues (1992) appear to have shown that tissue and blood acidosis per se fail to account for the ceiling of $\dot{V}O_2$max. High plasma lactate does not decrease maximum work or power (Seburn et al. 1992). Subjects exercising in hypoxia stop work before the working muscles are fatigued (unexcitable) except when work is confined to an arm or one leg (Beelen et al. 1995). Fatigue as defined by reduction in force accompanying a supermaximal electric stimulation is attributed to a change in excitation-contraction coupling, the cause of which remains unknown. Even if zero PO_2 is consistently shown to be the limiting value at $\dot{V}O_2$max, the cause of exhaustion would remain unclear. In slow-twitch muscle as used in steady exercise, due to accumula-

tion of reducing equivalents, the membrane electric potential may become sufficiently negative to make the muscle cell unexcitable or to stall the active relaxation process. Klausen et al. (1982) have shown that with one-leg cycling to exhaustion, that leg's O_2 consumption exceeds that of the same leg at $\dot{V}O_2$max with two-leg cycling. While these authors believe the difference to be due to greater blood flow and greater lactic acid excretion with its accompanying Bohr effect, exhaustion may be a complex function of additional unknown limits. Roca et al. (1992) noted this in their first tests of sedentary, untrained subjects.

Conclusions

Oxygen delivery appears to be the critical limitation to continuing work. O_2 consumption continues to be proportional to or slightly more than proportional to work rate even in the presence of maximal anaerobic lactate production. This indicates that steady-state work is essentially aerobic and that lactate production is incidental, but useful, in facilitating oxygen unloading. The ATP gained from lactate production and excretion contributes to the energy expended. However, it may be reduced or eliminated by blood loading with lactate without altering $\dot{V}O_2$max. The proposed hypothesis states that at $\dot{V}O_2$max, cytochrome PO_2 has reached zero, and this is the prime determinant of maximum work. New evidence of a relative work independence of myoglobin oxygen saturation implies that myoglobin facilitation of O_2 diffusion between blood and mitochondrial cytochrome aa_3 increases as O_2 consumption rises and myoglobin becomes desaturated.

References

Bangsbo, J., T. Graham, L. Johansen, S. Strange, C. Christensen, and B. Saltin. 1992. Elevated muscle acidity and energy production during exhaustive exercise in humans. *Am. J. Physiol.* 263:R891–R899.

Beelen, A., A.J. Sargeant, D.A. Jones, and C.J. de Ruiter. 1995. Fatigue and recovery of voluntary and electrically elicited dynamic force in humans. *J. Physiol.* 484:227–235.

Bogdanis, G.C., M.E. Nevill, and H.K. Lakony. 1994. Effects of previous dynamic arm exercise on power output during repeated maximal sprint cycling. *J. Sports Sci.* 12:363–370.

Gayeski, T.E.J., W.J. Federspiel, and C.R. Honig. 1988. A graphical analysis of the influence of red cell transit time, carrier-free layer thickness, and intracellular PO_2 on blood-tissue O_2 transport. *Adv. in Exp. Med. and Biol.* 222:25–35.

Gayeski, T.E., and C.R. Honig. 1991. Intracellular PO_2 in individual cardiac myocytes in cats, ferrets, rabbits and rats. *Am. J. Physiol.* 60:H522—H531.

Groebe, K., and G. Thews. 1988. Theoretical analysis of oxygen supply to contracted skeletal muscle. *Adv. in Exp. Med. and Biol.* 222:25–35.

Gutierrez, G., C. Marin, A.L. Acero, and N. Lund. 1990. Skeletal muscle PO_2 during hypoxemia and isovolemic anemia. *J. Appl. Physiol.* 68:2047–2053.

Hogan, M.C., J. Roca, P.D. Wagner, and J.B. West. 1988. Limitation of maximal O_2 uptake and performance by acute hypoxia in dog muscle in situ. *J. Appl. Physiol.* 65:815–821.

Honig, C.R., and T.E.J. Gayeski. 1993. Resistance to O_2 diffusion in anemic red muscle: Roles of flux density and cell PO_2. *Am. J. Physiol.* 265:H868–H875.

Honig, C.R., T.E.J. Gayeski., W. Federspiel, A. Clark, Jr., and P. Clark. 1984. Muscle O_2 gradients from hemoglobin to cytochrome: New concepts, new complexities. *Adv. Exp. Med. and Biol.* 169:23–28.

Jorfeldt, L., A. Juhlin-Dannfelt, and J. Karlsson. 1978. Lactate release in relation to tissue lactate in human skeletal muscle during exercise. *J. Appl. Physiol.* 44:350–352.

Klausen, K., N.H. Secher, J.P. Clausen, O. Hartling, and J. Trap-Jensen. 1982. Central and peripheral circulatory adaptations to one-leg exercise. *J. Appl. Physiol.* 52:976–983.

Knight, D.R., W. Schaffartzik, D.C. Poole, M.C. Hogan, D.E. Bebout, and P.D. Wagner. 1993. Effects of hyperoxia on maximal leg O_2 supply and utilization in man. *J. Appl. Physiol.* 75:2586–2594.

Li, Y., A. Naqui, T.G. Frey, and B. Chance. 1987. A new procedure for the purification of monodispersed highly active cytochrome c oxidase from bovine heart. *Biochem. J.* 242:417–423.

Mancini, D.M., L. Bolinger, H. Li, K. Kendrick, B. Chance, and J.R. Wilson. 1994. Validation of near-infrared spectroscopy in humans. *J. Appl. Physiol.* 77:2740–2747.

Naqui, A., B. Chance, and E. Cadenas. 1986. Reactive oxygen intermediates in biochemistry. *Ann. Rev. of Biochem.* 55:137–166.

Piiper, J. 1990. Unequal distribution of blood flow in exercising muscle of the dog. *Resp. Physiol.* 80:129–136.

Richardson, R.S., E.A. Noyszewski, K.F. Kendrick, J.S. Leigh, and P.D. Wagner. 1995. Myoglobin O_2 desaturation during exercise: Evidence of limited O_2 transport. *J. Clin. Invest.* 96:1916–1926.

Roca, J.A., G. Agusti, A. Alonso, D.C. Poole, C. Viegas, J.A. Barbera, R. Rodriguez-Roisin, A. Ferrer, and P.D. Wagner. 1992. Effects of training on muscle O_2 transport at VO_2max. *J. Appl. Physiol.* 73:1067–1076.

Roca, J.M., C. Hogan, D. Story, D.E. Bebout, P. Haab, R. Gonzalez, O. Ueno, and P.D. Wagner. 1989. Evidence for tissue diffusion limitation of VO_2max in normal humans. *J. Appl. Physiol.* 67:291–299.

Roth, D.A., and G. Brooks. 1990. Lactate and pyruvate transport is dominated by a pH- gradient-sensitive carrier in rat skeletal muscle sarcolemmal vescicles. *Arch. Biochem. Biophys.* 279:386-394.

Schaffartzik, W., E.D. Barton, D.C. Poole, K. Tsukimoto, M.C. Hogan, D.E. Bebout, and P.D. Wagner. 1993. Effect of reduced hemoglobin concentration on leg oxygen uptake during maximal exercise in humans. *J. Appl. Physiol.* 75:491–498.

Seburn, K.L., D.J. Sanderson, A.N. Belcastro, and D.C. McKenzie. 1992. Effect of manipulation of plasma lactate on integrated EMG during cycling. *Med. Sci. Sports Excerc.* 24:911–916.

Severinghaus, J.W. 1994. Exercise O_2 transport model assuming zero cytochrome PO_2 at VO_2max. *J. Appl. Physiol.* 77:671-678.

Stainsby, W.N., B. Snyder, and H.G. Welch. 1988. A pictographic essay on blood and tissue oxygen transport. *Med. Sci. Sports. Exerc.* 20:213–221.

Stanley, W.C., E.W. Geartz, J.A. Wisneski, D.L. Morris, R.A. Neese, and G. Brooks. 1985. Systemic lactate kinetics during graded exercise in man. *Am. J. Physiol.* 149:E595–E602.

Takahashi, E., and K. Doi. 1995. Visualization of oxygen level inside a single cardiac myocyte. *Am. J. Physiol.* 268:H2561–2568.

Wagner, P.D. 1988. An integrated view of the determinants of maximum oxygen uptake. In *O_2 transport to tissue XII*, eds. N.C. Gonzales and M.R. Fedde, Vol. 227, 245–256. New York: Plenum Press.

Wagner, P.D. 1992. Gas exchange and peripheral diffusion limitation. *Med. Sci. Sports Exerc.* 24:54–58.

Wagner, P.D., J. Roca, M.C. Hogan, D.C. Poole, D.E. Bebout, and P. Haab. 1990. Experimental support for the theory of diffusion limitation of maximum oxygen uptake. In *O_2 transport to tissue XII*, eds. J. Piiper, T.K. Goldstick, and M. Meyer, 825–833. New York: Plenum Press.

Wasserman, K. 1994. Coupling of external to cellular respiration during exercise: The wisdom of the body revisited. *Am. J. Physiol.* 266:E519–E539.

PART III

Environmental Factors in Cardiovascular Regulation

Chapter 14

Cardiovascular Regulation During Hypoxia

Robert C. Roach

When humans exercise while breathing hypoxic air or at high altitudes, the challenge to oxygen transport is met by a complex series of adjustments in the cardiopulmonary system. These cardiopulmonary responses and their regulation by the autonomic nervous system are the focus of this chapter.

Over sixty years ago Danish scientists lived together with colleagues from the United States for several months in the high mountains of northern Chile. They were the first to describe two related physiological responses during exercise (Christensen and Forbes 1937): the heart rate and lactate production during maximal exercise were diminished compared to sea level values, and these responses became more pronounced when the subjects stayed at a given altitude or went to higher altitudes. Today, the fundamental mechanisms for these responses remain a mystery. In this chapter we explore what is known about control of the cardiovascular system during exercise in hypoxia, with a focus on the role of the autonomic nervous system.

During progressive exercise at sea level, heart rate increases. Continuation to peak effort results in a reproducible increase in heart rate that is representative of the subject's training status, autonomic nervous system tone, and other inborn factors. At the onset of exercise, parasympathetic control is withdrawn and sympathetic activation then largely drives the heart rate up to maximal levels (Rowell, O'Leary, and Kellogg 1996). This entire picture changes in humans after they have spent several weeks to months in a hypoxic environment (Christensen and Forbes 1937). A much lower heart rate at maximal exercise is observed after acclimatization. Typically, peak heart rate reaches around 200 bpm at sea level, but maximal heart rates of only 120 bpm have been observed after a prolonged sojourn at 5000 m. These alterations may be explained by altered autonomic tone. Also, the changes are not permanent, as breathing high-oxygen gas mixtures instantly and nearly completely normalizes the exercise responses to sea level values.

After acclimatization to high altitudes, adaptive changes have been observed in the autonomic nervous system that could explain the marked reduction in peak heart rate and physical work capacity. In Operation Everest 2, a simulated ascent of Mount Everest, Sutton and colleagues (1988) observed a gradual deterioration in work performance, so that at an altitude equivalent to the summit of Mount Everest, the power output at maximal exercise was one-third that at sea level. The maximal oxygen uptake was only $1.2 \, L \cdot min^{-1}$, as compared to its sea level value of nearly $4 \, L \cdot min^{-1}$, and the plasma norepinephrine concentration dropped by 50% even though the subjects exercised to exhaustion. The maximal heart rate was only 127 compared to 178 at sea level. This suggests that either (a) the sympathetic nervous system was not activated centrally (decreased neuromuscular activation) because of the hypoxemia per se, or because of diminished feedback from the working muscles (Alam and Smirk 1938) or peripherally from receptors in the cardiovascular system; or (b) neurotransmission was somehow inhibited at the peripheral level (local modulation by metabolic factors). We use these observations as a basis for the following discussion of cardiovascular control during exercise in hypoxia.

Oxygen Transport System During Rest and Exercise in Hypoxia

Hypoxia causes a profound accommodation of the lungs, circulatory system, and blood in an attempt to maintain a steady oxygen supply to the central nervous system and working muscles. This response can be seen at rest at very low oxygen levels, but it is also apparent during exercise with much less severe levels of hypoxemia. The resting acclimatization process and associated pathophysiological responses to high altitudes have recently been reviewed (Hackett and Roach 1995). Here we focus on exercise responses to acute and chronic hypoxia and how this unique set of physiological responses can aid in our understanding of basic cardiovascular physiology.

Lung

The ventilatory response to hypoxia is a complex and well-studied phenomenon. Early in this century, Haldane and Priestley (1905) demonstrated that as P_IO_2 decreases (secondary either to a drop in the fraction of oxygen in the inspired air or a drop in barometric pressure), ventilation increases to minimize the drop in alveolar oxygen. This response starts within seconds of breathing hypoxic gas, but has a complicated path from this initial, and individually variable, response to the response encountered after days, weeks, and years at high altitudes. The initial increase in ventilation is not maintained as ventilation declines within the first 20 to 30 minutes of hypoxia (Weil 1986). A prolonged exposure of many hours to several days results in a secondary increase in ventilation. The ventilatory responses to hypoxia are thus complex, but largely adequate to increase alveolar PO_2 above the levels expected had no ventilatory acclimatization occurred.

Of the two peripheral chemoreceptors, the aortic and carotid bodies, it is the carotid chemoreceptors that account for most of the ventilatory stimulation during hypoxia. The hypoxic ventilatory response (HVR) describes the increase in chemoreceptor discharge to the central respiratory center as SaO_2 falls, the result being an increase in ventilation. An example of the increase in carotid sinus nerve firing as end tidal O_2 falls is shown from data on cats in figure 14.1a. In humans, a similarly shaped curve results when the increase in ventilation is plotted in place of carotid sinus nerve firing against barometric pressure (figure 14.1b). Note that resting $PaCO_2$ has dropped to less than 15 Torr at the highest altitude (figure 14.1c), from the sea level normal value of 40 Torr, reflecting the significant degree of ventilatory acclimatization achieved. Numerous studies have shown that a good ventilatory response to hypoxia enhances acclimatization and performance (Masson and Lahiri 1974; Masuyama et al. 1986; Schoene et al. 1984).

A persistent hyperventilation, secondary to carotid chemoreceptor activation, causes a profound respiratory alkalosis. The respiratory alkalosis, in turn, has a braking effect on the increase in ventilation with prolonged stay at high altitude (Huang et al. 1984). How ventilation can continue to increase in the presence of the inhibitory effect of an alkaline pH has not yet been solved. After several hours to days at altitude, Severinghaus and colleagues (1963) found that cerebrospinal fluid pH had adjusted to the respiratory alkalosis produced by the hyperventilation of acclimatization, and a new steady state was reached and maintained throughout the stay at high altitude. Others, however, studying animal and human models of ventilatory acclimatization to high altitude, found that cerebrospinal fluid pH was consistently alkaline (Forster, Dempsey, and Chosey 1974; Orr et al. 1975; Weiskopf, Gabel, and Fencl 1976), which would drive ventilation down. In a further attempt to clarify this question, humans were exposed for 7 days to a simulated altitude of 4300 m and brain nuclear magnetic resonance spectroscopy was completed before and after the altitude exposure. No demonstrable change in brain tissue pH was observed compared to pre-exposure values. These studies suggest that it is a factor other than $[H^+]$ that accounts for the hyperventilation observed during prolonged exposure to high altitude.

Heart

At rest, acute systemic hypoxia leads to an increase in cardiac output ($\dot{Q}$) and heart rate, with a variable response of stroke volume (Asmussen and Consolazio 1941; Hannon and Vogel 1977; Saltin et al. 1968; Vogel, Hansen, and Harris 1967). After several weeks to months of hypoxic exposure, especially above 4000 m ($P_IO_2 \sim 90$), there is a sustained increase in $\dot{Q}$ at rest (Saltin et al. 1968; Sutton et al. 1988). After 40 days of gradual decompression in Operation Everest 2, resting $\dot{Q}$ rose from 6 L · min^{-1} at sea level to almost 9 L · min^{-1} at the "summit" ($P_IO_2 = 43$ torr) (Sutton et al. 1988). At lower altitudes (~3000 m) it appears that $\dot{Q}$ returns to near sea level values after one or two weeks, due to a decrease in both stroke volume and heart rate (Hartley et al. 1967). Mean arterial pressure increases slightly from sea level values after two days at 4300 m, but begins to return to near sea level values by two weeks. Though slight, the elevation in chronic hypoxia of mean arterial pressure by 10–20 mmHg above normoxia or acute hypoxia values is a consistent finding (Bender et al. 1988; Wolfel 1993).

Blood

The relationship of oxygen binding to hemoglobin (Hb) is described by the sigmoidal shape of the oxygen-Hb dissociation curve (figure 14.2). At sea level, an arterial PO_2 (PaO_2) of 100 Torr results in Hb in arterial blood that is nearly completely saturated with oxygen ($SaO_2\%$ 96–98%). Moving down the

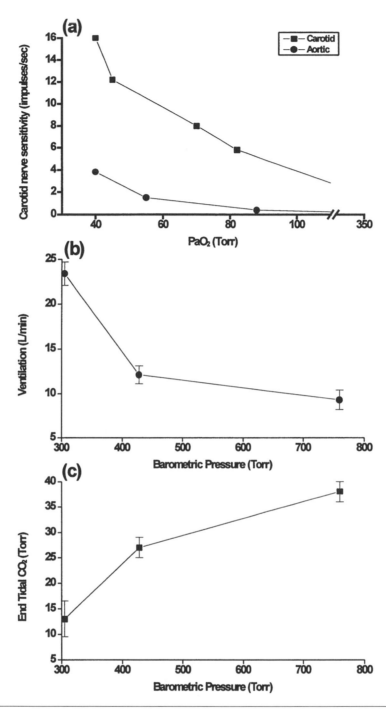

Figure 14.1   (*a*) Cat carotid and aortic chemoreceptor sensitivity to a drop in PO_2 (adapted from Fitzgerald and Lahiri 1986). (*b*) The rise in ventilation in humans with prolonged exposure to increasingly higher altitudes in Operation Everest 2. (*c*) The concomitant drop in end tidal CO_2 in the same study (adapted from Schoene et al. 1990).

curve (to the left), a drop in PaO_2 by one-half, to 50 Torr, results in a drop in SaO_2 of only 10%, to 87%. As the PaO_2 drops below about 40 Torr, the curve becomes much steeper, and ever smaller changes in the partial pressure of inspired oxygen (P_IO_2) result in larger decreases in SaO_2. As barometric pressure decreases with increasing altitude, P_IO_2 is lower, as calculated by the formula $P_IO_2 = [PB - (PH_2O @ 37$

°C)] $\times$ 0.2093 (table 14.1). From table 14.1, the effect on P_IO_2 of lowering the fraction of inspired air at constant PB is also apparent. Hence, throughout this chapter, especially where comparisons of cardiovascular responses are made between altitudes or levels of hypoxia, we will refer to the terrestrial altitude as meters above sea level (m) and frequently to the P_IO_2.

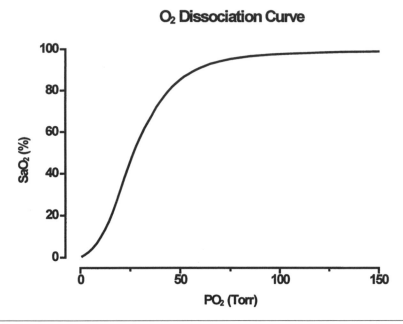

Figure 14.2   Oxygen dissociation curve for humans. Note the flat portion of the curve between a PO_2 of 100 and 150 torr. Contrast that with the region from 25 to 75 torr where a small drop in PO_2 results in a large drop in oxygen saturation.

Humans employ several strategies during their adjustment to hypoxia to maximize the blood-linked aspects of oxygen transport. In addition to aiding oxygen binding in the lungs by a left-shifted curve and aiding oxygen delivery in muscle by a right-shifted curve in the periphery, other blood-based responses also aid acclimatization. Since oxygen delivery is the product of arterial oxygen content × blood flow, any strategy that increases oxygen content with the same blood flow will increase the amount of oxygen available for muscular work (or maintenance of bodily functions). Initially, the plasma volume contracts in response to hypoxia, gradually returning to normal after several weeks (Alexander et al. 1967; Hannon and Vogel 1977; Hartley et al. 1967; Myhre et al. 1970). The 10–15% reduction in plasma volume is reflected in an apparent increased Hb concentration and thus also in CaO_2. The last of the blood-linked responses is an increase in total Hb concentration that occurs after several days to weeks of hypoxia, secondary to the stimulation of erythropoiesis by hypoxia (Jelkman 1992; Semenza et al. 1994).

Erythropoiesis is controlled by the induction of the glycoprotein erythropoietin. The basal production of erythropoietin is increased as much as 100- to 300-fold in response to hypoxic stress (Jelkman 1992; Semenza et al. 1994), reaching up to 17 ± 2 gm dl^{-1} at the simulated "summit" of Mount Everest (Sutton et al. 1988). Hypoxic induction of the erythropoietin gene depends primarily on increased transcription,

with markedly increased levels of erythropoietin mRNA observed following hypoxic stimuli to cultured cells (Bunn and Poyton 1996). The stimulus for erythropoietin mRNA production is likely hypoxia per se, as inhibition of oxidative phosphorylation does not induce erythropoietin RNA production or inhibit erythropoietin production under hypoxic conditions (Tan and Ratcliffe 1991). Another avenue to maximize oxygen transport during exercise at altitude is to increase alveolar PO_2 in the presence of a low inspired PO_2. This is accomplished largely by increases in pulmonary ventilation.

Exercise in Acute and Chronic Hypoxia

The most striking feature of exercise in hypoxia is the consistent and dramatic reduction in $\dot{V}O_2max$ (figure 14.3) due to the decrease in CaO_2, and hence, oxygen delivery. Also striking is that in spite of numerous physiological changes that enhance oxygen delivery, $\dot{V}O_2max$ does not increase appreciably with acclimatization (Buskirk et al. 1967; Faulkner et al. 1968). This fall in $\dot{V}O_2max$ is noticeable already at 2000 m (Consolazio et al. 1969) and continues a linear drop, up to the highest altitudes studied (Cymerman et al. 1989; Pugh et al. 1964), with a rate of decline above 1500 m of 2–3% per 300 m (Cymerman et al. 1989; Dill et al. 1931; Saltin et al. 1968).

The decrement of $\dot{V}O_2max$ appears related to aerobic fitness at sea level (Kjaer et al. 1988; Lawler,

Table 14.1 Barometric Pressure, Altitude, Inspired Oxygen (P_IO_2), and Fraction of Oxygen (F_IO_2), at Sea Level Barometric Pressure

PB^1	Alt (m)	Alt (ft)	$P_IO_2^2$	F_IO_2 at SL[3]	PB^1	Alt (m)	Alt (ft)	$P_IO_2^2$	F_IO_2 at SL[3]
759.6	0.0	0.00	149.15	0.209	409.2	5.2	17060.37	75.81	0.106
742.9	0.2	656.17	145.64	0.204	399.0	5.4	17716.54	73.67	0.103
726.4	0.4	1312.34	142.20	0.199	388.9	5.6	18372.70	71.56	0.100
710.2	0.6	1968.50	138.81	0.195	379.1	5.8	19028.87	69.50	0.097
694.3	0.8	2624.67	135.48	0.190	369.4	6.0	19685.04	67.48	0.095
678.7	1.0	3280.84	132.20	0.185	360.0	6.2	20341.21	65.51	0.092
663.3	1.2	3937.01	128.99	0.181	350.8	6.4	20997.38	63.58	0.089
648.2	1.4	4593.18	125.83	0.176	341.7	6.6	21653.54	61.68	0.087
633.4	1.6	5249.34	122.73	0.172	332.9	6.8	22309.71	59.83	0.084
618.8	1.8	5905.51	119.68	0.168	324.2	7.0	22965.88	58.02	0.081
604.5	2.0	6561.68	116.69	0.164	315.7	7.2	23622.05	56.25	0.079
590.5	2.2	7217.85	113.75	0.160	307.5	7.4	24278.22	54.51	0.076
576.7	2.4	7874.02	110.87	0.155	299.4	7.6	24934.38	52.82	0.074
563.2	2.6	8530.18	108.04	0.152	291.4	7.8	25590.55	51.16	0.072
549.9	2.8	9186.35	105.26	0.148	283.7	8.0	26246.72	49.54	0.069
536.9	3.0	9842.52	102.53	0.144	276.1	8.2	26902.89	47.95	0.067
524.1	3.2	10498.69	99.86	0.140	268.7	8.4	27559.06	46.40	0.065
511.6	3.4	11154.86	97.24	0.136	261.5	8.6	28215.22	44.89	0.063
499.3	3.6	11811.02	94.66	0.133	252.7	8.848	29028.87	43.06	0.060
487.2	3.8	12467.19	92.14	0.129	247.5	9.0	29527.56	41.96	0.059
475.4	4.0	13123.36	89.66	0.126	242.3	9.152	30026.25	40.88	0.057
463.8	4.2	13779.53	87.24	0.122	237.3	9.304	30524.93	39.82	0.056
452.4	4.4	14435.70	84.86	0.119	232.3	9.456	31023.62	38.78	0.054
441.3	4.6	15091.86	82.53	0.116	227.4	9.608	31522.31	37.76	0.053
430.4	4.8	15748.03	80.25	0.113	222.6	9.76	32021.00	36.76	0.052
419.7	5.0	16404.20	78.01	0.109					

[1] Calculated from model altitude equation [PB = exp $(6.63268 - 0.1112h - 0.00149h^2)$] where h is altitude in km. (West 1996).

[2] $P_IO_2 = [P_B - PH_2O$ at 37°C $\cdot$ 0 $\cdot$ 2093].

[3] F_IO_2 at sea level = [target $P_IO_2/760 - PH_2O$ at 37°C].

Powers, and Thompson 1988; Martin and O'Kroy 1993; Young, Cymerman, and Burse 1985). In a comparison of trained endurance athletes and untrained individuals, a high sea level $\dot{V}O_2max$ was related to the greatest decrease in $\dot{V}O_2max$ when breathing hypoxic gas (Lawler, Powers, and Thompson 1988) (r = 0.94, p < 0.001). The explanation offered was that the trained subjects had a larger decrease in $SaO_2\%$ during hypoxic exercise than the untrained subjects, thus decreasing oxygen delivery and hence $\dot{V}O_2max$.

The lack of increase of $\dot{V}O_2max$ with time in hypoxia is largely due to cardiovascular changes that

occur with acclimatization. Total blood volume is decreased early during altitude acclimatization due to a lower plasma volume that, initially at least, may account for the decrease in stroke volume. Stroke volume gradually returns to normoxia values and Hb increases. The peak heart rate, however, is decreased early in altitude exposure and remains depressed throughout an altitude stay above 4000 m, thus offsetting any gains in oxygen delivery due to the normalized stroke volume and increased Hb.

In spite of the lack of increase in $\dot{V}O_2max$ during acclimatization, endurance capacity gradually increases. In a study of eight men staying at 4300 m

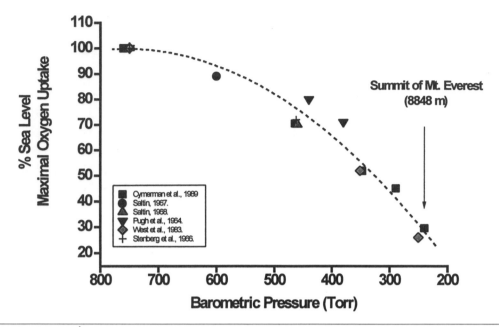

Figure 14.3   The drop in V̇O₂max with increasing altitude (or falling barometric pressure) is illustrated using data from several studies (see Cymerman et al. 1989; Saltin 1967; Saltin et al. 1968; Pugh et al. 1964; West et al. 1983; Stenberg, Ekblom, and Messin 1966).

($P_IO_2 = 86$) for 12 days, endurance time at 75% of sea level V̇O₂max increased 31% from day 2, when it was only 70 minutes, to 100 minutes by day 12. This increase in submaximal work capacity was accompanied by a decrease in lactate production during exercise (Maher, Jones, and Hartley 1974).

An increase in ventilation during exercise is key to maximizing performance in hypoxia. At submaximal exercise, ventilation is consistently elevated above values at similar workloads in normoxia, and is higher at maximal exercise in hypoxia compared to normoxia in spite of the large reduction in maximal power output (Åstrand 1954; Åstrand and Åstrand 1958) (figure 14.3). For example, ventilation at a V̇O₂max of 4 L · min⁻¹ was 105 l · min⁻¹ at sea level, 140 L · min⁻¹ at 2000 m, and increased to 160 L · min⁻¹ at 3000 m (Åstrand 1954; Åstrand and Åstrand 1958). The increased ventilatory drive during exercise serves to maintain SaO₂ during all but the heaviest work. West and colleagues (West, Boyer, et al. 1983) illustrated the importance of a brisk ventilatory response during exercise in extreme hypoxia (subjects living at 6300 m and exercising while breathing an F₁O₂ of 0.16) by contrasting the drop in SaO₂ during exercise between one climber who doubled his V̇ₑ when work was increased from 150 kpm to 300 kpm and consequently maintained his SaO₂ above 50%. In contrast, a second climber with a very poor drive to breathe increased his V̇ₑ by only 35% at 300 kpm and his SaO₂ dropped momentarily below 20% (oximeter reading).

One consequence of the large increases in pulmonary ventilation during hypoxic exercise is marked hypocapnia. For example, a $P_{ET}CO_2$ of 8 Torr was reported from one climber on the summit of Mount Everest (West, Hackett et al. 1983). Hypocapnia alters pulmonary ventilation but does not appear to have direct effects on cardiovascular responses to acute hypoxia. In 18 healthy men, hypoxia (F₁O₂ 0.08) caused a 76% increase in Q̇ and a 25% increase in heart rate (Richardson et al. 1966). When CO₂ was added and the level of hypoxia maintained, the Q̇ and heart rate responses were similar, indicating that hypocapnia is not responsible for the cardiovascular changes that accompany hypoxia, at least in the acute setting (Richardson et al. 1966).

During submaximal exercise at altitude, Q̇ may initially be increased (Asmussen and Consolazio 1941; Asmussen and Nielsen 1955; Pugh 1964) or similar to sea level values (Hartley et al. 1967), but all reports confirm a decrease from normoxic values in maximal Q̇ (Pugh 1964; Saltin et al. 1968; Sutton et al. 1988). After two weeks at 4300 m, peak Q̇ was 22% lower than sea level values (Saltin et al. 1968). Similar findings are reported at higher altitudes from studies in the field (Pugh 1964) and chamber (Sutton et al. 1988). In 4 men who spent 4–7 months at 4600 m or above, including 2–3 months at 5800, Pugh and colleagues observed a 30% decrease in peak Q̇ to 16 L · min⁻¹, from 23 L · min⁻¹ at sea level (Pugh et al. 1964). In Operation Everest 2, near peak Q̇ was 24 L · min⁻¹ at sea level and fell to 20 L · min⁻¹ with a P₁O₂ of 63 and further to 16 L · min⁻¹ at the "summit" with a P_IO_2 of 43 (Sutton et al. 1988). The $\dot{Q}$ response is largely determined by increased heart rate during submaximal exercise and a decrease at peak effort, with little change in stroke volume.

As first reported by Christensen and Forbes (1937; figure 14.4), at high altitude the peak heart rate falls from sea level values. In Operation Everest 2, peak heart rate was only 127 bpm at the simulated "summit," a 70% decrease from sea level values (Cymerman et al. 1989). When peak heart rate during maximal exercise first decreases in hypoxia has not yet been settled. During acute hypoxia, some studies show no consistent lowering of peak heart rate (Fagraeus et al. 1973; Lawler, Powers, and Thompson 1988; Saltin et al. 1968; Stenberg, Ekblom, and Messin 1966), while others show a consistent and significant decrease (Dill et al. 1966; Escourrou, Johnson, and Rowell 1984; Kjaer et al. 1988; Shephard et al. 1988). Several investigators report that sea level physical fitness determines the decrement in peak heart rate and $\dot{V}O_2$max in hypoxia (Grover et al. 1967; Kjaer et al. 1988; Martin and O'Kroy 1993; Shephard et al. 1988), and also determines, at least in part, the level of sympathetic activation during hypoxic exercise (Kjaer et al. 1988). In studies at simulated altitude or with hypoxic gas breathing, trained subjects have a consistently greater decrease in peak heart rate with acute hypoxia when compared to untrained subjects.

Stroke volume is slightly lower during submaximal and maximal exercise at altitudes from 3000 m to 5000 m (P_IO_2 103 to 78). The reduction in

stroke volume at submaximal work is less with acclimatization, but does not return to sea level values. Several explanations are possible for why stroke volume is reduced at high altitude. One likely mechanism of the reduction early in acclimatization is the reduction in total blood volume due to the contraction of the plasma volume. However, acute expansion of plasma volume with dextran infusion in 2 subjects exercising at 3100 m resulted in no increase in stroke volume in spite of a significant expansion of total blood volume (Alexander et al. 1967). Subsequent echocardiographic studies done at the same altitude ruled out left ventricular dysfunction as a possible cause of the lower stroke volume (Alexander and Grover 1983). Fatigue may play a role in lowering stroke volume at high altitude. In support of this, stroke volume at sea level is lower if it is measured in an exercise bout preceded by a previous, heavier workload (Hartley and Saltin 1968). When subjects exercise for 90 minutes at a high submaximal workload and then undergo a β-receptor agonist challenge, the results show significant β-receptor desensitization or down-regulation (Eysmann et al. 1996). Whether the β-receptor down-regulation noted after several days to weeks in hypoxia in animals (Kacimi et al. 1995; Maher et al. 1975) and humans (Antezana et al. 1994) also contributes to the lower stroke volume seen during altitude acclimatization is not known. Myocardial oxygen supply, which if limited would be detrimental to myocardial function, is well maintained during acute hypoxia (F_IO_2 0.12), largely due to an

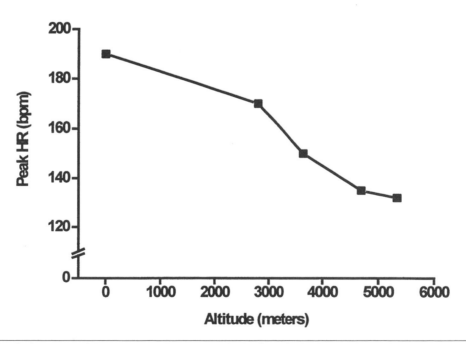

Figure 14.4 The decline in peak heart rate with ascent to high altitude. (Adapted from the 1937 report by Christensen and Forbes.)

oxygen demand-matching increase in myocardial blood flow (Kaijser, Grubbström, and Berglund 1993).

Exercise mean arterial pressure with acute hypoxia is similar to normoxia control values up to 200 watts (Bender et al. 1988). In contrast, after 3 weeks acclimatization to 4300 m, mean arterial pressure at rest and exercise was about 20 mmHg higher than sea level or acute hypoxia values. Whether mean arterial pressure is maintained at maximal exhaustive work with chronic hypoxia is not settled. Some authors report a decrease in mean arterial pressure at maximal exercise (Stenberg, Ekblom, and Messin 1966; Vogel, Hansen, and Harris 1967), while others report a slight increase (Cymerman et al. 1989; Reeves et al. 1987). In Operation Everest 2, mean arterial pressure was slightly elevated from 112 at sea level to 128 mmHg at the "summit" (P_IO_2 = 43 torr) during near maximal effort (97% $\dot{V}O_2$max) (Cymerman et al. 1989; Reeves et al. 1987).

That the peak heart rate to work output relationship during chronic exposure to hypoxia approaches sea level values has been used to argue that the heart rate to power output relationship is unaltered at altitude. For example, from the Operation Everest 2 data we can calculate the slope of the $\dot{V}O_2$max to watts relationship at sea level and at the "summit." The difference is a slight flattening at altitude of the

rise in $\dot{V}O_2$max per watt. Although the difference is slight, if the sea level $\dot{V}O_2$max to work output relationship was maintained, it would result in a 30% increase in $\dot{V}O_2$max at the "summit." Further support for departure from the expected exercise responses in chronic hypoxia comes from recent work on mountaineers studied at 5250 m. When subjects exercised with only one leg, near maximal heart rates were achieved at workloads that were twice as high as during two-legged exercise (Savard, Areskog, and Saltin 1995). This suggests that factors related to muscle mass (Lewis et al. 1985) or to the total oxygen supply/demand relationship may control maximal heart rate differently at altitude compared to sea level. Also, in many situations, including acute and chronic hypoxia, the heart rate to oxygen uptake relationship is a function of the relative work intensity regardless of the absolute work performed (Hermansen and Saltin 1967; Lewis et al. 1983).

Autonomic Nervous System

The primary pathway for autonomic regulation of cardiovascular responses to hypoxemia is via stimulation of the carotid bodies by a decrease in PaO_2, not CaO_2. As shown in figure 14.2 (page 180), the carotid

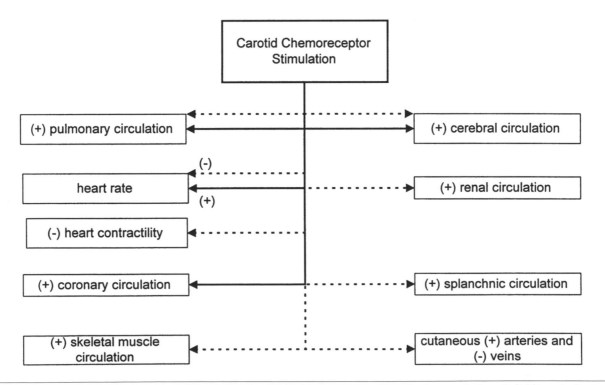

Figure 14.5 Graphic illustration of the role the carotid body plays in cardiovascular regulation. Solid lines indicate parasympathetic nerve fibers, and dashed lines indicate sympathetic nerve fibers. The plus and minus symbols indicate a rise or drop in nerve activity, respectively. (Adapted from Marshall 1994.)

body is notably insensitive to decreases in CaO_2 (Lahiri et al. 1979). The primary sympathetic and parasympathetic effects of carotid body activation on the cardiovascular system are illustrated in figure 14.5. The neurophysiology of carotid chemoreceptor activation has been extensively reviewed elsewhere (see Spyer 1990). In brief, studies in the cat and rat indicate that carotid chemoreceptor neural inputs project into several groups of neurons within the medulla that have sympathoexcitatory properties. Parasympathetic stimulation may occur via nerve projections from the carotid body to the vagal nucleus and nucleus ambiguus (Marshall 1994; Spyer 1990).

The sympathoadrenal response to hypoxia is characterized by an initial increase in adrenal medullary release of epinephrine, followed by a gradual increase in circulating norepinephrine and a diminution of epinephrine levels after several days exposure to hypoxia. Circulating plasma norepinephrine levels are the net result of synaptic norepinephrine release, reuptake, and clearance in the circulation. The extent to which these processes are altered in acute or chronic hypoxia has not been fully explored. It is now possible to directly record muscle sympathetic nerve activity (MSNA), and this technique has been used to unravel the relationships between sympathetic activation and cardiovascular regulation during hypoxia.

Sympathetic Regulation

Since the early work of Cannon and Hoskins showing that asphyxia stimulates medullary secretion in cats (1911), scientists have been interested in the role of activation of the sympathetic nervous system in the physiological responses to hypoxia. Epinephrine levels at rest with acute hypoxia are either increased (Kjaer et al. 1988; Mazzeo et al. 1991) or similar to normoxia values (Escourrou, Johnson, and Rowell 1984; Young et al. 1989). Also, Kjaer and colleagues found an elevation of resting epinephrine values only in trained subjects (1988). During exercise in acute hypoxia, however, epinephrine levels are uniformly elevated (Escourrou, Johnson, and Rowell 1984; Kjaer et al. 1988; Mazzeo et al. 1991). In 14 subjects breathing room air or an F_IO_2 of 0.16 during 1 hour of exercise at 65% of sea level $\dot{V}O_2$max, epinephrine and norepinephrine levels were elevated with exercise, but more so with hypoxic exercise (Strobel, Neureither, and Bärtsch 1995). Escourrou, Johnson, and Rowell (1984) reported similar catecholamine responses to hypoxic (F_IO_2 0.11–0.12) exercise when epinephrine was 80% higher than normoxic values.

Norepinephrine plasma levels are not consistently elevated with acute hypoxia, but begin to rise after several days in a hypoxic environment, and remain elevated during the initial days of acclimatization (Becker and Kreuzer 1968; Cunningham, Becker, and Kruezer 1965; Mazzeo et al. 1991). Cunningham, Becker, and Kruezer were the first to report an elevation of plasma norepinephrine concentration in man acclimatizing to chronic hypoxia (1965). They reported a 22% increase from sea level in plasma norepinephrine after 12 days at 4560 m (P_IO_2 83). In 5 men studied at sea level and after 21 days at 4300 m, arterial norepinephrine rose 174% from normoxia values during steady-state submaximal exercise (Mazzeo et al. 1991). After 7 days at 6542 m (P_IO_2 62), norepinephrine levels increased 3–4-fold during rest and submaximal exercise (Antezana et al. 1994) (figure 14.6a). As the altitude stay was prolonged to 3 weeks, norepinephrine values declined, particularly during exercise, although the values were still above those seen in normoxia (Antezana et al. 1994). In Operation Everest 2, the altitude was continually increased, so it is hard to see the effect of continued exposure to a fixed altitude on resting catecholamine values, except at one of the highest altitudes, at a pressure of about 280 torr (or a P_IO_2 of 49 Torr). At this altitude subjects spent almost a week while making short forays to higher altitudes. Their resting norepinephrine values at the start of their stay were increased 3-fold to 630 pg ml^{-1} from normoxia values, but decreased to 351 pg ml^{-1} after several days (figure 14.6b) (Young et al. 1989). Catecholamine samples taken immediately after an exhaustive exercise test in Operation Everest 2 revealed that maximal heart rate was markedly depressed, and accompanied by lower plasma levels of both epinephrine (–87%) and norepinephrine (–60%) at 282 torr (P_IO_2 49) compared to normoxia. Power output and time to exhaustion at this altitude were about 50% less than normoxia values. After sustained hypoxic exposure, these data suggest a marked attenuation of sympathetic activation during submaximal and maximal exercise (Antezana et al. 1994; Young et al. 1989).

Epinephrine kinetics during acute and chronic hypoxia remain largely unexplored, but recent work investigating norepinephrine kinetics may help explain the low norepinephrine levels observed with acute hypoxemia, and significantly influence interpretation of blood catecholamine levels in hypoxia. Skeletal muscle is the major source of circulating norepinephrine. Factors that increase muscle sympathetic activity increase norepinephrine spillover from muscle to plasma. A decrease in neuronal release of norepinephrine due to hypoxemia would

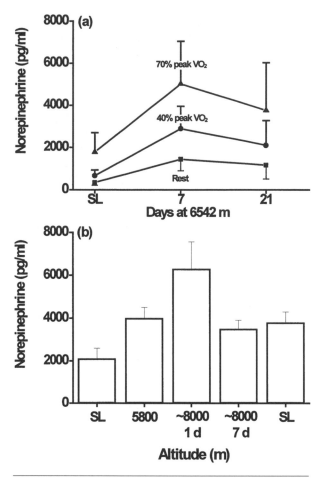

Figure 14.6 *(a)* Plasma norepinephrine during rest and submaximal exercise over 3 weeks at 6542 meters (adapted from Antezana et al. 1994); and *(b)* at rest from Operation Everest 2. Note that the norepinephrine level fell from the 1st to 7th day at ~8,000 m (adapted from Young et al. 1989).

account for the low norepinephrine values seen with acute hypoxia. Rowell and Blackmon (1986) used venomotor responses to hypoxia to test whether sympathetic vasoconstrictor mechanisms are intact during hypoxia. Veins, which are sensitive to small changes in sympathetic nerve activity, were consistently constricted with hypoxemia. This is supported by calculated norepinephrine release (leg blood flow $\times$ arterial-venous norepinephrine difference) that suggests norepinephrine switches from a pattern of net uptake at sea level to net release after 21 days of acclimatization (Mazzeo et al. 1991). Thus it seems that neuronal release of norepinephrine is not impaired by acute hypoxia. In studies using isolated perfused dog pulmonary arteries, hypoxia increased the norepinephrine release and overflow, with a marked reduction in neuronal reuptake and in intraneuronal degradation of norepinephrine by monoamine oxidase (Rorie and Tyce 1983). In hu-

mans breathing an F_IO_2 of 0.10 for 30 minutes, plasma norepinephrine clearance increased 20% compared to normoxia (Leuenberger et al. 1991). This meant that a 46% increase in norepinephrine spillover to plasma was seen as a 20% increase in plasma norepinephrine levels (figure 14.7a). This is surprising since other stress, such as high-intensity dynamic exercise, lower-body negative pressure, orthostatic stress, and heart failure are all associated with a decrease in norepinephrine clearance (see Leuenberger et al. 1993). An increase in norepinephrine clearance could reflect an increase in pulmonary or hepatomesenteric blood flow with hypoxia as these are the sites where norepinephrine is cleared from the circulation. Understanding that norepinephrine kinetics may be altered in acute and chronic hypoxia is important as we examine evidence of sympathetic activation from direct muscle sympathetic nerve recordings. Results from these studies show uniform activation of sympathetic nerves starting within minutes of the hypoxic stimulus, with persistent elevation even after 24 hours of hypoxia.

Recordings of MSNA reveal a graded increase in sympathetic activity with increasing hypoxia. In a study of 13 men exposed to 4000, 5000, and 6000 m in a hypobaric chamber, MSNA increased from sea level 34% at 4000 m, 44% at 5000 m, and 48% at 6000 m (Saito et al. 1988) (figure 14.7b). Prolonged exposure to the same degree of hypoxemia leads to greater sympathetic activation compared to a similar short exposure. For example, in a short exposure to 6000 m, MSNA was up 48%, while after breathing the equivalent hypoxia gas (F_IO_2 of 0.10) for 20 minutes, MSNA rose 260% (Rowell et al. 1989). Since lung inflation can have an attenuating effect on sympathetic outflow, MSNA responses were compared with a constant level of hypoxia but with or without adding carbon dioxide to the inspired gas (Somers et al. 1989a; Somers et al. 1989b). The addition of the carbon dioxide caused a greater chemoreceptor stimulation and subsequently a larger increase in ventilation. In turn, the increased ventilation attenuated the MSNA response, resulting in a 15% drop in MSNA compared to the response during hypocapnic hypoxia. Also of note from this study was that an F_IO_2 of 0.14, equivalent to about 3200 m, resulted in no change in MSNA (Somers et al. 1989a; Somers et al. 1989b).

To extend the above findings to include the interaction of the sympathetic stimuli from hypoxia and exercise, Seals and colleagues studied MSNA and plasma norepinephrine levels in young men made hypoxic while performing rhythmic handgrip exercise (Seals, Johnson, and Fregosi 1991). The MSNA and plasma norepinephrine levels rose more during

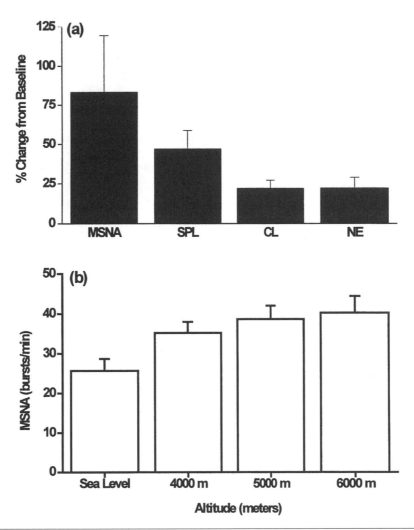

Figure 14.7 (*a*) Altered norepinephrine kinetics in the face of increased muscle sympathetic nerve activity (MSNA) (SPL = spillover of norepinephrine, CL = norepinephrine clearance, and NE = plasma norepinephrine. (Adapted from Leuenberger et al. 1991.) (*b*) Dose-response stimulus of MSNA by increasing levels of hypobaric hypoxia in humans. (Adapted from Saito et al. 1988.)

exercise with hypoxia than during normoxia. From this study it appears that hypoxia potentiates exercise-evoked sympathetic activation. Also, there may be a synergistic interaction between the two stimuli, because the hypoxic exercise MSNA responses were greater than the sum of the normoxic exercise and hypoxic rest trials.

To date, only one preliminary study has used MSNA recordings to examine sympathetic activation with prolonged hypoxia. Seven young men were exposed to a simulated altitude of 4300 m for 24 hours. MSNA was recorded in normoxia, and after 1 hour and 24 hours of hypoxia. Resting MSNA increased from 19 bursts · min^{-1} at sea level to 32 bursts · min^{-1} during acute hypoxia and remained elevated after 24 hours of hypoxia (Roach et al. 1996).

One result of the early and persistent elevation of circulating catecholamines in hypoxia is an appar-

ent down-regulation of the β-adrenergic receptor system. Animal and human studies have shown that β-receptor density is decreased, and response to a β-receptor agonist appears attenuated. A 5-week exposure of rats to 380 torr (P$_I$O$_2$ 70) caused a down-regulation of β-adrenergic receptors from right ventricles (Kacimi et al. 1992), and a similar study found such down-regulation in the left ventricle (Voelkel et al. 1981).

Indirect evidence from peripheral lymphocyte β-receptor populations suggest that β-receptor down-regulation also occurs in humans after several weeks of chronic hypoxia. Lymphocyte β-receptor density decreased by 45%, with no change in affinity for agonist (Antezana et al. 1994). In support of the physiological relevance of this observation is the decrease in chronotropic response to adrenergic activation observed after acclimatization. For example, during exercise in acclimatized men at 6542 m, a given plasma concentration of norepinephrine was associated with a lower heart rate, suggesting refractoriness to adrenergic stimulation. Responses to exogenous adrenergic agonist are also blunted after chronic hypoxia, both in humans (Antezana et al. 1994) and dogs (Maher et al. 1975). Isoproterenol infusion resulted in a lower change in heart rate at a given dose, and required a larger dose to increase the heart rate by 25 bpm, a consistent finding in studies conducted at 4350 and 6542 m (Antezana et al. 1994; Richalet et al. 1988; Richalet et al. 1989).

Attenuation of cardiovascular responsiveness to increases in β-adrenergic tone may occur quite rapidly (Hausdorff, Caron, and Lefkowitz 1990), even with acute hypoxia lasting only 30 minutes. Desensitization is essentially complete after exposure of cells to minute levels of agonist for 30 minutes. The short-term desensitization can disappear a few minutes after removal of the agonist. Marked desensitization induced by prolonged exposure to agonist (such as exposure for weeks to high circulating norepinephrine levels in chronic hypoxia) requires new protein synthesis for complete recovery, and thus is a process that may take several days to complete.

Parasympathetic Regulation

The usual increase in heart rate during exercise is linearly related to the oxygen requirements of the performed work and is mediated by parasympathetic withdrawal and sympathetic activation. The threshold for this response is thought to be in the range of 50–75% of $\dot{V}O_2$max. Evidence in humans of parasympathetic withdrawal during exercise is indirect, and largely based on lack of change in peak heart rate when atropine, a muscarinic antagonist, is given. Unlike our ability to directly measure sympathetic nerve impulses to skeletal muscle, no such direct techniques yet exist for measuring vagal nerve activity.

Marked elevations in peak heart rate at high altitude after atropine infusion has been consistently observed (Hartley, Vogel, and Cruz 1974; Richalet et al. 1990; Savard, Areskog, and Saltin 1995; Zhuang et al. 1993). In five men studied before and after several days acclimatization to 4600 m, peak heart rate was decreased 14 bpm from sea level values. After parasympathetic blockade with atropine, peak heart rate increased an average of 11 bpm, while $\dot{V}O_2$max was unchanged (figure 14.8a). In another study, of climbers at 5250 m, peak heart

rate was 21 bpm higher after atropine infusion (Savard, Areskog, and Saltin 1995) (figure 14.8b). Lifelong altitude residents had a significant increase in heart rate after atropine infusion at all workloads (Zhuang et al. 1993). In relative newcomers to altitude (only 1–2 years at 3658 m), heart rate after atropine was higher than control, but at the highest workload they experienced a slight decrease in heart rate after atropine infusion. Taken together these data suggest an unusually high degree of parasympathetic tone present during exercise in acclimatized individuals.

A noninvasive approach to examining autonomic balance during altitude acclimatization is available through a mathematical analysis of the variability of the R to R interval of the ECG. The heart rate variability has been interpreted as a reliable indicator of autonomic balance, particularly sensitive to changes in parasympathetic tone. Using heart rate variability analysis, Hughson and colleagues (1994) showed that the parasympathetic indicators were significantly increased early during altitude acclimatization, but that sympathetic indicators increased with further time at altitude, as parasympathetic activity decreased. The response of the sympathetic indicators fits well with known norepinephrine data from similar altitudes and similar times of exposure, and the parasympathetic data complement what has been observed with muscarinic blockade at altitude.

The increase in peak exercise heart rate after atropine infusion at altitude may be due to a direct effect of chronic hypoxia augmenting parasympathetic activation. On the other hand, cardiac parasympathetic activation may not be altered with chronic hypoxia; rather its effects may be unveiled because of the reduced sympathetic tone and lower absolute heart rate compared to normoxia (Savard, Areskog, and Saltin 1995). Recent animal work demonstrates a synergistic effect of vagal activation when it occurs in the presence of increased sympathetic tone. With increased sympathetic tone, the same degree of vagal nerve stimulation caused a much larger drop in heart rate than in the absence of such tone (Furukawa, Hoyano, and Chiba 1996; Levy 1971). In early altitude acclimatization, it seems reasonable that an interaction of increased cardiac parasympathetic and sympathetic tone plays an important role in determining the cardiovascular responses to exercise. The picture during chronic hypoxia, lasting several weeks to months in duration, is less clear. In this setting, sympathetic tone is probably lower than sea level values, and whether parasympathetic tone is different than in normoxia is not known.

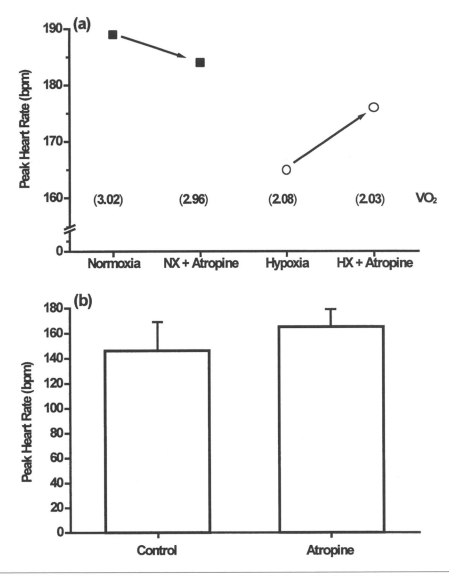

Figure 14.8 (*a*) Increased peak exercise heart rate after atropine infusion at high altitude in five men at 4600 m (adapted from Hartley, Vogel, and Cruz 1974); and (*b*) a similar rise in peak heart rate seen at 5300 m in base camp on Mount Everest (adapted from Savard, Areskog, and Saltin 1995).

Muscarinic Receptor Regulation

In hearts from guinea pigs exposed to hypoxia for 7 to 14 days, Crockatt and colleagues found increased choline acetyltransferase activity in the SA nodal region and a reciprocal decrease in muscarinic receptor number in many heart regions (Crockatt et al. 1981). In contrast, in rats made hypoxic for several days to weeks, cardiac muscarinic receptor density was increased (Kacimi, Crozatier, and Richalet 1992; Kacimi et al. 1995; Kacimi, Richalet, and Crozatier 1993; Wolfe and Voelkel 1983). Kacimi, Richalet, and Crozatier (1993) also demonstrated an increase in muscarinic receptor affinity for acetylcholine, which, in the rat, may reflect increased parasympathetic activity during chronic hypoxia.

An Integrated View of Cardiovascular Regulation During Exercise in Hypoxia

We now come back to the two ideas put forward to explain cardiovascular regulation during exercise in chronic hypoxia. We proposed that cardiovascular regulation in chronic hypoxia was controlled by a drop in central activation of the sympathetic nervous system due to a direct effect of hypoxia or diminished feedback from working muscles. Alternatively, we proposed that neurotransmission was modulated by local factors in the periphery. The scheme presented in figure 14.9 summarizes these ideas, with an emphasis on the role of arterial

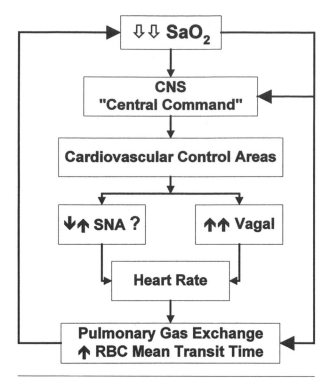

Figure 14.9 An integrated scheme of cardiovascular feedback and feed-forward mechanisms as they change with acclimatization to hypoxemia.

desaturation during exercise in cardiovascular regulation of exercise responses in chronic hypoxia. Support for arterial oxygen desaturation being causally related to cardiovascular regulation in the setting of chronic hypoxia is given in figure 14.10. Data

are shown in figure 14.10 from one individual exercising at 5800 m. During the first minute of exercise at a higher workload, the arterial oxygen saturation drops and exercise stops. Similar data linking desaturation in the early minutes of exercise in chronic hypoxia to a drop in heart rate are available from Operation Everest 2 (see figure 2 in Reeves et al. 1987). The scheme proposed in figure 14.9 accounts for known exercise responses in humans adapted to chronic hypoxia, and, in order to prompt future investigations, includes speculation about the involved physiological mechanisms.

References

Alam, M., and F.H. Smirk. 1938. Observations in man on a pulse accelerating reflex from the voluntary muscles of the legs. *J Physiol* 92:167–77.

Alexander, J. K., and R. F. Grover. 1983. Mechanism of reduced cardiac stroke volume at high altitude. *Clin Cardiol* 6:301–3.

Alexander, J.K., L.H. Hartley, M. Modelski, and R.F. Grover. 1967. Reduction of stroke volume during exercise in man following ascent to 3,100 m altitude. *J Appl Physiol* 23:849–58.

Antezana, A.M., R. Kacimi, J.L. Le Trong, M. Marchal, I. Abousahl, C. Dubray, and J.P. Richalet. 1994. Adrenergic status of humans during prolonged exposure to the altitude of 6542 m. *J Appl Physiol* 76:3001–5.

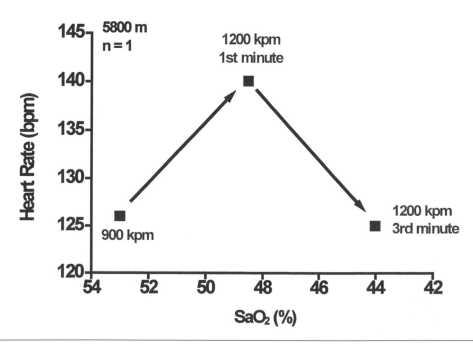

Figure 14.10 Decreased SaO$_2$ and a fall in heart rate at peak exercise in chronic hypoxia (5800 m). (Adapted from Pugh et al. 1964.)

Asmussen, E., and C.F. Consolazio. 1941. The circulation in rest and work on Mount Evans (4300 m). *Am J Physiol* 132:555–63.

Asmussen, E., and M. Nielsen. 1955. The cardiac output in rest and work at low and high oxygen pressures. *Acta Physiol Scand* 35:73–83.

Åstrand, P.O. 1954. The respiratory activity in man exposed to prolonged hypoxia. *Acta Physiol Scand* 30:343–68.

Åstrand, P.O., and I. Åstrand. 1958. Heart rate during muscular work in man exposed to prolonged hypoxia. *J Appl Physiol* 13:75–80.

Becker, E.J., and F. Kreuzer. 1968. Sympathoadrenal response to hypoxia. *Pflugers Arch* 304:1–10.

Bender, P.R., B.M. Groves, R.E. McCullough, R.G. McCullough, S.Y. Huang, A.J. Hamilton, P.D. Wagner, A. Cymerman, and J.T. Reeves. 1988. Oxygen transport to exercising leg in chronic hypoxia. *J Appl Physiol* 65:2592–97.

Bunn, H.F., and R.O. Poyton. 1996. Oxygen sensing and molecular adaptation to hypoxia. *Physiol Rev* 76:839–85.

Buskirk, E.R., J. Kollias, R.F. Akers, E.K. Prokop, and E.P. Reategui. 1967. Maximal performance at altitude and return from altitude in conditioned runners. *J Appl Physiol* 23:259–66.

Cannon, W.B., and R.G. Hoskins. 1911. The effects of asphyxia, hyperpnoea, and sensory stimulation on adrenal secretion. *Am J Physiol* 29:274–79.

Christensen, E.H., and W.H. Forbes. 1937. Der Kreislauf in grossen Höhen. *Skand Arch Physiol* 76:75–89.

Consolazio, C.F., L.O. Matoush, H.L. Johnson, H.J. Krzywicki, T.A. Daws, and G.J. Isaac. 1969. Effects of a high-carbohydrate diet on performance and clinical symptomology after rapid ascent to high altitude. *Fed Proc* 28:937–43.

Crockatt, L.H., D.D. Lund, P.G. Schmid, and R. Roskoski. 1981. Hypoxia-induced changes in parasympathetic neurochemical markers in guinea pig heart. *J Appl Physiol* 50:1017–21.

Cunningham, W.L., E.J. Becker, and F. Kruezer. 1965. Catecholamines in plasma and urine at high altitude. *J Appl Physiol* 20:607–10.

Cymerman, A., J.T. Reeves, J.R. Sutton, P.B. Rock, B.M. Groves, M.K. Malconian, P.M. Young, P.D. Wagner, and C.S. Houston. 1989. Operation Everest II: Maximal oxygen uptake at extreme altitude. *J Appl Physiol* 66:2446–53.

Dill, D.B., H.T. Edwards, A. Follings, S.A. Oberg, A.M. Pappenheimer, and J.H. Talbott. 1931. Adaptations of the organism to changes in oxygen pressure. *J Physiol (London)* 71:47–63.

Dill, D.B., L.G. Myhre, E.E. Phillips, and D.K. Brown. 1966. Work capacity in acute exposures to altitude. *J Appl Physiol* 21:1168–76.

Escourrou, P., D.G. Johnson, and L.B. Rowell. 1984. Hypoxemia increases plasma catecholamine concentrations in humans. *J Appl Physiol* 57:1507–11.

Eysmann, S.B., E. Gervino, D.E. Vatner, S.E. Katz, L. Decker, and P.S. Douglas. 1996. Prolonged exercise alters β-adrenergic responsiveness in healthy sedentary humans. *J Appl Physiol* 80:616–22.

Fagraeus, L., J. Karlsson, D. Linnarsson, and B. Saltin. 1973. Oxygen uptake during maximal work at lowered and raised ambient air pressure. *Acta Physiol Scand* 87:411–21.

Faulkner, J.A., J. Kollias, C.B. Favour, E.R. Buskirk, and B. Balke. 1968. Maximum aerobic capacity and running performance at altitude. *J Appl Physiol* 24:685–91.

Fitzgerald, R.S., and S. Lahiri. 1986. Reflex responses to chemoreceptor stimulation. In *Handbook of physiology. Section 3: The respiratory system*, ed. N.S. Cherniack and J.G. Widdicombe, 313–62. Bethesda, MD: American Physiological Society.

Forster, H.V., J.A. Dempsey, and L.W. Chosey. 1974. Incomplete compensation of CSF [H+] in man during acclimatization to high altitude (4300). *J Appl Physiol* 38:1067–72.

Furukawa, Y., Y. Hoyano, and S. Chiba. 1996. Parasympathetic inhibition of sympathetic effects on sinus rate in anesthetized dogs. *Am J Physiol* 271:H44–50.

Grover, R.F., J.T. Reeves, E.B. Grover, and J.E. Leathers. 1967. Muscular exercise in young men native to 3100 meter altitude. *J Appl Physiol* 22:555–64.

Hackett, P.H., and R.C. Roach. 1995. High-altitude medicine. In *Wilderness medicine*, ed. P.A. Auerbach, 1–37. St. Louis: Mosby.

Haldane, J.S., and J.G. Priestley. 1905. The regulation of the lung-ventilation. *J Physiol (London)* 32:225–66.

Hannon, J.P., and J.A. Vogel. 1977. Oxygen transport during early altitude acclimatization: A perspective study. *Eur J Appl Physiol* 36:285–97.

Hartley, L.H., J.K. Alexander, M. Modelski, and R.F. Grover. 1967. Subnormal cardiac output at rest and during exercise in residents at 3100 m altitude. *J Appl Physiol* 23:839–48.

Hartley, L.H., and B. Saltin. 1968. Reduction in stroke volume and increase in heart rate after a previous heavier submaximal workload. *Scand J Clin Lab Invest* 22:217–23.

Hartley, L.H., J.A. Vogel, and J.C. Cruz. 1974. Reduction of maximal exercise heart rate at altitude and its reversal with atropine. *J Appl Physiol* 36:362–65.

Hausdorff, W.P., M.G. Caron, and R.J. Lefkowitz. 1990. Turning off the signal: Desensitization of β-adrenergic receptor function. *FASEB J* 4:2881–89.

Hermansen, L., and B. Saltin. 1967. Blood lactate concentration during exercise at acute exposure to high altitude. In *Exercise at high altitude*, ed. R. Margaria, 48–53. New York: Excerpta Medica.

Huang, S.Y., J.K. Alexander, R.F. Grover, J.T. Maher, and R.E. McCullough. 1984. Hypocapnia and sustained hypoxia blunt ventilation on arrival at high altitude. *J Appl Physiol* 56:602–6.

Hughson, R.L., Y. Yamamoto, R.E. McCullough, J.R. Sutton, and J.T. Reeves. 1994. Sympathetic and parasympathetic indicators of heart rate control at altitude studied by spectral analysis. *J Appl Physiol* 77:2537–42.

Jelkman, W. 1992. Erythropoietin: Structure, control of production, and function. *Physiol Rev* 72:449–89.

Kacimi, R., B. Crozatier, and J.P. Richalet. 1992. Adenosinergic and muscarinic receptors coupling within rat heart in chronic hypoxia. *J Mol Cell Cardiol* 24:S27.

Kacimi, R., J.M. Moalic, A. Aldashev, D.E. Vatner, J.P. Richalet, and B. Crozatier. 1995. Differential regulation of G protein expression in rat hearts exposed to chronic hypoxia. *Am J Physiol* 38:H1865–73.

Kacimi, R., J.P. Richalet, A. Corsin, I. Abousahl, and B. Crozatier. 1992. Hypoxia-induced down-regulation of β-adrenergic receptors in rat heart. *J Appl Physiol* 73:1377–82.

Kacimi, R., J.P. Richalet, and B. Crozatier. 1993. Hypoxia-induced differential modulation of adenosinergic and muscarinic receptors in rat heart. *J Appl Physiol* 75:1123–28.

Kaijser, L., J. Grubbström, and B. Berglund. 1993. Myocardial lactate release during prolonged exercise under hypoxemia. *Acta Physiol Scand* 149:427–33.

Kjaer, M., J. Bangsbo, G. Lortie, and H. Galbo. 1988. Hormonal response to exercise in humans: Influence of hypoxia and physical training. *Am J Physiol* 23:R197–203.

Lahiri, S., E. Mulligan, T. Nishino, and A. Mokashi. 1979. Aortic body chemoreceptor responses to changes in PCO_2 and PO_2 in the cat. *J Appl Physiol* 47:858–66.

Lawler, J., S.K. Powers, and D. Thompson. 1988. Linear relationship between $\dot{V}O_2$max and $\dot{V}O_2$max decrement during exposure to acute hypoxia. *J Appl Physiol* 64:1486–92.

Leuenberger, U., K. Gleeson, K. Wroblewski, S. Prophet, R. Zelis, R. Zwillich, and L. Sinoway. 1991. Norepinephrine clearance is increased during acute hypoxemia in humans. *Am J Physiol* 261:H1659–64.

Leuenberger, U., L. Sinoway, S. Gubin, L. Gaul, D. Davis, and R. Zelis. 1993. Effects of exercise intensity and duration on norepinephrine spillover and clearance in humans. *J Appl Physiol* 75:668–74.

Levy, M.N. 1971. Sympathetic-parasympathetic interactions in the heart. *Circ Res* 29:437–45.

Lewis, S.F., P.G. Snell, W.F. Taylor, M. Hamra, R.M. Graham, W.A. Pettinger, and C.G. Blomqvist. 1985. Role of muscle mass and mode of contraction in circulatory responses to exercise. *J Appl Physiol* 58:146–51.

Lewis, S.F., W.F. Taylor, R.M. Graham, W.A. Pettinger, J.E. Schutte, and C.G. Blomqvist. 1983. Cardiovascular responses to exercise as functions of absolute and relative work load. *J Appl Physiol* 54:1314–23.

Maher, J.T., L.G. Jones, and L.H. Hartley. 1974. Effects of high altitude exposure on submaximal endurance capacity of men. *J Appl Physiol* 37:895–98.

Maher, J.T., S.C. Manchanda, A. Cymerman, D.L. Wolfe, and L.H. Hartley. 1975. Cardiovascular responsiveness to β-adrenergic stimulation and blockade in chronic hypoxia. *Am J Physiol* 228:477–81.

Marshall, J.M. 1994. Peripheral chemoreceptors and cardiovascular regulation. *Physiol Rev* 74:543–94.

Martin, D., and J. O'Kroy. 1993. Effects of acute hypoxia on the $\dot{V}O_2$max of trained and untrained subjects. *J Sports Sci* 11:37–42.

Masson, R.G., and S. Lahiri. 1974. Chemical control of ventilation during hypoxic exercise. *Respir Physiol* 22:242–62.

Masuyama, S., H. Kimura, T. Sugita, T. Kuriyama, K. Tatsumi, F. Kunitomo, S. Okita, H. Tojima, Y. Yuguchi, S. Watanabe, and Y. Honda. 1986. Control of ventilation in extreme-altitude climbers. *J Appl Physiol* 61:500–506.

Mazzeo, R.S., P.R. Bender, G.A. Brooks, G.E. Butterfield, B.M. Groves, J.R. Sutton, E.E. Wolfel, and J.T. Reeves. 1991. Arterial catecholamine response during exercise with acute and chronic high altitude exposure. *Am J Physiol* 261:E419–24.

Myhre, L.G., D.B. Dill, F.G. Hall, and D.K. Brown. 1970. Blood volume changes during three-week residence at high altitude. *Clin Chem* 16:7–14.

Orr, J.A., G.E. Bisgard, H.V. Forster, D.D. Buss, J.A. Dempsey, and J.A. Will. 1975. Cerebrospinal fluid alkalosis during high-altitude sojourn in unanesthetized ponies. *Respir Physiol* 23–37.

Pugh, L.G. 1964. Cardiac output in muscular exercise at 5800 m (19,000 ft). *J Appl Physiol* 19:441–47.

Pugh, L.G., M.B. Gill, S. Lahiri, J.S. Milledge, M.P. Ward, and J.B. West. 1964. Muscular exercise at great altitude. *J Appl Physiol* 19:431–40.

Reeves, J.T., B.M. Groves, J.R. Sutton, P.D. Wagner, A. Cymerman, M.K. Malconian, P.B. Rock, P.M. Young, and C.S. Houston. 1987. Operation Everest II: Preservation of cardiac function at extreme altitude. *J Appl Physiol* 63:531–39.

Richalet, J.P., P. Larmignat, C. Rathat, L.H. Hartleym, and R.L. Johnson. 1988. Decreased cardiac response to isoproterenol infusion in acute and chronic hypoxia. *J Appl Physiol* 65:1957–61.

Richalet, J.P., J.L. Le Trong, C. Rathat, P. Merlet, P. Bouissou, A. Keromes, and P. Veyrac. 1989. Reversal of hypoxia-induced decrease in human cardiac response to isoproterenol infusion. *J Appl Physiol* 67:523–27.

Richalet, J.P., C. Rathat, A. Keromes, and P. Larmignant. 1990. Effets de l'atropine sur la réponse adrenergique a l'exercise en hypoxia d'altitude (4350 m). *Science and Sport* 5:77–82.

Richardson, D.W., H.A. Kontos, W. Shapiro, and J.L. Patterson. 1966. Role of hypocapnia in the circulatory responses to acute hypoxia in man. *J Appl Physiol* 21:22–26.

Roach, R.C., S.F. Vissing, J.A.L. Calbet, G.K. Savard, and B. Saltin. 1996. Peak exercise heart rate after 24 hours at high altitude. *FASEB J* 10:A811.

Rorie, D.K., and G.M. Tyce. 1983. Effects of hypoxia on norepinephrine release and metabolism in dog pulmonary artery. *J Appl Physiol* 55:750–58.

Rowell, L.B., and J.R. Blackmon. 1986. Adrenergic activity during rest and exercise in hypoxemic humans. In *The sympathoadrenal system*, ed. N. J. Christensen, O. Hensriksen, and N. A. Lassen, 155–73. Copenhagen: Munksgaard.

Rowell, L.B., D.G. Johnson, P.B. Chase, K.A. Comess, and D.R. Seals. 1989. Hypoxemia raises muscle sympathetic activity but not norepinephrine in resting humans. *J Appl Physiol* 66:1736–43.

Rowell, L.B., D.S. O'Leary, and D.L. Kellogg. 1996. Integration of cardiovascular control systems in dynamic exercise. In *Exercise: Regulation and integration of multiple systems*, ed. L.B. Rowell and J. T. Shephard. New York: Oxford University Press.

Saito, M., T. Mano, S. Iwase, K. Koga, H. Abe, and Y. Yamazaki. 1988. Response in muscle sympathetic activity to acute hypoxia in humans. *J Appl Physiol* 65:1548–52.

Saltin, B. 1967. Aerobic and anaerobic work capacity at 2300 meters. *Med Thorac* 24:205–10.

Saltin, B., R.F. Grover, C.G. Blomqvist, L.H. Hartley, and R.L. Johnson. 1968. Maximal oxygen uptake and cardiac output after 2 weeks at 4300 m. *J Appl Physiol* 25:400–409.

Savard, G.K., N.H. Areskog, and B. Saltin. 1995. Cardiovascular response to exercise in humans following acclimatization to extreme altitude. *Acta Physiol Scand* 154:499–509.

Schoene, R.B., S. Lahiri, P.H. Hackett, R.M. Peters, Jr., J.S. Milledge, C.J. Pizzo, F.H. Sarnquist, S.J. Boyer, D.J. Graber, K.H. Maret, and J.B. West. 1984. The relationship of hypoxic ventilatory response to exercise performance on Mount Everest. *J Appl Physiol* 56:1478–83.

Schoene, R.B., R.C. Roach, P.H. Hackett, J.R. Sutton, A. Cymerman, and C.S. Houston. 1990. Operation Everest II: Ventilatory adaptation during gradual decompression to extreme altitude. *Medicine & Science in Sport & Exercise* 22:804–10.

Seals, D.R., D.G. Johnson, and R.F. Fregosi. 1991. Hypoxia potentiates exercise-induced sympathetic neural activation in humans. *J Appl Physiol* 71:1032–40.

Semenza, G.L., P.H. Roth, H.M. Fang, and G.L. Wang. 1994. Transcriptional regulation of genes encoding glycolytic enzymes by hypoxia-inducible factor 1. *J Biol Chem* 269:23757–63.

Severinghaus, J.W., R.A. Mitchell, B.W. Richardson, and M.M. Singer. 1963. Respiratory control at high altitude suggesting active transport regulation of CSF pH. *J Appl Physiol* 18:1155–66.

Shephard, R.J., E. Bouhlel, H. Vendewalle, and H. Monod. 1988. Peak oxygen uptake and hypoxia: Influence of physical fitness. *Intl J Sport Med* 9:279–83.

Somers, V.K., A.L. Mark, D.C. Zavala, and F.M. Abboud. 1989a. Contrasting effects of hypoxia and hypercapnia on ventilation and sympathetic activity in humans. *J Appl Physiol* 67:2101–6.

———. 1989b. Influence of ventilation and hypocapnia on sympathetic nerve responses to hypoxia in normal humans. *J Appl Physiol* 67:2095–2100.

Spyer, K.M. 1990. The central nervous organization of reflex cardiovascular control. In *Central regulation of autonomic function*, ed. A.D. Loewy and K.M. Spyer, 168–88. New York: Oxford University Press.

Stenberg, J., B. Ekblom, and R. Messin. 1966. Hemodynamic response to work at simulated altitude, 4000 m. *J Appl Physiol* 21:1589–94.

Strobel, G., M. Neureither, and P. Bärtsch. 1995. Effect of acute mild hypoxia during exercise on plasma free and sulphaconjugated catecholamines. *Eur J Appl Physiol* 73:82–87.

Sutton, J.R., J.T. Reeves, P.D. Wagner, B.M. Groves, A. Cymerman, M.K. Malconian, P.B. Rock, P.M. Young, S.D. Walter, and C.S. Houston. 1988. Operation Everest II. Oxygen transport during exercise at extreme simulated altitude. *J Appl Physiol* 64:1309–21.

Tan, C.C., and P.J. Ratcliffe. 1991. Effect of inhibitors of oxidative phosphorylation on erythropoietin mRNA in isolated perfused rat kidneys. *Am J Physiol* 261:F982.

Voelkel, N.F., L. Hegstrand, J.T. Reeves, I.F. McMurtry, and P.B. Molinoff. 1981. Effects of hypoxia on density of beta-adrenergic receptors. *J Appl Physiol* 50:363–66.

Vogel, J.A., J.E. Hansen, and C.W. Harris. 1967. Cardiovascular responses in man during exhaustive work at sea level and high altitude. *J Appl Physiol* 23:531–39.

Weil, J.V. 1986. Ventilatory control at high altitude. In *Handbook of physiology. The respiratory system*, ed. N.S. Cherniack and J.G. Widdicombe, 703–27. Bethesda, MD: American Physiological Society.

Weiskopf, R.B., R.A. Gabel, and V. Fencl. 1976. Alkaline shift in lumbar and intracranial CSF in man after 5 days at high altitude. *J Appl Physiol* 41:93–97.

West, J.B., S.J. Boyer, D.J. Graber, P.H. Hackett, K.H. Maret, J.S. Milledge, R.M. Peters, Jr., C.J. Pizzo, M. Samaja, F.H. Sarnquist, R.B. Schoene, and R.M. Winslow. 1983. Maximal exercise at extreme altitudes on Mount Everest. *J Appl Physiol* 55:688–98.

West, J.B., P.H. Hackett, K.H. Maret, J.S. Milledge, and R.M. Peters Jr. 1983. Pulmonary gas exchange on the summit of Mount Everest. *J Appl Physiol* 55:678–87.

West, J.B. 1996. Prediction of barometric pressures at high altitude with the use of model atmospheres. *J Appl Physiol* 81:1850–1854.

Wolfe, B.B., and N.F. Voelkel. 1983. Effect of hypoxia on atrial muscarinic cholinergic receptors and cardiac parasympathetic response. *Biochem Pharmacol* 32:1999–2002.

Wolfel, E.E. 1993. Sympatho-adrenal and cardiovascular adaptation to hypoxia. In *Hypoxia and molecular medicine*, ed. J.R. Sutton, C.S. Houston, and G. Coates. Burlington, VT: Queen City Press.

Young, A.J., A. Cymerman, and R.L. Burse. 1985. The influence of cardiorespiratory fitness on the decrement in maximal aerobic power at high altitude. *Eur J Appl Physiol* 54:12–15.

Young, P.M., M.S. Rose, J.R. Sutton, H.J. Green, A. Cymerman, and C.S. Houston. 1989. Operation Everest II: Plasma lipids and hormonal responses during a simulated ascent of Mount Everest. *J Appl Physiol* 66:1430–35.

Zhuang, J., T. Droma, J.R. Sutton, R.E. McCullough, R.G. McCullough, B. M. Groves, G. Rapmund, C. Janes, S. Sun, and L.G. Moore. 1993. Autonomic regulation of heart rate response to exercise in Tibetan and Han residents of Lhasa (3658 m). *J Appl Physiol* 75:1968–73.

Chapter 15

Hyperbaric Pressure

Dag Linnarsson

For a human diver, several factors may potentially influence the exercise performance. Along with the increase of the hydrostatic pressure of about 1 atmosphere (atm) (~1 bar, 0.1 MPa) per 10 m of water depth, parallel increases occur in the partial pressures of the inspired gas components and of the gas density for a given gas mixture. Table 15.1 summarizes the various environmental factors to which a diver is exposed, together with the corresponding manifestations and their assumed mechanisms. At moderate depths, the effects of the environment are reasonably well understood. Undesired effects can be avoided or ameliorated by protective equipment and synthetic breathing mixtures. At extreme depths and pressures, here defined as more than 90 m seawater (10 bar), biologic effects of the environ-ment are less well understood, and undesired effects are more difficult to avoid. Nevertheless, operational diving is routinely performed to depths on the order of 200–300 m. The highest pressure that a human has been exposed to in an experimental setting corresponds to a water depth of 701 m (Lafay et al. 1995; Medelli et al. 1993).

A number of inherent limitations are associated with studies of human physiology in the hyperbaric environment. Thus, changing the hydrostatic pressure of the environment is not possible without altering at least one of the partial pressures in the breathing gas. Effects of the hydrostatic pressure, therefore, are usually inferred from experiments with hyperbaric helium (He), a gas with minimal pharmacological effects due to its very low solubil-

Table 15.1 Diving Environmental Factors, Manifestations, and Their Assumed Mechanisms

Factor	Manifestation	Assumed mechanism
Hyperoxia, modest	Increased $\dot{V}O_2$max, airway irritation, impaired pulmonary gas exchange?	Increased arterial PO_2, O_2 toxicity
Hyperoxia, severe	Convulsions	O_2 toxicity
Increased gas density	CO_2 retention	Increased work of breathing
Increased N_2 partial pressure	Inert gas narcosis, impaired psychomotor performance	Physical solution of N_2 at hydrophobic sites in the CNS?
Increased hydrostatic pressure	High-pressure-nervous-syndrome (tremor and so on)	Slight volume decrease of hydrophobic sites in the CNS?
Inert gas loading	Impaired pulmonary gas exchange?	Tissue supersaturation during decompression, bubble formation
Unknown	Chouteau effect, cardiorespiratory failure	Hypoxia? Impaired respiratory control?

ity in biologic tissues. Dense gases, such as sulfur hexafluoride (SFl$_5$) and xenon (Xe) have higher narcotic potencies than lighter gases such as nitrogen (N$_2$) and helium. For a given gas mixture, density and narcotic effect increase proportionally with the hydrostatic pressure. For these two reasons, increasing the gas density above that of normal air is not possible without changing the narcotic effect in the same direction. The narcotic effect, however, can be increased without changing the density by inhaling gases with a high narcotic potency, for example nitrous oxide (N$_2$O).

Despite these limitations, the separate and combined effects of the various environmental factors have been gradually and successfully explored in humans and in relevant animal models. The additional effects of water immersion will not be described here since these are not specific for the hyperbaric environment.

Hyperoxia

Air is the most common breathing gas in recreational diving and in professional diving up to a depth of 50 m in seawater (6 bar). At 50 m, the inspired O$_2$ partial pressure is increased to 130% of a normal atmosphere. Parallel increases occur in alveolar and arterial O$_2$ partial pressures. In the hyperbaric air environment, a modest but statistically significant increase of the maximum O$_2$ uptake (V̇O$_2$max) has been found already at 1.4 bar (Fagraeus et al. 1973; Wyndham et al. 1970). However, no further increase occurred in V̇O$_2$max with higher air pressures (see figure 15.1). Instead, at 3 bar and 6 bar air pressure, V̇O$_2$max was no longer significantly increased compared with the surface (Fagraeus 1974; Linnarsson and Fagraeus 1976). When a 21% O$_2$ and 79% He mixture was respired, V̇O$_2$max at 3 bar ambient pressure was markedly increased by more than 10% (Fagraeus 1974; Linnarsson and Fagraeus 1976). These results are compatible with the notion that a large mass of exercising muscles can consume more O$_2$ than what can be provided by the cardiopulmonary system at maximal exercise during normal air breathing.

Gas Density

The work of breathing increases with the density of the breathing gas. The maximum voluntary ventilation is reduced roughly in proportion to the square root of the density increase (Hesser,

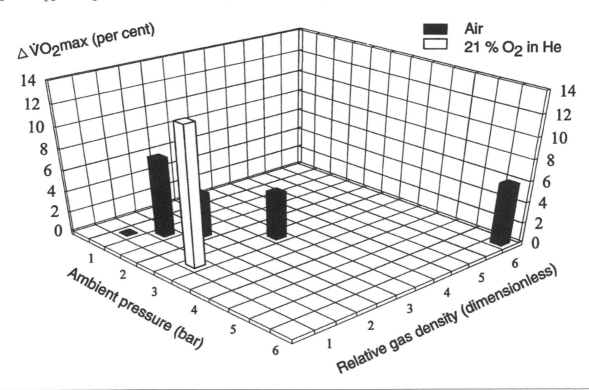

Figure 15.1 Changes in O$_2$ uptake (V̇O$_2$max) as functions of the ambient pressure and the relative density of the respired gas. Average control value at 1 bar air was 3.54 L/min standard temperature, pressure, and density (STPD). With air breathing, proportional increases occur in ambient pressure and gas density. Note the marked increase of V̇O$_2$max when hyperoxia was combined with normal gas density (3 bar, 21% O$_2$ in He). (Data from Fagraeus 1974; Fagraeus et al. 1973; Linnarsson and Fagraeus 1976.)

Linnarsson, and Fagraeus 1981; Lanphier and Camporesi 1993). Also, for very intense exercise, pulmonary ventilation is reduced with a high-density breathing gas, resulting in CO_2 retention (Fagraeus 1974; Fagraeus et al. 1973; Hesser, Linnarsson, and Fagraeus 1981; Linnarsson and Fagraeus 1976). Figure 15.2 shows changes in end-tidal CO_2 partial pressure during maximum exercise with air at 1, 1.4, 3, and 6 bar and with helium-oxygen at 3 bar. With normal gas density (1 bar air, 3 bar He/O_2), a typical hypocapnia (P_aCO_2 ~4 kPa) occurred as ventilation levels were increased out of proportion with the metabolic CO_2 production. With increased gas density (3 and 6 bar air), however, PCO_2 gradually increased due to a relative hypoventilation, reaching severely hypercapnic levels. This thereby added respiratory acidosis to the metabolic acidosis of exhaustive exercise. Researchers have proposed that this acidosis prevents the maximum aerobic metabolism to increase above 1 bar air control levels despite the increased availability of O_2 in the lungs and in the arterial blood (Fagraeus 1974; Linnarsson and Fagraeus 1976).

Further conclusions can be made from comparing data from 1 bar air, 3 bar air, and 3 bar 21% O_2/ 79% He (Fagraeus 1974). A small but significant elevation occurred in alveolar PCO_2 with 3 bar 21% O_2/79% He compared with 1 bar air with the same

gas density (see figure 15.2). This suggests that there normally is a component of PO_2-dependent respiratory drive during severe exercise, which was removed in the condition with three times greater PO_2. Thus, both the increased gas density and the hyperoxia of hyperbaric air contributed to reduce ventilation during maximum exercise. This is also true for the more modest reduction in ventilation seen when performing submaximal leg exercise in hyperbaric air (Fagraeus, Hesser, and Linnarsson 1974).

Inert Gas Narcosis

It is well established that hyperbaric N_2 impairs psychomotor performance (Bennett 1993). Nitrogen has a narcotic potency per unit partial pressure that is 30–40 times less than that for N_2O. Nitrogen also exerts its narcotic effect with a similar mechanism (Ostlund et al. 1992, 1994). Subjects breathing air at 7–10 bar have serious impairments of judgment, reaction time, manual dexterity, and short-term memory (Ostlund et al. 1994). Cardiopulmonary control functions, although severely impaired during deep levels of anesthesia, maintain their functional integrity in light-to-moderate inert gas narcosis. Thus, the response of the respiratory drive

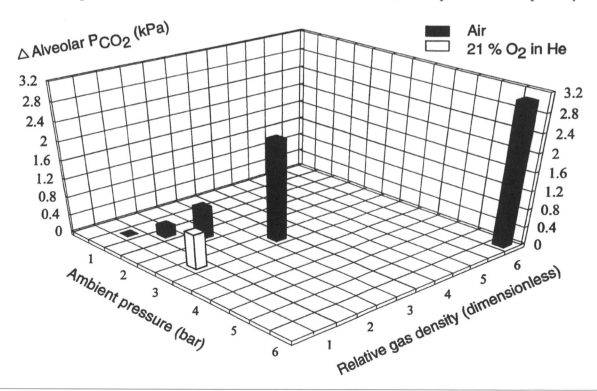

Figure 15.2   Alveolar PCO_2 at the end of supramaximal leg exercise, leading to exhaustion within 4–6 minutes. Data are differences from control at 1 bar air shown as functions of the ambient pressure and the relative density of the respired gas. Average control value at 1 bar air was 4.3 kPa (32 mmHg). (Data from Fagraeus 1974; Fagraeus et al. 1973; Linnarsson and Fagraeus 1976.)

to combined CO_2 stimulation and mechanical loading is not depressed during light inert gas narcosis (Linnarsson and Hesser 1978a, 1978b). Furthermore, light-to-moderate inert gas narcosis does not adversely influence cardiovascular functions during exercise (Bradley and Dickson 1976; Ciammaichella and Mekjavic 1994; Fothergill and Carlson 1994, 1995).

Inert gas narcosis is potentiated by CO_2 retention (Bennett 1993). This may contribute to limit exercise capacity in the hyperbaric air environment. Fagraeus (1974) and Linnarsson and Fagraeus (1976) observed that some of the subjects, while performing supramaximal leg work at 6 bar air, experienced a markedly increased sensation of intoxication as the work progressed. Upon cessation of exercise, these subjects reported that they ceased to work not because of fatigue but because the sensation of intoxication became so severe that they feared they would lose consciousness. In operational diving, therefore, the combination of heavy physical exertion and air breathing can be hazardous, especially when increased gas density or poor equipment increase the work of breathing.

Hydrostatic Pressure

Very high hydrostatic pressures alter the function of excitable tissues (Bennett and Rostain 1993). The section about extreme pressures will discuss this further. However, indications show that moderate levels of purely hydrostatic compression also affect the beating frequency and the contractility of isolated cardiac tissue, the heart rate, and the left ventricular dp/dt (first time derivative of pressure) in anaesthetized rats (Ask and Tyssebotn 1988; Risberg, Skei, and Tyssebotn 1995a, 1995b; Stuhr 1993). Also in the exercising human is a modest but reproducible bradycardia when breathing hyperbaric air (Fagraeus, Hesser, and Linnarsson 1974) and when breathing a normoxic, nonnarcotic helium-oxygen gas of normal density at a pressure of 5.5 bar (Linnarsson, Lind, and Hesser 1988). Since no other environmental differences occurred in comparison with air at normal pressure in this study, the researchers concluded that the elevated hydrostatic pressure must have caused the bradycardia. In addition to the hydrostatic bradycardia component, a further bradycardic component of a similar modest amplitude existed that could be seen when hydrostatic pressure was combined with moderate inert gas narcosis and increased gas density (Linnarsson, Lind, and Hesser 1988) and with hyperoxia (Fagraeus, Hesser, and Linnarsson 1974; Fagraeus and Linnarsson 1973). At pressures below

10 bar, the overall effect of these bradycardic influences is quantitatively modest and does not adversely influence the ability to perform submaximal (Fagraeus, Hesser, and Linnarsson 1974) or maximal (Fagraeus 1974; Fagraeus et al. 1973) dynamic exercise.

Exercise ventilation does not seem to be influenced by moderate levels of hydrostatic pressure. Lind, Linnarsson, and Hesser (1988) determined respiratory drive (inspiratory occlusion pressure), ventilatory timing, and volumes during light-to-severe exercise in subjects exposed to different combinations of hydrostatic pressure, inert gas narcosis, and gas density using normoxic mixtures of nitrogen, sulfur hexafluoride, and helium. They found no differences between 1 bar and 5.5 bar hydrostatic pressure when comparing conditions of equal gas density.

Hyperoxia and Decompression

When divers are exposed to extreme pressures, synthetic gas mixtures are used instead of air to avoid inert gas narcosis and to reduce density-induced respiratory effects. Respired oxygen partial pressures are usually kept in the range 25–50 kPa, i.e., slightly to moderately hyperoxic. This creates a balance between appropriate safety margins for hypoxia on the one hand and pulmonary oxygen toxicity (Clark 1993, 1994) on the other. Also, during the decompression phase, high partial pressures of O_2 are frequently respired for long periods in order to increase the rate of inert gas elimination. Following dives to extreme pressures and of long duration, decompression may take days. The cumulative exposure of the airways to elevated O_2 partial pressures will be substantial and may cause irritation of the airway mucosa, coughing, and decreased vital capacity (Clark 1993, 1994). Maximum exercise capacity, $\dot{V}O_2$max, and the lung diffusion capacity for carbon monoxide are reduced following deep saturation diving. Pulmonary oxygen toxicity has been suggested to be the cause (Thorsen et al. 1993). Decompression-induced bubbles in the pulmonary circulation have been suggested as an alternative or coexisting explanation (Thorsen et al. 1990, 1995).

Variations in Gas Density

With normoxic helium-oxygen mixtures, gas density will be above normal at pressures in excess of 5.5 bar. With hydrogen-oxygen, this is true above

11.4 bar. Thus, increased gas density is always present in deep diving regardless of gas mixture. In a typical diver working at the bottom of the continental shelf at a depth of 200 m and breathing 5–10% O_2 in He, the gas density is increased by a factor of 4–5. This diver, therefore, faces basically the same ventilatory limitations as an air-breathing diver at 30–40 m water depth. Therefore, very rigorous standards apply for flow-resistive pressure drops in breathing equipment for divers (Lanphier and Camporesi 1993).

Recently developed procedures to use when diving to depths in excess of 300 m include the addition of a narcotic gas component to the breathing in order to ameliorate pressure-induced symptoms from the nervous system (see below). When the narcotic inert gas component is 10% N_2, it further adds to the pressure-induced increase in gas density. In addition, gas density will be 10 and 14 times higher than in normal air at water depths of 400 m and 600 m, respectively. Even without flow-resistive pressure drops in the breathing equipment, the internal work of breathing imposes severe limitations on maximum expired flow rates due to dynamic airway compression (Lanphier and Camporesi 1993). Flook and Fraser (1989) also observed similar flow limitations during maximum inspirations. Compensatory changes in the breathing pattern include an elevated end-expired lung volume (Hesser, Lind, and Linnarsson 1990; Hesser, Linnarsson, and Fagraeus 1981; Spaur et al. 1977). Exercising divers frequently report inspiratory fatigue (Salzano et al. 1981). Decreased ventilatory CO_2 reactivity is also seen as an adaptive reaction to the increased work of breathing (Gelfand, Lainbertsen, and Peterson 1980; Gelfand and Peterson 1976).

The diffusivity of gases decreases with increasing gas density (Paganelli 1987; Van Liew, Paganelli, and Sponholtz 1982). Researchers have speculated that diffusive mixing between the inspired gas and the resident alveolar gas may become so slow that functionally important intra-alveolar partial pressure gradients for O_2 and CO_2 would occur in divers (for review, see Lanphier and Camporesi 1993). In a theoretical analysis, Hlastala, Ohlsson, and Robertson (1987) calculated that a diver breathing slightly hyperoxic *trimix* (10% N_2 in oxyhelium) at 47–66 bar would have intra-alveolar O_2 partial pressure gradients on the order of 1.5–2 kPa. Such a modest gradient would be easily compensated for by an equal elevation of the inspired O_2 partial pressure. This theoretical study and direct arterial blood gas determinations in exercising divers at very high pressure (Spaur et al. 1977) speak against intra-alveolar O_2 and CO_2 gradients as important respiratory problems in divers.

Hydrostatic Pressure and Inert Gas Narcosis

Vertebrates exposed to extreme pressures in the range of 50–100 bar develop motor disturbances, myoclonic jerks, and generated clonic or tonic-clonic convulsions (Brauer 1975). In humans, tremor and impaired motor performance have been observed in helium-oxygen breathing divers at 30 bar and above (Bennett and Rostain 1993). Such manifestations from the nervous system are called the high-pressure nervous syndrome (HPNS). Hydrostatic pressure is thought to cause them since similar manifestations can also be observed in aquatic animals compressed without concomitant elevation of the partial pressures of the gases physically dissolved in the water. Manifestations of HPNS occur earlier and with very rapid compression. An individual can avoid or postpone HPNS by gradually reducing the compression rate during the course of compression. Interestingly, the simultaneous administration of narcotic gases can increase the pressure threshold for HPNS manifestations. Inert gas narcosis can be reversed by hydrostatic compression in tissue and whole animal models (for reviews, see Brauer et al. 1982; Rostain 1994).

In addition, cardiac tissue is susceptible to high hydrostatic pressure. This reduces the beating frequency of spontaneously beating atrial preparations and enhances the twitch tension of preparations paced at a constant frequency (Ask and Tyssebotn 1988; Gennser and Omhagen 1989; Omhagen 1977; Stuhr 1993). As in nervous tissues, the presence of narcotic inert gases can counterbalance these effects of purely hydrostatic compression on the heart. Researchers have not, however, established that hydrostatic pressure can reverse narcotic effects on cardiac tissue (Gennser and Omhagen 1989).

Researchers have speculated that the opposite effects of hydrostatic pressure and narcotic inert gases represent an antagonism at a distinct molecular site in excitable tissues. However, observations in humans and higher animals suggest that inert gas narcosis may merely obscure, but not necessarily prevent, basic HPNS manifestations. Regardless of the underlying mechanism, HPNS symptoms become markedly reduced when narcotic gases such as N_2 and H_2 are added to the breathing mixture at great depths (Bennett and Rostain 1993).

The deepest dive performed by a human so far has been to a pressure of 71 bar (Lafay et al. 1995; Medelli et al. 1993). The breathing mixture then contained 20 bar, 0.4 bar O_2, and balanced He. If this very high H_2 partial pressure were acting alone without a concomitant increase of hydrostatic pressure, it would have a narcotic effect equivalent to the inhalation of 30–40% N_2O at normal atmospheric pressure (Brauer et al. 1982). One cannot exclude that the narcotic inert gas exerts adverse effects on the diver despite the very high hydrostatic pressure. For example, inhaling 30% N_2O at normal atmospheric pressure markedly lowers the threshold core temperature for thermogenic shivering and also depresses the shivering response per unit lowering of the core temperature (Mekjavic and Sundberg 1992). Researchers have not yet determined how well high hydrostatic pressure reverses this and other narcosis-induced impairments of essential control functions.

Dyspnea and Cardiorespiratory Problems of Unknown Origin

In a series of experiments exposing large mammals to very high pressures in a helium-oxygen atmosphere, Chouteau (1971) observed progressive signs of cardiorespiratory failure that could be reversed by increasing the PO_2 of the breathing gas. Although other researchers have had difficulty both explaining and reproducing these results (Imbert et al. 1981), an association may exist between these results in animals and subsequent observations of severe dyspnea in exercising humans at extreme pressures. Spaur et al. (1977) observed exercise-limiting dyspnea in divers performing light-to-moderate exercise at 50 bar while breathing an oxygen-helium mixture. They concluded that the dyspnea was of a mechanical rather than of a chemical origin since arterial blood gases remained in the normal range. Subsequently, Salzano et al. (1981) performed similar experiments at 47 bar. They confirmed that dyspnea limited the exercise capacity to what would correspond to submaximal intensities during surface control conditions. The same experiments were then repeated at 47 bar and 66 bar with the addition of up to 10% N_2 in the breathing mixture. Despite the substantially elevated gas density, the subjects reported much fewer dyspnea problems. These results suggest that an HPNS-related mechanism may be behind the dyspnea in addition to the problems caused by the very high gas density. One might speculate that the observations of cardiorespiratory failure in animals at very high pressure (Chouteau 1971) may have been due at least in part to HPNS-related problems in respiratory muscle function.

References

Ask, J.A., and I. Tyssebotn. 1988. Positive inotropic effect on the rat atrial myocardium compressed to 5, 10, and 30 bar. *Acta. Physiol. Scand.* 13:277–283.

Bennett, P.B. 1993. Inert gas narcosis. In *The physiology and medicine of diving*, eds. P.B. Bennet and D.H. Elliott, 4th ed., 170–193. London: W.B. Saunders.

Bennett, P.B., and J.C. Rostain. 1993. The high pressure nervous syndrome. In *The physiology and medicine of diving*, eds. P.B. Bennett and D.H. Elliott, 4th ed., 194–237. London: W.B. Saunders.

Bradley, M.E., and J.G. Dickson, Jr. 1976. The effects of nitrous oxide narcosis on the physiologic and psychologic performance of man at rest and during exercise. In *Underwater physiology V*, ed. C.J. Lambertsen, 617–626. Bethesda, MD: Faseb.

Brauer, R.W. 1975. The high pressure nervous syndrome: Animals. In *The physiology and medicine of diving and compressed air work*, eds. P.B. Bennett and D.H. Elliott, 2nd ed., 231–247. London: Bailliare Tindall.

Brauer, R.W., P.M. Hogan, M. Hugon, A.G. MacDonald, and K.W. Miller. 1982. Patterns of interaction of effects of fight metabolically inert gases with those of hydrostatic pressure as such. *Undersea. Biomed. Res.* 9:353–395.

Chouteau, J. 1971. Respiratory gas exchange in animals during exposure to extreme ambient pressures. In *Underwater physiology*, ed. C.J. Lambertsen, 385–397. New York: Academic Press.

Ciammaichella, R., and I.B. Mekjavic. 1994. The effect of narcosis on maximal aerobic performance. *Undersea & Hyperbar. Med.* 21 (Suppl.):39.

Clark, J.M. 1993. Oxygen toxicity. In *The physiology and medicine of diving*, eds. P.B. Bennett and D.H. Elliott, 4th ed., 121–169. London: W.B. Saunders.

Clark, J.M. 1994. Oxygen toxicity. In *Hyperbaric medicine practice*, ed. E.P. Kindwafl, 33–43. Flagstaff, AZ: Best Publishing.

Fagraeus, L. 1974. Maximal work performance at raised air and helium-oxygen pressures. *Acta Physiol. Scand.* 91:545–556.

Fagraeus, L., C.M. Hesser, and D. Linnarsson. 1974. Cardiorespiratory responses to graded exercise at increased ambient air pressure. *Acta Physiol. Scand.* 91:259–274.

Fagraeus, L., J. Karisson, D. Linnarsson, and B. Saltin. 1973. Oxygen uptake during maximal work at lowered and raised ambient air pressures. *Acta Physiol. Scand.* 87:411–421.

Fagraeus, L., and D. Linnarsson. 1973. Heart rate in hyperbaric environment after autonomic blockade. *Forsvarsmedicin (Swedish Journal of Defense Medicine)* 9:260–264.

Flook, V., and I.M. Fraser. 1989. Inspiratory flow limitation in divers. *Undersea Biom. Res.* 16:305–311.

Fothergill, D.M., and N.A. Carlson. 1994. The influence of inert gas narcosis on psychophysical responses to exercise with inspiratory resistive loading. *Undersea & Hyperbar. Med.* 21 (Suppl.):38–39.

Fothergill, D.M., and N.A. Carlson. 1995. The roles of gas density, high pressure, and inert gas narcosis on exercise tolerance and psychophysical perceptions of exertion. *Undersea & Hyperbar. Med.* 22 (Suppl.):70.

Gelfand, R., C.J. Lainbertsen, and R.E. Peterson. 1980. Human respiratory control at high ambient pressures and inspired gas densities. *J. Appl. Physiol.* 48:528–539.

Gelfand, K., and K. Peterson. 1976. The effects on CO_2 reactivity of breathing crude neon, helium and nitrogen at high pressure. In *Underwater physiology V*, ed. C.J. Lambertsen, 603–615. Bethesda, MD: Faseb.

Gennser, M., and H.C. Omhagen. 1989. Effects of hydrostatic pressure, H_2, N_2, and He, on beating frequency of rat atria. *Undersea Biomed. Res.* 16:153–164.

Hesser, C.M., F. Lind, and D. Linnarsson. 1990. Significance of airway resistance for the pattern of breathing and lung volumes in exercising humans. *J. Appl. Physiol.* 68:1875–1882.

Hesser, C.M., D. Linnarsson, and L. Fagraeus. 1981. Pulmonary mechanics and work of breathing at maximal ventilation and raised air pressure. *J. Appl. Physiol.* 50:747–753.

Hlastala, M.P., J. Ohlsson, and H.T. Robertson. 1987. Alveolar gas-phase diffusion limitation in the hyperbaric environment. In *Underwater and hyperbaric physiology IX*, eds. A.A. Bove, A.J. Bachrach, and L.J. Greenbaum, Jr., 457–464. Bethesda, MD: Undersea and Hyperbaric Medical Society.

Imbert, G., Y. Jammes, N. Narald, J.C. Dufiot, and C. Grimaud. 1981. Ventilation, pattern of breathing, and activity of respiratory muscles in awake cats during oxygen-helium simulated dives (1000 msw). In *Underwater physiology VI*, eds. A.J. Bachrach and M.M. Matzen, 273–282. Bethesda, MD: Undersea Medical Society.

Lafay, V., P. Barthelemy, B. Comet, Y. Frances, and Y. Jammes. 1995. ECG changes during the experimental human dive HYDRA 10 (71 atm/7,200 kPa). *Undersea & Hyperbar. Med.* 22:51–60.

Lanphier, E.H., and E.M. Camporesi. 1993. Respiration and exertion. In *The physiology and medicine of diving*, eds. P.B. Bennett and D.H. Elliott, 4th ed., 77–120. London: W.B. Saunders.

Lind, F., D. Linnarsson, and C.M. Hesser. 1988. Abstract. Moderately increased hydrostatic pressure does not influence ventilatory control during exercise in humans. *Undersea & Med. Res.* 15 (Suppl.):33.

Linnarsson, D., and L. Fagraeus. 1976. Maximal work performance in hyperbaric air. In *Underwater physiology V*, ed. C.J. Lambertsen, 55–60. Bethesda, MD: Faseb.

Linnarsson, D., and C.M. Hesser. 1978a. Dissociated ventilatory and central respiratory responses to CO_2 at raised N_2 pressure. *J. Appl. Physiol.* 45:756–761.

Linnarsson, D., and C.M. Hesser. 1978b. Effect of hyperbaric nitrogen on central respiratory response to CO_2. In *Underwater physiology VI*, eds. C.W. Shilling and M.W. Becket, 139–144. Bethesda, MD: Faseb.

Linnarsson, D., F. Lind, and C.M. Hesser. 1988. Abstract. Effects of moderate hydrostatic pressure on exercise heart rate in man. *Undersea Biomed. Res.* 15 (Suppl.):87–88.

Medelli, J., H. Jullien, B. Boulferache, J.Y. Massimelli, M. Comet, and B. Gardette. 1993. Exercises during Hydra 10 until 701 msw: Cardiac and ventilatory monitoring. *Undersea & Hyperbar. Med.* 20 (Suppl.):33–34.

Mekjavic, I.B., and C.J. Sundberg. 1992. Human temperature regulation during narcosis induced by inhalation of 30% nitrous oxide. *J. Appl. Physiol.* 73:2246–2254.

Omhagen, H. 1977. Hyperbaric bradycardia and arrhythmia. Experiments in liquid breathing mice and sinus node preparations. Ph.D. diss., University of Lund, Sweden.

Ostlund, A., D. Linnarsson, F. Lind, and A. Sporrong. 1994. Relative narcotic potency and mode of action of sulfur hexafluoride and nitrogen in humans. *J. Appl. Physiol.* 76:439–444.

Ostlund, A., A. Sporrong, D. Linnarsson, and F. Lind. 1992. Effects of sulfur hexafluoride on psychomotor performance. *Clin. Physiol.* 12:409–418.

Paganelli, C.V. 1987. Gas-phase diffusion of O_2 in helium and nitrogen under pressure. In *Underwater and hyperbaric physiology IX*, eds. A.A. Bove, A.J. Bachrach, and L.J. Greenbaum, Jr., 439–446. Bethesda, MD: Undersea and Hyperbaric Medical Society.

Risberg, J., S. Skei, and I. Tyssebotn. 1995a. Effects of gas density and ambient pressure on myocardial contractility in the rat. *Av. Space Envir. Med.* 66:1159–1168.

Risberg, J., S. Skei, and I. Tyssebotn. 1995b. Pressure increase myocardial contractility independent of breathing gas density in anesthetized rats. *Undersea & Hyperbar. Med.* 22 (Suppl.):30.

Rostain, J.C. 1994. Nervous system at high pressure. In *Basic and applied high pressure biology*, eds. P.B. Bennet and R.E. Marquis, 157–172. Rochester, NY: University of Rochester Press.

Salzano, J.V., B.W. Stolp, R.E. Moon, and E.M. Camporesi. 1981. Exercise at 47 and 66 ATA. In *Underwater physiology VII*, eds. A.J. Bachrach and M.M. Matzen, 181–196. Bethesda, MD: Undersea Medical Society.

Spaur, W.H., L.W. Raymond, M.M. Knott, J.C. Crothers, W.R. Braithwaite, E.D. Thaltnann, and D.F. Uddin. 1977. Dyspnea in divers at 49.5 ATA: Mechanical, not chemical in origin. *Undersea Biomed. Res.* 4:183–198.

Stuhr, L.E.B. 1993. Cardiovascular changes during moderate hyperbaric exposures. Effect of acute, and repeated hyperbaric exposure, and pharmacological interventions on cardiac functions. Ph.D. diss., University of Bergen, Norway.

Thorsen, E., J. Hjefie, K. Segadal, and A. Gulsvik. 1990. Exercise tolerance and pulmonary gas exchange after deep saturation dives. *J. Appl. Physiol.* 68:1809–1814.

Thorsen, E., J. Risberg, K. Segadal, and A. Hope. 1995. Effects of venous gas microemboli on pulmonary gas transfer function. *Undersea & Hyperbar. Med.* 22:347–353.

Thorsen, E., K. Segadal, J.W. Reed, C. Elhott, A. Gulsvik, and J.O. Hjelle. 1993. Contribution of hyperoxia to reduced pulmonary function after deep saturation dives. *J. Appl. Physiol.* 75:657–662.

Van Liew, H.D., C.V. Paganelli, and D.K. Sponholtz. 1982. Estimation of gas-phase diflusivities in hyperbaric environments. *Undersea Biomed. Res.* 9:175–181.

Wyndham, C.H., N.B. Strydom, A.J. van Rensburg, and G.G. Rogers. 1970. Effects on maximal oxygen intake of acute changes in altitude in a deep mine. *J. Appl. Physiol.* 29:552–555.

Chapter 16

Microgravity

Gunnar Blomqvist

Galileo and Newton introduced the physical concept of gravity during the 17th century. Recognition of its biologic role came much later. This began during the early 19th century with the demonstration of gravitational effects on the growth pattern of plants (geotropism) and the immediate effects of changes in posture on the human cardiovascular system. However, progress was slow until Gagarin's first brief space flight on 12 April 1961. During the 35 years that have elapsed since then, frequent access to space has provided opportunities to examine the physical and biologic effects of microgravity (μG). The range of experimental techniques that can be utilized in space has been slowly but steadily expanded. Research has shown that although gravity affects most biologic systems, the impact of μG varies greatly across species and systems. The effects of μG on the human cardiovascular, musculoskeletal, and vestibular systems are particularly striking.

The transition to μG has major effects on the distributions of pressures and volumes that prevail in the human cardiovascular system in the dominant upright posture at 1 G. However, little evidence shows major circulatory impairment in space. Following an early central fluid shift with increased cardiac filling, adaptation to μG includes a return to the hemodynamic conditions that normally exist in the upright posture at 1 G. This process sets the stage for major dysfunction on return to earth. This chapter will discuss this process in detail.

Unloading of skeletal muscle produces atrophy, particularly of the muscle groups that normally control posture, support body weight, and include large slow-twitch fiber populations. μG also induces an unrelenting loss of calcium from the skeleton that can approach clinically significant cumulative levels during long flights. Urinary tract stone formation is a secondary major hazard of demineralization. Only very recently have promising pharmacological countermeasures been identified (Ruml, Hw, and Pak 1996). Significant neurosensory prob-

lems also occur, particularly vestibular dysfunction early after arrival at μG. Finally, radiation exposure (which is a feature of the space environment rather than of μG) is the most significant biologic hazard during interplanetary space travel. Exposure in low earth orbits is limited, and the levels reached are only slightly above the terrestrial norm.

Current detailed reviews of these important areas are available in *The Handbook of Physiology* in a recent section about environmental physiology, edited by Greenleaf (1996). It includes a chapter about the cardiovascular system in μG by Watenpaugh and Hargens (1996). General information about space medicine can be found in a text by Nicogossian, Huntoon, and Pool (1994). Atkov and Bednenko (1992) have published an extensive review of Russian work. The present chapter deals primarily with the adaptation of the human cardiovascular system to actual and simulated μG and its readaptation to 1 G.

Simulated Microgravity

The following sections discuss simulated microgravity. Specifically, they deal with methodology and physiological effects.

Methodology

Many of the major μG-induced effects on the cardiovascular and musculoskeletal systems can be reproduced by strict bed rest. However, the fidelity of the model has not been rigorously tested by exposing the same individuals to spaceflight and bed rest of equal duration. With respect to human physiology, bed rest shares two major effects with μG. First, a marked reduction occurs in the magnitude of the hydrostatic gradients that normally exist in the dominant upright posture, which causes a redistribution of intra- and extravascular fluid volumes. Second, the overall level of physical activity decreases. Technically, more accurate μG models are available than bed rest, e.g., controlled free fall from drop towers or during parabolic airplane flights. However, these modalities can provide only brief μG equivalents (< 1 min) and are of limited value in life sciences research. Upright water immersion has been used, but it produces pressure and volume distributions that differ significantly from actual μG conditions.

Kakurin et al. (1976) introduced bed rest with head-down tilt (HDT) as a more appropriate model of μG than horizontal bed rest. HDT at an angle of 4–8° (which minimizes all hydrostatic gradients within the human body) rapidly became the standard tool for μG simulation during human cardiovascular studies. Nevertheless, it is now evident that μG causes truly unique physiological effects that cannot be recreated at 1 G. Most important—from a clinical and operational point of view—is μG-induced vestibular dysfunction. This condition can be severe during the first few days in space and be present to a lesser degree early after return to 1 G. The exact neurophysiological mechanisms are still unknown. The transition from 1 G to μG likely produces a sensory conflict. The afferent nerve traffic from the vestibular system and the sensory receptors in skin and skeletal muscle is likely to be markedly altered at μG, whereas the visual input will remain intact and accurate (Daunton 1996).

Physiological Effects

The physiological effects of prolonged bed rest were first described in the 1920s. Important clinical cardiovascular and metabolic consequences, including decreased blood volume and post-bed rest orthostatic intolerance, skeletal muscle atrophy, and loss of calcium from the skeleton were recognized during the 1940s. (For references, see Blomqvist and Stone 1983 and Fortney, Schneider, and Greenleaf 1996.) Saltin et al. (1968) provided the first data about the magnitude of the full range of physiological changes that long-term changes in posture and physical activity can induce in healthy subjects. They examined a group of five healthy young men before and after prolonged bed rest and also after a period of intense physical training that immediately followed the bed rest phase. A three-week bed rest period reduced stroke volume, cardiac output, and maximal oxygen uptake by one-third, whereas maximal heart rate remained unchanged. Previously untrained subjects responded to an eight-week period of intensive physical training by an increase in maximal oxygen uptake to one-third above baseline levels before bed rest. Half of the improvement was attributable to an increase in

oxygen extraction (as reflected by the systemic arteriovenous oxygen difference) and half to an increase in maximal cardiac output. The increase in cardiac output was due to an increase in heart size and stroke volume without any change in maximal heart rate.

The development of human spaceflight programs in the United States and Russia prompted extensive use of bed rest as a method to simulate μG. HDT was employed in a majority of the studies performed after the report published by Kakurin et al. (1976). Nicogossian et al. (1979) tabulated results from more than 500 simulation experiments. Bed rest with HDT has remained an important adjunct in the study of space physiology to supplement the limited access to space. Work done in several laboratories, including ours (Gaffney et al. 1985; Nixon et al. 1979), has demonstrated early but transient increases in cardiac filling beyond normal supine levels during HDT. This has been documented by echocardiographic estimates of left ventricular end-diastolic dimensions and by measurements of central venous pressure (CVP). Stroke volume also increases beyond supine levels early during HDT without any change in contractile state. Within a few hours, a negative fluid balance produces a decrease in total body weight and intravascular volume. After 24 hours, cardiac filling and stroke volume approach pretilt upright levels. As expected, orthostatic tolerance is impaired on return to the upright body position. In general, these data seem to confirm Gauer's view (Gauer and Thron 1965) that the dominant upright body position defines the normal operating conditions for the human cardiovascular system. If the upright hemodynamic pattern is perturbed, compensatory mechanisms are activated to restore baseline conditions (Baisch et al. 1992; Fortney, Schneider, and Greenleaf 1996; Sandler and Vernikos 1986). However, the achievement of this collection has significant negative side effects on return to the normally dominant upright posture.

Human Cardiovascular Adaptation to the Microgravity of Space

Pre-Gagarin concerns about human tolerance of the μG of space focused on the cardiovascular system. However, the catastrophic failure predicted by many experts did not occur. Strong evidence now shows that the functional capacity of the cardiovascular system is relatively well maintained at μG but that the return to earth uncovers significant dysfunction as does the return to the upright body posture after prolonged bed rest. The following review is based primarily on our findings from three recent Spacelab missions, two U.S. flights, Spacelab Life Sciences-1 or SLS-1 (1991) and SLS-2 (1993), and a German flight, D-2 (also 1993). D-2 supported many different scientific disciplines with participation from the European Space Agency and NASA.

Cardiovascular Responses to μG

Early clinical observations in space documented a rapid shift of body fluids toward the head. This manifested visually as distended neck veins and edematous facial features. As during HDT, a negative body fluid balance is evident during the first few days in space, associated with a decrease in intravascular volume and total body mass (Fortney, Schneider, and Greenleaf 1996). Serial in-flight echocardiographic data have been consistent with these observations and also with findings from ground-based cardiovascular HDT simulators. Charles and Lathers (1991) reported (based on M-mode measurements of left ventricular internal short axis diameter) a small increase of left ventricular size beyond preflight supine levels on flight day 1 followed by a decrease to preflight upright dimensions. Together, these data about body fluid shifts and cardiac dimensions made monitoring human cardiac filling pressures important. It is preferably performed by using actual left ventricular filling pressures but also by using the most accurate indicator available with right heart catheterization, i.e., pulmonary capillary wedge pressure (PCW), during the transition from 10 to actual μG. However, only measurement of CVP is currently feasible in the space flight environment.

CVP (measured with careful placement of the catheter tip and the transducer reference level under fluoroscopic control) closely reflects PCW over a wide range of hemodynamic conditions at rest in healthy human subjects (Levine et al. 1989). We obtained CVP data in three crew members during the flights of SLS-1 and SLS-2 in 1991 and 1993 (Buckey et al. 1993; Buckey, Gaffney, et al. 1996). Foldager et al. (1994) measured CVP in one crew member during the D-2 flight in 1993 by using a different type of instrumentation. The findings of both groups were similar. In our series, mean CVP in the space shuttle before launch (measured with the crew members supine with elevated legs) was approximately 10 mmHg. CVP increased significantly during launch in direct proportion to front-to-back acceleration forces up to 3.5 +Gx and then fell precipitously on main engine cutoff and arrival in μG. CVP remained at levels between 5 and 0 mmHg for

as long as measurements were made, i.e., respectively for 4, 5, and 40 hours on orbit. Measurements during CVP catheter removal verified a true zero level. These results were totally unexpected and must be interpreted in the context of other hemodynamic measurements.

Heart rates (and arterial blood pressure in one crew member) tended to be slightly elevated (80–115 bpm) but only during the first 20–30 min of orbit. The low CVP was not caused by a greatly elevated cardiac output. Cardiac output (estimated from heart rate and echocardiographic left ventricular dimensions) did increase but only from a preflight mean of 4.0 L/min to 5.0 L/min. Most importantly, early in-flight echocardiographic measurements of the three CVP subjects showed an increase in cardiac filling beyond preflight supine levels without appreciable changes in heart rate or measurements reflecting contractile left ventricular performance. This involves end-systolic diameter and ejection fraction, i.e., the ratio between end-systolic and end-diastolic left ventricular volume and the velocity of circumferential fiber shortening. The initial increase in left ventricular size was followed within 48 hours by a significant decrease relative to preflight supine dimensions.

Figure 16.1 presents sequential echocardiographic data from one of the crew members instrumented with CVP. Contractile state (as reflected by left ventricular ejection fraction and end-systolic volume) did not change significantly at μG.

Serial Echocardiographic Data on EDV, ESV, and EF

Prisk et al. (1993) obtained independent data about cardiac output and stroke volume during the flight of SLS-1. They used the foreign gas rebreathing technique. Their measurements provided additional information consistent with our echocardiographic data about left ventricular dimensions. The stroke volume data, acquired after two days in space, approximated the 1 G supine data. Measurements performed after five days or later approached but did not quite reach preflight upright levels.

Thus, the combined serial data about CVP, left ventricular dimensions, and stroke volume suggest that arrival at μG caused a transient increase in cardiac filling slightly beyond 1 G supine conditions at μG. By inference, a change also occurred in the complex pressure distributions within the chest that altered the quantitative relationship between CVP and effective transmural ventricular filling pressure that exists at 1 G. At 1 G, gravity and the physical properties of extracardiac tissues (e.g., the

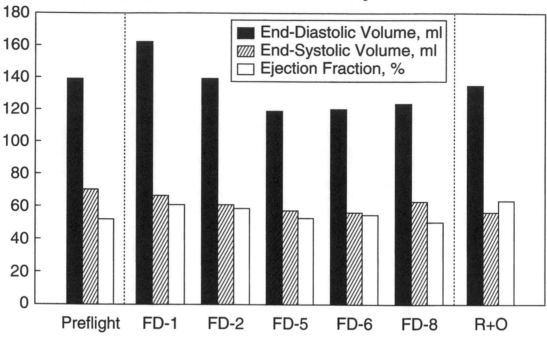

Figure 16.1 Indices of left ventricular function measured before (preflight), during (FD1 –FD8), and after return (R + O) from spaceflight. Filled bars are end-diastolic volume; hatched bars are end-systolic volume; and open bars are the derived ejection fractions (%).

pericardium and the lungs) most likely contribute to the definition of ventricular pressure-volume characteristics. The rapidity of the change effectively eliminates the possibility that the change in pressure-volume characteristics was caused by a change in intrinsic myocardial mechanical properties rather than being a consequence of μG. The initial increase in cardiac filling in space is followed by restoration of the predominant systemic cardiovascular state pattern at 1 G, i.e., heart size and hemodynamics approximating the normal terrestrial conditions in the upright.

We also considered the possibility that μG may change fundamental operating characteristics of cardiovascular neurohumoral control mechanisms in a manner that has not been predicted from 1 G-based knowledge. Maximal exercise involving large muscle groups is an important means of probing integrated cardiovascular function. Maximal oxygen uptake at μG can be maintained at preflight levels only by preserving the integrity of both cardiac pump function and the multiple neurohumoral control mechanisms that normally mediate regulatory exercise responses. Measurements (see Schykoff et al. 1996) in six SLS-1 and SLS-2 crew members during bicycle ergometer exercise before, during, and after spaceflight (Levine et al. 1996) showed no change in maximal oxygen uptake or maximal heart rate after seven to eight days in space (figure 16.2). A significant mean reduction in maximal oxygen uptake, –22%, was documented on landing day within hours after return to earth. This decrease was attributable to reductions in maximal stroke volume and cardiac output that, in turn, can be linked to a decrease in total blood volume. Maximal heart rate, arterial blood pressure, and estimated systemic arteriovenous oxygen difference were unchanged.

Maximal Exercise Results From SLS-1 and SLS-2

These in-flight data about maximal exercise extend informal observations made during the flight of Skylab 4 (Michel et al. 1977). The results provide evidence against any defect that significantly affects cardiac pump function and/or globally impairs cardiovascular regulation at μG. The exact mechanisms underlying the decrease in postflight maximal exercise performance are yet to be defined. The extent to which the hemodynamic effects that can be linked simply to the hypovolemia (that develops early in orbit and persists for several hours after return to 1 G) has not been rigorously tested. Recent data from bed rest experiments by Levine et al. (1996) suggest that changes in left ventricular dia-

SLS-1 and SLS-2 Maximal Exercise
(N=6)

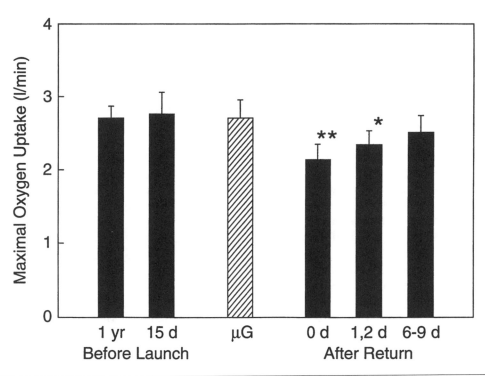

Figure 16.2   Maximal oxygen uptake measured before spaceflight launch, during in-flight of Skylab 4 (μG), and after spaceflight return. Significant reductions in maximal oxygen uptake were noted only after return from microgravity, but not during the spaceflight period.

stolic pressure-volume characteristics develop over a two-week period of bed rest in healthy subjects and produce a ventricle that is stiffer over the normal physiological range of filling pressures. This change would amplify the effects of postflight hypovolemia on ventricular filling.

Orthostatic Intolerance

Postflight orthostatic intolerance remains a very significant clinical problem. Growing evidence shows that the cardiovascular dysfunction early after return to 1 G is not a simple consequence of μG-induced hypovolemia. This raises the possibility that μG induces fundamental changes in neurohumoral circulatory regulation. Eckberg, Fritsch, and their associates (Fritsch et al. 1992, 1994) have examined the characteristics of the human carotid-cardiac baroreflex over a wide range of physiological and pathological conditions, including actual and simulated μG. They used a computer-controlled pneumatic neck collar to produce beat-to-beat changes in carotid artery transmural pressures and have reported μG-induced changes in the stimulus-

response characteristics of the carotid baroreflex. The change in R-R interval for a given change in carotid transmural pressure was significantly attenuated. These data suggest that μG changes the properties of one of the mechanisms that regulate arterial pressure in a direction that may contribute to postflight orthostatic intolerance.

We have examined a wider range of potential μG effects on cardiovascular regulatory mechanisms by comparing preflight and postflight hemodynamic responses to a stand test. Detailed measurements were obtained in 14 crew members from three Spacelab flights with durations of 9, 10, and 14 days. Studies were performed within four hours after landing (Blomqvist et al. 1994; Buckey, Lane, et al. 1996). In the analyses of the stand test data, we viewed arterial blood pressure as the triple product of heart rate, stroke volume, and peripheral resistance. As expected, mean body weight and total blood volume decreased significantly at μG by 1.1 kg and by 9% (Alfrey et al. personal communication). A 12% postflight decrease occurred in stroke volume, but hemodynamics at supine rest were otherwise similar preflight and postflight. Cardiac output was kept at preflight levels by an increase in heart rate. During preflight, all subjects completed our stand test. However, almost two-thirds, or nine of 14 participants, were unable to stand for the required 10 minutes postflight. Comparisons of preflight and postflight hemodynamic mean data, based on the last measurement obtained during standing in each crew member, showed that—as expected—postflight stroke volume was significantly lower and heart rate higher. Analysis of the continuous finger blood pressure and heart rate recordings showed that the most common failure mode (four of nine) was a progressive fall in arterial pressure with little or no further change in heart rate. Two crew members had low initial pressure and episodic hypotension. Only one of the nine had a classical cardioinhibitory failure. Furthermore, two of our subjects were symptomatic with dizziness, lightheadedness, and sweating but had arterial pressures at or above 100/75.

Postflight orthostatic intolerance commonly occurs. It has usually been attributed to reduced stroke volume and cardiac output in the upright position caused by central hypovolemia (with or without excessive peripheral pooling). The principal consequence of the reduced cardiac output is then arterial hypotension with inadequate cerebral blood flow. Analysis of our data suggest that the pathophysiology is more complex. A comparison of responses in finishers and nonfinishers (those who were or were not able to complete the 10-minute standing period) clearly identifies an inadequate systemic vasoconstrictor response in nonfinishers as the principal difference (Blomqvist et al. 1994; Buckey, Gaffney, et al. 1996). The heart rate response to standing was actually greater in nonfinishers than in finishers of the stand test (figure 16.3). This finding does not exclude the presence of a subtle defect in heart rate regulation but makes it a very unlikely primary cause of the impaired orthostatic tolerance. The magnitude of the orthostatic venous pooling and of the related decrease in stroke volume was also similar in finishers and nonfinishers, which leaves inadequate vasoconstriction as the principal apparent cause of failure. Mulvagh et al. (1991) have presented data derived from echocardiographic recordings suggesting that limited systemic vasoconstriction is an important factor.

Stand Test Responses in Finishers and Nonfinishers

The mechanisms underlying the inadequate systemic vasoconstrictor response remain to be identified. Whitson et al. (1995) noted markedly elevated postflight plasma catecholamine levels relative to preflight with only a minor increase in the systemic vasoconstrictor response. This suggests decreased adrenergic responsiveness. However, individual data were not related to stand test responses. We

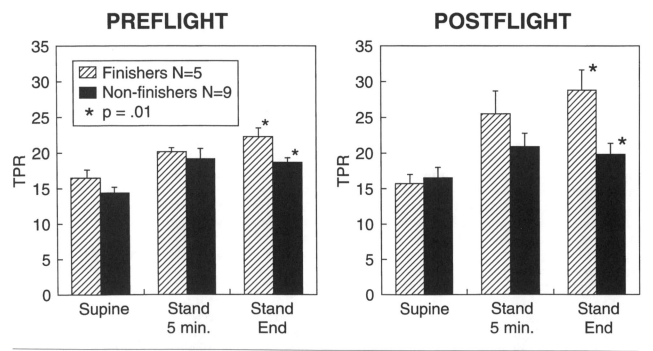

Figure 16.3   Total peripheral resistance (TPR) measured in the supine and standing positions before and after spaceflight. TPR is taken to reflect levels of systemic vasoconstrictor tone. Comparisons were made in the supine position and during an orthostatic challenge manifest by upright standing for 10 minutes.

found no differences between preflight and post-flight cardiovascular results (to be published). However, these studies were performed 24–48 hours after return from space, and negative results are inconclusive.

Two of the returning astronauts in our series were unable to continue standing, but they maintained low normal arterial pressures. Recent data from bed rest studies by Levine et al. (1996) have identified a likely cause. They showed that simulated μG can produce an impairment of cerebral autoregulation of blood flow manifest at low systemic arterial pressures. Vestibular dysfunction is a possible alternative cause. Together, these data suggest that adaptation to μG causes a degradation of neurohumoral cardiovascular control mechanisms, but the specific site of the defect(s) is unknown. μG may produce functionally important defects in afferent pathways, central integration, and/or efferent traffic. Alternatively, the dynamic range of the mechanisms that control vasoconstriction and cerebral blood flow during orthostatic stress may be an inborn characteristic of the individual. A limited range that is adequate at ordinary 1 G conditions may become inadequate in the hypovolemic condition early after return from space. The two alternatives are not mutually exclusive. These concepts will be examined during a future series of space flights, including the last Spacelab flight, Neurolab, scheduled for 1998.

Summary

Much is still unknown about the biologic effects of μG. However, during the last few years, increasingly complex and sophisticated methods and procedures have been applied to life sciences research in space. New information has been acquired about the μG responses of biologic systems ranging from subcellular to intact humans. However, much remains to be done, including integration of data across all levels of biologic complexity.

Acknowledgments

The work upon which much of this paper is based was supported by the National Aeronautics and Space Administration (Flight Experiment Contracts NAS 9-16044 and NAS 9-18139; Grant NAGW 3582 NASA Center of Research and Training in Integrated Physiology).

Credit for the scientific work belongs the U.T. Southwestern flight experiment team, including Jay C. Buckey, Jr., M.D., F. Andrew Gaffney, M.D., Lynda D. Lane, M.S., R.N., Benjamin D. Levine, M.D., Geralyn M. Meny, M.D., Donald E. Watenpaugh, Ph.D., Sheryl J. Wright, M.D., Dan Meyer, M.D., Clyde W. Yancy, M.D. Invaluable support was provided by Carolyn Donahue, Willie E. Moore, and Boyce Moon.

References

Atkov, O.Y., and V.S. Bednenko. 1992. *Hypokinesia and weightlessness: Clinical and physiological aspects.* Madison, CT: International Universities Press.

Baisch, F., L. Beck, C.G. Blomqvist, J.M. Karemaker, and B. Saltin. 1992. Headdown tilt bed rest. HDT '88—an international collaborative effort in integrated systems physiology. *Acta Physiological Scandinavica* 144(Suppl. 604):1–141.

Blomqvist, C.G., J.C. Buckey, F.A. Gaffney, L.D. Lane, B.D. Levine, and D.E. Watenpaugh. 1994. Mechanisms of post-flight orthostatic intolerance. *Journal of Gravitational Physiology* 1:122–124.

Blomqvist, C.G., and H.L. Stone. 1983. Cardiovascular adjustments to gravitational stress. In *Handbook of physiology. The cardiovascular system peripheral circulation and organ blood flow,* eds. J.T. Shepherd and F.M. Abboud, Sect. 2, Pt. 2, 1025–1063. Bethesda, MD: American Physiology Society.

Buckey, J.C, F.A. Gaffney, L.D. Lane, B.D. Levine, D.E. Watenpaugh, and C.G. Blomqvist 1993. Central venous pressure in space (Letter). *New England Journal of Medicine* 328:1853–1854.

Buckey, J.C., Jr., F.A. Gaffney, L.D. Lane, B.D. Levine, D.E. Watenpaugh, S.J. Wright, C.W. Yancy, Jr., D.M. Meyer, and C.G. Blomqvist. 1996. Central venous pressure in space. *Journal of Applied Physiology* 81:19–22.

Buckey, J.C., Jr., L.D. Lane, B.D. Levine, D.E. Watenpaugh, S.J. Wright, W.E. Moore, Jr., F.A. Gaffney, and C.G. Blomqvist. 1996. Orthostatic intolerance after spaceflight. *Journal of Applied Physiology* 81:7–18.

Charles, J.B., and C.M. Lathers. 1991. Cardiovascular adaptation to spaceflight. *Journal of Clinical Pharmacology* 31:1010–1023.

Daunton, A.C. 1996. Adaptation of the vestibular system to microgravity. In *Handbook of physiology: Environmental physiology,* eds. M.J. Fregly and C.M. Blatteis, Sect. 4, Vol. I, 765–784. New York and Oxford: Oxford University Press.

Foldager, N., T.A.E. Andersen, F.B. Jessen, P. Ellegard, C. Stadaeger, R. Videbaek, and P. Norsk. 1994.

Central venous pressure during weightlessness in humans. In *Proceedings of the Nordeney symposium on scientific results of the German Spacelab mission D-2*, eds. P.R. Sahm, M.H. Keller, and B. Schieve, 695–696. Germany: Wissenschaftliche.

Fortney, S.M., V.S. Schneider, and J.F. Greenleaf. 1996. The physiology of bed rest. In *Handbook of physiology: Environmental physiology*, eds. M.J. Fregly and C.M. Blatteis, Sect. 4, Vol. I, 889–942. New York and Oxford: Oxford University Press.

Fritsch, J.M., J.B. Charles, B.S. Bennett, M.M. Jones, and D.L. Eckberg. 1992. Short-duration spaceflight impairs human carotid-cardiac reflex responses. *Journal of Applied Physiology* 73:664–671.

Fritsch, J.M., J.B. Charles, M.M. Jones, L.A. Beightol, and D.L. Eckberg. 1994. Spaceflight alters autonomic regulation of arterial pressure in humans. *Journal of Applied Physiology* 77:1776–1783.

Gaffney, F.A., J.V. Nixon, E.S. Karlson, W. Campbell, A.B.C. Dowdy, and C.G. Blomqvist. 1985. Cardiovascular deconditioning produced by 20-hour bedrest with head-down (–5 degree) tilt in middle aged men. *American Journal of Cardiology* 56:634–638.

Gauer, O., and H.L. Thron. 1965. Postural changes in the circulation. In *Handbook of physiology*, ed. W.F. Hamilton, Sect. 2, Vol. III, 2409–2439. Bethesda, MD: American Physiological Society.

Greenleaf, J.E. 1996. The gravitational environment. In *Handbook of physiology: Environmental physiology*, eds. M.J. Fregly and C.M. Blatteis, Sect. 4, Vol. I and H, 631–975. New York and Oxford: Oxford University Press.

Kakurin, L.I., V.I. Lobachi, V.M. Mickhailov, and Y.A. Senkevich. 1976. Antiorthostatic hypokinesia as a method of weightlessness simulation. *Aviation Space and Environmental Medicine* 47:1083–1086.

Levine, B.D., J.C. Buckey, D.B. Friedman, L.D. Lane, D.E. Watenpaugh, and C.G. Blomqvist. 1989. Right atrial pressure (RA) vs. pulmonary capillary wedge pressure (PCW) in normal man. *Circulation* 80 (Suppl. H):250.

Levine, B.D., L.D. Lane, D.E. Watenpaugh, F.A. Gaffney, J.C. Buckey, and C.G. Blomqvist. 1996. Maximal exercise performance after adaptation to microgravity. *Journal of Applied Physiology*.

Michel, E.L., J.A. Rummel, C.F. Sawin, M.C. Buderer, and J.D.F. Lem. 1977. In *Biomedical results from Skylab NASA SP-377*, eds. R.S. Johnston and L.F. Dictlein, 372–387. Washington, DC: Scientific and Technical Information Office, National Aeronautics and Space Administration.

Mulvagh, S.L., J.B. Charles, J.M. Riddle, T.L. Rehbein, and M.W. Bungo. 1991. Echocardiographic evaluation of the cardiovascular effects of short-duration space flight. *Journal of Clinical Pharmacology* 31:1024–1026.

Nixon, J.V., R.G. Murray, C. Bryant, R.L. Johnson, Jr., J.H. Afitchell, O.B. Holland, C. Gomez-Sanchez, P. Vergne-Marini, and C.G. Blomqvist. 1979. Early cardiovascular adaptation to simulated zero gravity. *Journal of Applied Physiology* 46:541–548.

Prisk, G.K., H.J.B. Guy, A.R. Elliott, R.A. Deitschman, and J.B. West. 1993. Pulmonary diffusing capacity, capillary blood volume, and cardiac output during sustained microgravity. *Journal of Applied Physiology* 75:15–26.

Nicogossian, A.E., C.L. Huntoon, and S.L. Pool, eds. 1994. *Space physiology and medicine*. Philadelphia: Lea & Febiger.

Nicogossian, A.E., H. Sandler, A.A. Whyte, C.S. Leach, and P.C. Rambaut. 1979. *Chronological summaries of United States, European, and Soviet bed rest studies*. Washington, DC: National Aeronautics and Space Administration Biotechnology Office of Life Sciences.

Ruml, L., K. Hw, and C.Y.C. Pak. 1996. Effect of prolonged bed rest on the urinary saturation of stone-forming salts. Abstract. VII International Symposium on Urolithiasis at Dallas, TX.

Saltin, B., C.G. Blomqvist, J.H. Mitchell, R.J. Johnson, Jr., K. Wildenthal, and C.B. Chapman. 1968. Response to exercise after bed rest and training. A longitudinal study of adaptive changes in oxygen transport and body composition. *Circulation* 37–39(Suppl. VII):1–78.

Sandler, H., and J. Vemikos. 1986. *Inactivity: Physiological effects*. Orlando and London. Academic Press.

Schykoff, B.E., L.W. Farhi, A.J. Olzowka, D.R. Pendergast, M.A. Roldkta, C.G. Eisenhardt, and R.A. Morin. 1996. Cardiovascular response to submaximal exercise in sustained microgravity. *Journal of Applied Physiology* 81:26–32.

Watenpaugh, D.E., and A.R. Hargens. 1996. The cardiovascular system in microgravity. In *Handbook of physiology: Environmental physiology*, ed. M.J. Fregly and C.M. Blatters, Sect. 4, 631–674. New York and Oxford: Oxford University Press.

Whitson, P.A., J.B. Charles, W.J. Williams, and N.M. Cintron. 1995. Changes in sympathoadrenal response to standing in humans after space flight. *Journal of Applied Physiology* 79:428–433.

Chapter 17

Heat and Cold

Bodil Nielsen Johannsen

During dynamic exercise, the increase in metabolic requirements of the exercising muscle is met by an increased blood flow and O_2 extraction. The increase in muscle blood flow is accomplished by an increase in cardiac output and a redistribution of blood flow away from the visceral organs proportional to the intensity of the exercise ($\%\dot{V}O_2$max). An additional need is the thermoregulatory requirement for skin circulation (SkBF), particularly when exposed to heat stress. The skin surface temperature varies with the environmental temperature, while core temperature is largely determined by the intensity of the exercise.

In cold conditions, the core-to-skin temperature gradient is large, i.e., 10–15 °C. However, this gradient decreases with increasing environmental temperature to become only about 1.0–1.5 °C at 35–40 °C. This means that for a given exercise intensity and heat liberation (H), an increasingly greater blood flow is needed to transport the heat to the warmer environment, as seen from the equation:

$$H_{skin} = \text{skin blood flow} \times c \times (T_a - T_v),$$

where H_{skin} is the heat transported to the skin, c the heat capacity of blood, and T_a and T_v the temperature of the arterial and venous blood reaching and leaving the skin, respectively. These can be taken to equal deep body temperature and skin surface temperature, respectively.

As skin blood flow increases with exercise in warm environments, the vessels in the skin fill. A large volume of blood pools in the skin, displacing the blood from the thorax. Therefore, central blood volume and the filling of the heart may be reduced (Rowell 1986). In this way, environmental conditions affect cardiac output and blood pressure regulation.

In resting subjects, SkBF can be roughly estimated to vary between 0.2–0.5 L/min in a thermoneutral environment and increasing up to 7–8 L/min during extreme heat stress (Rowell 1986, 1993). However, during exercise in hot environments, the maximal SkBF is lower and may reach only half of the mentioned value. The balance between the competing needs for metabolic and thermoregulatory blood flow during exercise in different environments is still a matter of discussion and will be taken up in the following.

Hot Conditions

In hot environments, compared with cool, the displacement of blood flow and volume to the periphery, as mentioned above, may cause a reduction in central blood volume and stroke volume (SV) (Rowell 1974). However, cardiac output is still maintained, or even increased, through a compensatory increase in heart rate (HR). Discussion is still ongoing as to whether the blood flow to the exercising muscles, thermoregulatory skin circulation, or both are compromised when cardiac output must be shared. Several studies have addressed the problem of muscle blood flow during exercise in the heat when core temperature is increasing toward critical high values (Febbrario et al. 1994; Fink, Costill, and van Handel 1975; Nielsen et al. 1990). A further question of discussion can be raised. Are fatigue and exhaustion as they occur during exercise in hot environments due to circulatory and metabolic factors or associated with high core temperatures (Nielsen et al. 1990, 1993)?

Central Circulation

During exercise in the heat as compared with the cold, cardiac output (Q) is mostly found to increase (Rowell 1986; Williams et al. 1962). This is due to an increase in heart rate (HR), which more than compensates for the fall in stroke volume as long as the maximal HR and Q have not been reached. It depends, however, on the exercise intensity, type of exercise (upright, supine, bicycling, walking), the state of acclimation, and training status (reviewed in Nielsen et al. 1993). Furthermore, the hydration level plays a significant role in whether heat stress leads to an increase, no change, or a decrease in cardiac output. Sawka et al. (1992) observed a decreased Q after 4% loss of body water during running at 70% of $\dot{V}O_2$max in hot conditions. Nadel, Fortney, and Wenger (1980) used diuretics to reduce plasma volume. They found a decrease in Q during cycling exercise at 50% of maximum at 35 °C as compared with a euhydrated control condition. Furthermore, during prolonged cycle ergometer exercise at 35 °C, leading to up to 4-5% body weight loss, Q was found to decline 3-4 L/min, which represented a 14% lower Q than in the fully hydrated control condition where Q was maintained (see figure 17.1) (González-Alonso et al. 1995, 1998).

Thus, the central circulation is affected in hot environments by two factors. First, the acute blood volume shifts due to the increase in skin circulation. Second, plasma volume gradually changes due to thermoregulatory sweat loss during prolonged exercise, as a following section will further discuss.

Skin Circulation

The vessels in the skin respond to direct heating and cooling, to reflexes from the thermoregulatory center due to changes in core and skin temperature, as well as to nonthermal stimuli, exercise-related reflexes, and baroreceptor reflexes in the control of blood (reviewed by Kenney and Johnson 1992). Studies of skin blood flow in hot environments have used both laser Doppler flowmetry and venous occlusion plethysmography on the forearm (FBF). These studies show that with increasing exercise intensity in hot conditions, the rise in FBF per °C rise in core temperature is decreased, the core temperature threshold for vasodilation is increased, and the rate of rise is attenuated (see figure 17.2). A level of FBF is reached at T_{core} of about 38 °C, which is not surpassed even if the core temperature keeps rising (Brengelmann et al. 1977; Johnson, Rowell, and Brengelmann 1974; Nadel et al. 1979).

If central blood volume increases by having subjects exercise in 35 °C warm water up to the xiphoid, FBF as well as cardiac output can increase further than with the same skin temperature (35 °C) in hot air at the same exercise intensity (see figure 17.3) (Nielsen, Rowell, and Bonde-Petersen 1984). Dehydration or reductions in blood volume leads to a serious decline in forearm blood flow (see figure 17.1) (González-Alonso, Calbet, and Nielsen 1998; González-Alonso et al. 1995; Montain and Coyle 1992; Nadel et al. 1979; Nadel, Fortney, and Wenger 1980).

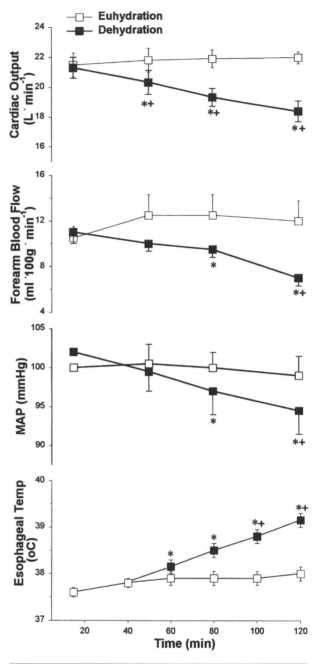

Figure 17.1 Cardiac output, forearm blood flow, mean arterial pressure (MAP), and esophageal temperature responses to prolonged exercise in 35 °C environmental temperature with euhydrated condition and with dehydration. Subjects either maintained euhydration by oral fluid replacement or became progressively dehydrated. (Modified from González-Alonso et al. 1995.)

Both arterial and cardiopulmonary reflexes are involved in the reduction in FBF elicited with heat and exercise stress. The study by Mack, Nose, and Nadel (1988) illustrated the role of cardiopulmonary receptors. They showed that negative pressure

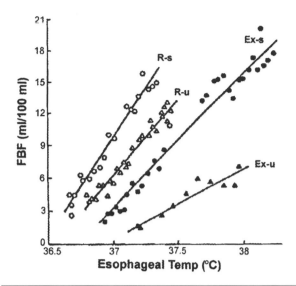

Figure 17.2   Effect of exercise intensity on the forearm blood flow (FBF)/T_{es} relationship where T_{es} equals core temperature. The threshold shifts to the right and the slope is reduced with increasing activity level. R-s supine rest; R-u upright rest; Ex-s supine exercise; Ex-u upright exercise. (From Johnson, Rowell, and Brengelmann 1974, with permission.)

breathing led to increased FBF even though mean arterial pressure (MAP) was reduced during the negative pressure breathing.

However, peripheral displacement of blood volume may not be the only cause for the fall in SV with prolonged exercise in warm conditions. SV keeps falling and HR increases during continued exercise in spite of the leveling out of FBF in hot conditions where T_{core} is increasing until exhaustion (González-Alonso, Calbet, and Nielsen 1997; González-Alonso et al. 1999).

We measured FBF in subjects exercising at 150 W, 40%, and FBF and leg blood flow (LBF) at 220 W, 70% of $\dot{V}O_2$max in 40 °C dry heat. The rates of heat storage and rise in core temperature differed. A core temperature (T_{es}) of approximately 40 °C was reached at exhaustion in either 60 min or 30 min, respectively. In our study, the rate of change in FBF per °C change in T_{es} was similar. The plateau FBF values attained (18 ml/100 ml/min) were equal for the two exercise intensities, seated bicycling. Q was 2.5–3.0 L/min higher at the high work intensity. Thus, the skin circulation did not increase although Q was higher with the high exercise intensity. The LBF was measured only at the high intensity in these subjects. However, in a previous study at the low intensity in the same experimental conditions and with a similar Q (17.8 L/min), the LBF in one leg was 5.7 L/min at the 150 W intensity. These values can be compared with the measurements at high intensity, 70% $\dot{V}O_2$max in the present study where Q was 20.5 L/min and

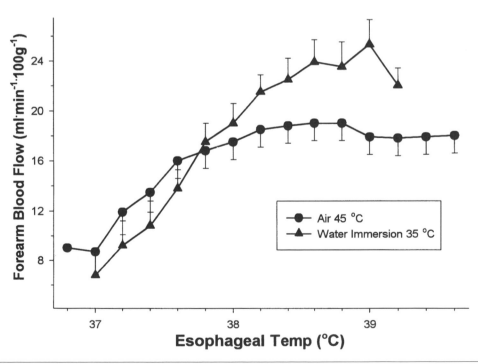

Figure 17.3   Forearm blood flow plotted against esophageal temperature. Results during exercise in air and immersed in water to the level of the xiphoid. Skin temperature 35 °C in both conditions (number of data shown if less than eight). Indicates significant differences between air and water experiments. (From Nielsen, Rowell, and Bonde-Petersen 1984, with permission.)

LBF was 6.8 L/min. Therefore, the approximately 2.5 L/min greater Q in high exercise intensity seems to have been directed to the more intensely exercising legs at the expense of skin blood flow. Evidently, the heat dissipation to the environment was not increased compared with the low workload. This resulted in the higher rate of heat storage and rate of core temperature rise (Nielsen et al. unpublished).

Leg Blood Flow

The metabolic requirements of the working muscles during exercise in the heat have been studied in various ways: biopsy studies, measurements of blood concentrations of metabolites (lactate), and lately by direct blood flow measurements. Fink, Costill, and van Handel (1975) observed an increased glycogen breakdown in the working leg in a hot (43 °C) environment compared with a cold (9 °C) environment after exercise at 70–85% $\dot{V}O_2$max. They took this as an indication that muscle blood flow declines in the hot condition compared with the cold condition. In later biopsy studies comparing exercise in hot and cool environments, Febbrario et al. (1994) also found a higher glycogen usage in a hot (40 °C) environment compared with a cool (20 °C) environment. This altered metabolism could be explained by a reduced muscle blood flow, an increased muscle temperature, and/or a heat stress-induced elevated catecholamine concentration. The independent effects of increased epinephrine and elevated muscle temperature on muscle glycogenolysis during heat stress were recently confirmed (Febbrario et al. 1996, 1998).

In the same light, Young et al. (1995) had subjects exercise and train while submerged to the neck in 20 °C and 35 °C water to separate rectal and muscle temperature in the two conditions. They found no difference in glycogen utilization or in blood or muscle lactate in the hot compared with the cold water. Therefore, they stated that the increase in body temperature due to exercise is not an important stimulus to the metabolic adaptations to training and heat stress. They also stated that this increase in body temperature is not an important stimulus to the increased glycogen usage during exercise in heat stress.

By using direct measurements, Nielsen et al. (1990) studied leg blood flow during exercise in cool conditions and with heat stress. During 30 min treadmill exercise at a work intensity of 55–60% $\dot{V}O_2$max in cool conditions of 18–20 °C, cardiac output was 15.2 L/min. After shifting to a 40 °C environmental temperature and continuing work at the same speed and inclination in the heat, final cardiac output had increased to 18.4 L/min. Leg blood flow was the same, 6.1 L/min ± 0.8 vs. 6.0 ± 0.9 L/min. Leg vascular conductance did not change in the heat, although T_{es} surpassed 39.6 °C (see figure 17.4). The extra Q was likely used to increase skin blood flow during the exercise at 40 °C compared with 18–20 °C.

Even at higher exercise intensities (70–75% $\dot{V}O_2$max) in seated cycling at 40 °C in which condition T_{es} and femoral venous blood temperature increases to > 39.5 °C (in 30 min), LBF was maintained unchanged in spite of increasing core temperature when, as here, Q was maintained during exercise. The skin blood flow of the forearm, measured both with plethysmography and laser Doppler flowmetry, already reached peak values at 38 °C T_{es}. The same is probably the case for the thigh skin. Therefore, if reduction in muscle blood flow does take place during prolonged exercise in heat stress, the work intensity probably has to be even higher than 75% $\dot{V}O_2$max, at least if the subject is not dehydrated.

Dehydration

Prolonged exercise in the heat leads to loss of sweat. This results in dehydration with hypovolemia combined with hyperosmolarity. Both of these interfere with thermoregulatory as well as cardiovascular functions (Nadel, Fortney, and Wenger 1980; Nielsen 1974). Saltin (1964) found no significant change in Q during submaximal and maximal cycle ergometer exercise in cool conditions. However, in warm and hot conditions, the effects of dehydration on cardiovascular function occur (Montain and Coyle 1992; Sawka et al. 1992). González-Alonso et al. (1995) had subjects exercising for 2 h at 62% $\dot{V}O_2$max in 35 °C T_a. On one occasion, the subjects maintained euhydration by ingesting fluids. During another trial, they experienced a gradual dehydration to 5% of body weight (3.6 L) (see figure 17.1). With dehydration, cardiac output as well as MAP decreased significantly. This was due to a 27% fall in stroke volume in spite of a 13% rise in heart rate.

Note that the SV decrease was not linked with an increased skin blood flow and forearm venous volume. Skin blood flow actually decreased with the gradual dehydration in spite of a rising core temperature (up to approximately 39.3 °C). Plasma catecholamines (NA significantly more) also rose, an indication of increased sympathetic activity. In such conditions, a reduction in muscle blood flow in defense of blood pressure might be expected. This was demonstrated by direct measurements comparing exercise at 65% $\dot{V}O_2$max at 35 °C in a euhydrated condition and during progressive dehydration. Leg blood flow started to decline after 20 min and be-

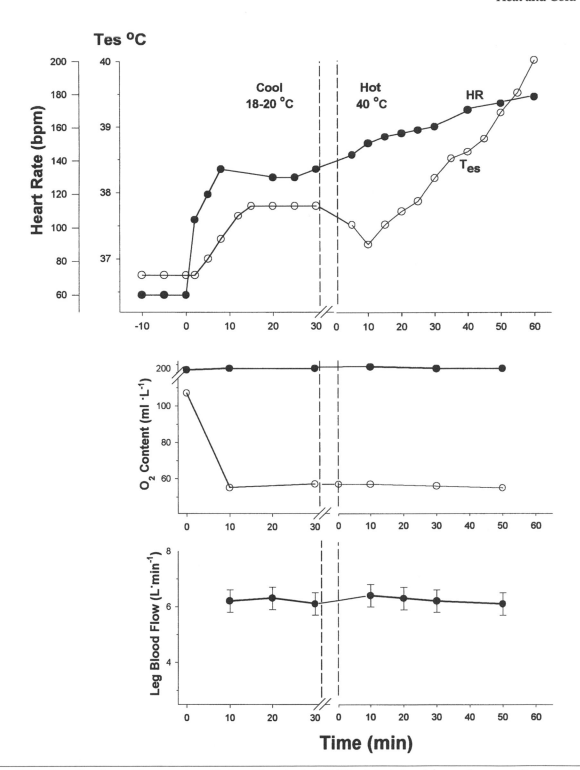

Figure 17.4 Top: Esophageal temperature (Tes) and heart rate. Middle: Arterial and femoral venous O_2 content. Bottom: Leg blood flow measured in one leg. All were measured during uphill walking for 30 min in cool conditions that continued in 40 °C dry heat until exhaustion. (Modified from Nielsen et al. 1990.)

came significantly reduced at exhaustion with dehydration after 134 min. Q and FBF were also reduced at this stage, while these variables tended to increase in the euhydrated control condition (González-Alonso, Calbet, and Nielsen 1998).

Fatigue and Circulation

Fatigue and exhaustion occur after prolonged exercise in hot conditions. In figure 17.5, the subject exercised in a hot environment until exhaustion on

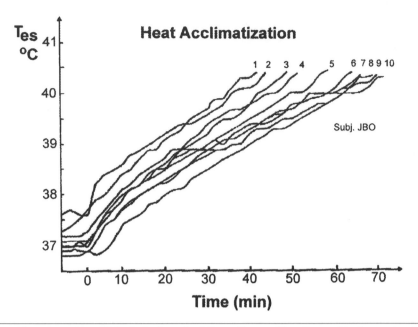

Figure 17.5 Esophageal temperature (T$_{es}$) plotted against time. One acclimating subject exercised at 40 °C until exhaustion during 10 consecutive days. (From Nielsen et al. 1993, with permission.)

10 consecutive days. During this time, he became gradually acclimatized, as seen from the increasing performance time. However, exhaustion always coincided with the same very high core temperature. This fatigue could result from a reduction in muscle blood flow and an altered muscle metabolism due to the high temperature. However, direct measurements of muscle blood flow in subjects exercising at intensities from 40–75% of $\dot{V}O_2$max until exhaustion in hot environments showed no decline in blood flow as measured with thermal dilution in the exercising leg. Muscle metabolism had not changed, as seen in figure 17.6. No lactate accumulation was observed, and glycogen stores were not depleted (Nielsen et al. 1990, 1993; Savard et al. 1988). The conclusion of these studies is that the high core temperature attained at exhaustion ultimately caused the fatigue.

However, in conditions where dehydration is also a problem, cardiac output reaches its ceiling for the given intensity or even becomes reduced (González-Alonso et al. 1997). In this condition, the blood flow to the exercising muscles also declines. The metabolic consequences of this may be a factor for the fatigue during exercise in the dehydrated state. Hargreaves et al. (1996) found similar results. Those biochemical changes in the working muscle associated with peripheral muscular fatigue were studied by looking at the glycogen and lactate concentrations at exhaustion obtained in biopsies (González-Alonso, Calbet, and Nielsen 1997). Glycogen concentrations had decreased significantly,

and lactate increased from resting values. However, at exhaustion, the values were far from those reported to be involved in muscular fatigue during intense exercise in cool conditions.

Exhausted subjects who were unable to continue exercise in the hot environment were nevertheless able to perform the same maximal voluntary contraction at this stage as before exercise (Nielsen et al. 1993). Thus, the transmission from the motor areas to the muscles are not fatigued. This leads to the conclusion that the fatigue is of central origin. High core temperature in itself could cause fatigue, exerting its action directly in the brain. This could perhaps occur on a central command for motor pattern activation in the brain but not on neural transmission or on the circulatory or biochemical events in the muscles (González-Alonso et al. 1999). In fact, in a recent study, we have observed changes in brain electric activity (EEG) over the frontal lobe, changes in frequency and amplitude. An increase occurred in the ratio of A to B wave power spectrum. The sleep index rose proportional to core temperature, indicating a decline in motivation with increasing T$_{es}$.

Conclusions

In hot environments, the additional stress induces an increase in cardiac output and a decrease in splanchnic and visceral blood flow. The skin blood flow increases over that in cool conditions but is lower during exercise than during rest in the same

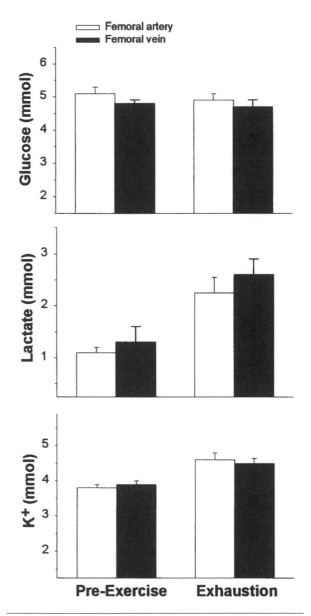

Figure 17.6 Blood glucose, lactate, and potassium ion (K⁺) concentrations in the femoral artery and vein before and after exercise until exhaustion at 40 °C. Mean ±SE for eight subjects. (From Nielsen et al. 1993.)

thermal stress. Apparently, blood flow to the exercising muscles has a priority over the visceral as well as skin blood flow even in the face of severe hyperthermia. However, we may consider that the changes observed in muscle metabolism during heat and exercise stress in some studies are perhaps not due to circulatory failure, i.e., oxygen lack caused by reduced muscle blood flow. They may be due to hormonal changes elicited by the high temperature, which interferes with muscle metabolism. Likewise, the exhaustion due to prolonged exercise in a hot environment appears to be a direct effect of the

high core temperature on the brain and not caused by circulatory or metabolic shortcomings.

Cold Conditions

The effects of cold environments on the circulation is far less studied than the stress of hot environments. What are cold environments? Cold may be defined as a condition in which physiological reactions reducing the heat loss from the body and increasing heat production are activated. Obviously, therefore, for exercising subjects, cold conditions refer to ambient temperatures other than cold during rest.

The primary effect of cold is that it elicits heat conservation. Heat loss from the body surface is reduced by vasoconstriction in the skin vessels when the skin surface is cooled. Further cooling, when body core temperature tends to decrease, will activate shivering and nonshivering thermogenesis. These classical observations were already discussed thoroughly more than 100 years ago (Burton and Edholm 1955; Cannon et al. 1926; Johansson 1897; Sjöström 1913). The increase in metabolic heat liberation prevents or delays the fall in deep body temperature. An increase in metabolic rate due to shivering of more than 150 W has been reported both in resting subjects (Burton and Bazett 1936; Hardy and Söderström 1938) and during exercise (Erikson et al. 1956; Nadel et al. 1974; Nielsen 1976). The contribution of nonshivering thermogenesis to the metabolic response to cold is much less in humans, an increase of 25–30% in basal metabolic rate (Jessen, Rabøl, and Winkler 1980; Nielsen et al. 1993). The recent review by Jansky (1995) discusses the nature and control of nonshivering thermogenesis.

Cold stress further causes a hemoconcentration. This is classically explained by an increased diuresis caused by low-pressure baroreflex inhibition of antidiuretic hormone. This could be elicited by the increase in central blood volume resulting from the peripheral vasoconstriction. However, other mechanisms may also be involved. Vogelaere, Deklunder, et al. (1992) showed that a 15% reduction in plasma volume resulting from cold exposure in subjects resting reclined at 1 °C air for 120 min would be reduced to 3% after 60 min recovery in a thermoneutral environment. The urinary water loss equaled the 3% change observed after the recovery. They proposed, therefore, that a transient shift of plasma fluid into the interstitial space takes place due to an increase in blood pressure during cold stress.

Whole-body cooling can be studied in cold air and during immersion in cold water. The cooling power of water is much stronger due to the higher heat capacity and heat conductivity. However, the effect of cold on the heart and circulation may differ in air and water, because the water environment has specific effects on the circulation. These include abolishing the hydrostatic effects of gravity so the central venous volume and the filling of the heart increases (Nielsen, Rowell, and Bonde-Petersen 1984).

During exercise, the critical temperature—the temperature at which the metabolic rate starts to increase due to the cold exposure—increasingly lowers with an increasing rate of exercise. However, shivering thermogenesis adds to the metabolic heat liberation (Nielsen 1976). Thus, $\dot{V}O_2$ is higher during exercise in a cold environment rather than in a warm environment. However, in a given cold environment, the difference from neutral conditions becomes smaller the higher the work intensity and heat production (McArdle et al. 1976). The problem to be discussed is how the circulation is matched to this increased metabolic rate in cold conditions and how cold affects the flow distribution.

Central Circulation, Responses to Cold

In resting subjects, Raven et al. (1970) measured cardiac output, stroke volume, $(a-\bar{v})O_2$ difference, and heat production during cold stress: 2 h supine rest at 5 °C air temperature. Q increased gradually and linearly with increases in $\dot{V}O_2$, reaching 95% above the control levels obtained at 28 °C prior to the cold exposure. However, $\dot{V}O_2$ increased 173%. Thus, the increase in $\dot{V}O_2$ was partly achieved through a widening of the $(a-\bar{v})O_2$ difference. This observation was later confirmed over a range of air temperatures, 5–28 °C, where the $(a-\bar{v})O_2$ difference increased linearly with decreasing temperature (Raven et al. 1975). SV had increased (+78%) and was responsible for most of the increase in Q since HR increased only insignificantly. Both systolic and diastolic blood pressure increased, while total peripheral resistance fell gradually. Raven et al. (1970) explained the latter as an effect of tissue level oxygen lack in the splanchnic region. On the other hand, Vogelaere, Savourey, et al. (1992) found that in 1 °C air, the increased Q was due only to the higher HR.

Wagner and Horvath (1985b) also measured Q in resting subjects (males and females, age 20–72 years) exposed for 2 h to air between 10 °C and 28 °C. This is a cold stress severe enough to cause a decline in rectal temperature toward the end of the exposure

(Wagner and Horvath 1985a). Q and SV increased with time in cold, both in males and females, while HR decreased in males but not in the females (see figure 17.7). Minimal increases were observed in systolic and diastolic blood pressure in the younger, larger increases occurred in the older subjects. Again the $(a-\bar{v})O_2$ difference increased with the increasing Q, mostly in the older subjects, but well inside their extraction capacity.

Studies about circulation and cold exposure during exercise are few. They are mostly done in cold water since obtaining low enough air temperatures in laboratory experiments to produce whole-body cooling during exercise in air is difficult.

Pirnay, Deroanne, and Petit (1977) had subjects exercise on a cycle ergometer at 30% and 60% $\dot{V}O_2$max while immersed to the neck in water between 20 °C and 40 °C. Water temperature did not significantly influence $\dot{V}O_2$. However, HR, rectal temperature, and muscle temperature were higher the warmer the water. $\dot{V}O_2$max was lower at 20 °C than at 25 °C and 35 °C but decreased again at 40 °C. The maximal HR was 175 bpm at 20 °C versus 201 bpm at 40 °C. The maximal $\dot{V}O_2$ coincided with a muscle temperature of 39 °C. It was reduced at 38 °C T_{muscle} at 20 °C water and again when T_{muscle} reached 40.2 °C at the highest water temperature. The results speak against a circulatory failure in cold water but point to a biochemical limitation both at low and high muscle temperatures. Bergh and Ekblom (Bergh 1980) also determined $\dot{V}O_2$max but in air at 34.5–39 °C muscle and core temperatures obtained by swimming in cold water. Their results also indicate that the reduced $\dot{V}O_2$max, total work performance, and lower maximum HR are caused by a direct effect of low temperature on the muscles and heart, not circulatory limitations.

During steps of increasing exercise intensities, McArdle et al. (1976) measured $\dot{V}O_2$, Q, and HR in subjects immersed (to the level of the first thoracic vertebrae) in 18, 25, and 33 °C water and in air (25–27 °C) during exercise on the same arm-leg bicycle ergometer. $\dot{V}O_2$ was higher in 18 °C than in either 33 °C water or 25–27 °C air, the latter two being equal. The difference (23%) was higher at rest, but less and less with increasing $\dot{V}O_2$. At 120 W and 3.3 L $\dot{V}O_2$/min, the metabolic rate at 18 °C and 33 °C water became equal. In the cold water, the stroke volume was higher while HR was lower. However, in their study, Q was equal for the same $\dot{V}O_2$. Thus, the $(a-\bar{v})O_2$ difference was also the same for the same $\dot{V}O_2$ (in contrast with the above-mentioned measurements during rest made by Raven et al. (1970) and Wagner and Horvath (1985a). The authors discuss that an increase in stroke volume compensates

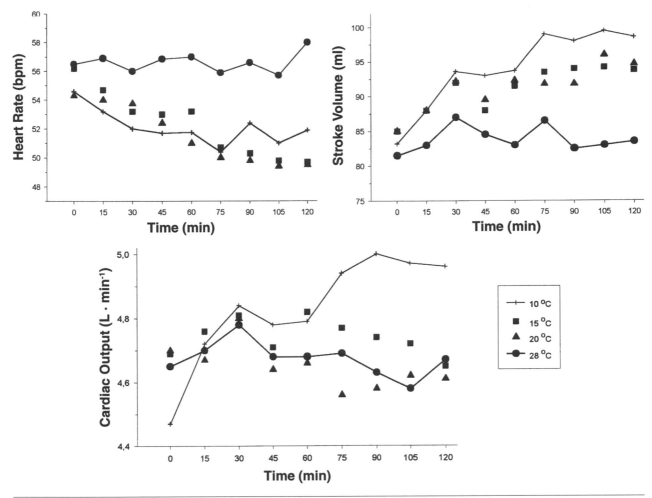

Figure 17.7   Cardiac output (20 male and 17 female subjects, 20–72 years old), heart rate, and stroke volume (only the 20 male subjects shown) measured during 2 h seated rest while exposed to ambient temperatures ranging from 10 °C to 28 °C. (From Wagner and Horvath 1985a, with permission.)

for the bradycardia in cold water, so Q remains unchanged. The increased cutaneous vasoconstriction would result in an increase in venous return and SV, the resulting baroreceptor stimulus would slow the HR.

Stevens, Graham, and Wilson (1987) studied subjects, males and females, at rest and during exercise of 50, 100, and 150 W in 21 °C and 5 °C air. The results from the men confirm the previous studies. However, the women did not react to cold stress by increased SV and reduced HR. This demonstrated a gender difference to cold stress.

In Young et al.'s study (1995) with exercise in hot (35 °C) and cold (20 °C) water while bicycling upright immersed to the neck at 60% $\dot{V}O_2$max 2 L/min, the oxygen uptake was the same in the cold and warm water. Heart rate was highest in the hot water (approximatley 20 bpm), and stroke volume was highest in cold water. Water temperature, though, did not affect cardiac output. However, the cold water in their study can hardly be interpreted as a cold stress in the sense of the previously mentioned definition, because core temperature also rose in the 20 °C water, only less so than in the hot.

In the study of swimming at 18, 26, and 33 °C (Nadel et al. 1974), increases in $\dot{V}O_2$ of 0.5 L/min were observed in the cold at two submaximal work intensities. The blood lactate (venous) after swimming was increased proportionally to the relative $\dot{V}O_2$ and did not differ from the relationship lactate/$\dot{V}O_2$ in bicycling. In other words, the increase in the cold reflected the increased $\dot{V}O_2$ and probably did not indicate a reduced muscle blood flow or aerobic metabolism due to cold muscles. HR was linearly correlated with $\dot{V}O_2$ for increasing swimming speeds but was shifted to a higher level in warm water. No measurements of Q were made in this or other studies about swimming in cold water with hypothermia and increased metabolism due to shivering (Galbo et al. 1979; Nielsen 1976).

Skin Circulation

The primary defense against cooling is the peripheral vasoconstriction, which is elicited from thermoreceptors in the skin. This has been studied by measuring/calculating the conductance of peripheral tissues and by venous occlusion plethysmography (Burton and Edholm 1955; Hardy and Söderström 1938; Winslow, Herrington, and Gagge 1937). The minimal conductance during rest is lower than during exercise. This is probably due to the reduced peripheral insulation with leg exercise when heat is produced in the limbs (Bazett et al. 1948). The minimal conductance, heat flux when the skin is maximally constricted, is further determined by the body fat content (Carlson et al. 1958).

Bittel et al. (1988) observed in resting subjects that minimal conductance and skin circulation occurred at 10 °C air temperature. However, conductance increased at further cold stress, at 5 °C and 1 °C, probably due to the increase in the muscle blood flow due to shivering activity. Wagner and Horvath (1985b) measured finger and forearm blood flow by occlusion plethysmography in resting subjects, males and females, 20–72 years. The finger and forearm blood flow decreased, and maximal vasoconstriction was completed within 15 min both at 10 °C and 15 °C air temperature in all subjects.

Another characteristic vascular response to cold is the countercurrent arrangement of the return of venous blood from cold hands (and feet) adjacent to the arteries in a way that reduces the cooling since arterial blood is precooled. A gradient of > 10 °C was measured over 15 cm of the radial artery in resting subjects (Bazett et al. 1948). At very low skin temperatures with fingers immersed in ice water, the skin temperature of fingers after a drop to almost 0 °C fluctuated, presumably due to variation in blood flow—the hunting phenomenon (Lewis 1930). This has also been attributed to reactions in the arteriovenous anastomosis, a cold vasodilatation, studied by, for example, Greenfield, Shepherd, and Whelan (1951) and Aschoff (1944ab). Greenfield, Shepherd, and Whelan (1951) demonstrated that nervous pathways were not essential for the hunting reaction.

Vanggaard (1975) showed that during rest in 8–9 °C air, the arteriovenous anastomoses in hands and fingers are able to vasoconstrict so violently during local cooling that the cooling rate of the arm is the same as with arrested arterial flow, a cuff around the upper arm. A recent study by Daanen (1997) reviews the effect of cold on finger blood flow.

To what degree the peripheral blood flow is reduced during whole-body exercise in cold conditions is less well studied. The calculated conductance during swimming in 18 °C cold water (20, respectively, 29 W/m^2/°C) were higher than the reported minimum values even though core temperature was decreasing (Nadel et al. 1974; Nielsen 1976). This could be due to a reduction in the peripheral insulation, caused by perfusion of a large exercising muscle mass in swimming, or that maximal vasoconstriction cannot be achieved during this type of exercise in spite of falling body core temperature. This remains to be studied.

Conclusions

The cardiovascular effects of cold stress during exercise are a peripheral vasoconstriction and an increased stroke volume compared with neutral conditions. This occurs due to the shift of blood volume centrally. The heart rate is reduced, probably due to baroreceptor reflexes (starting mechanism). Arterial blood pressure will tend to rise. Shivering and nonshivering thermogenesis are activated in conditions where core temperature tends to fall during exercise, adding to the rate of oxygen consumption. It is not clear if cardiac output increases to the same extent as $\dot{V}O_2$ or if $(a-\bar{v})O_2$ difference increases, which has been shown in cold stress during rest. The maximal $\dot{V}O_2$ during cold stress is reduced and reached with a lower maximal HR compared with the neutral condition. This could be due to four possible factors. First, a limitation in the capacity of the heart could be due to temperature effects on the biochemical and mechanical function. Second, perhaps the maximal work capacity of the heart (= SV × HR × MAP) has been reached. Third, peripheral circulatory factors, such as reduction in blood flow to the exercising muscles because of the strong vasoconstriction, could cause this. Fourth, the reduction in the rate of the metabolic processes in the cold muscles could be the cause. Which factor or factors are the actual cause remain to be elucidated.

References

Aschoff, J. 1944a. Die Vasodilation einer Extremität bei örtlicher Kälteeinwirkung. *Pflügers Archiv* 248:178–182.

Aschoff, J. 1944b. Über die Kältedilatation der Extremität des Menschen in Eiswasser. *Pflügers Archiv* 248:183–196.

Bazett, H.C., L. Love, M. Neuton, R.D. Eisenberg, and R. Forster. 1948. Temperature change in blood flow in the arteries and veins in man. *Journal of Applied Physiology* 1:3–19

Bergh, U. 1980. Human power at subnormal body temperatures. *Acta Physiologica Scandinavica Supplementum* 478:1–39.

Bittel, J.H.M., C. Nonotte-Varly, G.H. Livecchi-Gonnot, G.L.M.J. Savourey, and A.M. Hanniquet. 1988. Physical fitness and thermoregulatory reactions in a cold environment in men. *Journal of Applied Physiology* 45:1984–1989.

Brengelmann, G.L., J.M. Johnson, L. Hermansen, and L.B. Rowell. 1977. Altered control of skin blood flow during exercise at high internal temperatures. *Journal of Applied Physiology* 43:790–794.

Burton, A.C., and H.C. Bazett. 1936. A study of the average temperature of the tissues, of the exchanges of heat and vasomotor responses in man by means of a bath calorimeter. *American Journal of Physiology* 177:36–53.

Burton, A.C., and O.G. Edholm. 1955. *Man in a cold environment.* London: Edward Arnold Ltd.

Cannon, W.B., A. Querido, S.W. Britton, and E.M. Bright. 1926. Studies on condition of activity in endocrine glands: The role of adrenal secretion in the chemical control of body temperatures. *American Journal of Physiology* 79:466–507.

Carlson, L.D., A.C.L. Hsieh, F. Fullington, and R.W. Elsner. 1958. Immersion in cold water and body tissue isolation. *Journal of Aviation Medicine* 29:145–152.

Daanen, H. 1997. *Central and peripheral control of finger blood flow in the cold.* Ph.D. diss., Vrije Universitet te Amsterdam.

Erikson, H., J. Krog, K. Lange Andersen, and P.F. Scholander. 1956. The critical temperature in naked man. *Acta Physiologica Scandinavica* 37:35–39.

Febbrario, M.A., M.F. Carey, R.J. Snow, C.G. Stathis, and M. Hargreaves. 1996. Influence of elevated muscle temperature on metabolism during intense dynamic exercise. *American Journal of Physiology (Regulatory, Integrative and Comparative Physiology)* 271:R1251–R1255.

Febbrario, M.A., D.L. Lambert, R.L. Starkie, J. Proietto, and M. Hargreaves. 1998. Effect of epinephrine on muscle glycogenolysis during exercise in trained men. *Journal of Applied Physiology* 82:465–470.

Febbrario, M.A., R.J. Snow, M. Hargraves, C.G. Stathis, I.K. Martin, and M.F. Carey. 1994. Muscle metabolism during exercise and heat stress in trained men: Effect of acclimation. *Journal of Applied Physiology* 76:589–597.

Fink, W.J., D.L. Costill, and P.J. van Handel. 1975. Leg muscle metabolism during exercise in the heat and cold. *European Journal of Applied Physiology* 34:183–190.

Galbo, H., M.E. Houston, N.J. Christensen, J.J. Holst, B. Nielsen, E. Nygaard, and Y. Suzuki. 1979. The effect of water temperature on the hormonal response to prolonged swimming. *Acta Physiologica Scandinavia* 105:326–337.

González-Alonso, J., J.A.L. Calbet, and B. Nielsen. 1997. Metabolic alterations with dehydration-induced reductions on muscle blood flow in exercising humans. Abstract, Second Annual Congress of the European College of Sport Science, Copenhagen, 492–493.

González-Alonso, J., J.A.L. Calbet, and B. Nielsen. 1998. Muscle blood flow is reduced with dehydration during prolonged exercise in humans. *Journal of Physiology* 513:895–905.

González-Alonso, J., R. Mora-Rodríguez, P.R. Below, and E.F. Coyle. 1995. Dehydration reduces cardiac output and increases systemic and cutaneous vascular resistance during exercise. *Journal of Applied Physiology* 79:1487–1496.

González-Alonso, J., R. Mora-Rodríguez, P.R. Below, and E.F. Coyle. 1997. Dehydration markedly impairs cardiovascular function in hyperthermic endurance athletes during exercise. *Journal of Applied Physiology* 82:1229–1236.

González-Alonso, J., C. Teller, S.L. Andersen, F.B. Jensen, T. Hyldig, and B. Nielsen. Influence of body temperature on the development of fatigue during prolonged exercise in the heat. *J. Appl. Physiol.* 86:1032–1039, 1999.

Greenfield, A.D.M., J.T. Shepherd, and R.F. Whelan. 1951. The loss of heat from the hands and from the fingers immersed in cold water. *Journal of Physiology* 112:459–475.

Hardy, J.D., and G.F. Söderström. 1938. Heat loss from the nude body and peripheral blood flow at temperatures of 22–35 °C. *Journal of Nutrition* 16:493–510.

Jansky, L. 1995. Humoral thermogenesis and its role in maintaining energy balance. *Physiological Reviews* 75:237–259.

Jessen, K., A. Rabøl, and K. Winkler. 1980. Total body and splanchnic thermogenesis in curarized man during a short exposure to cold. *Acta Anaesthetica Scandandinavica* 24:339–344.

Johansson, J.E. 1897. Ueber den Einfluss der Temperature in der Umgebung auf die Kohlensäureabgabe des menschlichen Körpers. *Skandinavische Archiv für Physiologie* 7:123–177.

Johnson, J.M., L.B. Rowell, and G.L. Brengelmann. 1974. Modification of the skin blood flow-body temperature relationship by upright exercise. *Journal of Applied Physiology* 37:880–886.

Kenney, W.L., and J.M. Johnson. 1992. Control of skin blood flow during exercise. *Medicine and*

Science in Sports and Exercise 24:303–312.

Lewis, T. 1930. Observations upon the reactions of the vessels of the human skin to cold. *Heart* 15:177–208.

Mack, G.W., H. Nose, and E.R. Nadel. 1988. Role of cardiopulmonary baroreflexes during dynamic exercise. *Journal of Applied Physiology* 65:1827–1832.

McArdle, W.D., J.R. Magel, G.R. Lesmes, and G.S. Pechar. 1976. Metabolic and cardiovascular adjustment to work in air and water at 18, 25, and 33 °C. *Journal of Applied Physiology* 40:85–90.

Montain, S.J., and E.F. Coyle. 1992. Fluid ingestion during exercise increases skin blood flow independent of increases in blood volume. *Journal of Applied Physiology* 73:903–910.

Nadel, E.R., E. Cafarelli, M.F. Roberts, and C.B. Wenger. 1979. Circulatory regulation during exercise in different ambient temperatures. *Journal of Applied Physiology* 46:430–437.

Nadel, E.R., S.M. Fortney, and C.B. Wenger. 1980. Effect of dehydration state on circulatory and thermal regulation. *Journal of Applied Physiology* 49:715–721.

Nadel, E.R., I. Holmér, U. Bergh, P.-O. Åstrand, and J.A.J. Stolwijk. 1974. Energy exchanges of swimming man. *Journal of Applied Physiology* 36:465–471.

Nielsen, B. 1974. Effects of changes in plasma volume and osmolarity on thermoregulation during exercise. *Acta Physiologica Scandinavia* 90:725–730.

Nielsen, B. 1976. Metabolic reactions to changes in core and skin temperature in man. *Acta Physiologica Scandinavia* 97:129–138.

Nielsen, B., A. Astrup, P. Samuelsen, H. Wengholt, and N.J. Christensen. 1993. Effect of physical training on thermogenic responses to cold and ephedrine in obesity. *International Journal of Obesity* 17:383–390.

Nielsen, B., J.R.S. Hales, S. Strange, N. Juel Christensen, J. Warberg, and B. Saltin. 1993. Human circulatory and thermoregulatory adaptations with heat acclimation and exercise in a hot, dry environment. *Journal of Physiology* 460:467–485.

Nielsen, B., L.B. Rowell, and F. Bonde-Petersen. 1984. Cardiovascular responses to heat stress and blood volume displacements during exercise in man. *European Journal of Applied Physiology* 52:370–374.

Nielsen, B., G. Savard, E.A. Richter, M. Hargreves, and B. Saltin. 1990. Muscle blood flow and muscle metabolism during exercise and heat stress. *Journal of Applied Physiology* 69:1040–1046.

Pirnay, F., R. Deroanne, and J.M. Petit. 1977. Influence of water temperature on thermal, circulatory and respiratory responses to muscular work. *European Journal of Applied Physiology* 37:129–136.

Raven, P.B., I. Niki, T.E. Dahms, and S.M. Horvath. 1970. Compensatory cardiovascular responses during an environmental cold stress, 5 °C. *Journal of Applied Physiology* 29:417–421.

Raven, P.B., J.E. Wilkerson, S.M. Horvath, and N.W. Bolduan. 1975. Thermal, metabolic, and cardiovascular responses to various degrees of cold stress. *Canadian Journal of Physiology and Pharmacology* 53:293–298.

Rowell, L.B. 1974. Human cardiovascular adjustments to exercise and thermal stress. *Physiological Review* 54:75–159.

Rowell, L.B. 1986. *Human circulation regulation during physical stress.* New York, Oxford: Oxford University Press.

Rowell, L.B. 1993. *Human cardiovascular control.* New York, Oxford: Oxford University Press.

Rowell, L.B., J.A. Murray, G.L. Brengelmann, and K.K. Kraning. 1969. Human cardiovascular adjustments to rapid changes in skin temperature during exercise. *Circulation Research* 24:711-724.

Saltin, B. 1964. Circulatory response to submaximal and maximal exercise after thermal dehydration. *Journal of Applied Physiology* 19:1125–1132.

Savard, G.K., B. Nielsen, J. Laszczynska, B.E. Larsen, and B. Saltin. 1988. Muscle blood flow is not reduced in humans during moderate exercise and heat stress. *Journal of Applied Physiology* 82:649-657.

Sawka, M.N., A.J. Young, W.A. Latzka, P.D. Neufer, M.D. Quicley, and K.B. Pandolf. 1992. Human tolerance to heat strain during exercise: influence of hydration. *Journal of Applied Physiology* 73:368-375.

Sjöström, L. 1913. Über den Einfluss der Temperature der umgebenden Luft auf die Kohlensäureabgabe beim Menschen. *Skandinavische Archiv für Physiologie* 30:1–72.

Stevens, G.H.J., T.E. Graham, and B.A. Wilson. 1987. Gender differences in cardiovascular and metabolic responses to cold and exercise. *Canadian Journal of Physiology and Pharmacology* 65:165-171.

Vanggaard, L. 1975. Physiological reactions to wet-cold. *Aviation, Space and Environmental Medicine* 46:33–36.

Vogelaere P., G. Deklunder, J. Lecroart, G. Savourey, and J. Bittel. 1992. Factors enhancing cardiac output in resting subjects during cold exposure in air environment. *Journal of Sports Medicine, Physiology and Fitness* 32:378–386.

Vogelaere, P., G. Savourey, G. Deklunder, J. Lecroart, M. Brasseur, S. Bekaert, and J. Bittel. 1992. Reversal of cold induced hemoconcentration. *European Journal of Applied Physiology* 64:244–249.

Wagner, J.A., and S.M. Horvath. 1985a. Cardiovascular reactions to cold exposures differ with age and gender. *Journal of Applied Physiology* 58:187–192.

Wagner, J.A., and S.M. Horvath. 1985b. Influences of age and gender on human thermoregulatory responses to cold exposures. *Journal of Applied Physiology* 58:180–186.

Williams, C.G., G.A.G. Bredell, C.H. Wyndham, N.B. Strydom, J.F. Morrison, J. Peter, and P.W. Fleming. 1962. Circulatory and metabolic reactions to work in heat. *Journal of Applied Physiology* 17:625–638.

Winslow, G., L.P. Herrington, and A.P. Gagge. 1937. Physiological reactions of the human body to varying environmental temperatures. *American Journal of Physiology* 120:1–22.

Young, A.J., M.N. Sawka, L. Levine, P.W. Burgoon, W.A. Latzka, R.R. Gonzalez, and K.B. Pandolf. 1995. Metabolic and thermal adaptations from endurance training in hot or cold water. *Journal of Applied Physiology* 78:793–801.

Chapter 18

Cardiovascular Regulation With Endurance Training

Robert Boushel, Peter Snell, and Bengt Saltin

Endurance training provokes structural and regulatory adaptations in the heart and peripheral circulations, enhancing the delivery of oxygen to muscle and improving physical work capacity and endurance performance. This chapter reviews training-induced adaptations intrinsic to the heart and vasculature as well as changes in autonomic neural control of the cardiovascular response to exercise.

Changes in autonomic neural tone and a reduced sinus node automaticity contribute to the notable bradycardia associated with endurance training. Plasma volume expansion and increased diastolic filling result in an enlargement of the stroke volume and of maximal cardiac output. In the peripheral vasculature, enhanced vasodilatory capacity combined with reduced sympathetic vasoconstriction account for a lowering of vascular resistance. These concurrent adaptations facilitate increases in preload and decreases in afterload contributing to the enhancement of both cardiac output and muscle perfusion during exercise. The primary outcome of these adaptations is an improvement in oxygen transport, the predominant mechanism by which oxygen uptake and performance are enhanced by endurance training.

Oxygen Transport and Uptake

Maximal oxygen uptake ($\dot{V}O_2$max) is the principal measure of an individual's ability to perform dynamic exercise and is determined largely by the capacity of the cardiovascular system to deliver oxygen to skeletal muscle (Taylor, Buskirk, and Henschel 1955; Mitchell, Sproule, and Chapman 1958; Åstrand et al. 1964; Saltin 1985). In previously sedentary individuals, $\dot{V}O_2$max increased approximately 15–25% in response to endurance training. The magnitude of the increase depended on the initial fitness level: the lower the initial fitness level, the higher the gain in $\dot{V}O_2$max, and visa versa (Blomqvist and Saltin 1983; Saltin and Rowell 1980).

Adaptability to endurance training is also determined by genetic factors (Keul et al. 1996) as suggested by heterogeneous responses to a training regimen that are associated with genotype differences (Prud'Homme et al. 1984; Bouchard et al. 1986). During submaximal exercise, both cardiac output and O_2 uptake are unaltered by training (Saltin et al. 1969; Hartley et al. 1969; Clausen et al. 1973; Clausen 1977). However, the distribution of cardiac output is more efficient. Increases in maximal oxygen uptake after training result from an elevation of cardiac output (Hartley et al. 1969; Clausen et al. 1973; Klausen et al. 1982), from improvements in the distribution of blood toward the muscles most activated during exercise, and from a greater extraction and utilization of O_2 by muscle (Rowell, Blackmon, and Bruce 1964; Grimby, Häggendal, and Saltin 1967; Clausen et al. 1973; Saltin et al. 1976; Saltin 1986; Keul et al. 1996; Delp 1998). Of these factors, oxygen delivery to muscle is the most important contributor to the increase in O_2 uptake, as reflected by proportional increases in maximal cardiac output and O_2 uptake (figure 18.1).

Cardiac Function

The elevation of maximal cardiac output and systemic O_2 transport with endurance training are attributed to the maintenance of a larger stroke volume stemming from improved diastolic filling and an enlargement of the end-diastolic heart volume (Bevegard, Holmgren, and Johnsson 1963; Shapiro and Smith 1983; Voutilainen et al. 1991; Schairer et al. 1992; Huonker, Halle, and Keul 1996). This response develops from

- expansion of the plasma volume (Fortney et al. 1981; Kanstrup and Ekblom 1978; Convertino, Mack, and Nadel 1991; Fellmann 1992; Kanstrup, Marving, and Høilund-Carlsen 1992; Shoemaker et al. 1998; Krip et al. 1997);

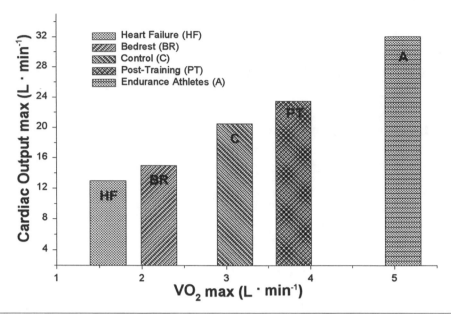

Figure 18.1 The relationship between increases in cardiac output and maximal oxygen uptake ($\dot{V}O_2$max) with varying degrees of functional capacity. HF: responses in heart failure patients (Sullivan et al. 1989); BR: healthy subjects after 20 days of bedrest (Saltin et al. 1968); C: same subjects before bedrest (Saltin et al. 1968); PT: same subjects after ~50 days of endurance training (Saltin et al. 1968); A: endurance athletes (Åstrand et al. 1964).

- a more rapid early diastolic filling associated with improved venous return by the muscle pump, and a prolongation of the period of diastole, (Bevegard, Holmgren, and Johnsson 1963; Shapiro and Smith 1983); and
- a greater ventricular compliance and enlargement of the ventricular volume (Douglas et al. 1986; Levine 1993; Vanoverschelde et al. 1993; George et al. 1995; Huonker, Halle, and Keul 1996) (see chapter 20).

As a result of an enlarged end-diastolic volume, left ventricular systolic performance is enhanced by way of the Frank-Starling mechanism (Jensen-Urstad et al. 1998; Di Bello et al. 1996; Mier et al. 1997). Most studies indicate either no change in contractility and ejection fraction during both submaximal and maximal exercise after training (Kronenberg et al. 1988; Nixon et al. 1991; Levine 1993; Kanstrup, Marving, and Høilund-Carlsen 1992; Brandao et al. 1993; Huonker et al. 1998) but increases have also been reported (Jensen-Urstad et al. 1998).

During submaximal exercise, the work of the heart is reduced due to a lower heart rate at a given cardiac output as well as a reduced afterload attributed to lower peripheral resistance (Blomqvist and Saltin 1983; Saltin 1986; Clausen 1976; Scheuer and Tipton 1977; Klausen et al. 1982). A lowering of myocardial oxygen demand and consumption also accompany the reduction in cardiac work (Keul et al. 1996). The reduced afterload and maintenance of

end-diastolic volume via the muscle pump ensure that stroke volume is maintained during submaximal through maximal exercise, as compared to the sedentary in which stroke volume may fall as maximal exertion is approached (Blomqvist and Saltin 1983; Saltin 1986; Martin et al. 1987; Spina et al. 1992; Di Bello et al. 1996; Mier et al. 1997) (see chapter 20).

Systemic Hemodynamics and Sympathetic Activation

There are conflicting reports on the effect of endurance training on blood pressure and sympathetic neural outflow at rest. Some longitudinal studies have shown a decrease in resting blood pressure and sympathetic nerve activity after training (Grassi et al. 1994; see chapter 23), while others show little change (Svedenhag et al. 1984; Sheldahl et al. 1994; Somers et al. 1992) or an increase (Ng et al. 1994).

During exercise, a close relationship develops among the levels of sympathetic activation, vascular resistance, and mean arterial blood pressure. At given submaximal exercise loads, blood pressure and vascular resistance are reduced after training (Clausen et al. 1973; Eckblom, Kilbom, and Soltysiak 1973; Lehman et al. 1981; Klausen et al. 1982; Seals and Chase 1989; Martin et al. 1990; Sinoway et al. 1987; Shenberger et al. 1990; Martin et al. 1991) in association with diminished sympathetic outflow to the heart and vasculature, as well as lower circulating plasma norepinephrine (Häggendal, Hartley, and Saltin 1970; Cousineau et al. 1977; Winder et al. 1979; Lehman et al. 1981; Svedenhag et al. 1986; Parker et al. 1994; Ray, Pace, and Clary 1992; Somers et al. 1992; Negrao et al. 1993; Sinoway et al. 1996; McAllister 1998; Johnson 1998; Kjaer 1998). Additionally, endurance training reduces the activity of plasma vasopressin and renin, potent vasoconstrictor hormones, during exercise (Convertino, Keil, and Greenleaf 1983; Shoemaker et al. 1998). When comparisons are made of the trained and untrained at the same relative intensities, sympathetic activation appears similar in the trained and untrained as reflected by similar plasma norepinephrine concentrations (Häggendal, Hartley, and Saltin 1970; Winder et al. 1979; Peronnet et al. 1981; Lehman et al. 1981) although there may be regional differences as determined by direct microneurography recordings (Ray and Hume 1998).

Mechanisms of Reduced Sympathetic Activity With Training

During dynamic exercise engaging a large muscle mass, total peripheral vascular resistance decreases progressively to reach a nadir when cardiac output and mean arterial pressure approach maximum. Evidence suggests that muscle vascular conductance can exceed the heart's ability to maintain blood pressure as maximal exercise is approached

(Secher et al. 1977; Klausen et al. 1982; Saltin 1985), and in both the trained and untrained, sympathetic vasoconstriction curtails vascular conductance even in active muscles in order to prevent a fall in blood pressure (see chapter 12).

At maximal exercise in the trained, vascular resistance is reduced in proportion to the elevation of cardiac output (Clausen et al. 1973; Gleser 1973; Klausen et al. 1982; Blomqvist and Saltin 1983; Levine 1993; Reading et al. 1993). The close coupling between changes in vascular resistance and cardiac output largely underlies the adjustments in sympathetic neural activation after training. Since cardiac output contributes to blood pressure, without a drop in peripheral resistance, blood pressure would be markedly elevated and the higher afterload would in turn impair stroke volume (Miura et al. 1993). Thus, sympathetic neural activation regulates vascular resistance and tissue blood flow in relation to cardiac pumping dynamics (figure 18.2) (see chapter 12). Vascular resistance is reduced for a given submaximal cardiac output, and the relationship between these variables is widened: the higher the maximal cardiac output, the higher the vascular conductance (Reading et al. 1993; Clausen et al. 1973; Klausen et al. 1982; DiCarlo and Bishop 1988, 1990; Levine et al. 1991; Raven and Pawelczyk 1993; Levine 1993).

The diminished sympathetic drive at a given workload and level of cardiac output after training is attributed in part to the larger stroke volume, which maintains activation of the cardiopulmonary

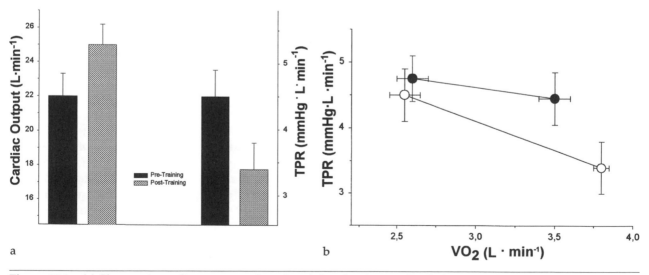

a   b

Figure 18.2 (a) Changes in cardiac output and total peripheral resistance (TPR) during cycling exercise after endurance training of the legs. Note that TPR is attenuated in parallel with the increase in cardiac output. (b) Relationship between total peripheral resistance (TPR) and oxygen uptake ($\dot{V}O_2$) during cycle exercise before (filled circles) and after endurance training (open circles). Note that TPR is lowered at a given oxygen uptake and at maximal $\dot{V}O_2$. (Adapted from Klausen et al. 1982.)

afferents resulting in diminished sympathetic out-flow (see chapter 3). Thus, vasoconstriction to maintain systemic perfusion pressure is delayed during graded absolute work due to a larger cardiac reserve, since cardiac output is an important contributor to maintaining or elevating arterial pressure. In highly trained humans and animals, cardiopulmonary baroreflex-mediated increases in arterial pressure (Pawelczyk, Kenney, and Kenney 1988; Raven and Pawelczyk 1993), renal and mesenteric sympathetic nerve activity (DiCarlo and Bishop 1988; Meredith et al. 1992; Negrao et al. 1993; Convertino, Keil, and Greenleaf 1983), and regional vascular resistance (Mack et al. 1987; DiCarlo and Bishop 1990; Mack, Convertino, and Nadel 1993; Green et al. 1994; Shi et al. 1996) are attenuated. Consistent with this adaptation, both splanchnic and skin blood flow are elevated in both humans and animals at a given absolute work rate after an endurance training regimen (Rowell, Blackmon, and Bruce 1964; Clausen et al. 1973; Fellmann 1992; Lash 1998; McAllister 1998; Johnson 1998). The cardiac afferents have also been shown to attenuate sympathetic activity induced by muscle metaboreflex activation (Collins and DiCarlo 1993).

Another mechanism contributing to the lowering of vascular resistance during exercise in the trained is diminished muscle metaboreflex-mediated vasoconstriction (Somers et al. 1992; Sinoway et al. 1996; Piepoli et al. 1996). This adaptation is linked to improved muscle perfusion, improved microcirculatory transport capacity, and greater muscle oxidative capacity. A more efficient blood flow distribution that enhances oxygen delivery more selectively to type I activated muscle diminishes the reliance on anaerobic metabolism and fibers, and with improved muscle pump activity results in greater clearance of metabolites (Saltin 1977; Winder et al. 1979; Brooks 1991; Green et al. 1991; MacRae et al. 1992; Delp 1998).

Altered metabolite accumulation reduces the activation of chemosensitive muscle afferent nerves and attenuates efferent sympathetic nerve activity (Winder et al. 1979; Somers et al. 1992; Sinoway et al. 1996; Piepoli et al. 1996; Ray and Hume 1998). Supporting the concept of attenuated muscle reflex activity is the finding that heart rate and metabolic responses to exercise are attenuated when a trained leg is engaged in exercise compared to when an untrained, contralateral leg is involved (Saltin 1977). Thus, with training, enhanced cardiac pumping capacity and improvements in muscle oxygen delivery and utilization contribute to alterations in both central and peripheral mechanisms regulating cardiac function and peripheral vascular tone. In turn, the outcome of these adaptations is an optimization of blood flow distribution and oxygen delivery to match habitual oxygen demand.

Structural and Functional Adaptations in the Vasculature

While the capacity of the systemic circulation to increase its conductance for blood flow is primarily a function of autonomic regulatory mechanisms affecting arteriolar diameter, some evidence suggests local adaptations in the vasculature. The stiffness of the major arteries is an important determinant of systemic compliance and constitutes an important determinant of arterial pressure and ventricular afterload. Endurance training results in enhanced arterial compliance (Mohiaddin et al. 1989; Vaitkevicius et al. 1993; Cameron and Dart 1994; Stewart et al. 1998), which is also linearly related to increases in maximal oxygen uptake (Cameron and Dart 1994). The arterial pulse amplitude affects arterial resistance (and therefore arterial pressure), which is greater after training as a result of a higher stroke volume. The increase in arterial compliance offsets the pressure elevation associated with an increase in pulse amplitude, so that exercise blood pressure in the trained is not elevated compared to the sedentary.

Studies in which autonomic neural influences have been excluded as a factor contributing to vascular resistance reveal the importance of local vasoregulatory adaptations with training. Reactive hyperemia in a limb following tourniquet ischemia is thought to elicit maximal muscle blood flow and to represent the intrinsic capacity of vessels to vasodilate (Sinoway et al. 1986; Snell et al. 1987; Martin et al. 1990). In this model, blood flow is enhanced specifically in the trained muscles, although some crossover effects occur in inactive tissues (Kingwell et al. 1997; Yasuda and Miyamura 1983; Silber, McLaughlin, and Sinoway 1991; Lash 1998). In addition to enhanced perfusion, expansion of resting and maximally dilated arterial diameter has been demonstrated, indicating structural and functional alterations in the vasculature with training (Delp 1998; Keul et al. 1996; Huonker et al. 1998; Shenberger et al. 1990; Laughlin and McAllister 1992; Haskell et al. 1993).

Endurance training has been found to increase the cross-sectional area of medium and small peripheral arteries (Shenberger et al. 1990; Laughlin and McAllister 1992; Huonker, Halle, and Keul 1996; Huonker et al. 1998; Keul et al. 1996). Vessel size is correlated with both training status and peak pulmonary oxygen uptake (Mann et al. 1972; Kool et al. 1992; Huonker et al. 1998; Radegran and Saltin unpublished findings). These findings indicate that structural increases in vessel diameter serve to increase maximal perfusion, oxygen delivery, and oxygen uptake. Often, it has been suggested that endurance training increases the size of the coronary arteries and improves collateral circulation in humans. The evidence to support this hypothesis is still inconclusive due to the difficulty of performing longitudinal studies on healthy individuals. Most of the data has been obtained from studies of patients with cardiomyopathy and from cross-sectional or autopsy studies of athletes (Currens and White 1961; Mann et al. 1972; Fisher et al. 1989). Haskell et al. (1993) reported that the combined cross-sectional area of four coronary arteries determined by angiography was similar at rest in runners and control sedentary subjects. However, when the coronary arteries were dilated with nitroglycerine, the total cross-sectional area in the vessels of the runners was double that of the controls, indicating superior vasodilator reserve. Alterations in functional vasodilatory capacity have been observed after as little as one week of training (George et al. 1995; Stewart et al. 1998), while structural changes require a more prolonged period of several weeks to months (Wyatt and Mitchell 1978).

Experimental limitations prevent a detailed understanding of the mechanisms regulating adaptations in the vasculature of humans with training, yet isolated tissue studies in animals have been informative. Most of the mechanistic work in this area has focused on the function of the coronary vessels. Using isolated coronary vascular tissue preparations as well as intact vessels, flow-mediated adaptations in coronary vasculature with training have been described (Laughlin et al. 1996; Laughlin, Oltman, and Bowles 1998; Kingwell et al. 1997; Lash 1998), including altered responsiveness to pharmacological interventions and greater sensitivity of the vascular endothelium and smooth muscle to circulating and or locally released vasoactive substances (Rogers et al. 1991; Lash 1998; Laughlin, Oltman, and Bowles 1998; Delp 1998). Training-induced adaptations in vascular responsiveness to vasoactive substances are highly specific to the size and region of the vascular segments and their location in the branching pattern (Parker et al. 1994; Laughlin, Oltman, and Bowles 1998; Lash 1998). In isolated, conduit-sized, proximal coronary vessels, vasodilatation is enhanced in response to a given level of nitroglycerine, adenosine, and acetylcholine with a reduced sensitivity to norepinephrine (Stewart et al. 1998; Lash 1998). In the smaller resistance arteri-

oles, enhanced relaxation in response to bradykinin and shear stress is exhibited (Muller, Myers, and Laughlin 1994). This response supports the concept of a major role of nitric oxide in the enhanced vasodilator capacity response in the microcirculatory vessels (Delp 1998). Furthermore, increases in nitric oxide synthase gene expression and nitric oxide release eliciting cyclic GMP-mediated relaxation of vascular smooth muscle have been demonstrated in response to training (Sessa et al. 1994; George et al. 1995; Stewart et al. 1998; Laughlin, Oltman, and Bowles 1998). In skeletal muscle, contraction-induced vasodilatation is enhanced after training (Mackie and Terjung 1983; Lash 1998) in association with increased responsiveness to vasoactive substances which are specific to the size of the vessels. Greater response to acetylcholine has been demonstrated in arterioles, while gradual increases in sensitivity to nitroprusside are observed in terminal feed arteries (Lash 1998). An additional feature of the vasculature after training is the more rapid vasodilatation of skeletal muscle terminal arterioles in the transition from rest to exercise with corresponding increases in blood flow (Lash 1998).

Myogenic reactivity of the vascular smooth muscle in response to increases in pressure is also enhanced after training (Muller, Myers, and Laughlin 1993; Lash 1998). This response may be linked to altered control of sarcoplasmic reticular calcium release in vascular smooth muscle consistent with structural changes in calcium channels (Underwood, Laughlin, and Sturek 1994; Stewart et al. 1998). Thus, changes in both the vascular endothelium and smooth muscle cells appear to be coordinated to balance vasoregulatory responses.

Significance of Vascular Adaptations

The importance of altered vasodilatory capacity of the vasculature with training is obviated by adaptations in capillary flow and substrate exchange characteristics observed in animal studies (Overholser, Laughlin, and Bhatte 1994; Laughlin et al. 1996; Laughlin, Oltman, and Bowles 1998; Delp 1998). Measurements of capillary flow and exchange products under both maximally dilated and controlled perfusion conditions reveal that capillary transport capacity and functional surface area for nutrient exchange is enhanced at a given level of vascular resistance. This response occurs in the absence of anatomical evidence for angiogenesis or enhanced capillary permeability (Laughlin et al. 1996), and implies improved flow distribution, either by increased perfusion in areas of high exchange capacity or enhanced flow to previously under-utilized regions. It is not clear, however, whether training decreases flow heterogeneity or increases flow dispersion in skeletal muscle (Laughlin et al. 1996). Another unanswered question is how vascular adaptations in upstream vessels are coordinated with specific adaptations in flow at the capillary level. Nonetheless, the enhancement of capillary surface exchange area with training appears to be linked to a lowering of upstream vascular resistance.

In endurance-trained muscles, both capillary flow and surface area available for transport are elevated at a given arterial perfusion pressure (Laughlin et al. 1996; Delp 1998). The significance of this adaptation is that for a given perfusion pressure, capillary recruitment and surface area for nutrient exchange and metabolic by-product removal expands. Presumably, this pattern would arise only by selective distribution of perfusion increases to the most metabolically active fibers (Delp 1998). Muscle pump activity becomes more efficient as a larger proportion of the muscle blood volume perfuses the contracting fibers. These intrinsic adaptations underlie the functional advantage to muscle of a reduced vascular resistance with training and explain in part the close relationship between increases in oxygen uptake and local and systemic vascular resistance. Additionally, local vascular adaptations, which favor a greater vasodilatory sensitivity, may explain the observation that in the trained, blood flow is maintained at a higher vascular conductance despite a lower metabolite accumulation. This adaptation would not only ensure metabolic vasodilatation but also would result in a reduction in metabolically triggered afferent neural signals eliciting vasoconstriction.

Muscle Perfusion and O_2 Delivery

In the endurance-trained subject at rest, muscle blood flow is unaltered. However, in anticipation of exercise, blood flow to the muscles engaged in the training-specific activity is elevated (Delp 1998). The mechanisms underlying this response in humans likely involves reductions in sympathetic adrenergic vasoconstrictor activity or decreased sensitivity to norepinephrine (Callister, Ng, and Seals 1994). In some animal species, evidence suggests cholinergic vasodilatation (Bolme and Novotny 1969). During submaximal dynamic exercise at a given absolute work rate, total limb perfusion is unchanged or is slightly reduced by training. However, training alters the distribution of limb flow whereby flow is partitioned toward the muscles

most activated by the specific training exercise (Clausen and Trap-Jensen 1970; Varnauskas et al. 1970; Saltin et al. 1976; Saito, Matsui, and Miyamura 1980; Mackie and Terjung 1983; Armstrong and Laughlin 1984; Delp 1998). This adaptation is associated with a more efficient muscle recruitment pattern favoring the activation of oxidative type I fibers, coupled with local vasoregulatory adaptations enhancing vasodilatation in these fibers (Varnauskas et al. 1970; Mackie and Terjung 1983; Armstrong and Laughlin 1984; Bergman et al. 1973; Clausen et al. 1973; Delp 1998). Peak muscle perfusion during exercise is enhanced by endurance training as a result of a higher maximal cardiac output (Hartley et al. 1969; Clausen 1977; Klausen et al. 1982) and a greater diversion of the total blood volume toward active muscle (Clausen and Trap-Jensen 1970; Delp 1998; Armstrong and Laughlin 1984; Mackie and Terjung 1983; Musch et al. 1987; Magnusson et al. 1997). Local structural changes also contribute to the enhancement of peak muscle perfusion. Training increases muscle capillary density and volume resulting in a greater capacity of the muscles to accommodate blood flow (Saltin et al. 1986; Saltin and Gollnick 1983; Magnusson et al. 1996). This results in an expansion of the surface area for O_2 exchange within muscle tissue, especially in the type I oxidative fibers. The myoglobin content, the mitochondrial volume, and the activity of rate-limiting enzymes of oxidative phosphorylation are also enhanced (Saltin and Gollnick 1983; Somers et al. 1992; Magnusson et al. 1996). The outcome of these peripheral adaptations is that capillary transit time is preserved to maintain or increase local diffusion and O_2 extraction at a higher muscle perfusion rate (table 18.1) (Saltin 1985; Saltin et al. 1986). Thus, training evokes peripheral adaptations to allow for effective utilization of the increase in O_2 delivery resulting in a higher O_2 diffusional conductance of O_2 in muscle (Saltin and Rowell 1980; Blomqvist and Saltin 1983; Wagner 1992).

Training Bradycardia

The most obvious cardiovascular adaptation with endurance training is a lowering of the heart rate at rest and during submaximal exercise, termed *training bradycardia*. This phenomenon is mediated by alterations in the autonomic nervous system and by changes in the intrinsic automaticity of the sinus node and right atrial myocytes (Robinson et al. 1966; Eckblom et al. 1972; Barnard, Corre, and Cho 1976; Hughson et al. 1977; Lewis, Nylander, Gad, et al. 1980; Smith et al. 1989; Nylander and Areskog 1982;

Table 18.1   Relationship Among Muscle Blood Flow, Mean Transit Time, Muscle Capillary Density, and Oxygen Extraction ([a-v̄] O_2 diff)

Maximal muscle blood flow (ml · 100g⁻¹)	Mean transit time (msec)	Capillary density (n/mm²)	a-v O_2 diff (vol %)
~100	650	350	16
	850	450	17.5

The effect of endurance training can be observed as a greater capillary density allows for a longer transit time at a given blood flow and also a higher (a-v̄) O_2 diff.

Areskog 1985; Bolter, Banister, and Singh 1986; Schaefer et al. 1992; Negrao et al. 1992). Several experimental models have been employed to identify the neural and intrinsic components involved including studies using pharmacological autonomic neural blockade, power spectrum analysis of heart rate variability, cardiac denervation, and isolated cardiac tissue preparations.

Both cross-sectional and longitudinal studies involving pharmacological autonomic blockade and power spectrum analysis of heart rate variability have established that increased parasympathetic (vagal) tone to the heart contributes importantly to resting bradycardia (Eckblom, Kilbom, and Soltysiak 1973; Smith et al. 1989; Seals and Chase 1989; Billman and Dujardin 1990; Goldsmith et al. 1992; Dixon et al. 1992; De Meersman 1993; Sacknoff et al. 1994; Shi, Stevens, Foresman, et al. 1995; Gregoire et al. 1996). It is still debatable whether sinus node automaticity (intrinsic heart rate) is reduced with training, but this adaptation may require an intensive training regimen (Nylander and Areskog 1982; Areskog 1985; Katona et al. 1982; Smith et al. 1989; Nylander and Dahlström 1984; Bolter, Banister, and Singh 1986; Shi, Potts, Raven, et al. 1995; Bonaduce et al. 1998). Findings conflict on whether the sympathetic influence on the heart at rest reduces after training (Eckblom, Kilbom, and Soltysiak 1973; Smith et al. 1989; Meredith et al. 1992). The chronic increase in parasympathetic tone occurs within a few weeks after beginning regular training, independent of a lower intrinsic heart rate. Thus, increased vagal activity accounts for the initial reduction in resting heart rate (Kenney 1985; Swaine, Linden, and Mary 1994; Bonaduce et al. 1998) while further latent reductions in heart rate are attributed to a lowering of intrinsic automaticity (Barnard, Corre, and Cho 1976; Bonaduce et al. 1998). The mechanisms underlying the increase in vagal tone and lower intrinsic heart rate have not been clearly defined, but animal and human experiments suggest redundant influences contributing to these adaptations.

Intrinsic Heart Rate

Despite extensive study of the phenomenon of training bradycardia, the debate still rages over whether a lowering of intrinsic heart rate is a true adaptation to endurance training. In humans, a lowering of intrinsic heart rate has been established mainly from cross-sectional studies (Lewis, Thompson, Areskog, et al. 1980; Katona et al. 1982; Smith et al. 1989). This has raised the notion that differences in intrinsic heart rate may represent a genetic difference between sedentary individuals and those who engage in regular training (Lewis, Thompson, Areskog, et al. 1980; Raven and Pawelczyk 1993; Shi, Stevens, Foresman, et al. 1995). However, an intensive training period may be the critical requisite for inducing a lower intrinsic heart rate (Barnard, Corre, and Cho 1976; Hughson et al. 1977; Areskog 1985; Bolter, Banister, and Singh 1986; Raven and Pawelczyk 1993; Shi, Stevens, Foresman, et al. 1995; Bonaduce et al. 1998).

Little is known about the mechanisms underlying the reduction in intrinsic heart rate. This adaptation is associated with a prolonged duration of the atrial action potential suggesting structural and metabolic changes in the pacemaker cells (Brorson et al. 1976). Based on animal studies, an intact nervous system (at least the sympathetic nerves) may be necessary for the response. Chronic endurance training in sympathectomized rats (Sigvardsson, Svanfeldt, and Kilbom 1977; Nylander and Areskog 1982) and denervated dogs (Ordway et al. 1982) produces no reduction in intrinsic heart rate. Similarly, chronic electrically evoked pacing of isolated right atrial cells where the sinus node pacemaker cells are situated fails to lower intrinsic automaticity (Hughson et al. 1977). In contrast, spontaneously-beating right atrial cells isolated from rats that had previously undergone an intensive endurance training regimen does display the intrinsic heart rate reduction compared to sedentary controls when studied under pharmacological beta receptor and parasympathetic blockade (Schaefer et al. 1992). These same trained animals also showed no alterations in β-1 receptor activity (Williams 1985; Davidson, Banerjee, and Liang 1986; Dickhuth et al. 1987; Hammond, Ransnas, and Insel 1988).

Studies in heart transplant patients, who are without cardiac innervation, have shown discrepant findings. In an early study in these patients, exercise training produced no reduction in resting heart rate (Squires et al. 1983), but in a later report by Kavanaugh et al. (1988), a prolonged training period of 36 months did result in a lower resting heart rate. Changes in cardiac β-cell receptor response to circulating catecholamines has been proposed as the mechanism underlying this reduction (Tohmeh and Cryer 1980; Bristow et al. 1982; Borow et al. 1985). While intact sympathetic nerves appear important for the lowering of intrinsic heart rate, this adaptation has nonetheless been demonstrated in rats trained under beta blockade (Nylander and Areskog 1982; Sigvardsson, Svanfeldt, and Kilbom 1977; Nylander and Dahlström 1984; Schaefer et al. 1992). This would suggest that if the intrinsic heart rate reduction is sympathetically-mediated, then it occurs by a mechanism independent of β receptor stimulation and is not dependent on the magnitude of heart rate during training bouts (Nylander and Areskog 1982; Nylander 1985). This is supported by the finding that sedentary rats exposed to long-term β-receptor stimulation that induced high heart rates showed no reduction in intrinsic heart rate in the absence of exercise training (Nylander and Dahlström 1984). However, rats trained under parasympathetic blockade demonstrate an even greater reduction in intrinsic heart rate compared to control trained animals when the right atria are isolated and studied under β and muscarinic receptor blockade (Hughson et al. 1977). This would suggest that the parasympathetic neural branch does not appear to be necessary to induce a lowering of intrinsic heart rate, and the magnitude of heart rate during training may be a factor influencing the degree of intrinsic pacemaker cell automaticity.

Recently, circuit neurons in the right atria, which stain for nitric oxide synthase (NOS) and are activated by NO, have been identified in the dog. These neurons modulate parasympathetic and sympathetic activity both centrally and locally and may play a role in adaptive responses (Armour et al. 1995). A specific group of these neurons enhance heart rate when activated, while a smaller group of NO-sensitive neurons are cardioinhibitory. When the intrinsic neurons are dissected from central innervation and stimulated by NO, their cardioinhibitory and excitatory responses are attenuated (Armour et al. 1995). These neurons also possess α-1,2, and β-1,2 adrenoreceptors and are tonically activated by β agonists. Their role in lowering intrinsic heart rate is unclear, but they may mediate the autonomic neural adjustments, and perhaps explain the presumed necessity of intact sympathetic nerves for the response. Further work is required to elucidate the relation between cardiac innervation, local circuit neurons, and intrinsic heart rate.

The observation that primates with larger hearts have lower intrinsic heart rates underlies the hypothesis that training-induced cardiac enlargement accounts for the lower intrinsic heart rate with training. The development of hypertrophy is not a prerequisite (Barnard, Corre, and Cho 1976; Nylander and Areskog 1982; Dickhuth et al. 1987), but heart volume enlargement per se has been proposed as the mechanism underlying this adaptation (Lewis, Nylander, Gad, et al. 1980; Saltin and Blomqvist 1983; Shi, Stevens, Foresman, et al. 1995). Atrial pacemaker firing rate increases in response to stretch (Hughson et al. 1977), and training has been shown to increase ventricular compliance and distensibility (Levine et al. 1991). Thus, one plausible mechanism for reduced intrinsic heart rate is that atrial chamber volume enlargement reduces the stretch-depolarization stimulus through structural changes in the myocyte membrane, therefore lowering the spontaneous depolarization rate.

Parasympathetic Tone and Bradycardia

Three suggested triggers for increased vagal tone in trained individuals are blood volume-baroreceptor activation due to an increase in blood volume and stroke volume, opiod modulation of autonomic activity, and the stimulation of the dopaminergic system.

Blood Volume-Baroreceptor Activation

Evidence suggests that the mechanism for the training-induced increase in vagal tone involves a greater activation of the cardiac baroreceptors in response to the enlargement of blood volume and stroke (Pawelczyk, Kenney, and Kenney 1988; Convertino, Mack, and Nadel 1991; Raven and Pawelczyk 1993; Levine 1993; Negrao et al. 1993; Shi, Stevens, Foresman, et al. 1995). Detectable hypervolemia is observed within minutes or hours postexercise, and a 9–25% increase in plasma volume has been found in response to training (Convertino, Keil, and Greenleaf 1983; Fellmann 1992; Krip et al. 1997). Recent studies suggest that following an acute bout of exercise blood volume expansion is triggered by a reduction of central blood volume due to pooling of blood in peripherally dilated vessels (Hanel et al. 1997). With a reduced central blood volume, right atrial pressure is lowered, and there is a pronounced reduction in atrial natriuretic peptide (ANP) (Hanel et al. 1997). ANP inhibits the fluid-retention influence of aldosterone, vasopressin, and plasma proteins in the kidney. Therefore, the lower ANP allows for greater sodium and water retention by the kidney, and successive exercise training bouts, therefore, are likely to elicit a chronic stimulus for expansion of blood volume. The link between blood volume and training bradycardia is that stroke volume is enlarged, stretching the atrial and/or ventricular mechanoreceptors. Right atrial receptor activation by plasma volume expansion reduces sympathetic nerve activity, and the parasympathetic influence is enhanced (Seals and Chase 1989; Levine et al. 1991; Negrao et al. 1993; Shi et al. 1993, 1996; Shi, Stevens, Foresman, et al. 1995; Raven and Pawelczyk 1993). The elevation of stroke volume in the trained would therefore exert a chronic increase in vagal tone.

Opioids

Opioids are known to modulate both sympathetic and parasympathetic control of heart rate at the neuronal, prejunctional level (Cherdchu, Robinson, and Hexum 1987; Musha et al. 1989; Caffrey et al. 1995; Baron, Laughlin, and Gwirtz 1997). Arterial infusion of enkephalins attenuates sympathetic neural suppression of vagal bradycardia by reducing acetylcholine release in the atria (Caffrey et al. 1995). Consistent with this pattern is the finding that enkephalin inhibition by dinorphin attenuates epinephrine release in sympathetic preganglionic junctions, enhancing the effects of a given level of vagal tone (Caffrey et al. 1995). Peripherally, activation of opioid receptors modulates the transmission of group III and IV muscle afferent nerve activity through the L-7 dorsal horn of the cat (Meintjes et al. 1995), influencing muscle reflex signals affecting sympathetic discharge. In humans, plasma catecholamine levels during intense dynamic exercise are altered by opioid receptor blockade by naloxone (Angelopoulos et al. 1995).

In dogs, endurance training has been shown to down-regulate enkephalin receptors in the heart, providing the mechanism for enhancing vagal tone (Baron, Laughlin, and Gwirtz 1997). Indirect evidence for a role of the opioids in training bradycardia in humans comes from studies on orthostatic tolerance. Training lowers tolerance to lower-body negative pressure, unloading the cardiopulmonary baroreceptors. This response is alleviated by opioid receptor blockade by naloxone (Madsen et al. 1995). Thus, down-regulation of enkephalin receptors in response to endurance training is a plausible mechanism contributing to enhanced parasympathetic influence and a lower heart rate.

Dopaminergic System

Excitation of dopamine D_2 receptors in the nucleus ambiguus induces bradycardia by activation of dopamine D_2 receptors present on vagal preganglionic cardioinhibitory neurons controlling heart rate (Chitravanshi and Calaresu 1992). Endurance training activates and increases the number of binding sites of dopamine receptors (subtype DA_2) located in central and peripheral nerves, and elevated resting and exercise plasma concentrations of dopamine have been reported (MacRae et al. 1987; Slavik and LaPointe 1993; Bove, Dewey, and Tyce 1984; Gilbert 1995). This pattern is associated with reductions in both resting and exercise heart rate. Thus, it has been postulated that training up-regulates the dopaminergic system, enhancing the effect of parasympathetic activity contributing to resting bradycardia.

Relative Bradycardia During Exercise

During exercise in the trained, a given increase in cardiac output requires less increase in heart rate due to the maintenance of a larger stroke volume. Studies focusing on autonomic and endocrine responses to training indicate that heart rate is reduced during submaximal exercise (absolute load) in the trained due to a lower intrinsic heart rate, a reduction in sympathetic activity, circulating catecholamines, and a greater vagal influence (Eckblom, Kilbom, and Soltysiak 1973; Lewis, Nylander, Gad, et al. 1980; Peronnet et al. 1981; Cousineau et al. 1977; Winder et al. 1979; Ishida and Okada 1997; Tulppo et al. 1998). The chronotropic response of the heart to catecholamines and autonomic neural activity appears unchanged (Svedenhag et al. 1986; Bolter, Banister, and Singh 1986; Davidson, Banerjee, and Liang 1986; Hatfield et al. 1998), but whether cardiac β cell-mediated signal transduction is altered is inconclusive (Hughson et al. 1977; Lehman et al. 1981; Butler et al. 1982; Dickhuth et al. 1987; Svedenhag et al. 1986; Davidson, Banerjee, and

Liang 1986; Williams 1985; Hammond, Ransnas, and Insel 1988). The reduced sympathetic influence on the heart activity is due in part to diminished afferent signals from muscle metaboreflexes (Stegemann and Kenner 1971). The baroreflexes are thought to exert a significant role in preserving a lower heart rate during submaximal exercise. Training diminishes cardiopulmonary baroreflex cardio-acceleration as the higher stroke volume maintains the activation of atrial and ventricular mechanoreceptors (Smith et al. 1988; Levine et al. 1991; Levine 1993; Chen, DiCarlo, and Scislo 1995). This influence is thought to be mediated through inhibition of carotid and aortic baroreceptor-mediated sympathetic activity (see chapters 1 and 3).

Rheoreceptors located in the carotid sinus may also exert an influence on autonomic neural adjustments with training. These receptors are sensitive to flow and shear stress. When activated, they reduce efferent sympathetic outflow. In dogs, increases in stroke volume produce proportional increases in carotid sinus afferent nerve activity, suggesting a modulating influence on arterial baroreflex-mediated sympathetic activity (Levine 1993). Thus, the training-induced elevation of flow amplitude asso-

ciated with a higher stroke volume is thought to contribute to reflex inhibition of sympathetic activity during exercise after training.

Summary

Young sedentary individuals are able to increase their oxygen utilization 10–15 times that of rest in response to the demand of heavy dynamic exercise by increasing their cardiac output four- to five-fold, coupled with a reduction in peripheral resistance that enhances muscle perfusion. Oxygen diffusion from the capillaries to mitochondria is accelerated so that even with the large increase in muscle blood flow, oxygen uptake by the muscles results in a wider arterio-venous oxygen difference. With regular endurance exercise training, these responses are enhanced by structural and regulatory adaptations that result in a 15–25% increase in oxygen transport and utilization. This chapter has focused on the neural and hormonal influences that may be responsible for the development of characteristics of the heart and circulation that allow for the impressive improvement in exercise capacity. The key com-

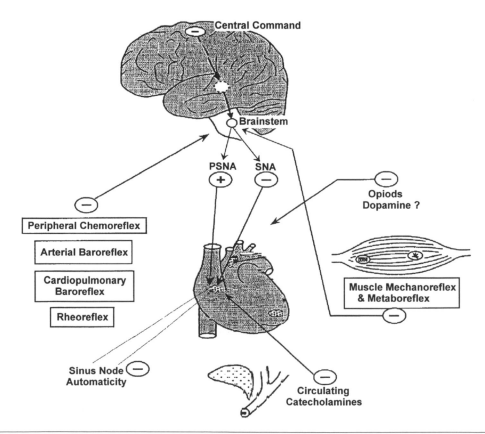

Figure 18.3   Schematic showing the multiple mechanisms likely to be involved in the lowering of heart rate with endurance training.

ponents include expansion of the plasma volume and intrinsic and neurally mediated adjustments in the heart and vasculature, which improve blood flow distribution and enhance diastolic filling of the heart, stroke volume, and maximal cardiac output. Hormonal and autonomic neural activity and intrinsic cardiac pacemaker activity constitute redundant influences contributing to the reduction of heart rate at rest and during exercise after training. In response to an increase in plasma volume and stroke volume, the baroreceptors maintain greater vagal tone to the heart. This influence is thought to be enhanced by the up-regulation of dopaminergic receptors in the heart. This pattern is further supported by down-regulation of enkephalin receptors, resulting in less attenuation of vagal activity in the atria. Aortic rheoreceptor activation by an enhanced stroke volume, together with diminished reflex signals from skeletal muscle contribute to a reduction of sympathetic outflow to the heart. The mechanism underlying the lowering of intrinsic heart rate remains to be elucidated, but is likely to involve structural adaptations within the pacemaker cells of the right atria (figure 18.3).

References

Angelopoulos, T.J., B.G. Denys, C. Weikart, S.G. Dasilva, T.J. Michael, and R.J. Robertson. 1995. Endogenous opioids may modulate catecholamine secretion during high intensity exercise. *European Journal of Applied Physiology* 70(3):195–199.

Areskog, N.H. 1985. Effects and adverse effects of autonomic blockade in physical exercise. *American Journal of Cardiology* 55(10):132D–134D.

Armour, J.A. 1997. Intrinsic cardiac neurons involved in cardiac regulation possess alpha-1, alpha-2, beta-1, and beta-2 adrenoreceptors. *Canadian Journal of Cardiology* 13(3):277-284.

Armour, J.A., F.M. Smith, A.M. Losier, H.H. Ellenberger, and D.A. Hopkins. 1995. Modulation of intrinsic cardiac neuronal activity by nitric oxide donors induces cardiodynamic changes. *American Journal of Applied Physiology* 268: R403–R413.

Armstrong, R.B., and M.H. Laughlin. 1984. Exercise blood flow patterns within and among rat muscles after training. *American Journal of Physiology* 264:H59–H68.

Åstrand, P-O., T.E. Cuddy, B. Saltin, and J. Stenberg. 1964. Cardiac output during submaximal and maximal work. *Journal of Applied Physiology* 19:268–274.

Barnard, R.J., K. Corre, and H. Cho. 1976. Effect of training on the resting heart rate of rats. *European Journal of Applied Physiology* 35(4):285–289.

Baron, B.A., M.H. Laughlin, and P.A. Gwirtz. 1997. Exercise effect on canine and miniswine cardiac catecholamines and enkephalins. *Medicine and Science in Sports and Exercise* 29(10):1338–1343.

Bergman, H., P. Björntorp, T.B. Conradson, M. Fahlén, J. Stenberg, and E. Varnauskas. 1973. Enzymatic and circulatory adjustments to physical training in middle aged men. *European Journal of Clinical Investigation* 3:414–418.

Bevegard, S., A. Holmgren, and B. Johnsson. 1963. Circulatory studies in well trained athletes at rest and during heavy exercise with special reference to stroke volume and the influence of body position. *Acta Physiologica Scandinavica* 57:26.

Billman, G.E., and J.P. Dujardin. 1990. Dynamic changes in cardiac vagal tone as measured by time-series analysis. *American Journal of Physiology* 258:H896–H902.

Blomqvist, G., and B. Saltin. 1983. Cardiovascular adaptations to physical training. *Annual Review of Physiology* 45:169–189.

Bolme, P., and J. Novotny. 1969. Conditional reflex activation of the sympathetic cholinergic vasodilator nerves in skeketal muscle. *Acta Physiologica Scandinavica* 21:123–132.

Bolter, C.P., E.W. Banister, and A.K. Singh. 1986. Intrinsic rates and adrenergic responses of atria from rats on sprinting, endurance, and walking exercise programs. *Australian Journal of Experimental Biology and Medical Science* 64(3):251–256.

Bonaduce, D., M. Petretta, V. Cavallaro, C. Apicella, A. Ianniciello, M. Romano, R. Breglio, and F. Marciano. 1998. Intensive training and cardiac autonomic control in high level athletes. *Medicine and Science in Sports and Exercise* 30(5):691–696.

Borow, K.M., A. Neumann, F.W. Arensman, and M.H. Yacoub. 1985. Clinical evidence for differential sensitivity of alpha and beta adrenergic receptors after cardiac transplantation. *Circulation* 71:866–872.

Bouchard, C., R. Lesage, G. Lortie, J.A. Simoneau, P. Hamel, M.R. Boulay, L. Perusse, G. Theriault, and C. Leblanc. 1986. Aerobic performance in brothers, dizygotic, and monozygotic twins. *Medicine and Science in Sports and Exercise* 18:639–646.

Bove, A.A., J.D. Dewey, and G.M. Tyce. 1984. Increased conjugated dopamine in plasma after exercise training. *Journal of Laboratory Clinical Medicine* 104(1):77–85.

Brandao, M.U.P., M. Wajngarten, E. Rondon, C.P. Giorgi, F. Hironaka, and C.E. Negrao. 1993. Left ventricular function during dynamic exercise in untrained and moderately trained subjects. *Journal of Applied Physiology* 75:1989–1995.

Bristow, M.R., R. Ginsburg, W. Minobe, et al. 1982. Decreased catecholamine sensitivity and β-receptor density in failing human hearts. *New England Journal of Medicine* 307:205–211.

Brooks, G.A. 1991. Current concepts in lactate exchange. *Medicine and Science in Sports and Exercise* 23:895–906.

Brorson, L., T.B. Conradson, B. Olsson, and E. Varnauskas. 1976. Right atrial monophasic action potential and effective refractory periods in relation to physical training and maximal heart rate. *Cardiovascular Research* 10(2):160–168.

Butler, J., M. O'Brian, K. O'Malley, and J.G. Kelle. 1982. Relationship of β-receptor density to fitness in athletes. *Nature* 298:60.

Caffrey, J.L., Z. Mateo, L.D. Napier, J.F. Gaugl, and B.A. Barron. 1995. Intrinsic cardiac enkephalins inhibit vagal bradycardia in the dog. *American Journal of Physiology* 268:H848–H855.

Callister, R., A.V. Ng, and D.R. Seals. 1994. Arm muscle sympathetic nerve activity during preparation for and initiation of leg cycling exercise in humans. *Journal of Applied Physiology* 77:1403–1410.

Cameron, J.D., and A.M. Dart. 1994. Exercise training increases total systemic arterial compliance. *American Journal of Physiology* 266:H693–H701.

Chen, C., S.E. DiCarlo, and T.J. Scislo. 1995. Daily spontaneous running attenuated the central gain of the arterial baroreflex. *American Journal of Physiology* 268: H662–H669.

Cherdchu, C., L.A. Robinson, and T.D. Hexum. 1987. Proenkephalin-A derived peptides do not modulate cardiovascular effects of epinephrine on the isolated rat atrial preparations. *Neuropeptides* 10(3):299–312.

Chitravanshi, V.C., and F.R. Calaresu. 1992. Dopamine microinjected into the nucleus ambiguus elicits vagal bradycardia in spinal rats. *Brain Research* 583(1–2):308–311.

Clausen, J.P. 1976. Circulatory adjustments to dynamic exercise and effects of physical training in normal subjects and in patients with coronary artery disease. *Progress in Cardiovascular Disease* 18:459–495.

Clausen, J.P. 1977. Effect of physical training on cardiovascular adjustments to exercise in man. *Physiological Reviews* 57(4):779–814.

Clausen, J.P., K. Klausen, B. Rasmussen, and J. Trap-Jensen. 1973. Central and peripheral circulatory changes after training of the arms or legs. *American Journal of Physiology* 225:675–682.

Clausen, J.P., and J. Trap-Jensen. 1970. Effects of training on the distribution of cardiac output in patients with coronary artery disease. *Circulation* (17):611–624.

Collins, H., and S. DiCarlo. 1993. Cardiac afferents attenuate the muscle metaboreflex in the rat. *Journal of Applied Physiology* 75:114–120.

Convertino, V.A., L.C. Keil, and J.E. Greenleaf. 1983. Plasma volume, renin, and vasopressin responses to graded exercise after training. *Journal of Applied Physiology* 54:508–514.

Convertino, V.A., G.A. Mack, and E.R. Nadel. 1991. Elevated central venous pressure: A consequence of exercise training-induced hypervolemia. *American Journal of Physiology* 29: R273–R277.

Cousineau, D., R.J. Ferguson, J. De Champlain, P. Gauthier, P. Cote, and M. Bourassa. 1977. Catecholamines in coronary sinus during exercise in man before and after training. *Journal of Applied Physiology* 43:801–806.

Currens, J.H., and P.D. White. 1961. Half century of running: Clinical, physiologic, and autopsy findings in the case of Clarence de Mar "Mr. Marathoner". *New England Journal of Medicine* 265:988–993.

Davidson, W.R., Jr., S.P. Banerjee, and C.S. Liang. 1986. Dodutamine-induced cardiac adaptations: Comparisons with exercise-trained and sedentary rats. *American Journal of Physiology* 250:H725–H730.

Delp, M.D. 1998. Differential effects of training on the control of skeletal muscle perfusion. *Medicine and Science in Sports and Exercise* 30(3):361–374.

De Meersman, R.E. 1993. Heart rate variability and aerobic fitness. *American Heart Journal* 126:726–731.

Di Bello, V., G. Santoro, L. Talarico, C. Di Muro, M.T. Caputo, D. Giorgi, A. Bertini, M. Bianchi, and C. Giusti. 1996. Left ventricular function during exercise in athletes and in sedentary men. *Medicine and Science in Sports and Exercise* 28(2):190–196.

DiCarlo, S.E., and V.S. Bishop. 1988. Exercise training attenuates baroreflex regulation of nerve activity in rabbits. *American Journal of Physiology* 255:H974–H979.

DiCarlo, S.E., and V.S. Bishop. 1990. Regional vascular resistance during exercise: Role of cardiac afferents and exercise training. *American Journal of Physiology* 258:H842–H847.

Dickhuth, H-H., M. Lehman, W. Auch-Schwelk, T. Meinertz, and J. Keul. 1987. Physical training, vegetative regulation, and cardiac hypertrophy. *Journal of Cardiovascular Pharmacology* 10(suppl 6):S71–S78.

Dixon, E.M., M.V. Kamath, N. McCartney, and E.L. Fallen. 1992. Neural regulation of heart rate variability in endurance athletes and sedentary controls. *Cardiovascular Research* 26:713–719.

Douglas, P.A., M.L. O'Toole, D.B. Hiller, and N. Reichek. 1986. Left ventricular structure and function by echocardiography in ultraendurance athletes. *American Journal of Cardiology* 58:805–809.

Eckblom, B., A.N. Goldbarg, Å. Kilbom, and P.O. Åstrand. 1972. Effects of atropine and propranolol on the oxygen transport system in man. *Scandinavian Journal of Clinical Laboratory Investigation* 30:35–42.

Eckblom, B., Å. Kilbom, and J. Soltysiak. 1973. Physical training, bradycardia, and autonomic nervous system. *Scandinavian Journal of Clinical Laboratory Investigation* 32:251–256.

Fellmann, N. 1992. Hormonal and plasma volume alterations following endurance exercise. *Sports Medicine* 13:3749.

Fisher, A.G., T.D. Adams, F.G. Yanowitz, J.D. Ridges, and G. Orsmond. 1989. Noninvasive evaluation of world class athletes engaged in different modes of training. *American Journal of Cardiology* 63:337–341.

Fortney, S.M., E.R. Nadel, C.B. Wenger, and J.R. Bove. 1981. Effect of acute alteration of blood volume on circulatory performance in humans. *Journal of Applied Physiology* 50:292–298.

George, K.P., L.A. Wolfe, G.W. Burggraf, and R. Norman. 1995. Electrocardiographic and echocardiographic characteristics of female athletes. *Medicine and Science in Sports and Exercise* 27(10):1362–1370.

Gilbert, C. 1995. Optimal physical performance in athletes: Key roles of dopamine in a specific neurotransmitter/hormonal mechanism. *Mechanisms of Ageing and Development* 84(2):83–102.

Gleser, M.A. 1973. Effects of hypoxia and physical training on hemodynamic adjustments to one-legged exercise. *Journal of Applied Physiology* 34:655–659.

Goldsmith, R.L., J.T. Bigger, Jr., R.C. Steinman, and J.L. Fleiss. 1992. Comparison of 24 hour parasympathetic activity in endurance-trained and untrained men. *Journal of the American College of Cardiology* 20:552–558.

Grassi, G., G. Seravalle, D.A. Calhoun, and G. Mancia. 1994. Physical training and baroreceptor control of sympathetic nerve activity in humans. *Hypertension* 23:294–301.

Green, D.J., N.T. Cable, C. Fox, J.M. Rankin, and R.R. Taylor. 1994. Modification of forearm resistance vessels by exercise training in young men. *Journal of Applied Physiology* 77:1829–1833.

Green, H.J., S. Jones, M. Ball-Burnett, and I. Fraser. 1991. Early adaptations in blood substrates, metabolites, and hormones to prolonged exercise training in man. *Canadian Journal of Physiology and Pharmacology* 69:1222–1229.

Gregoire, J., S. Tuck, Y. Yamamoto, and R.L. Hughson. 1996. Heart rate variability at rest and exercise: Influence of age, gender, and physical training. *Canadian Journal of Applied Physiology* 21(6):455–470.

Grimby, G., E. Häggendal, and B. Saltin. 1967. Local xenon 133 clearance from the quadriceps muscle during exercise in man. *Journal of Applied Physiology* 22:305–310.

Häggendal, J., J.U.H. Hartley, and B. Saltin. 1970. Arterial noradrenaline concentration during exercise in relation to the relative work levels. *Scandinavian Journal of Clinical and Laboratory Investigation* 26(4):337–342.

Hammond, H.K., L.A. Ransnas, and P.A. Insel. 1988. Noncoordinate regulation of cardiac Gs protein and beta-adrenergic receptors by a physiological stimulus, chronic dynamic exercise. *Journal of Clinical Investigation* 82(6):2168–2171.

Hanel, B., I. Teunissen, A. Rabol, J. Warberg, and N.H. Secher. 1997. Restricted postexercise pulmonary diffusing capacity and central blood volume depletion. *Journal of Applied Physiology* 83:11–17.

Hartley, L.H., H.G. Grimby, A. Kilbom, et al. 1969. Physical training in sedentary middle-aged and older men. III. Cardiac output and gas exchange at submaximal and maximal exercise. *Scandinavian Journal of Clinical Lab Investigation* 24:335–344.

Haskell, W.L., C.S. Sims, J. Myll, W.M. Bortz, F.G. St. Goar, and E.L. Alderman. 1993. Coronary artery size and dilating capacity in ultra-distance runners. *Circulation* 87:1076–1082.

Hatfield, B.D., T.W. Spaulding, D. Laine Santa Maria, S.W. Porges, J.T. Potts, E.A. Byrne, E.B. Brody, and A.D. Mahon. 1998. Respiratory sinus during exercise in aerobically trained and untrained men. *Medicine and Science in Sports and Exercise* 30(2):206–214.

Hughson, R.L., J.R. Sutton, J.D. Fitzgerald, and N.L. Jones. 1977. Reduction of intrinsic sinoatrial frequency and norepinephrine response of the exercised rat. *Canadian Journal of Physiology and Pharmacology* 55(4):813–820.

Huonker, M., M. Halle, and J. Keul. 1996. Structural

and functional adaptations of the cardiovascular system by training. *International Journal of Sports Medicine* 17:S164–S172.

Huonker, M., A. Schmid, S. Sorichter, A. Schmidt-Trucksab, P. Mrosek, and J. Keul. 1998. Cardiovascular differences between sedentary and wheelchair-trained subjects with paraplegia. *Medicine and Science in Sports and Exercise* 30(4):609–613.

Ishida, R., and M. Okada. 1997. Spectrum analysis of heart rate variability for the assessment of training effects. *Rinsho Byori* 45(7):685–688.

Jensen-Urstad, M., F. Bouvier, M. Nejat, B. Saltin, and L.A. Brodin. 1998. Left ventricular function in endurance runners during exercise. *Acta Physiologica Scandinavica* 164(2):167–172.

Johnson, J.M. 1998. Physical training and the control of skin blood flow. *Medicine and Science in Sports and Exercise* 30(3):382–386.

Kanstrup, I-L., and B. Ekblom. 1978. Influence of age and physical activity on central hemodynamics and lung function in active adults. *Journal of Applied Physiology* 45(5):709–717.

Kanstrup, I-L., J. Marving, and P.F. Høilund-Carlsen. 1992. Acute plasma expansion: Left ventricular hemodynamics and endocrine function during exercise in healthy subjects. *Journal of Applied Physiology* 73:1791–1796.

Katona, P.G., M. McLean, D.H. Dighton, and A. Guz. 1982. Sympathetic and parasympathetic cardiac control in athletes and non-athletes at rest. *Journal of Applied Physiology* 52:1652–1657.

Kavanaugh, T., M.H. Yacoub, D.J. Mertens, J. Kennedy, and R.B. Campbell. 1988. Cardiorespiratory responses to exercise training after orthotopic cardiac transplantation. *Circulation* 77:162–171.

Kenney, W.L. 1985. Parasympathetic control of resting heart rate: Relationship to aerobic power. *Medicine and Science in Sports and Exercise* 17:451–455.

Keul, J., D. Konig, M. Huonker, M. Halle, B. Wohlfahrt, and A. Berg. 1996. Adaptation to training and performance in elite athletes. *Research Quarterly for Exercise and Sport* 67(suppl):S29–S36.

Kingwell, B.A., B. Sherrard, G.L. Jennings, and A.M. Dart. 1997. Four weeks of cycle training increases basal production of nitric oxide from the forearm *American Journal of Physiology* 272(3):H1070–H1077.

Kjaer, M. 1998. Adrenal medulla and exercise training. *European Journal of Applied Physiology* 77(3):195–199.

Klausen, K., N.H. Secher, J.P. Clausen, O. Hartling,

and J. Trap-Jensen. 1982. Central and regional circulatory adaptations to one-leg training. *Journal of Applied Physiology* 52:976–983.

Kool, M.J., H.A. Struyker-Boudier, J.A. Wijnen, A.P. Hoeks, and L.M. Van Bortel. 1992. Effects of diurnal variability and exercise training on properties of large arteries. *Journal of Hypertension* 10(suppl 6):S49–S52.

Krip, B., N. Gledhill, V. Jamnik, and D. Warburton. 1997. Effect of alterations in blood volume on cardiac function during maximal exercise. *Medicine and Science in Sports and Exercise* 29(11):1469–1476.

Kronenberg, M.W., J.P. Uetrecht, W.D. Dupont, M.H. Davis, B.K. Phelan, and G.C. Friesinger. 1988. Intrinsic left ventricular contractility in normal subjects. *American Journal of Cardiology* 61:621–627.

Lash, J.M. 1998. Training-induced alterations in contractile function and excitation-contraction coupling in vascular smooth muscle. *Medicine and Science in Sports and Exercise* 30(1):60–66.

Laughlin, M.H., R.J. Korthuis, D.J. Duncker, and R.J. Backe. 1996. Control of blood flow to cardiac and skeletal muscle during exercise. In *Handbook of Physiology, Exercise: Regulation and Integration of Multiple Systems* edited by L.B. Rowell and J.T. Shepherd. Bethesda, MD: American Physiological Society. 705–769.

Laughlin, M.H., and R.M. McAllister. 1992. Exercise training-induced coronary vascular adaptation. *Journal of Applied Physiology* 73:2209–2225.

Laughlin, M.H., C. Oltman, and D. Bowles. 1998. Exercise training-induced adaptations in the coronary circulation. *Medicine and Science in Sports and Exercise* 30(3):352–360.

Lehman, M., J. Keul, G. Huber, and M. Da Prada. 1981. Plasma catecholamines in trained and untrained volunteers during graduated exercise. *International Journal of Sports Medicine* 2(3):143–147.

Levine, B. 1993. Regulation of central blood volume and cardiac filling in endurance athletes: The Frank-Starling mechanism as a determinant of orthostatic tolerance. *Medicine and Science in Sports and Exercise* 23:727–732.

Levine, B., J.C. Buckey, J.M. Fritsch, C.W. Yancy, Jr., D.E. Watenpaugh, P.G. Snell, L.D. Lane, D. Eckberg, and C.G. Blomqvist. 1991. Physical fitness and cardiovascular regulation: Mechanisms of orthostatic intolerance. *Journal of Applied Physiology* 70:112–122.

Lewis, S.F., E. Nylander, P. Gad, and N.H. Areskog. 1980. Non-autonomic component in bradycardia of endurance trained men at rest and during exercise. *Acta Physiologica Scandinavica* 109:297–305.

Lewis, S.F., P. Thompson, N.H. Areskog, et al. 1980. Endurance training and heart rate control studied by combined parasympathetic and beta adrenergic blockade. *International Journal of Sports Medicine* 1:42–49.

Mack, G.W., V.A. Convertino, and E.R. Nadel. 1993. Effect of exercise training on cardiopulmonary baroreflex control of forearm vascular resistance in humans. *Medicine and Science in Sports and Exercise* 25:722–726.

Mack, G.W., S. Xiangrong, H. Nose, A. Tripathi, and E. Nadel. 1987. Diminished baroreflex control of forearm vascular resistance in physically fit humans. *Journal of Applied Physiology* 63:105–110.

Mackie, G.B., and R.L. Terjung. 1983. Influence of training on blood flow to different skeletal muscle fiber types. *Journal of Applied Physiology* 55:1072–1078.

MacRae, H.H-S., S.C. Dennis, A.N. Bosch, and T.D. Noakes. 1992. Effects of training on lactate production and removal during progressive exercise in humans. *Journal of Applied Physiology* 72(5):1649–1656.

MacRae, P.G., W.W. Spirduso, G.D. Cartee, R.P. Farrar, and R.E. Wilcox. 1987. Endurance training and striatal dopamine metabolite levels. *Neuroscience Letters* 79(1-2):138–144.

Madsen, P., M. Klokker, H.L. Olesen, and N.H. Secher. 1995. Naloxone-provoked vaso-vagal response to head-up tilt in man. *European Journal of Applied Physiology* 70(3):246–251.

Magnusson, G., A. Gordon, L. Kaijser, C. Sylven, B. Isberg, J. Karpakka, and B. Saltin. 1996. High intensity knee extensor training, in patients with chronic heart failure: Major skeletal muscle improvement. *European Heart Journal* 17(7):1048–1055.

Magnusson, G., L. Kaijser, C. Sylven, K.E. Karlberg, B. Isberg, and B. Saltin. 1997. Peak skeletal muscle perfusion is maintained in patients with heart failure when only a small muscle mass is exercised. *Cardiovascular Research* 33(2):297–306.

Mann, G.V., A. Spoerry, M. Gray, and D. Jarashow. 1972. Atherosclerosis in the Masai. *American Journal of Epidemiology* 95:26–37.

Martin, W., J. Montgomery, P.G. Snell, J.R. Corbett, J.J. Sokolov, J.C. Buckey, D.A. Maloney, and C.G. Blomqvist. 1987. Cardiovascular adaptations to intense swim training in sedentary middle aged men and women. *Circulation* 75:323–330.

Martin, W.H., W.M. Kohrt, M.T. Malley, E. Korte, and S. Stolz. 1990. Exercise training enhances leg vasodilatory capacity of 65-yr-old men and women. *Journal of Applied Physiology* 69:1804–1809.

Martin, W.H., T. Ogawa, W.M. Kohrt, M.T. Malley, E. Korte, P.S. Keifer, and K.B. Schechtman. 1991. Effects of aging, gender, and physical training on peripheral vascular function. *Circulation* 84:654–664.

McAllister, R.M. 1998. Adaptations in control of blood flow with training: Splanchnic and renal blood flows. *Medicine and Science in Sports and Exercise* 30(3):375–381.

Meintjes, A.F., A.C. Nobrega, I.E. Fuchs, A. Ally, and L.B. Wilson. 1995. Attenuation of the exercise pressor reflex: Effect of opioid agonist on substance P release in L-7 dorsal horn of cats. *Circulation Research* 77(2):326–334.

Meredith, I.T., P. Friberg, G.L. Jennings, T. Ogawa, W.M. Kohrt, M.T. Malley, E. Korte, P.S. Kieffer, W.H. Martin, and M.D. Esler. 1992. Regular endurance training lowers resting renal but not cardiac sympathetic activity. *Hypertension* 18:575–582.

Mier, C.M., M.J. Turner, A.A. Ehsani, and R.J. Spina. 1997. Cardiovascular adaptations to 10 days of cycle exercise. *Journal of Applied Physiology* 83(6):1900–1906.

Mitchell, J.H., B.J. Sproule, and C.B. Chapman. 1958. The physiological meaning of the maximal oxygen uptake test. *Journal of Clinical Investigation* 37:538–547.

Miura, T., V. Bhargava, B.D. Guth, K.S. Sunnerhagen, S. Miyazaki, C. Indolfi, and K.L. Peterson. 1993. Increased afterload intensifies asynchronous wall motion and impairs ventricular relaxation. *Journal of Applied Physiology* 75(1):389–396.

Mohiaddin, R.H., S.R. Underwood, H.G. Bogren, D.N. Firmin, R.S.O. Reese, and D.B Longmore. 1989. Regional aortic compliance studied by magnetic resonance imaging: The effect of age, training, and coronary artery disease. *British Heart Journal* 62:90–96.

Muller, J.M., P.R. Myers, and M.H. Laughlin. 1993. Exercise training alters myogenic responses in porcine coronary resistance arteries. *Journal of Applied Physiology* 75:2677–2682.

Muller, J.M., P.R. Myers, and M.H. Laughlin. 1994. Vasodilator responses of coronary resistance arteries of exercise-trained pigs. *Circulation* 89:2308–2314.

Musch, T.I., G.C. Haidet, G.A. Ordway, J.C. Longhurst, and J.H. Mitchell. 1987. Training effects on regional blood flow response to maximal exercise in foxhounds. *Journal of Applied Physiology* 62:1724–1732.

Musha, T., E. Satoh, H. Koyanagawa, T. Kimura, and S. Satoh. 1989. Effects of opioid agonists on sympathetic and parasympathetic transmission to the dog heart. *Journal of Pharmacology and Experimental Therapy* 250(3):1087–1091.

Negrao, C.E., M.C. Irigoyen, E.D. Moreira, P.C. Brum, P.M. Freire, and E.M. Krieger. 1993. Effect of exercise training on RSNA, baroreflex control, and blood pressure responsiveness. *American Journal of Physiology* 265:R365–R370.

Negrao, C.E., E.D. Moreira, M.C.L.M. Santos, V.M.A. Farah, and E.M. Krieger. 1992. Vagal function impairment after exercise training. *Journal of Applied Physiology* 72:1749–1753.

Ng, A.V., R. Callister, D.G. Johnson, and D.R. Seals. 1994. Endurance training is associated with increased basal sympathetic nerve activity in healthy older humans. *Journal of Applied Physiology* 77:1366–1374.

Nixon, J.V., A.R. Wright, T.R. Porter, V. Roy, and J.A. Arrowood. 1991. Effects of exercise on left ventricular diastolic performance in trained athletes. *American Journal of Cardiology* 68:945–949.

Nylander, E. 1985. Training-induced bradycardia in rats on cardio-selective and non-selective beta receptor blockade. *Acta Physiologica Scandinavica* 123(2):147–149.

Nylander, E., and N.H. Areskog. 1982. New aspects on training bradycardia. *Ups Journal of Medical Science* 87(1):1–10.

Nylander, E., and U. Dahlström. 1984. Influence of long-term beta-receptor stimulation with prenalterol on intrinsic heart rate in rats. *European Journal of Applied Physiology* 53(1):48–52.

Ordway, G.A., J.B. Charles, D.C. Randall, G.E. Billman, and D.R. Wekstein. 1982. Heart rate adaptation to exercise training in cardiac-denervated dogs. *Journal of Applied Physiology* 52:1586–1590.

Overholser, K.A., H.M. Laughlin, and M.J. Bhatte. 1994. Exercise training-induced increase in coronary transport capacity. *Medicine and Science in Sports and Exercise* 26:1239–1244.

Parker, J.L., C.L. Oltman, J.M. Muller, P.R. Myers, H.R. Adams, and H. Laughlin. 1994. Effects of exercise training on regulation of tone in coronary arteries and arterioles. *Medicine and Science in Sports and Exercise* 26:1252–1261.

Pawelczyk, J., W. Kenney, and P. Kenney. 1988. Cardiovascular responses to head-up tilt after an endurance exercise program. *Aviation Space Environmental Medicine* 59(2):107–112.

Peronnet, F., J. Cleroux, H. Perrault, D. Cousineau, J. de Champlain, and R. Nadeau. 1981. Plasma norepinephrine response to exercise before and after training in humans. *Journal of Applied Physiology* 51(4):812–815.

Piepoli, M., A.L. Clark, M. Volterrani, S. Adamopoulos, P. Sleight, and A.J. Coats. 1996. Contribution of muscle afferents to the hemodynamic, autonomic, and ventilatory responses to exercise in patients with chronic heart failure: Effects of physical training. *Circulation* 93(5):940–952.

Prud'homme, D., C. Bouchard, C. Leblanc, F. Landry, and E. Fontaine. 1984. Sensitivity of maximal aerobic power to training is genotype-dependent. *Medicine and Science in Sports and Exercise* 16(5):489–493.

Raven, P.B., and J. Pawelczyk. 1993. Chronic endurance exercise training: A condition of inadequate blood pressure regulation and reduced tolerance to LBNP. *Medicine and Science in Sports and Exercise* 25:713–721.

Ray, C.A., and K.M. Hume. 1998. Sympathetic neural adaptations to exercise training in humans: Insights from microneurography. *Medicine and Science in Sports and Exercise* 30(3):387–391.

Ray, C.A., W.M. Pace, and M.P. Clary. 1992. Sympathetic and cardiovascular adaptations to one-legged exercise training (abstract). *Physiologist* 34:235.

Reading, J.L., J.M. Goodman, M.J. Plyley, J.S. Floras, P.P. Liu, P.R. McLaughlin, and R.J. Shephard. 1993. Vascular conductance and aerobic power in sedentary and active subjects and heart failure patients. *Journal of Applied Physiology* 74:567–573.

Robinson, B.F., S.E. Epstein, G.D. Beiser, and E. Braunwald. 1966. Control of heart rate by the automatic nervous system. *Circulation Research* 19:400–411.

Rogers, P.J., T.D. Miller, B.A. Bauer, J.M. Brum, A.A. Bove, and P.M. Vanhoute. 1991. Exercise training and responsiveness of isolated coronary arteries. *Journal of Applied Physiology* 71:2346–2351.

Rowell, L.B., J.R. Blackmon, and R.A. Bruce. 1964. Indocyanine green clearance and estimated hepatic blood flow during mild to maximal exercise in upright man. *Journal of Clinical Investigation* 43:1677–1690.

Sacknoff, D.M., D.W. Gleim, N. Stachenfeld, and N.L. Coplan. 1994. Effect of athletic training on heart rate variability. *American Heart Journal* 127:1275–1278.

Saito, M., H. Matsui, and M. Miyamura. 1980. Effects of training on the calf and thigh blood flows. *Japanese Journal of Physiology* 30:955–959.

Saltin, B. 1977. The interplay between peripheral and central factors in the adaptive response to exercise and training. *Annals of the New York Academy of Sciences* 301:224–231

Saltin, B. 1985. Hemodynamic adaptions to exercise. *American Journal of Cardiology* 55:42D–47D.

Saltin B. 1986. Physiological adaptation to physical conditioning: Old problems revisited. *Acta Medica Scandinavica* suppl 711:11–24.

Saltin, B., and G. Blomqvist. 1983. Cardiovascular adaptations to physical training. *Annual Review of Physiology* 45:169–189

Saltin, B., G. Blomqvist, J.H. Mitchell, R.L. Johnson, Jr., K. Wildenthal, and C.B. Chapman. 1968. Response to exercise after bedrest and after training. *Circulation* 38(5 suppl. 7):1–55.

Saltin, B., and P.D. Gollnick. 1983. Skeletal muscle adaptability: Significance for metabolism and performance. In *Handbook of Physiology: Skeletal Muscle* edited by L. D. Peachey. New York: The American Physiological Society, Oxford University Press. 555–631.

Saltin, B., L.H. Hartley, A. Kilbom, et al. 1969. Physical training in sedentary middle-aged and older men. II. Oxygen uptake, heart rate and blood lactate concentration at submaximal and maximal exercise. *Scandinavian Journal of Clinical Lab Investigation* 24:323–334.

Saltin, B., B. Kiens, G. Savard, and P.K. Pedersen. 1986. Role of hemoglobin and capillarization for oxygen delivery and extraction in muscular exercise. *American Journal of Physiology* 128:21–32.

Saltin, B., K. Nazar, D.L. Costill, E. Stein, E. Jansson, B. Esson, and P.D. Gollnick. 1976. The nature of the training response: Peripheral and central adaptations to one-legged exercise. *Acta Physiologica Scandinavica* 96:289–305.

Saltin, B., and L.B. Rowell. 1980. Functional adaptations to physical activity and inactivity. *Federation Proceedings* 39:1506–1513.

Schaefer, M.E., J.A. Allert, H.R. Adams, and M.H. Laughlin. 1992. Adrenergic responsiveness and intrinsic sioatrial automaticity of exercise-trained rats. *Medicine and Science in Sports and Exercise* 24:887–894.

Schairer, J.R., P.D. Stein, S. Keteyian, F. Fedel, J. Ehrmann, M. Alam, J.W. Henry, and T. Shaw. 1992. Left ventricular response to submaximal exercise in endurance-trained athletes and sedentary adults. *American Journal of Cardiology* 70:930–933.

Scheuer, J., and C.M. Tipton. 1977. Cardiovascular adaptations to physical training. *Annual Review of Physiology* 39:221–251.

Seals, D.R., and P.B. Chase. 1989. Influence of physical training on heart rate variability and baroreflex circulatory control. *Journal of Applied Physiology* 66:1886–1895.

Secher, N.H., J.P. Clausen, K. Klausen, I. Noer, and J. Trap-Jensen. 1977. Central and regional effects of adding arm exercise to leg exercise. *Acta Physiologica Scandinavica* 100:288–297.

Sessa, W.C., K. Pritchard, N. Seyedi, J. Wang, and T.H. Hintze. 1994. Chronic exercise in dogs increases coronary vascular nitric oxide production and endothelial cell nitric oxide synthase gene expression. *Circulation Research* 74:349–353.

Shapiro, L.M., and R.G. Smith. 1983. Effect of training on left ventricular structure and function. *British Heart Journal* 50:534–539.

Sheldahl, L.M., T.J. Ebert, B. Cox, and F.E. Tristani. 1994. Effect of aerobic training on baroreflex regulation of cardiac and sympathetic function. *Journal of Applied Physiology* 76:158–165.

Shenberger, J.S., G.J. Leaman, M.M. Neumyer, T.I. Musch, and L.I. Sinoway. 1990. Physiologic and structural indices of vascular function in paraplegics. *Medicine and Science in Sports and Exercise* 22:96–101.

Shi, X., J.M. Andresen, J.T. Potts, B.H. Foresman, S.A. Stern, and P.B. Raven. 1993. Aortic baroreflex control of heart rate during hypertensive stimuli: Effect of fitness. *Journal of Applied Physiology* 74(4):1555–1562.

Shi, X., K.M. Gallagher, S.A. Smith, K.H. Bryant, and P.B. Raven. 1996. Diminished forearm vasomotor response to central hypervolemic loading in aerobically fit individuals. *Medicine and Science in Sports and Exercise* 28(11):1388–1395.

Shi, X., J.T. Potts, P.B. Raven, and B.H. Foresman. 1995. Aortic-cardiac reflex during dynamic exercise. *Journal of Applied Physiology* 78(4):1569–1574.

Shi, X., G.H.J. Stevens, B. Foresman, S. Stern, and P.B. Raven. 1995. Autonomic nervous system control of the heart: Endurance exercise training. *Medicine and Science in Sports and Exercise* 27(10):1406–1413.

Shoemaker, J.K., H.J. Green, M. Ball-Burnett, and S. Grant. 1998. Relationships between fluid and electrolyte hormones and plasma volume during exercise with training and detraining. *Medicine and Science in Sports and Exercise* 30(4):497–505.

Sigvardsson, K., E. Svanfeldt, and Å. Kilbom. 1977. Role of adrenergic nervous system in development of training-induced bradycardia. *Acta Physiologica Scandinavica* 101:481–488.

Silber, D., D. McLaughlin, and L. Sinoway. 1991. Leg exercise conditioning increases peak forearm blood flow. *Journal of Applied Physiology* 71:1568–1573.

Sinoway, L.I., T.I. Musch, J.R. Minotti, and R. Zelis.

1986. Enhanced maximal metabolic vasodilation in the dominant forearms of tennis players. *Journal of Applied Physiology* 61:673–678.

Sinoway, L.I., J. Shenberger, G. Leaman, et al. 1996. Forearm training attenuates sympathetic responses to prolonged rhythmic forearm exercise. *Journal of Applied Physiology* 81:1778–1784.

Sinoway, L.I., J. Shenberger, J. Wilson, D. McLaughlin, T. Musch, and R. Zelis. 1987. A 30-day forearm work protocol increases maximal forearm blood flow. *Journal of Applied Physiology* 62:1063–1067.

Slavik, K., and J. LaPointe. 1993. Involvement of inhibitory dopamine-2 receptors in resting bradycardia in exercise-conditioned rats. *Journal of Applied Physiology* 74:2086–2091.

Smith, M.L., H.M. Graitzer, D.L. Hudson, and P.B. Raven. 1988. Baroreflex function in endurance- and static-exercise trained men. *Journal of Applied Physiology* 64(2):585–591.

Smith, M.L., D.L. Hudson, H.M. Graitzer, and P.B. Raven. 1989. Exercise training bradycardia: The role of autonomic balance. *Medicine and Science in Sports and Exercise* 21(1):40–44.

Snell, P.G., W.H. Martin, J.C. Buckey, and C. Gunnar Blomqvist. 1987. Maximal vascular leg conductance in trained and untrained men. *Journal of Applied Physiology* 62:606–610.

Somers, V.K., K.C. Leo, R. Shields, M. Cleary, and A. Mark. 1992. Forearm endurance training attenuates sympathetic nerve response to isometric handgrip in normal humans. *Journal of Applied Physiology* 72:1039–1043.

Spina, R.J., T. Ogawa, W.H. Martin, A.R. Coggan, J. Holloszy, and A.A. Ehsani. 1992. Exercise training prevents decline in stroke volume during exercise in young healthy subjects. *Journal of Applied Physiology* 72:2458–2462.

Squires, R.W., P.R. Arthur, G.T. Gau, A. Muri, and W.B. Lambert. 1983. Exercise after cardiac transplantation: A report of two cases. *Journal of Cardiac Rehabilitation* 3:570–574.

Stegemann, J., and T. Kenner. 1971. A theory of heart rate control by muscular metabolic receptors. *Archives Kreislauffosch* 64(3):185–214.

Stewart, J.M., X. Xu, M. Ochoa, and J.H. Hintze. 1998. Exercise decreases epicardial coronary artery wall stiffness: Roles of cGMP and cAMP. *Medicine and Science in Sports and Exercise* 30(2):220–228.

Sullivan, M.J., J.D. Knight, M.B. Higgenbotham, and F.R. Cobb. 1989. Relation between central and peripheral hemodynamics during exercise in patients with chronic heart failure: Muscle blood flow is reduced with maintenance of arterial perfusion pressure. *Circulation* 80:769–781.

Svedenhag, J., A. Martinsson, B. Eckblom, and P. Hjemdahl. 1986. Altered cardiovascular responsiveness to adrenaline in endurance-trained subjects. *Acta Physiologica Scandinavica* 126:539–550.

Svedenhag, J., B.G. Wallin, G. Sundlof, and J. Henriksson. 1984. Skeletal muscle sympathetic activity at rest in trained and untrained subjects. *Acta Physiologica Scandinavica* 120:499–504.

Swaine, I.L., R.J. Linden, and D.A.S.G. Mary. 1994. Loss of exercise training-induced bradycardia with continued improvement in fitness. *Journal of Sport Sciences* 12:477–481.

Taylor, H.L., R. Buskirk, and A. Henschel. 1955. Maximal oxygen uptake as an objective measure of cardiorespiratory performance. *Journal of Applied Physiology* 8:73–80.

Tohmeh, J.F., and P.E. Cryer. 1980. Biphasic adrenergic modulation of beta adrenergic receptors in man. *Journal of Clinical Investigation* 65:836–840.

Tulppo, M.P., T.H. Makikallio, T. Seppanen, R.T. Laukkanen, and H.V. Huikuri. 1998. Vagal modulation of heart rate during exercise: Effects of age and physical fitness. *American Journal of Physiology* 274:H424–H429.

Underwood, F.B., M.H. Laughlin, and M. Sturek. 1994. Altered control of calcium in coronary smooth muscle cells by exercise training. *Medicine and Science in Sports and Exercise* 26:1230–1238.

Vaitkevicius, P.V., J.L. Fleg, J.H. Engel, F.C. O'Conner, J.G. Wright, L.E. Lakatta, F.C.P Yin, and E.G. Lakatta. 1993. Effects of age and aerobic capacity on arterial stiffness in healthy adults. *Circulation* 88:1456–1462.

Vanoverschelde, J.L.J., B. Essamri, R. Vanbutsele, A.M. D'Honte, J.R. Cosyns, J.M.R. Detry, and J.A. Melin. 1993. Contribution of left ventricular diastolic function to exercise capacity in normal subjects. *Journal of Applied Physiology* 74:2225–2233.

Varnauskas, E., P. Bjorntorp, M. Fahlen, I. Prerovsky, and J. Stenberg. 1970. Effects of physical training on exercise blood flow and enzymatic activity in skeletal muscle. *Cardiovascular Research* 4:418–422.

Voutilainen, S., M. Kupari, M. Hippelainen, K. Karpinnen, M. Ventila, and J. Heikkila. 1991. Factors influencing Doppler indexes of left ventricular filling in healthy persons. *American Journal of Cardiology* 68:653–659.

Wagner, P.D. 1992. Gas exchange and peripheral

diffusion limitation. *Medicine and Science in Sports and Exercise* 24:54–58.

Williams, R.S. 1985. Role of receptor mechanisms in the adaptive response to habitual exercise. *American Journal of Cardiology* 55(10):68D–73D.

Winder, W.W., R.C. Hickson, J.M. Hagberg, A.A. Ehsani, and J.A. McLane. 1979. Training-induced changes in hormonal and metabolic responses to submaximal exercise. *Journal of Applied Physiology* 46:766–771.

Wyatt, H.L., and J.H. Mitchell. 1978. Influences of physical conditioning and deconditioning on coronary vasculature of dogs. *Journal of Applied Physiology* 45:619–625.

Yasuda, Y., and M. Miyamura. 1983. Cross transfer effects of muscular training on blood flow in the ipsilateral and contralateral forearms. *European Journal of Applied Physiology* 51:321–329.

Chapter 19

Aging: Can the Effects Be Prevented?

Frederick R. Cobb, Martin J. Sullivan, Michael B. Higginbotham, and Dalane Kitzman

Assessment of the effects of age on cardiovascular function is confounded by the interaction of the normal aging process, the progressive development of cardiovascular disease with increasing age (Eiveback and Lie 1984; White, Edwards, and Dry 1950), and lifestyle factors that include the level of habitual physical activity. Although studies of the effects of age on cardiovascular function have attempted to exclude cardiovascular disease and physically inactive subjects, such screening may select a study group that does not adequately represent the typical aged population.

The following will provide an overview of the effects of age on

- cardiovascular anatomy,
- cardiovascular function at rest,
- cardiovascular function during exercise,
- exercise leg blood flow and metabolism,
- autonomic nervous system activity,
- the interactive effects of age and hypertension on cardiovascular function, and
- the extent to which regular exercise and aggressive treatment of hypertension may prevent age-related effects on cardiovascular function.

Effects of Age on Cardiovascular Anatomy

Postmortem studies have shown that with increasing age, cardiac interstitial collagen and elastin increase. This is accompanied by diffuse foci of fibrosis located in the subendocardium and myocardium (Hutchins 1980; Kitzman and Edwards 1990; Lenkiewicz, Davies, and Rosen 1972), a decrease in a number of myocytes, an increase in myocyte size (Anversa et al. 1986; Pomerance 1975), and the appearance of lipofuscin pigment (Strehier et al.

1959) and amyloid deposits (Glenner 1980). Echocardiographic studies from the Baltimore Longitudinal Study on Aging (BLSA) (Gerstenblith et al. 1977; Stone and Norris 1966), which included a group of active community volunteers screened for absence of cardiovascular disease and for blood pressure < 141/90, demonstrated modest age-related left ventricular hypertrophy in men age 25–80 years controlled for body surface area. As illustrated in figure 19.1, left ventricular diastolic posterior wall thickness measured by an M-mode echocardiography and indexed to body size increased approximately 30% over this age range; left ventricular diastolic and systolic dimensions did not change. Kitzman and Edwards (1990), in an autopsy study, observed an increase in the indexed ventricular septal thickness in men and women. No significant increase occurred in right and left ventricular free wall thickness. The left ventricular hypertrophy was associated with mild left atrial enlargement. Fleg (1986) has suggested that the age-related development of left ventricular hypertrophy is related to the age-related increase in systolic blood pressure observed in multiple populations (see figure 19.2).

Effects of Age on Rest Left Ventricular Function

Rest systolic left ventricular function, as assessed by ejection fraction by radionuclide angiography (Port et al. 1980) or fractional shortening by echocardiography (Gerstenblith et al. 1977), is not altered by increasing age. In contrast, diastolic function as measured by an increase in early diastolic closure rate (E-F slope) of the mitral valve (Gerstenblith et al. 1977) by echocardiography and a decrease in early left ventricular filling velocity and increase in atrial filling velocity by Doppler echocardiography

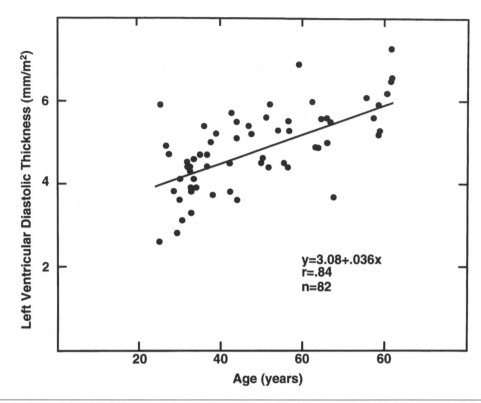

Figure 19.1   The effect of age on left ventricular diastolic posterior wall thickness determined by echocardiography in 62 healthy, normotensive men from the Baltimore Longitudinal Study on Aging. (Reproduced from Gerstenblith et al. 1977 with permission.)

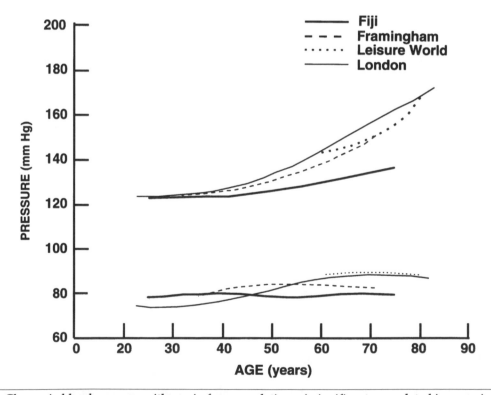

Figure 19.2   Change in blood pressure with age in four populations. A significant age-related increase in systolic but not in diastolic pressure occurs in the three industrialized populations. The effects are blunted in the nonindustrialized island of Fiji. (Reproduced from Berman 1982 with permission.)

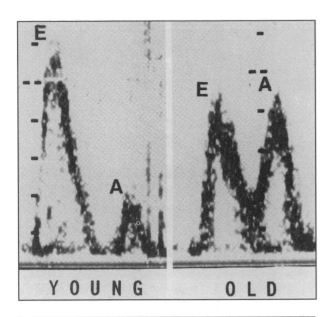

Figure 19.3 Noninvasive Doppler recording of early left ventricular filling velocity (E) and late filling velocity (A). (Reproduced from Kitzman, Sheikh, et al. 1991 with permission.)

decline in VO₂max from approximately 55 ml/kg/min to 20–25 ml/kg/min. This approximately represented a 60% total or 10% per decade reduction in maximum exercise performance with age (see figure 19.4).

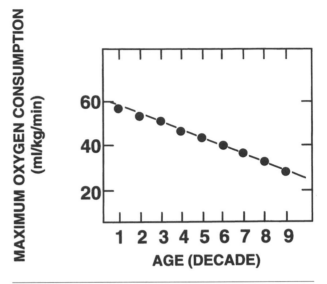

Figure 19.4 Effects of age on VO$_2$max. Data are recalculated from 17 studies and 700 observations from the literature, subjects age 10 to 90. Data are normalized for body weight. The data demonstrate an age-related decrease in VO$_2$max beginning in childhood and decreasing uniformly throughout life. (Reproduced from Dehn and Bruce 1972 with permission.)

(Kitzman, Sheikh, et al. 1991) is altered by increasing age. Figure 19.3 shows a noninvasive Doppler recording of early left ventricular filling velocity (E) and late filling velocity (A) that coincides with atrial systole (Kitzman, Sheikh, et al. 1991). In the younger subject (left), filling rate peaks in early diastolic and has a relatively low rate in late diastolic. In the older subject (right), early filling velocity is decreased and late filling velocity is increased consistent with a decrease in left ventricular diastolic relaxation and decrease in compliance. Multiple studies of rest cardiac function have demonstrated no change or a slight increase in heart rate, no change or a mild decrease in cardiac output and stroke volume, and an increase in systolic blood pressure and vascular resistance with increasing age (Brandponbrener, Landowne, and Shock 1955; Granath, Jonsson, and Strandell 1964; Higginbotham, Morris, Williams, Coleman, et al. 1986; Kutkka and Lansimies 1982; Rodeheffer et al. 1984).

Effects of Age on Exercise Cardiovascular Function

Dehn and Bruce (1972) combined data from 17 cross-sectional studies and 700 subjects to assess the relationship between maximum exercise cardiovascular function assessed by VO$_2$ (ml/kg/min) and age (10–90 years). They observed a progressive

What factors contribute to the age-related decrease in exercise and cardiovascular function? The determinants of VO$_2$max can be assessed by considering the components of the Fick equation during exercise:

VO = cardiac output (CO) × arteriovenous oxygen difference (AVO)

CO = stroke volume (SV) × heart rate

SV = end-diastolic volume (EDV) × ejection fraction (EF)

SV = EDV – end-systolic volume (ESV)

Our laboratory has evaluated these determinants of rest and exercise cardiovascular performance by using various techniques. These include breath-by-breath analyses of expired gases, right heart and pulmonary artery catheterization techniques to assess cardiac output, stroke volume by Fick equation, and thermodilution techniques. We also used pulmonary artery and capillary wedge pressure while simultaneously measuring radionuclide angiography to assess left ventricular ejection fraction (LVEF) and to calculate ventricular volume, EDV =

SV (Fick)/EF (RNA), during upright maximum bicycle ergometer exercise (Granath, Jonsson, and Strandell 1964; Higginbotham, Morris, Williams, McHale, et al. 1986; Kitzman et al. 1989; Sullivan, Cobb, and Higginbotham 1991). Measurements of cardiac output and stroke volume by Fick equation and thermodilution techniques and LVEF by radionuclide angiographic techniques are accurate and reproducible. However, we have observed excessive variability in the noninvasive assessment of absolute ventricular volumes by radionuclide angiography. In 24 healthy male volunteers age 20–50 years (Higginbotham, Morris, Williams, McHale, et al. 1986), we observed that VO_2 increased from 0.33 L/min to 2.55 L/min (a 7.7-fold increase) during maximum exercise. The increase in VO_2 resulted from a 3.2-fold linear increase in cardiac index from 3.0–9.7 L/min·M² (see figure 19.5a) and 2.5-fold increase in AVO_2 difference from 5.8 Vol% to 14.1 Vol%. The increase in cardiac output resulted from a 2.5-fold increase in heart rate and a 1.4-fold increase in stroke volume. Stroke volume increased during low levels of exercise primarily from an increase in end-diastolic volume, the Frank Starling mechanism (see figure 19.5b), and small increases in pulmonary capillary wedge pressure. Exercise end-diastolic volume increased to peak values at approximately 40% of VO_2max and then remained unchanged or decreased slightly to peak exercise (see figure 19.5b). Stroke volume also reached a plateau at approximately 50% of VO_2max. It was maintained during higher levels of exercise by a progressive increase in LVEF and a decrease in end-systolic volume, indicating a progressive increase in contractility.

In subsequent studies in female subjects, we observed comparable changes in the determinants of VO_2max (Sullivan, Cobb, and Higginbotham 1991). We have now performed studies in a total of 104 healthy subjects, age 20–76 years, by also using right heart catheterization and radionuclide angiographic techniques (Kitzman et al. 1989). In these studies, VO_2max decreased 48% over the six decades, or approximately 8% per decade, similar to that observed by Dehn and Bruce (1972). Over the six decades studied, maximum arteriovenous oxygen difference remained unchanged. However, cardiac index decreased 45% (see figure 19.6). The age-related decrease in cardiac index resulted from a 26% decrease in heart rate and a 24% decrease in stroke volume. The age-related decrease in stroke volume resulted from a decrease in maximum ejection fraction without a compensatory increase in left ventricular end-diastolic volume (see figure 19.7). These data are consistent with studies by other

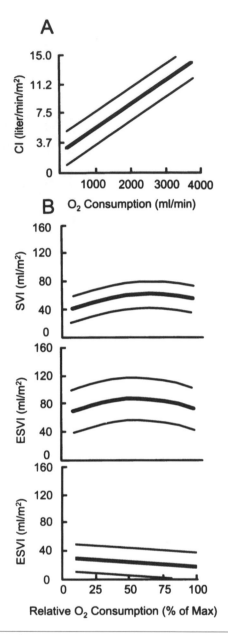

Figure 19.5 (*a*) Cardiac index (CI) during staged maximum upright exercise in 24 healthy male subjects, age 20–50 years, as a function of oxygen consumption. (*b*) Stroke volume index (SVI), end-diastolic volume index (EDVI), and end-systolic volume index (ESVI) as a function of normalized oxygen consumption. Each variable is plotted as a computer-generated regression with 95% confidence limits.

investigators who also observed age-related decreases in cardiac output during bicycle exercise due to a decrease in heart rate and stroke volume (Granath, Jonsson, and Strandell 1964).

In contrast, Rodeheffer et al. (1984) measured end-diastolic volume, stroke volume, and cardiac output by using noninvasive radionuclide angiographic techniques in 61 subjects, age 25–84

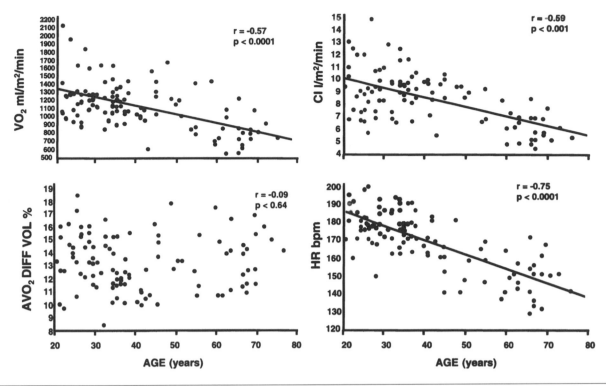

Figure 19.6 Maximal oxygen consumption (VO₂), cardiac index (CI), arteriovenous oxygen difference (O₂ diff), and heart rate (HR) related to age in 104 healthy subjects age 21–76 years. The age-related decrease in VO₂ was related to decreases in maximal cardiac output and heart rate but not to arteriovenous O₂ difference. (Reproduced from Kitzman et al. 1989 with permission.)

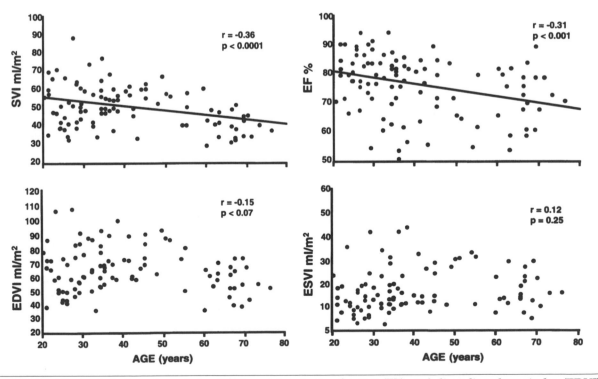

Figure 19.7 Maximal stroke volume index (SVI), exercise ejection fraction (EF), end-diastolic volume index (EDVI), and end-systolic volume index (ESVI) in 104 healthy subjects age 21–76 years. The age-related decrease in stroke volume index was accounted for by a decrease in maximal ejection fraction without a compensatory increase in EDVI. (Reproduced from Kitzman et al. 1989 with permission.)

years. They reported no age-related decrease in cardiac output or stroke volume. This occurred due to an apparent compensatory increase in end-diastolic volume and despite an age-related decrease in ejection fraction. This was surprising since the investigators did observe the expected age-related decline in maximum workload and heart rate. These studies did not measure expired gases. In early studies from our laboratory, we also observed age-related decreases in exercise left ventricular ejection fraction (Port et al. 1980). The age-related decrease in heart rate and possibly ejection fraction appear to be related to a decrease in responsiveness to catecholamines with advancing age (Lakatta 1979), as reviewed in the following section.

Effects of Age on Leg Blood Flow and Metabolism

Wahren et al. (1974) compared local circulatory adaptation with leg exercise in well-trained men.

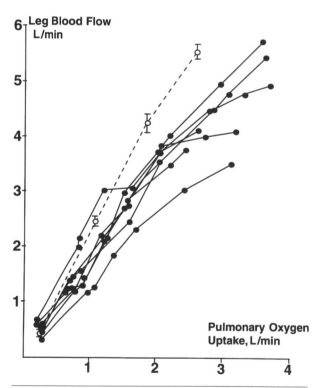

Figure 19.8 Individual data for leg blood flow in relationship to pulmonary oxygen uptake at rest and during exercise in seven healthy men age 52–59 years (closed circles and lines). The open circles and dashed lines illustrate the mean values for a group of subjects 25–30 years old. The slopes of the regression of leg blood flow on oxygen uptake is significantly steeper in the younger subjects. (Reproduced from Wahren et al. 1974 with permission.)

The two groups were based on age, 52–59 years versus 25–30 years. Peak leg blood flow was lower and tended to level off at the heaviest workloads in the older subjects (see figure 19.8). A larger femoral arterial venous oxygen difference during exercise compensated for the reduced blood flow. The researchers concluded that blood flow to the leg in middle-aged men rises in a curvilinear rather than in a linear manner in response to increasing exercise intensity. They also concluded that leg circulation during exercise becomes relatively hypokinetic with age. Kiessiing et al. (1974) observed insignificant increases in mitochondrial volume in middle-aged men after training as compared with a doubling of volume in younger men after training. This suggests a negative effect of age on the ability of muscles to increase their work capacity. Saltin, Wahren, and Pernow (1974) observed, however, that phosphagen (ATP and CP) and carbohydrate metabolism in leg muscle during exercise did not greatly differ in highly trained, middle-aged men age 52–59 as compared with younger, trained subjects.

Effects of Age on the Autonomic Nervous System

Aging is associated with augmentation of sympathetic nervous system activity (Ziegler, Lake, and Kopin 1976). This includes an increase in plasma norepinephrine concentration, an increase in sympathetic nerve firing (Ng et al. 1993; Yamada et al. 1989), an increase in the rate of norepinephrine spillover, and an impairment of neuronal reuptake of neurotransmitters by sympathetic nerves (Esier et al. 1981; Morrow et al. 1987). In contrast, Esier, Kaye, et al. (1995) observed two groups of subjects age 20–30 years versus 50–75 years. They found that plasma concentrations of epinephrine and appearance and clearance of epinephrine were reduced in the older group (Esier, Kaye, et al. 1995). In the younger men subjected to mental stress and isometric and dynamic exercise, epinephrine secretion doubled or tripled but increased only 44% of the response in older men (Esier, Kaye, et al. 1995). Others (Esier, Thompson, et al. 1995) have observed no difference in total norepinephrine spillover in younger men vs. older men in response to mental, isometric, or dynamic stress. However, cardiac norepinephrine spillover was increased to these stimuli in older men. These investigators observed that during mental stress and isometric exercise, heart rate increased comparably in older and younger men, 17 ± 3 beats/min versus 18 ± 4 beats/min. However heart rate increased less during dynamic exercise in older as compared with younger subjects, 18 ± versus 48 ± 4 beats/min, respectively. Mean arterial pressure increased comparably to mental stress in older men versus younger men, 21 ± 4 mmHg and 17 ± 3 mmHg. Several factors, thus, may influence the reduced heart rate response during dynamic exercise in the elderly. These include reduced epinephrine secretion (Esier, Kaye, et al. 1995), a reduction in the capacity for vagal withdrawals, a decrease in B-adrenergic responsiveness (Lakatta 1979; White et al. 1994), and anatomic changes in the sinoatrial node (Davies 1976). Thus, aging alters neural, humeral, and adrenergic receptor mechanisms. This contributes to altered heart rate, blood pressure, and possibly systolic cardiac function in response to dynamic and isometric exercise.

Interaction Effects of Age and Hypertension on Cardiovascular Function

Age and hypertension have similar but independent effects on the myocardium and vasculature. These changes include myocyte hypertrophy, increased left ventricular mass, decreased early diastolic filling rates, increased left atrial contribution to filling, increased arterial stiffness, increased aortic pulse wave velocity, and decreased cardiovascular responsiveness to catecholamines (Lakatta 1989). Age and hypertension produce strikingly similar changes in cardiac muscle excitation-contraction coupling mechanisms (Lakatta 1991; Lompre et al. 1991). In advanced age normotensive rats as compared with young hypertensive rats, isometric contractions, transmembrane action potentials, cytosolic calcium transients, and the resultant contractions were prolonged. The rate at which the sarcoplasmic reticulum pumps calcium was reduced. Myosin isoenzyme composition was shifted to predominantly the V3 form. Myosin ATPase was decreased in both aged and young hypertensive rats.

As previously noted (see figure 19.2), systolic blood pressure progressively increases with increasing age in a broad range of populations (Fleg 1986). This age-related increase in systolic blood pressure is likely a contributing factor to the age-related development of left ventricular hypertrophy. Clinical hypertension, defined as systolic pressure > 140 mmHg and diastolic pressure > 90 mmHg, may be expected to act in concert with the changes that occur during the normal aging process to promote the development of left ventricular hypertrophy. Left ventricular hypertrophy is a major risk factor for development of heart failure. Heart failure may result from myocardial hypertrophy and diastolic dysfunction and/or myocardial ischemia, myocyte loss, interstitial fibrosis, and, ultimately, systolic dysfunction.

Topol, Traill, and Fortuin (1985) have described a hypertensive hypertrophic cardiomyopathy in an elderly population, mean age 73, range 59–91 years. Of the 20 subjects studied, 16 were women and 15 were African American. All subjects had mild-to-moderate hypertension. Mean systolic blood pressure at the time of study entry was 161 ± 12 mmHg. Patients presented with pulmonary congestion, angina, or syncope. As compared with age-matched normotensive control subjects, ejection fraction was increased 79% ± 4% versus 59% ± 5%. End-diastolic and end-systolic dimensions were reduced, and wall thickness was increased 1.8 ± 0.3 cm versus 0.9 ± cm. In addition, diastolic function was reduced as measured by a delay in mitral valve opening, reduced peak rate of diastolic filling, and a prolonged early diastolic filling period.

Our laboratory has studied systemic and cardiac hemodynamics during upright exercise in a group of adults with heart failure due to left ventricular hypertrophy, diastolic dysfunction, and normal ejection fraction (Kitzman, Higginbotham, et al. 1991). Most subjects were elderly and had a history of hypertension. As compared with age-matched healthy controls, VO_2max was reduced 48% in the heart failure group. The reduction in VO_2 primarily resulted from a 41% reduction in cardiac output reserve with peak AVO_2 difference reduced 13% as compared with age-matched healthy controls (see figure 19.9). The reduced cardiac output resulted from an inability to increase stroke volume. Although end-diastolic volume, end-systolic volume, stroke volume, and ejection fractions were normal at rest, the heart failure group demonstrated an inability to increase exercise stroke volume. This occurred because the group had an inability to increase end-diastolic volume despite an abnormal increase in left ventricular filling pressure as measured by pulmonary capillary wedge pressure (see figure 19.10). These observations support the view that the heart failure resulted from diastolic dysfunction that severely limited use of the Starling mechanism to increase stroke volume during physical activity (see figure 19.11). The interaction of the normal aging process with hypertension to potentiate the development of left ventricular hypertrophy and subsequent heart failure supports an aggressive approach to the treatment of even mild degrees of hypertension in the elderly.

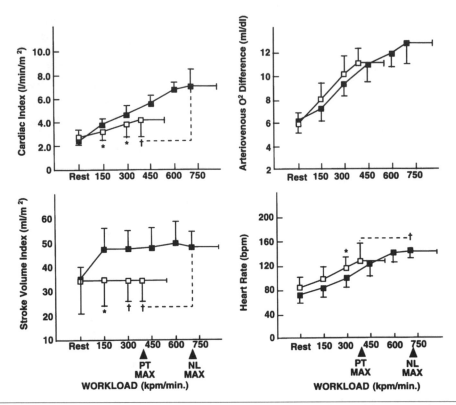

Figure 19.9 Cardiac index, arteriovenous oxygen difference, stroke volume index, and heart rate during progressive upright exercise in healthy subjects (N = 10; age 61 ± 8 years) (closed symbols) and patients (open symbols) with diastolic heart failure (N = 7; age 65 ± 12). * indicates a significant difference during submaximal workloads; † indicates a significant difference between maximum values. (Reproduced from Kitzman, Higginbotham, et al. 1991 with permission.)

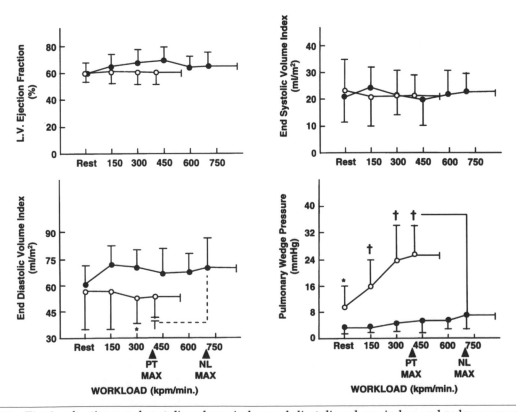

Figure 19.10 Ejection fractions, end-systolic volume index, end-diastolic volume index, and pulmonary wedge pressure at rest and during progressive upright exercise in healthy subjects and in patents with diastolic heart failure. Symbols are as in figure 19.9. (Reproduced from Kitzman, Higginbotham, et al. 1991 with permission.)

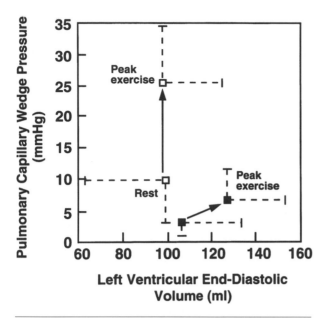

Figure 19.11 Relationship between pulmonary capillary wedge pressure and left ventricular end-diastolic volume at rest and during peak exercise in healthy subjects (closed symbols) and in patients with diastolic heart failure (open symbols). Patient failed to increase end-diastolic volume despite marked increases in pulmonary capillary wedge pressure, indicating a failure to use the Starling mechanism. (Reproduced from Kitzman, Higginbotham, et al. 1991 with permission.)

Can the Effects of Age on Cardiovascular Function Be Prevented?

Figure 19.12 illustrates data from three longitudinal studies of the relationship between age and VO_2max in inactive versus habitually active men (Dehn and Bruce 1972). Dill, Robinson, and Ross (1967) studied champion runners. Holimann (1965) and Dehn and Bruce (1972) studied healthy, middle-aged men. Although VO_2max declined with age in both the inactive and habitually active subjects, the rate of decline in VO_2max was substantially less in the active subjects in each study. This demonstrated the beneficial effect of a habitual physical activity program.

Stratton et al. (1994) compared the effects of six months of endurance training on cardiovascular response to graded supine exercise in younger (24–32 years), healthy men and in older (60–82 years), healthy men. Exercise training increased maximum VO_2 by 21% in the older and 17% in the younger subjects. Although these investigators observed a reduced exercise heart rate, stroke volume, and cardiac output in the older as compared with the younger subjects at baseline, the effects of endurance training were comparable in the younger and older populations. These effects included significant increases in exercise end-diastolic volume,

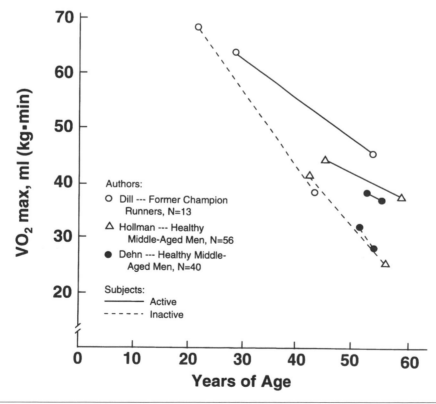

Figure 19.12 Longitudinal decrease in VO_2max as a function of habitual physical activity status. Habitual physical activity reduced the rate of age-related declines in VO_2max. (Reproduced from Dehn and Bruce 1972 with permission.)

ejection fraction, stroke volume, and cardiac output. The researchers concluded that despite significant changes in cardiovascular function as men age, older and younger men demonstrated similar cardiovascular adaptations to training.

Levy et al. (1993) measured diastolic filling parameters by radionuclide angiography in younger (24–32 years), healthy men and older (60–82 years), healthy men before and after six months of endurance training. Although the older men had a reduced early diastolic filling at rest and during exercise as compared with the younger men, endurance training increased early diastolic filling at rest and during exercise in both the older and younger. Beere et al. (1991) measured cardiac output and limb blood flow before and after a 12-week training period in younger (age 20–40) and older (age 60–80) sedentary subjects. Age-related decreases in VO_2max were due to reduced cardiac output and proportionately lower limb blood flow in the older subjects. VO_2max increased comparably after training in younger and older subjects via increases in the distribution of blood flow in exercising muscles. No significant increases occurred in cardiac output in the younger and older subjects.

Habitual exercise, therefore, attenuates, but does not prevent, the age-related decline in exercise performance. On the other hand, the cardiovascular system in the aged appears to respond to exercise training in a comparable fashion with that observed in younger subjects. Exercise training results in comparable improvement in overall cardiovascular function.

Summary

In summary, increasing age is associated with anatomic changes within the myocardium and vasculature. These changes reduce diastolic rather than systolic ventricular function. They interact with altered autonomic nervous system activity to reduce cardiac output progressively via reduction in heart rate. They also cause reduced ability to increase end-diastolic volume, stroke volume, and ejection fraction during exercise. These factors contribute to decreased limb perfusion and exercise performance with increasing age. Exercise training in the elderly improves exercise performance by a similar magnitude in younger and older subjects by apparently similar mechanisms. Habitual exercise reduces but does not prevent an age-related decline in exercise performance.

Age-related effects on the heart and vasculature are similar to effects caused by hypertension at any age. Clinical hypertension thus interacts with age effects to promote myocardial hypertrophy, diastolic dysfunction, reduced exercise tolerance, and development of heart failure. These observations support the role of habitual exercise for enhancement of cardiovascular function and physical performance and for aggressive treatment of hypertension at all ages.

Acknowledgments

From the Department of Medicine, Division of Cardiology, Duke University Medical Center, Duke Center for Living and Durham Veterans Administrative Medical Center, Durham, North Carolina, and Bowman Gray School of Medicine, Winston-Salem, North Carolina. Supported by grant HL-54314-02 SUB #1 from the National Heart, Lung and Blood Institute, Bethesda, Maryland.

References

Anversa, P., B. Hiler, R. Ricci, G. Giancario, and G. Olivetti. 1986. Myocyte cell loss and myocyte hypertrophy in the aging rat heart. *J. Am. Coll. Cardiol.* 8:1441–1448.

Beere, P.A., D.W. Kitzman, M. Morey, and M.B. Higginbotham. 1991. Peripheral adaptations to exercise training compensate for age related differences in aerobic capacity in healthy men. *Circulation* 84:186.

Berman, N.D. 1982. *Geriatric cardiology.* Lexington, MA: Collamore Press.

Brandponbrener, M., M. Landowne, and N.W. Shock. 1955. Changes in cardiac output with age. *Circulation* 12:557.

Davies, M.J. 1976. Pathology of the conducting system. In *Cardiology in old age,* eds. F.L. Caird, J.L.C. Dall, and R.D. Kennedy. New York: Plenum.

Dehn, M.M., and R.A. Bruce. 1972. Longitudinal variations in maximal oxygen intake with age and activity. *J. Appl. Physiol.* 33:805–807.

Dill, D.B., S. Robinson, and J.C. Ross. 1967. A longitudinal study of 16 champion runners. *J. Sports Med.* 7:4–32.

Eiveback, L., and J.T. Lie. 1984. Continued high incidence of coronary artery disease at autopsy in Oimsted County, Minnesota, 1950 to 1979. *Circulation* 70:345–349.

Esier, M., D. Kaye, J. Thompson, G. Jennings, H. Cox, A. Turner, G. Lambert, and D. Seals. 1995. Effects of aging on epinephrine secretion and regional release of epinephrine from the human heart. *J. Clin. Endocrinol. Metab.* 8:435–442.

Esier, M., H. Skews, P. Leonard, G. Jackson, A. Cobik, and P. Korner. 1981. Age-dependence of noradrenaline kinetics in normal subjects. *Clin. Sci.* 60:217–219.

Esier, M.D., J.M. Thompson, D.M. Kaye, A.G. Turner, G.L. Jennings, H.S. Cox, G.W. Lamber, and D.R. Seals. 1995. Effects of aging on the responsiveness of the human cardiac sympathetic nerves to stressors. *Circulation* 91:351–358.

Fleg, J.L. 1986. The aging heart anatomic and physiologic alterations. In *Coronary heart disease in the elderly*, eds. N.K. Wenger, C.D. Furbery, and E. Pitt, 253–273. New York: Elsevier.

Gerstenblith, G., J. Frederiksen, F.D. Yin, et al. 1977. Echocardiographic assessment of a normal adult aging population. *Circulation* 56:273–278.

Glenner, G.G. 1980. Amyloid deposits and amyloidosis: The B-fibritioses. *N. Eng. J. Med.* 302:1333–1343.

Granath, A., B. Jonsson, and T. Strandell. 1964. Circulation in healthy old men studied by right heart catheterization at rest and during exercise in supine sitting position. *Acta Med. Scand.* 176:425.

Higginbotham, M.B., K.G. Morris, R.S. Williams, R.E. Coleman, and F.R. Cobb. 1986. Physiologic basis for the age-related decline in aerobic work capacity. *Am. J. Cardiol.* 57:1374–1379.

Higginbotham, M.B., K.G. Morris, R.S. Williams, P.A. McHale, R.E. Coleman, and F.R. Cobb. 1986. Regulation of stroke volume during submaximal and maximal exercise in normal men. *Cir. Res.* 58:261–291.

Holimann, W. 1965. *Korperliches Training als Pravention von Herz-Kreislauf-Krankheiten.* Stuttgart, West Germany: Hippokrates-Verlag.

Hutchins, G.M. 1980. Structure of the aging heart. In *The aging heart: Its function and response to stress*, ed. M.I. Weisfeidt. New York: Raven Press.

Kiessiing, K.-H., L. Pilstrom, A.-Ch. Bylund, B. Saltin, and K. Piehi. 1974. Enzyme activities and morphometry in skeletal muscle of middle-aged men after training. *Scand. J. Clin. Lab. Invest.* 33:63–69.

Kitzman, D.W., and W.D. Edwards. 1990. Age-related changes in the anatomy of the normal human heart. *J. Gerontology Medical Sciences* 45:M33–M39.

Kitzman, D.W., M.B. Higginbotham, F.R. Cobb, K.H. Sheikh, and M.J. Sullivan. 1991. Exercise intolerance in patients with heart failure and preserved left ventricular systolic function: Failure of the Frank-Starling mechanism. *J. Am. Coll. Cardiol.* 17:1065–1072.

Kitzman, D.W., K.H. Sheikh, P.A. Beere, J.L. Philips, and M.B. Higginbotham. 1991. Age-related alterations of Doppler left ventricular filling indexes in normal subjects are independent of left ventricular mass, heart rate, contractility, and loading conditions. *J. Am. Coll. Cardiol.* 19:1243–1250.

Kitzman, D.W., M.J. Sullivan, F.R. Cobb, and M.B. Higginbotham. 1989. Exercise cardiac output declines with advancing age in normal subjects. *J. Am. Coll. Cardiol.* 13:241A.

Kutkka, J.T., and E. Lansimies. 1982. Effect of age on cardiac index, stroke index, and left ventricular ejection fraction at rest and during exercise as studied by radiocardiography. *Acta Physiol. Scand.* 114:339.

Lakatta, E.G. 1979. Age related alterations in the cardiovascular system that occur in advanced age. *Fed. Proc.* 38:163–167.

Lakatta, E.G. 1989. Mechanisms of hypertension in the elderly. *J. Am. Geriatr. Soc.* 37:780–790.

Lakatta, E.G. 1991. Regulation of cardiac muscle function in the hypertensive heart. In *Cellular and molecular mechanisms of hypertension*, ed. R.H. Cox, 149–173. New York: Plenum Press.

Lenkiewicz, J.E., M.J. Davies, and D. Rosen. 1972. Collagen in human myocardium as a function of age. *Cardiovasc. Res.* 6:549.

Levy, W.C., M.D. Cerqueira, I.B. Abrass, R.S. Schwartz, and J.R. Stratton. 1993. Endurance exercise training augments diastolic filling at rest and during exercise in healthy young and older men. *Circulation* 88:116–126.

Lompre, A.M., F. Lambert, E.G. Lakatta, B. Nadai-Ginard, and K. Schwartz. 1991. Expression of sarcoplasmic reticulum Ca 21-ATPase and calsequestrin genes in rat heart during ontogenic development and aging. *Circ. Res.* 9:1380–1388.

Morrow, L.A., O.A. Linares, T.J. Hill, J.A. Sanfield, M.A. Supiano, S.G. Rosen, and J.B. Hatier. 1987. Age difference in the plasma clearance mechanism for epinephrine and norepinephrine in humans. *J. Clin. Endocrinol. Metab.* 65:508–511.

Ng, A.V., R. Callister, D.G. Johnson, and D.R. Seals. 1993. Age and gender influence muscle sympathetic nerve activity at rest in healthy humans. *Hypertension* 21:498–503.

Pomerance, A. 1975. Aging and degenerative changes. In *The pathology of the heart*, eds. A. Pomerance and M.J. Davies MJ, 49–79. London: Blackwell Scientific.

Port, E., F.R. Cobb, R.E. Coleman, and R.H. Jones. 1980. Effect of age on the response of the left ventricular ejection fraction to exercise. *New Eng. J. Med.* 303:1133–1137.

Rodeheffer, R.J., G. Gerstenblith, L.C. Becker, et al. 1984. Exercise cardiac output is maintained with advancing age in health human subjects: Cardiac dilatation and increased stroke volume compensate for a diminished heart rate. *Circulation* 69:202–213.

Saltin, B., J. Wahren, and B. Pernow. 1974. Phosphagen and carbohydrate metabolism during exercise in trained middle-aged men. *Scand. J. Clin. Lab. Invest.* 33:71–77.

Stone, J.L., and A.H. Norris. 1966. Activities and attitudes of participants of the Baltimore Longitudinal Study. *J. Gerontology.* 21:575–580

Stratton, J.R., W.C. Levy, M.D. Cerqueira, R.S. Schwartz, and I.B. Abrass. 1994. Cardiovascular responses to exercise. Effects of aging and exercise training in healthy men. *Circulation* 89:1648–1655.

Strehier, B.L., D.D. Mark, A.S. Miidvan, and M.V. Gee. 1959. Rate and magnitude of age pigment accumulation in the human myocardium. *J. Gerontol.* 14:430–439.

Sullivan, M.J., F.R. Cobb, and M.B. Higginbotham. 1991. Stroke volume increases by similar mechanisms during upright exercise in normal men and women. *Am. J. Cardiol.* 13:241A.

Topol, E.K., T.A. Traill, and N.J. Fortuin. 1985. Hypertensive hypertrophic cardiomyopathy of the elderly. *N. Engl. J. Med.* 312:277–283.

Wahren, J., B. Saltin, L. Jorfeldt, and B. Pernow. 1974. Influence of age on the local circulatory adaptation to leg exercise. *Scand. J. Clin. Lab. Invest.* 33:79–86.

White, N.K., J.E. Edwards, and T.J. Dry. 1950. The relationship of the degree of coronary atherosclerosis with age in men. *Circulation* 1:1345–1354.

White, M., R. Reden, W. Minobe, F. Khan, P. Larrabee, M. Wolimering, J.D. Prot, F. Anderson, D. Campbell, A.M. Feldman, and M.R. Bristow. 1994. Age-related changes in B-adrenergic neuroeffector systems in the human heart. *Circulation* 90:1225–1238.

Yamada, Y., E. Miyajima, A.T. Matsukawa, and M. Lshii. 1989. Age-related changes in muscle sympathetic nerve activity in essential hypertension. *Hypertension* 13:870–877.

Ziegler, M.G., C.R. Lake, and I.J. Kopin. 1976. Plasma norepinephrine increases with age. *Nature* 261:333–335.

PART IV

Cardiovascular Regulation in Disease

Chapter 20

Heart Failure

Lawrence I. Sinoway and Jere H. Mitchell

Congestive heart failure (CHF) is a common, disabling disease affecting millions of individuals worldwide. This disease is earmarked by low cardiac outputs for a given level of physical exertion as well as by elevated left ventricular filling pressures. These pathophysiological responses initiate a cascade of compensatory responses that maintain performance. However, a slow process of decompensation occurs, ultimately resulting in multiple symptoms. Two of the most important are shortness of breath and fatigue upon exertion (Zelis and Sinoway 1992). In addition to cardiac dysfunction, other pathophysiological events importantly contribute to symptomatology (Zelis and Sinoway 1992). For example, Cheyne-Stokes ventilation, a periodic breathing pattern seen in many metabolic diseases, is a major contributor to the sensation of shortness of breath in severely decompensated heart failure patients (Harrison et al. 1934). With regards to fatigue, reduced cardiac output during exercise leads to a reduction in blood flow delivery to the exercising muscle (LeJemtel et al. 1986; Longhurst, Gifford, and Zelis 1976; Magnusson et al. 1997; Sullivan and Cobb 1991). However, this may not be sufficient to explain entirely the severe exercise limitation seen in the disease. Impaired vessel dilation (Zelis, Mason, and Braunwald 1968) and altered muscle metabolism (Minotti et al. 1991) have been suggested as potential contributors to the reduced exercise capacity.

Recently, interest has focused on control of the circulation in heart failure and how autonomic adjustments contribute to the symptoms seen in heart failure. However, understanding the impact of autonomic function on circulatory control in heart failure is difficult when so much still needs to be learned about the role of this system in healthy subjects. This chapter will briefly describe circulatory responses to exercise in heart failure. It will also review the patterns and mechanisms of altered autonomic neural control during exercise.

Systemic and Regional Circulatory Responses

Heart failure patients have a reduced exercise capacity due to a systemic circulatory limitation as a consequence of a reduced left ventricular ejection fraction and local muscle function impairment (Magnusson et al. 1997; Minotti et al. 1991; Sullivan and Cobb 1991). Research has demonstrated that peak leg blood flow is reduced during cycle ergometer exercise, and similar values are reached with the engagement of either one or two legs (LeJemtel et al. 1986; Sullivan and Cobb 1991). This finding has provoked the question of whether the reduced peak limb flow in heart failure is due to limited local flow capacity or because the muscle mass engaged is too large to be perfused by a limited cardiac output. Evidence in support of local muscle limitations stems from findings of a reduced vasodilator capacity (Sinoway and Prophet 1990; Zelis and Sinoway 1992), muscle wasting, a greater percentage of type II fibers, reduced capillary density, and lower mitochondrial enzyme activities (Drexler et al. 1992; Lipkin et al. 1988; Magnusson et al. 1997; Mancini et al. 1989; Shafiq et al. 1972).

In a recent study, Magnusson et al. (1997) sought to determine the relative importance of the heart and skeletal muscle for the reduced exercise capacity in heart failure. In contrast with others (Franciosa, Parker, and Levine 1981; Francis, Goldsmith, and Cohn 1982), Magnusson and colleagues found a strong correlation between maximal oxygen uptake and ejection fraction. The smaller cross-sectional area of the quadriceps muscle accounted for over half the reduction in muscle strength. A lower endurance capacity was associated with reduced enzyme activity and capillary density. However, the maximal oxygen uptake measured during dynamic exercise with one and two legs was most closely related to a limited cardiac output. They concluded that both the heart and skeletal muscle contributed to impaired dynamic exercise capacity, but the limited pumping capacity of the heart was the most significant factor. This premise is supported by the finding that when heart failure patients undergo high-intensity one-leg knee extensor training where the circulatory limits of the heart are not encroached upon, significant improvements in muscle enzyme activity, capillary density, and performance are observed (Magnusson et al. 1997). While some impairment in muscle adaptability may occur at the genetic level, recent studies suggest that altered sympathetic neural control of vascular resistance may be a more primary influence than factors intrinsic to the muscle in heart failure patients.

To explore the contribution of central circulatory factors further, Magnusson et al. (1997) recently compared cardiac output, leg blood flow, and nonleg blood flow responses in healthy and heart failure subjects during one-leg and two-leg knee extension. They found that when only the quadriceps of one leg was engaged in dynamic work, patients with moderate heart failure achieved an equally high peak perfusion as healthy age-matched controls. However, when both legs were engaged, they observed a lower peak muscle perfusion and a reduced blood flow to inactive vascular beds, while the controls maintained the blood flow in both active muscle and inactive regions. Thus, in spite of a normal capacity to vasodilate, the reduced blood flow in heart failure patients was related to a limited capacity to increase cardiac output. The arterial norepinephrine as well as norepinephrine spillover from muscle was significantly higher in heart failure patients in both one- and two-legged exercise. This supports the role of the sympathetic nervous system for regulating blood pressure and muscle perfusion pressure during one-leg exercise and for attenuating muscle blood flow during two-leg exercise. Thus, sympathetic vasoconstriction elevates vascular resistance to maintain blood pressure when cardiac output fails to support blood pressure.

Sullivan et al. (1989) and Sullivan and Cobb (1991) had similar findings. They observed a higher vascular resistance in the legs during cycling exercise in heart failure patients compared with the controls, while they saw no differences in mean blood pressure. They concluded that with a limited cardiac output in heart failure, vascular conductance is dynamically regulated by the sympathetic nervous system to maintain blood pressure. The sympathetic response to a low cardiac output is a pronounced compensatory adaptation since even cerebral perfusion is threatened during exercise in heart failure (Hellström et al. 1996).

Sympathoexcitatory Systems in Healthy Human Subjects

Figure 20.1 presents a schematic representation of neural activation during exercise in healthy subjects. As demonstrated, two systems may contribute to the increased sympathetic tone accompanying exercise. The first system, central command, involves the parallel activation of α motor neurons and sympathetic efferents. This is a feed-forward system requiring no afferent feedback for the full expression of the sympathoexcitatory response to exercise. This system may be linked to skeletal muscle metabolism need via motor neuron recruitment patterns. Specifically, as exercise progresses and some muscle fibers fatigue, new motor units are recruited, perhaps with a commensurate increase in sympathetic discharge (Gandevia et al. 1993; Mitchell, Kaufman, and Iwamoto 1983; Victor, Pryor et al. 1989).

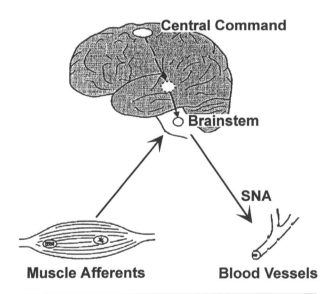

Figure 20.1 Schematic representation of neural control systems contributing to autonomic adjustments seen during exercise.

The sympathetic nervous system also engages via activation of a muscle-based reflex (Coote, Hilton, and Perez-Gonzalez 1971; McCloskey and Mitchell 1972; Rowell, Hermansen, and Blackmon 1976). Two receptor types responsive to different stimuli have been demonstrated within the muscle. The first type, termed muscle mechanoreceptors, are responsive to the mechanical deformation of the free afferent nerve endings (Kaufman et al. 1983; Sinoway et al. 1993; Victor, Rotto, et al. 1989). These afferents increase their discharge frequency in response to stretch or contraction. In general, finely myelinated group III muscle afferents subserve this role (Kaufman and Forster 1996).

In 1937, Alam and Smirk provided important evidence for the presence of a skeletal muscle metabolite-sensitive reflex (Alam and Smirk 1937). These authors demonstrated that postexercise muscle ischemia caused blood pressure to remain elevated above baseline values. They hypothesized that ischemic muscle released metabolites, stimulating afferent nerves and initiating the reflex. A number of important reports have suggested that unmyelinated, slowly conducting group IV fibers are the predominant fiber group that subserve metaboreceptor function (Kaufman and Forster 1996; Kaufman et al. 1983, 1984; Mense and Stahnke 1983). The free nerve endings for these receptors are in muscle interstitium in close proximity to small blood vessels. This allows rapid responses to changes in the local chemical milieu (Von During and Andres 1990). This system is clearly activated during intense lower-limb exercise (Sinoway and Prophet 1990). Important studies have suggested that muscle metaboreceptor stimulation is responsible for sympathoexcitation during "near-fatiguing," vigorous static handgrip exercise (Saito, Mano, and Iwase 1989; Seals and Enoka 1989). Mark and colleagues were among the first to use the microneurographic technique to record sympathetic nerve traffic during bouts of handgrip exercise (Mark et al. 1985). This technique has allowed investigators to make major advances in the understanding of sympathetic regulation during exercise.

A number of substances have been suggested as potential muscle metaboreceptor stimulants. The list of substances includes lactic acid, H^+, adenosine, arachidonic acid metabolites, diprotonated phosphate, and bradykinin (see chapter 2 "The Exercise Pressor Reflex: Afferent Mechanisms, Medullary Sites, and Efferent Sympathetic Responses").

One way of determining the specific substances evoking the metaboreflex in humans would be to record the activity of some efferent sympathetic index while examining exercising skeletal muscle metabolism. For example, one prior study used ^{31}P nuclear magnetic resonance (NMR) spectroscopy to measure forearm muscle pH and high energy phosphate metabolism. This same study also used strain gauge plethysmography to measure calf vascular resistance during static handgrip exercise. In these experiments, calf vascular resistance increased during handgrip and remained elevated during postexercise forearm circulatory arrest. As mentioned earlier, during posthandgrip circulatory arrest (PHGCA), muscle mechanoreceptors and cen-

tral command are not engaged. Therefore, the increased calf vascular resistance noted during PHGCA must have been due to metabolite-sensitive muscle afferent stimulation. The major finding of this report was that the intracellular skeletal muscle pH curve mirrored the pattern of calf vascular resistance changes noted during the paradigm. Additional work suggested that calf vascular resistance did not rise unless muscle pH fell. Victor et al. (1988) also used ^{31}P NMR during forearm work and found an impressive inverse relationship between muscle pH and increases in peroneal nerve muscle sympathetic nerve activity.

These studies were followed by additional experiments demonstrating that lactic acid generated during muscle work played some role in evoking the exercise reflex. Three strategies were employed. In the first, studies were performed in McArdle's syndrome, a disease where subjects are incapable of producing lactic acid. These reports suggested a decreased ability of the subjects to increase muscle sympathetic nerve activity (MSNA) during handgrip. However, recent preliminary work suggests that MSNA responses may be intact in these subjects if the bouts of exercise are near fatiguing. Furthermore, microdialysis and NMR measurements suggest that lactate and cellular acidosis are not necessary to evoke these MSNA responses during bouts of static exercise (Vissing et al. 1997).

The second strategy employed glycogen depletion techniques to reduce lactic acid responses to lower-limb exercise. Sinoway, Wroblewski, et al. (1992) demonstrated that reductions in plasma lactate levels due to glycogen depletion during supine leg exercise led to reduced vasoconstrictor drive directed to dilated forearm muscle beds.

In the third strategy, lactic acid formation was reduced chemically by reducing the conversion of pyruvate to lactate by intravenously administering dichloroacetate (DCA) to healthy volunteers. This drug inhibits the inhibitor of pyruvate dehydrogenase. Thus, for any given level of pyruvate produced during exercise, a greater percentage will enter the Krebs cycle, and less lactate will be produced. DCA lowered venous efferent lactate and H^+ levels during handgrip exercise. This reduction in lactate production was associated with attenuated muscle sympathetic nerve activity during forearm exercise (Ettinger et al. 1991).

When the results of all of these reports are considered together, one can conclude that muscle acidity plays some role in engaging the muscle reflex. However, we must emphasize that these studies do not provide evidence that lactic acid is a direct muscle afferent stimulant. They also do not support the concept that lactic acid is the sole muscle afferent stimulant.

To explore further the role H^+ plays in evoking muscle reflexes, experiments were performed examining the effects of chronic exercise on muscle reflex responses to handgrip. In these studies, the researchers hypothesized that chronic bodybuilding training would lead to decreased muscle H^+ and lactate production, thereby reducing muscle metaboreceptor stimulation during bouts of handgrip. The rationale for studying bodybuilders was that these subjects perform repetitive bouts of moderate-intensity exercise to fatigue. The researchers felt that this type of work would decrease metabolite production and in the process, modify the expression of the exercise pressor reflex. In this prior report, MSNA increased less during 30% maximal voluntary contraction (MVC) in the bodybuilders than in the control subjects (Sinoway, Rea, et al. 1992) despite the fact that forearm muscle mass and tension generation were greater in the bodybuilders. Both muscle mass and tension generation influence the magnitude of the sympathoexcitatory response to exercise (Seals and Victor 1991). Handgrip exercise in the bodybuilders was associated with less cellular acidosis than in the controls.

These results would imply a linkage between acid production and sympathoexcitation. However, in this report, alternate explanations were explored for the reduced MSNA response in the bodybuilders. These explanations included the possibility that conditioning evoked a desensitizing process such that the relationship between metabolite production and reflex responses were reduced. In an effort to explore this issue, additional experiments were performed designed to generate similar levels of muscle acidosis in the controls and bodybuilders. If under these conditions the bodybuilders and controls developed similar levels of sympathetic nerve traffic, the data would then be consistent with the hypothesis that the hydrogen ion concentration is directly linked to the magnitude of the sympathetic nerve response to static exercise. On the other hand, if acid generation was similar in the two groups but the MSNA responses were attenuated in the bodybuilders, factors aside from the hydrogen ion concentration would contribute to the different MSNA responses seen in the two groups.

Accordingly, individuals performed ischemic rhythmic handgrip. During ischemic exercise, the level of cellular acidosis at fatigue should be similar in control and conditioned individuals. This should occur since ATP would have to be generated anaerobically, and the enzymes of anaerobic glycolysis are far less affected by conditioning stimuli than the enzymes of the Krebs cycle and those involved in

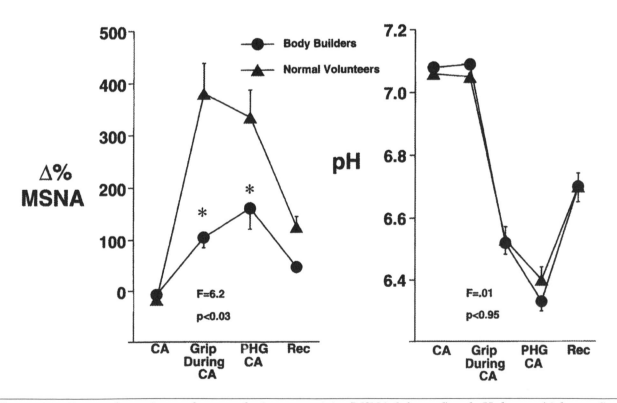

Figure 20.2   Percent change in muscle sympathetic nerve activity (MSNA; left panel) and pH changes (right panel) during ischemic handgrip exercise in bodybuilders and controls. pH responses were similar for the two groups, yet the MSNA response was lower in the bodybuilders.

the oxidation of lipids. This paradigm led to similar levels of cellular acidosis in the controls and bodybuilders. However, muscle sympathetic nerve activity was less in the bodybuilders than in the controls (see figure 20.2). These results were interpreted as supporting the hypothesis that rhythmic exercise led to a reduced production of muscle metabolites and also reduced responsiveness of metabolite-sensitive muscle afferents.

In addition to lactic acid, another substance that increases with muscle work is diprotonated phosphate. This is an intriguing potential stimulant because its concentration depends on both phosphate concentration as well as the proton concentration. Assuming that interstitial concentrations of diprotonated phosphate would show a large relative increase during exercise would therefore be reasonable. This substance has been suggested as a vasodilator as well as an important determinant of muscle fatigue (Hilton, Hudlicka, and Marshall 1978; Nosek, Fender, and Godt 1987).

In these studies, subjects performed two bouts of ischemic static exercise with an intervening freely perfused 1 min rest period. These studies measured forearm muscle metabolism (^{31}P NMR spectroscopy) and peroneal nerve recordings of muscle sympathetic nerve activity. The researchers demonstrated that diprotonated phosphate correlated better with muscle sympathetic nerve activity than did H^+. In animal experiments, bolus injections of $H_2PO_4^-$ (pH 6.0 phosphate) into the arterial supply of the triceps surae muscle evoked far greater reflex increases in blood pressure than did equimolar infusions of lactate or pH 7.5 phosphate (predominantly monoprotonated phosphate). These responses were due to a muscle reflex since outlining the sciatic nerve prevented the response (Sinoway et al. 1994). This report as well as preliminary work examining the discharge of group III afferents suggests that $H_2PO_4^-$ may play an important role in evoking the muscle reflex (Dzwonczyk, Gray, and Sinoway 1994).

Recently, Boushel et al. (1998) measured intracellular as well as venous $H_2PO_4^-$ and pH during handgrip and postexercise forearm ischemia. They found that the elevation of blood pressure was most closely related to intracellular $H_2PO_4^-$. Blood pressure was also predicted by both intracellular and venous pH. The significant finding was that no increase occurred in venous phosphate, and therefore no relation existed between the venous $H_2PO_4^-$ and blood pressure. This raised the question of whether phosphate is released from muscle or is rapidly taken up by muscle during contraction. However, recent work by MacLean et al. (1998) suggest that phoshate does increase in the interstitium where the afferent nerves are located.

Studies have also suggested that adenosine may be a muscle afferent stimulant in humans. A report by Costa and Biaggioni (1994) has suggested that infusions of adenosine into the forearm arterial supply evokes an increase in sympathetic nerve activity. Additionally, the increase in MSNA during handgrip was reduced after giving the intrabrachial adenosine receptor antagonist theophylline. However, recent observations by MacLean et al. (1998) suggest that femoral artery injections of adenosine do evoke an increase in MSNA but with a surprisingly long onset latency (16 ± 1 s). In separate trials, no increase in MSNA was observed when the adenosine was injected arterially but distal to a femoral artery occlusion cuff inflated to suprasystolic levels. Upon release of this cuff, MSNA increased rather dramatically. These results would suggest that the majority of the adenosine effect on MSNA is due to stimulation of afferents not in skeletal muscle. Also note that prior animal work suggests that adenosine does not stimulate group III and IV muscle afferents in cats (Rotto and Kaufman 1988).

In summary, a number of substances have been suggested as being the metabolic stimulant. However, to date, no specific clear-cut agreement exists about the key specific metaboreceptor stimulant. Clearly, accurate measurements of interstitial concentrations of the many substances during contraction are necessary. Intracellular and intravenous measurements may not tell the entire story. Finally, we must emphasize that this lack of agreement as to the metaboreceptor stimulant makes statements regarding this issue in heart failure problematic.

Recent work from our laboratory has employed the microdialysis method during twitch contractions in a decerebrate cat model. In these experiments, microdialysis probes were placed in the triceps surae muscle of the cat. We measured H^+, lactate, phosphate, and K^+. We observed that K^+ most closely matched the pattern of blood pressure and heart rate rise seen with contraction (MacLean et al. 1998). Interstitial phosphate increased during contraction. However, the similar increases at two different intensities could not explain the significant differences in blood pressure. The findings that K^+ follows the pattern of autonomic responses to exercise are consistent with a recent report by Fallentin et al. (1992). They measured venous blood concentrations of K^+ and found that the time course of K^+ and blood pressure closely tracked one another. K^+ remains a strong candidate because it evokes discharge of the afferents. In addition, the exercise intensity-related changes in the blood pressure parallel the changes in K^+.

What is yet to be resolved about K^+ is the common observation in humans that venous potassium typically increases to reach a plateau early during exercise. Sympathetic nerve activity, however, is characterized by a latent onset. It as well as blood pressure may rise progressively during moderate and intense exercise. Thus, while K+ likely plays a role in activating the muscle reflex, other factors also appear to be involved. Further experiments measuring interstitial metabolites during exercise are needed to elucidate these factors.

Neural Responses During Exercise in Heart Failure

A large body of work conducted over the years has examined many aspects of the blood flow delivery problem in heart failure subjects. However, little work has examined the muscle reflex responses in these patients.

Based on the bodybuilder work described earlier, Sterns et al. (1991) speculated that heart failure subjects would also exhibit metaboreceptor desensitization. In an effort to examine this hypothesis, CHF and age-matched controls performed static exercise at 30% of maximum voluntary contraction for 2 min followed by a 2-min period of PHGCA while measuring peroneal nerve recordings of sympathetic nerve traffic (Sterns et al. 1991). Sympathetic nerve activity was far higher in the heart failure subjects at rest than in the control subjects. This observation confirmed prior findings by Leimbach et al. (1986).

During exercise, the increase in MSNA was similar in the heart failure and control groups. However, during the posthandgrip ischemic phase when muscle metaboreceptors were selectively engaged, MSNA responses returned to baseline values in the heart failure subjects but remained elevated in the control subjects. NMR spectroscopy of the forearm muscle during bouts of handgrip suggested that the attenuated MSNA responses during ischemia in the heart failure subjects was not due to the reduced production of ischemic metabolites (Sterns et al. 1991). These findings suggested that the metaboreceptor contribution to the MSNA response during static handgrip is attenuated in heart failure. Moreover, if neural responses during handgrip itself were not decreased in CHF but the metaboreceptor contribution was reduced, some other neural system must have then been activated to a greater degree in heart failure than in control subjects (see figure 20.3). Specifically, central command and/or muscle mechanoreceptors must be activated to a greater degree during static handgrip in heart failure subjects.

Based on these results, a series of experiments were begun to examine the muscle mechanoreflex in healthy human subjects. The rationale was that if evidence were gathered demonstrating that muscle mechanoreceptors contributed to MSNA responses in humans, studies examining this system in heart failure would then be justified. Prior animal experiments have suggested that external pressure and/or muscle compression can engage muscle mechanore-

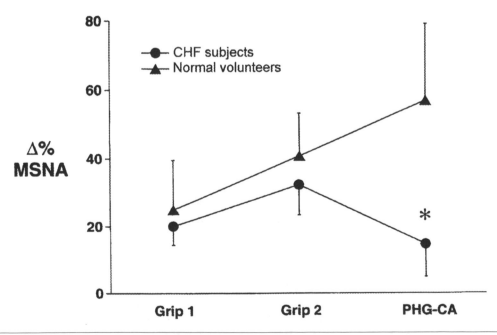

Figure 20.3 MSNA response in control and CHF subjects at baseline and during posthandgrip–circulatory arrest (PHG–CA). MSNA responses are attenuated in CHF during PHG–CA.

ceptors and evoke muscle reflexes (Stebbins et al. 1988). With the prior work by Stebbins et al. (1988) and Sterns et al. (1991) in mind, McClain, Hardy, and Sinoway (1994) had healthy individuals perform static handgrip exercise with and without external forearm compression. These individuals ischemically performed bouts of exercise to exclude an effect of forearm compression on blood flow. External forearm compression of 110 mmHg evoked far greater increases in sympathetic outflow than was observed during ischemic exercise without forearm compression (see figure 20.4) (McClain, Hardy, and Sinoway 1994). Interestingly, 90 mmHg of external forearm compression did not enhance the MSNA response to ischemic static handgrip. Neither 90 mmHg nor 110 mmHg of external compression raised MSNA at rest. At the end of the 110 mmHg ischemic exercise paradigm, the external forearm compression was removed. However, posthandgrip ischemia was continued. During this portion of the experiment, no difference occurred in sympathetic outflow between the two trials. Thus, this group of experiments suggests that muscle mechanoreceptors play a role in evoking MSNA responses to exercise. However, the stimulus necessary to engage this system, at least for static handgrip, needs to be relatively large (McClain, Hardy, and Sinoway 1994). Recent studies by Williamson et al. (1994) also provide support for the concept that leg muscle mechanoreceptors can evoke pressor responses.

A series of experiments were performed to examine muscle sympathetic nerve responses to 30 min of rhythmic handgrip at 25% MVC (12 contractions/min, each 2 s in length). The rationale for performing these studies was that 30 min of exercise at a nonfatiguing workload should not engage muscle metaboreceptors. Therefore, any increase in MSNA would activate some other reflex system. This rhythmic exercise paradigm led to a slow, gradual increase in sympathetic outflow not associated with a postexercise ischemic MSNA response (Batman et al. 1994). These results suggested that the rhythmic handgrip paradigm evoked sympathetic discharge by a nonmetaboreceptor-mediated process. Presumably, this type of exercise activated muscle mechanoreceptors.

To explore further the potential role muscle mechanoreflexes might play in pathophysiological conditions such as heart failure, acute limb congestion was employed as a potential way of selectively engaging muscle mechanoreflexes. It is well-known that individuals with heart failure have elevated central venous pressure values leading to increased interstitial volumes and eventually to peripheral edema. McClain et al. (1993) speculated that peripheral muscle would augment mechanoreceptor activity and increase the sympathetic nerve response to handgrip exercise.

To address this issue, healthy individuals performed handgrip exercise under two experimental

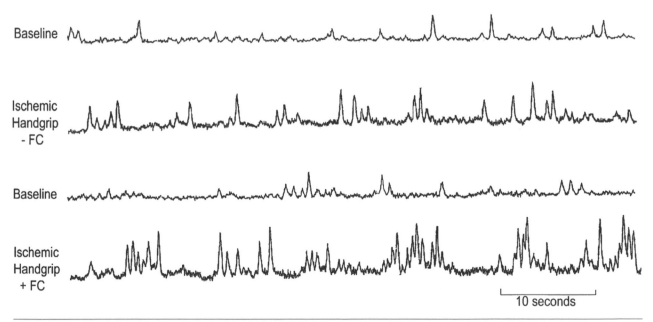

Figure 20.4   MSNA response at baseline and during ischemic handgrip without and with forearm compression (–FC and +FC, respectively). External forearm compression markedly increased the MSNA response to ischemic handgrip. Not shown in this figure are similar levels of MSNA during circulatory arrest periods following the two bouts of handgrip. These findings provide evidence that forearm compression augmented MSNA via a nonmetaboreceptor mediator process.

conditions. During the first group of experiments, healthy volunteers performed static handgrip (2 min/40% MVC) followed by postexercise ischemia with and without first inducing limb congestion. Limb congestion was induced by a 5-min period of forearm venous occlusion that preceded the bout of handgrip. Venous occlusion to 90 mmHg caused a 5% increase in forearm volume. This intervention increased the MSNA noted during the first minute of handgrip. Based on prior work, it seemed unlikely that muscle metaboreceptors were contributing significantly to the MSNA response seen during the first minute of handgrip (Mark et al. 1985; Victor et al. 1988). Limb congestion also increased MSNA seen at end exercise but did not augment the posthandgrip ischemic MSNA. These results provide evidence that limb congestion does not enhance the muscle metaboreceptor contribution to MSNA.

A second group of limb congestion experiments were performed under ischemic conditions. If limb congestion augments MSNA independent of changes in flow (and engagement of the muscle metaboreflex), the effects of congestion should then be present during congested ischemic handgrip. Venous congestion augmented the MSNA response during the second minute of ischemic handgrip. However, it had no effect on MSNA responses during the postexercise circulatory arrest phase.

To examine if limb congestion increased central command, skin sympathetic nerve activity (SSNA) was measured during control and congested handgrip. Vissing, Scherrer, and Victor (1991) had previously provided data that skin sympathetic nerve activity is an index of central command during handgrip. Accordingly, if limb congestion augmented central command, an effect of venous congestion on SSNA would be implicated. In these experiments, limb congestion had no effect on skin sympathetic nerve activity during static handgrip. In separate experiments, the effects of limb congestion on MSNA responses to involuntary biceps contractions were obtained. During involuntary contraction, no activation of central command would occur. One minute of involuntary biceps contractions did not increase MSNA above baseline values. When involuntary contraction was coupled with limb congestion, a significant increase in MSNA was observed (McClain et al. 1993). These findings provide evidence that forearm compression augmented MSNA via a nonmetaboreceptor mediator process. In summary, acute limb congestion increases sympathetic outflow during handgrip. This increase in sympathetic outflow does not appear to be due to metaboreceptor activation or to increased central command. These observations are most likely due to an effect of limb congestion on mechanically sensitive muscle afferents.

Recent work has focused on the effects of rhythmic exercise on the muscle reflex. Work by Piepoli et al. (1996) has demonstrated that patients with heart failure engage the muscle reflex to a greater degree than control subjects. In these studies, upright seated subjects performed rhythmic handgrip exercise until fatigue. Vascular resistance was measured in the dependent calf and was used as an index of efferent sympathetic drive. In heart failure subjects, postexercise ischemia led to sustained levels of calf vascular resistance and blood pressure. In control subjects, however, vascular resistance and blood pressure fell toward baseline values during postexercise ischemia. Interestingly, minute ventilation was increased above baseline in both groups, but the magnitude of the ventilatory response was greater in the heart failure subjects. These results suggest that the muscle reflex is engaged to a greater degree during rhythmic exercise. That the enhanced reflex responses observed in the heart failure subjects were attenuated by forearm exercise conditioning is interesting (Piepoli et al. 1996).

Recent preliminary observations using a rhythmic handgrip paradigm (25% MVC, 12 contractions/min, 2 s each) have examined MSNA responses in CHF and age-matched controls. As opposed to the healthy volunteers, the heart failure subjects studied were unable to complete the 20 min of rhythmic exercise. This early onset of fatigue was associated with a large increase in MSNA due to engagement of the muscle metaboreflex since MSNA stayed elevated during postexercise forearm circulatory arrest. Separate experiments measured SSNA during rhythmic handgrip in CHF subjects and in controls. SSNA responses increased in a similar fashion in the two groups. However, during the postexercise muscle ischemia period, SSNA remained elevated in the CHF subjects but, as expected, fell toward baseline values in the control studies. These results were interpreted as suggesting that in CHF, SSNA is not an index of central command but, rather, is due to muscle metaboreceptor engagement. NMR experiments during this rhythmic exercise paradigm demonstrated far larger levels of muscle acidosis in the CHF subjects. These observations provide evidence for the first time that rhythmic exercise potently engages the metaboreflex because of profoundly greater production of cellular metabolites. When engaged, this reflex evokes an increase in sympathetic activity directed to both skin and muscle. This is very different than the responses seen in control subjects (Leuenberger, Silber, and Sinoway 1995; Sutliff, Sinoway, and Leuenberger 1994). Clearly, data are needed to determine if metaboreflex activation in CHF causes vasoconstriction within active vascular beds.

In summary, muscle reflexes play an important role in regulating the circulation during exercise. However, many questions regarding this system remain unanswered. Further work in healthy volunteers and heart failure subjects is necessary to understand better the mechanisms and role this reflex system plays in health and disease. From what is presently known, the muscle metaboreflex is apparently not effectively engaged during static exercise in heart failure. Associated with this reduced level of metaboreceptor activation is heightened activity of some other reflex system. Enhanced discharge of muscle mechanoreceptors may be important in heart failure where increased interstitial volume is commonly seen. During rhythmic exercise, the situation differs. Here the production of muscle metabolites is dramatically increased in CHF as compared with control subjects. This leads to potent metaboreceptor activation. Interestingly, this potent metaboreceptor activation evokes increases in both MSNA and SSNA. The importance of these sympathetic drives in regulating blood flow delivery to various circulations in humans remains to be explored.

Summary

Heart failure patients possess a limited exercise capacity due to a limited pumping capacity of the heart and also to a lesser degree due to peripheral limitations in muscular vasodilatory reserve and metabolic capacity. During dynamic exercise with a small muscle mass (e.g., 1-leg knee extension), heart failure patients can achieve peak muscle perfusion values similar to healthy controls, because the pumping capacity of the heart is not threatened. However, they do so with enhanced sympathetic activation. When the active muscle mass is large, however, peak muscle perfusion is markedly reduced by enhanced sympathetic vasoconstriction, which is apparently linked directly to impaired left ventricular function and reduced cardiac output.

Handgrip exercise in humans stimulates muscle afferents evoking the exercise pressor reflex. The increase in blood pressure is associated with increased sympathetic discharge directed to skeletal muscle. The muscle metaboreceptors, group IV unmyelinated nerve fibers, are thought to play a prominent role in evoking this response. Lactic acid,

H+, diprotonated phosphate, adenosine, and the by-products of prostaglandin synthesis have been suggested as potential afferent stimulants.

In heart failure subjects, the metaboreceptor contribution to muscle sympathetic nerve activity during static handgrip is reduced. This effect may be due to afferent desensitization. Despite this presumed metaboreceptor desensitization in congestive heart failure, sympathoexcitatory responses during static forearm exercise are preserved. This suggests that some other neural system aside from muscle metaboreceptors must be activated to a greater degree during static handgrip in heart failure. We speculate that in heart failure, the activity of muscle mechanoreceptors may be increased. Data from healthy subjects utilizing acute forearm venous congestion suggests that limb edema can increase the discharge of muscle mechanoreceptors, thereby evoking nonmetaboreceptor-mediated increases in sympathetic discharge.

During rhythmic handgrip, CHF subjects (as compared with controls) produce much larger quantities of cellular by-products, and they prematurely fatigue. This process leads to an impressive activation of the muscle metaboreflex. Surprisingly, this activation evokes increases in both MSNA and SSNA.

References

Alam, M., and F.H. Smirk. 1937. Observations in man upon a blood pressure raising reflex arising from the voluntary muscles. *J. Physiol. (Lond.)* 89:372–383.

Batman, B.A., J.C. Hardy, U.A. Leuenberger, M.B. Smith, Q.X. Yang, and L.I. Sinoway. 1994. Sympathetic nerve activity during prolonged rhythmic forearm exercise. *J. Appl. Physiol.* 76:1077–1081.

Boushel, R., P. Madsen, B. Quistorff, and N.H. Secher. 1998. Contribution of pH, potassium and diprotonated phosphate for the reflex increase in blood pressure during handgrip. *Acta Physiol. Scand.* 164:269–275.

Coote, J.H., S.M. Hilton, and J.F. Perez-Gonzalez. 1971. The reflex nature of the pressor response to muscular exercise. *J. Physiol. (Lond.)* 215:789–804.

Costa, F., and I. Biaggioni. 1994. Role of adenosine in the sympathetic activation produced by isometric exercise in humans. *J. Clin. Invest.* 93:1654–1660.

Drexler, H., U. Reide, T. Munzel, H. Konig, E. Funke, and H. Just. 1992. Alterations of skeletal muscle in chronic heart failure. *Circulation* 85:1751-1759.

Dzwonczyk, T.D., K.S. Gray, and L.I. Sinoway. 1994. Diprotonated phosphate is an important muscle afferent stimulant. *Circulation* 90:74–81.

Ettinger, S., K. Gray, S. Whisler, and L. Sinoway. 1991. Dichloroacetate reduces sympathetic nerve responses to static exercise. *Am. J. Physiol.* 261:H1653–H1658.

Fallentin, N., B.R. Jensen, S. Byström, and G. Sjøgaard. 1992. Role of potassium in the reflex regulation of blood pressure during static exercise in man. *J. Physiol.* 451:643-651.

Franciosa, J.A., M. Parker, and T.B. Levine. 1981. The lack of correlation between exercise capacity and indexes of resting ventricular performance in heart failure. *Am. J. Cardiol.* 47:33-39.

Francis, G.S., S.R. Goldsmith, and J.N. Cohn. 1982. Relationship of exercise capacity to resting left ventricular performance and basal plasma norepinephrine levels in patients with congestive heart failure. *Am. Heart J.* 104:725-731.

Gandevia, S.C., K. Killian, D.K. McKenzie, M. Crawford, G.M. Allen, R.B. Gorman, and J.P. Hales. 1993. Respiratory sensations, cardiovascular control, kinesthesia and transcranial stimulation during paralysis in humans. *J. Physiol.* 470:85–107.

Harrison, T.R., C.E. King, J.A. Calhoun, and W.G. Harrison. 1934. Congestive heart failure: Cheyne-Stokes respiration as the cause of paroxysmal dyspnea at the onset of sleep. *Arch. Intern. Med.* 53:891–910.

Hellström, G., G. Magnusson, N.G. Wahlgren, A. Gordon, C. Sylven, and B. Saltin. 1996. Physical exercise may impair cerebral perfusion in patients with chronic heart failure. *Cardiol Elderly* 4:191-194.

Hilton, S.M., O. Hudlicka, and J.M. Marshall. 1978. Possible mediators of functional hyperemia in skeletal muscle. *J. Physiol.* 282:131–147.

Kaufman, M.P., and H.V. Forster. 1996. Reflexes controlling circulatory, ventilatory and airway responses to exercise. In *Handbook of physiology, section 12, exercise: Regulation and integration of multiple systems*, ed. L.B. Rowell and J.T. Shepherd, 381–447. New York: Oxford University Press.

Kaufman, M.P., J.C. Longhurst, K.J. Rybicki, J.H. Wallach, and J.H. Mitchell. 1983. Effects of static muscular contraction on impulse activity of group IH and IV afferents in cats. *J. Appl. Physiol.* 55:105–112.

Kaufman, M.P., K.J. Rybicki, T.G. Waldrop, and G.A. Ordway. 1984. Effect of ischen-iia on responses of group III and IV afferents to contraction. *J. Appl. Physiol.* 57:644–650.

Leimbach, W.N., Jr., B.G. Wallin, R.G. Victor, P.E. Aylward, G. Sundlöf, and A.L. Mark. 1986. Direct evidence from intraneural recordings for increased central sympathetic outflow in patients with heart failure. *Circulation* 73:913–919.

LeJemtel, T.H., C.S. Maskin, D. Lucido, and B.J. Chadwick. 1986. Failure to augment maximal limb blood flow in response to one-leg versus two-leg exercise in patients with severe heart failure. *Circulation* 74:245–251.

Leuenberger, U.A., D.H. Silber, and L.I. Sinoway. 1995. Differential control of skin sympathetic nerve activity during rhythmic handgrip exercise in heart failure and controls. *Circulation* 92:I655.

Lipkin, D.P., D.A. Jones, J.M. Round, and P.A. Poole-Wilson. 1988. Abnormalities of skeletal muscle in patients with chronic heart failure. *Int J. Cardiol.* 18:187-195.

Longhurst, J., W. Gifford, and R. Zelis. 1976. Impaired forearm oxygen consumption during static exercise in patients with congestive heart failure. *Circulation* 54:477–480.

MacLean, D.A., B. Saltin, G. Rådegran, and L. Sinoway. 1998. The femoral arterial injection of adenosine in humans elevates MSNA via central but not peripheral mechanisms. *J. Appl. Physiol.* 83:1045-1053.

Magnusson, G., L. Kaijser, C. Sylvèn, K.-E. Karlberg, B. Isberg, and B. Saltin. 1997. Peak skeletal muscle perfusion is maintained in patients with chronic heart failure when only a small muscle mass is exercised. *Cardiovasc. Res.* 33:297-306.

Mancini, D.M., E. Coyle, A. Coggan, J. Beltz, N. Ferraro, S. Montain, and J.R. Wilson. 1989. Contribution of intrinsic skeletal muscle changes to 31P NMR skeletal muscle metabolic abnormalities in patients with chronic heart failure. *Circulation* 80:1338-1346.

Mark, A.L., R.G. Victor, C. Nerhed, and B.G. Wallin. 1985. Microneurographic studies of the mechanisms of sympathetic nerve responses to static exercise in humans. *Circ. Res.* 57:461–469.

McClain, J., C. Hardy, B. Enders, M. Smith, and L. Sinoway. 1993. Limb congestion and sympathoexcitation during exercise: Implications for congestive heart failure. *J. Clin. Invest.* 92:2353–2359.

McClain, J., J. Hardy, and L. Sinoway. 1994. Forearm compression during exercise increases sympathetic nerve traffic. *J. Appl. Physiol.* 77:2612–2617.

McCloskey, D.I., and J.H. Mitchell. 1972. Reflex cardiovascular and respiratory responses originating in exercising muscle. *J. Physiol. (Lond.)* 224:173–186.

Mense, S., and M. Stahnke. 1983. Responses in muscle afferent fibres of slow conduction velocity to contractions and ischaemia in the cat. *J. Physiol.* 243:383–397.

Minotti, J.R., I. Christoph, R. Oka, M.W. Weiner, L. Wells, and B.M. Massie. 1991. Impaired skeletal muscle function in patients with congestive heart failure. Relationship to systemic exercise performance. *J. Clin. Invest.* 88:2077–2082.

Mitchell, J.H., M.P. Kaufman, and G.A. Iwamoto. 1983. The exercise pressor reflex: Its cardiovascular effects, afferent mechanism, and central pathways. *Ann. Rev. Physiol.* 45:229–242.

Nosek, T., K. Fender, and R. Godt. 1987. It is diprotonated inorganic phosphate that depresses force in skinned skeletal muscle fibers. *Science Wash. DC* 236:191–192.

Piepoli, M., A.L. Clark, M. Volterrani, S. Adamopoulos, P. Sleight, and A.J.S. Coats. 1996. Contribution of muscle afferents to the hemodynamic, autonomic, and ventilatory responses to exercise in patients with chronic heart failure. Effects of physical training. *Circulation* 93:940–952.

Rotto, D., and M. Kaufman. 1988. Effect of metabolic products of muscular contraction on the discharge of group III and IV afferents. *J. Appl. Physiol.* 64:2306–2313.

Rowell, L.B., L. Hermansen, and J.R. Blackmon. 1976. Human cardiovascular and respiratory responses to graded muscle ischemia. *J. Appl. Physiol.* 41:693–701.

Saito, M., T. Mano, and S. Iwase. 1989. Sympathetic nerve activity related to local fatigue sensation during static contraction. *J. Appl. Physiol.* 67:980–984.

Seals, D.R., and R.M. Enoka. 1989. Sympathetic activation is associated with increases in EMG during fatiguing exercise. *J. Appl. Physiol.* 66:88–95.

Seals, D.R., and R.G. Victor. 1991. Regulation of muscle sympathetic nerve activity during exercise in humans. In *Exercise & sport sciences reviews,* ed. J.O. Holloszy, vol. 19, 313–349. Baltimore: Williams and Wilkins.

Shafiq, S.A., V. Askanas, S.A. Aseidu, and A.T. Milhorat. 1972. Structural changes in human and chicken muscular dystrophy. *Muscle Biol.* 1:255–272.

Sinoway, L.I., J.M. Hill, J.G. Pickar, and M.P. Kaufman. 1993. Effects of contraction and lactic acid on the discharge of group III muscle afferents in cats. *J. Neurophysiol.* 69:1053–1059.

Sinoway, L., and S. Prophet. 1990. Skeletal muscle metaboreceptor stimulation opposes peak metabolic vasodilation in humans. *Circ. Res.* 66:1576–1584.

Sinoway, L.I., R.F. Rea, T.J. Mosher, M.B. Smith, and A.L. Mark. 1992. Hydrogen ion concentration is not the sole determinant of muscle metaboreceptor responses in humans. *J. Clin. Invest.* 89:1875–1884.

Sinoway, L.I., M.B. Smith, B. Enders, U. Leuenberger, T. Dzwonczyk, K. Gray, S. Whisler, and R.L. Moore. 1994. Role of diprotonated phosphate in evoking muscle reflex responses in cats and humans. *Am. J. Physiol. (Heart Circ. Physiol.)* 267(36):H770–H778.

Sinoway, L.I., K.J. Wroblewski, S.A. Prophet, S.M. Ettinger, K.S. Gray, S.K. Whisler, G. Miller, and R.L. Moore. 1992. Glycogen depletion-induced lactate reductions attenuate reflex responses in exercising humans. *Am. J. Physiol. (Heart Circ. Physiol.)* 263(32):H1499–H1505.

Stebbins, C.L., B. Brown, D. Levin, and J.C. Longhurst. 1988. Reflex effect of skeletal muscle mechanoreceptor stimulation on the cardiovascular system. *J. Appl. Physiol.* 65:1539–1547.

Sterns, D.A., S.M. Ettinger, K.S. Gray, S.K. Whisler, T.J. Mosher, M.B. Smith, and L.I. Sinoway. 1991. Skeletal muscle metaboreceptor exercise responses are attenuated in heart failure. *Circulation* 84:2034–2039.

Sullivan, M.J., and F.R. Cobb. 1991. Dynamic regulation of leg vasomotor tone in patients with chronic heart failure. *J. Appl. Physiol.* 71:1070–1075.

Sullivan, M.J., J.D. Knight, M.B. Higginbotham, and F.R. Cobb. 1989. Relation between central and peripheral hemodynamics during exercise in patients with chronic heart failure. *Circulation* 80:769–781.

Sutliff, G.M., L.I. Sinoway, and U.A. Leuenberger. 1994. Rhythmic exercise leads to muscle metaboreflex activation in heart failure but not in controls. *Circulation* 90:I-539.

Victor, R.G., L. Bertocci, S. Pryor, and R. Nunnally. 1988. Sympathetic nerve discharge is coupled to muscle cell pH during exercise in humans. *J. Clin. Invest.* 82:1301–1305.

Victor, R.G., S.L. Pryor, N.H. Secher, and J.H. Mitchell. 1989. Effects of partial neuromuscular blockade on sympathetic nerve responses to static exercise in humans. *Circ. Res.* 65:468–476.

Victor, R.G., D.M. Rotto, S.L. Pryor, and M.P. Kaufman. 1989. Stimulation of renal sympathetic activity by static contraction: Evidence for mechanoreceptor-induced reflexes from skeletal muscle. *Circ. Res.* 64:592–599.

Vissing, S.F., U. Scherrer, and R.G. Victor. 1991. Stimulation of skin sympathetic nerve discharge by central command. Differential control of sympathetic outflow to skin and skeletal muscle during static exercise. *Circ. Res.* 69:228–238.

Vissing, J., S. Vissing, D.A. MacLean, B. Saltin, B. Quistorff, and R.G. Haller. 1997. Sympathetic activation in exercise is not dependent on muscle acidosis: Evidence from studies in metabolic myopathies. *Neurology* in press, (American Academy of Neurology Spring Meeting, April, 1997): Abstract.

Von During, M., and K.H. Andres. 1990. Topography and ultrastructure of group III and IV nerve terminals of cat's gastrocnemius-soleus muscle. In *The primary afferent neuron: A survey of recent morpho-functional aspects*, ed. W. Zenker and W.L. Neuhuber, 35–41. New York: Plenum Press.

Williamson, J.W., J.H. Mitchell, H.L. Olesen, P.B. Raven, and N.H. Secher. 1994. Reflex increase in blood pressure induced by leg compression. *J. Physiol.* 475:351–357.

Zelis, R., D.T. Mason, and E. Braunwald. 1968. A comparison of peripheral resistance vessels in normal subjects and in patients with congestive heart failure. *J. Clin. Invest.* 47:960–969.

Zelis, R., and L.I. Sinoway. 1992. Pathophysiology of congestive heart failure. In *Textbook of internal medicine*, ed. W.N. Kelley, 2d ed., 104–112. Philadelphia: Lippincott.

Chapter 21

Circulatory Regulation in Muscle Disease

Ronald G. Haller and John Vissing

Skeletal muscle requires the continuous delivery of oxygen and oxidizable substrate and the removal of metabolic end products to meet muscle metabolic demand. From rest to peak exercise, the rate of muscle oxidative phosphorylation may increase 50-fold or greater, a requirement range for O_2 and substrate that is unmatched by any tissue (Astrand and Rodahl 1986). To meet this exercise increase in muscle metabolic demand, systemic O_2 delivery (cardiac output) may increase more than fourfold from rest to peak exercise. The proportion of cardiac output specifically directed to active skeletal muscle may increase almost 10-fold (Mitchell and Blomqvist 1971). Thus, the increase in skeletal muscle oxidative demand represents the primary necessity for circulatory adjustments that accompany exercise. The high metabolic requirement of skeletal muscle alone taxes the full range of circulatory capacity.

The regulatory mechanism responsible for matching the delivery of oxygen and substrate to the wide range of muscle oxidative requirements in exercise is not known. Central (central command) and peripheral mechanisms may be involved. An important regulatory mechanism involves metaboreceptors, probably in the form of free nerve endings. These metaboreceptors are stimulated by metabolites that accumulate in exercising muscle (Mitchell 1990). Impulses from these receptors are transmitted by groups III and IV afferent nerves to the central nervous system. These impulses mediate efferent sympathetic neural responses that increase systemic blood pressure, heart rate, and myocardial contractility and, thus, ultimately increase systemic O_2 transport (cardiac output).

A central feature of circulatory regulation in exercise is a virtually 1:1 relationship between changes in O_2 utilization and delivery. This relationship is expressed by the ratio of increase in O_2 transport relative to the increase in O_2 utilization, i.e. $\Delta Q / \Delta(a-v)O_2$. In healthy humans, this ratio is approximately 5-6 as determined by the fact that arterial blood in individuals with normal O_2-carrying capacity contains approximately 200 ml O_2 per liter of blood. The

1:1 relationship between O_2 delivery and utilization (normally corresponding to about 5 L of Q per 1 L of $\dot{V}O_2$) is maintained irrespective of age, sex, or level of physical conditioning (Faulkner, Heigenhauser, and Schork 1977). Decreased O_2-carrying capacity due to anemia or hypoxia results in an increased ratio of Q relative to $\dot{V}O_2$. However, the ratio of O_2 delivery to O_2 utilization remains 1:1. This is consistent with the hypothesis that O_2 delivery, rather than the volume of cardiac output or blood flow per se, is regulated.

Studies of inborn errors of muscle metabolism provide unique insights into the probable link between muscle metabolism and circulatory regulation. Specifically, such studies indicate that muscle metabolic disorders that severely limit muscle oxidative phosphorylation, including metabolic defects impairing electron transport, tricarboxylic acid (TCA) cycle function, or causing complete blocks in muscle glycolysis or glycogenolysis (see figure 21.1), disrupt the normal coupling between O_2 utilization and delivery in exercise. These disorders then cause exaggerated exercise circulatory responses relative to metabolic rate. In contrast, metabolic myopathies that preserve muscle oxidative phosphorylation, including partial glycolytic defects and defects in muscle fatty acid metabolism, demonstrate a normal circulation in exercise (see figure 21.1). These findings indicate that normal muscle oxidative metabolism is central to the normal coupling of O_2 utilization and delivery in exercise. They also suggest that the signal to increase O_2 delivery is a function of muscle oxidative demand as expressed by muscle metabolites that reflect the level of muscle oxidative phosphorylation.

Circulatory Response to Exercise in Muscle Disease

The following sections discuss the circulatory response to exercise in muscle disease. Specifically, they review the Larsson-Linderholm syndrome,

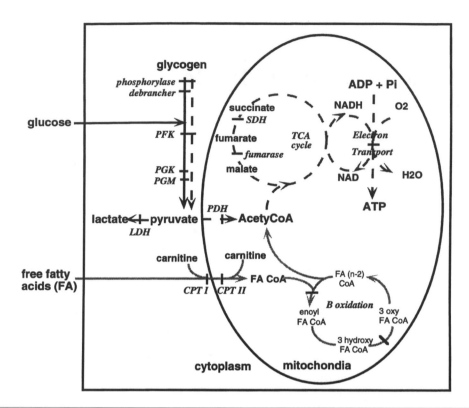

Figure 21.1   Outline of muscle metabolic pathways indicating some of the known metabolic myopathies associated with specific enzyme deficiencies. The dashed line, encompassing complete blocks of muscle glycogenolysis/ glycolysis, the tricarboxylic acid cycle, and the electron transport chain, indicates the location of metabolic blocks which result in exaggerated O_2 transport during exercise.

respiratory chain defects, muscle glycolytic defects, and circulatory regulation in other muscle disorders.

Larsson-Linderholm Syndrome

The classic model of disordered circulatory regulation in exercise in metabolic myopathy is the familial muscle disease described by Larsson et al. (1964) and Linderholm et al. (1969). These investigators reported about 14 patients from five families in northern Sweden with an autosomal recessive disorder associated with severe lifelong exercise intolerance punctuated by episodes of increased muscle fatigue, weakness, muscle swelling, and myoglobinuria. Oxygen utilization and circulation were normal at rest. Exercise was limited by low oxidative capacity. Exercise was associated with dramatic increases in lactate and pyruvate relative to workload and with a low peak ratio of lactate/pyruvate. $\dot{V}O_2$max was reduced to approximately 25% of that of healthy subjects despite the fact that peak levels of cardiac output were fully normal (Linderholm et al. 1969). The low oxidative capacity was thus related to a striking limitation in O_2 extraction. In contrast with the normal threefold increase in (a-v) O_2 difference from rest to peak exercise in healthy subjects, virtu-

ally no increase occurred in (a-v)O_2 difference in patients (Linderholm et al. 1969) (see figure 21.2). This marked limitation in O_2 extraction by working muscle was associated with a hyperkinetic circulation. In this type of circulation, systemic O_2 delivery (cardiac output) and blood flow to working muscle were greatly exaggerated relative to muscle work and metabolic rate. Instead of the normal 5-6 L increase in cardiac output and in muscle blood flow per liter of increased O_2 utilization (i.e, $\Delta Q/\Delta \dot{V}O_2 =$ 5–6), patients with this disorder had levels of cardiac output that were fourfold to sixfold higher (i.e., $\Delta Q/ \Delta \dot{V}O_2 =$ 20–30) (see figure 21.2).

The underlying muscle metabolic abnormality has been identified as deficiency of several iron-sulfur proteins of the TCA cycle and respiratory chain in skeletal muscle, including aconitase and succinate dehydrogenase (SDH) (Drugge et al. 1995; Hall et al. 1993; Haller et al. 1991; Linderholm, Essøn-Gustavsson, and Thomell 1990). The metabolic defect is apparently restricted to skeletal muscle as indicated by normal histochemical activity for SDH in muscle blood vessels (Haller et al. 1991), normal activities of aconitase and SDH in liver (Drugge et al. 1995), and sparing of cardiac metabolism as indicated by a normal capacity for cardiac

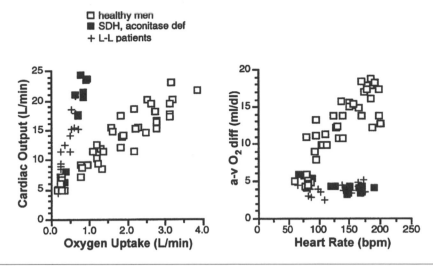

□ healthy men
■ SDH, aconitase def
+ L-L patients

Figure 21.2 Combined data from Linderholm et al. (1969) (+ = patients) and Haller et al. (1991) (closed square = patient, open square = control men) illustrating the hyperkinetic circulatory response to exercise and striking limitation in O_2 extraction in the Larsson-Linderholm syndrome. Note the exaggerated cardiac output relative to oxygen uptake in the patients com-pared to healthy subjects and the lack of increase in systemic arteriovenous O_2 difference in exercise in pati-ents in contrast to the three fold increase in healthy men.

work. Available evidence suggests that the rate-limiting muscle oxidative impairment in this disorder resides at the level of complex II (SDH) and/or aconitase. This causes deficient carbon flux via the TCA cycle with a low rate of production of reducing equivalents by the TCA cycle, thus limiting peak rates of electron transport and oxidative phosphorylation (Hall et al. 1993; Haller et al. 1991).

Linderholm and coworkers attributed the hyper-kinetic circulation to exaggerated metabolic vasodilation in active muscle related to the abnormally large accumulation of lactate and pyruvate during exercise in this condition. However, despite a marked fall in total peripheral resistance, mean arterial pressure in the patients increased on average more than 20 mmHg with an average cycle workload of 25 W (Linderholm et al. 1969). This indicated an exaggerated exercise pressor response. This is consistent with the view that the excessive increase in cardiac output in this metabolic myopathy is driven by an exaggerated "muscle-heart reflex" in exercise rather than by baroreceptor activation. Furthermore, a hyperkinetic exercise circulatory response also occurs in disorders that block muscle glycogenolysis and limit pyruvate-dependent oxidative metabolism. Thus, increased muscle glycogenolysis is not a common denominator of exaggerated circulatory responses to exercise in metabolic myopathies. Apparently, the exaggerated circulatory response to exercise in the Larsson-Linderholm syndrome, with a disproportionate increase in O_2 delivery relative to muscle metabolic rate, is a direct consequence of the block in muscle oxidative metabolism.

Respiratory Chain Defects

Respiratory chain disorders are attributable to a deficiency of the enzyme complexes of the mito-chondrial inner membrane that link the oxidation of reducing equivalents (NADH, reduced flavin nucle-otide) and the reduction of molecular O_2 to H_2O (complexes I–IV) to the phosphorylation of ADP via ATP synthase (complex V) (Morgan-Hughes 1994). The fundamental oxidative limitation in respiratory chain disorders is a block in electron transport with the accumulation of reducing equivalents behind the block. A typical metabolic feature of respiratory chain defects is a steep increase in blood lactate and in the lactate/pyruvate ratio due to a disproportionate increase in lactate relative to pyruvate. This reflects a steep increase in muscle cytoplasmic and mitochondrial redox. This contrasts with SDH-deficient and aconitase-deficient patients in whom pyruvate increases in parallel with lactate.

Detailed evaluations of physiological responses to exercise are few in persons with respiratory chain defects. However, consistent findings include an absent or strikingly attenuated increase in extraction of oxygen from circulating blood in exercise. This is indicated by simultaneous determination of cardiac output and oxygen uptake (Haller et al. 1989; Haller and Lewis 1990; Vissing, Galbo, and Haller 1996a) or by near-infrared (NIR) spectroscopy monitoring of oxyhemoglobin levels in working muscles (Bank and Chance 1994). As in SDH and aconitase deficiency, the circulation is normal at rest but is hyperkinetic during exercise with steep increases in heart rate with trivial exercise (Elliot et al. 1989). In addition, increases that are 3-4-fold normal occur in cardiac output in relation to increased oxygen uptake in exercise (Haller et al. 1989; Haller and Lewis 1990; Vissing, Galbo, and Haller 1996a).

Muscle blood flow to working muscle has not been systematically evaluated. However, NIR data indicate that the delivery of oxygenated blood is exaggerated relative to O_2 extraction in working muscle (Bank and Chance 1994). A hyperadrenergic response to exercise, indicated by a greatly exaggerated increase in norepinephrine and epinephrine compared with that in healthy subjects working at the same workload, is typical. This finding suggests that the hyperdynamic circulation in exercise results from exaggerated reflex activation of sympathetic outflow (Vissing, Galbo, and Haller 1996a).

Muscle Glycolytic Defects

The third major category of muscle oxidative defects associated with an exaggerated circulatory response to exercise is complete blocks in muscle glycogenolysis. This is attributable to muscle phosphorylase deficiency, also known as McArdle's disease (MD), and to muscle phosphofructokinase (PFK) deficiency (Tarui disease). While glycolytic disorders have classically been exclusively considered as defects of muscle anaerobic metabolism (Layzer 1985), researchers now recognize that a fundamental limitation of energy production during exercise in MD and PFK deficiency is a block in aerobic glycogenolysis/glycolysis. As a result, affected patients have maximal oxidative capacities of approximately half that of age-matched, healthy, sedentary individuals (Carroll et al. 1979; Elliot et al. 1989; Hagberg et al. 1982; Haller et al. 1983; Kissel et al. 1985; Lange, Lund-Johansen, and Clausen 1969). As in muscle respiratory chain defects and TCA cycle defects, oxygen utilization in exercise is limited by an attenuated increase in systemic (a-v)O_2 difference (Brechue et al. 1994; Haller et al. 1985; Haller and Lewis 1991). Instead of the normal three-fold increase in (a-v)O_2 difference, only an approximately 50% increase occurs in O_2 extraction in affected individuals. Correspondingly, NIR monitoring of changes in oxyhemoglobin reveal impaired O_2 extraction in working muscle (Bank and Chance 1994).

The basis of limited oxidative capacity in MD and PFK deficiency is substrate-limited oxidative phosphorylation (state 2 oxidative phosphorylation). The apparent mechanism is a pyruvate-dependent limitation in substrate flux via the TCA cycle. This results in an attenuated rate of production of reducing equivalents by the TCA cycle relative to the capacity of the respiratory chain for electron transport (Lewis and Haller 1986). Consistent with this hypothesis is an absence of the normal increase in mitochondrial NADH with exhausting exercise in McArdle's patients (Sahlin et al. 1990). The role of pyruvate in limiting TCA cycle function may be twofold. First, the generation of substrate for the TCA cycle, i.e., acetyl-CoA, is impaired as indicated by an absence of the normal increase in muscle levels of acetylcarnitine (which is in equilibrium with acetyl-CoA via carnitine acetyltransferase) during exhausting exercise (Sahlin et al. 1990). In addition, anaplerotic reactions involving pyruvate normally play an important role in increasing mitochondrial levels of four-carbon TCA cycle intermediates, such as malate, that likely are necessary to spark maximal rates of TCA cycle flux. Heavy exercise in patients with MD cause greatly attenuated increases in malate, fumarate, and citrate, consistent with impaired pyruvate-mediated anaplerosis (Sahlin et al. 1995).

Oxidative defects attributable to complete blocks in glycogenolysis have a normal circulation at rest. However, they display exaggerated increases in systemic O_2 transport and in limb blood flow during exercise. Cardiac output increases more than 10 L per liter of increase in O_2 utilization in exercise (Lewis et al. 1984; Lewis, Vora, and Haller 1991). Mean arterial pressure increases, and total peripheral resistance falls steeply relative to the change in metabolic rate in exercise (Lewis and Haller 1986). Blood levels of norepinephrine are high relative to work rate (Vissing et al. 1992; Vissing, Galbo, and Haller 1996b), consistent with an exaggerated exercise pressor response that outstrips metabolic vasodilation in working muscle. Exaggerated delivery of O_2 to working muscle relative to O_2 utilization has been documented by using near-infrared spectroscopy (Bank and Chance 1994). This is consistent with reports of greater-than-normal increases in limb blood flow during exercise (Barcroft et al. 1966; McArdle 1951; Porte et al. 1966).

A central feature of complete deficiency of muscle phosphorylase or PFK is a fluctuation in exercise capacity according to the availability of extramuscular substrate. At one extreme, patients experience a second wind in which they can easily tolerate previously exhausting exercise (Pearson, Rimer, and Mommaerts 1961; Pernow, Havel, and Jennings 1967; Porte et al. 1966). This improved exercise capacity may occur spontaneously or be induced by administration of fuels that bypass the metabolic block. It is apparently due to a substrate-mediated increase in oxidative capacity attributable to an increased rate of electron transport and oxidative phosphorylation (Bertocci, Haller, and Lewis 1993; Lewis and Haller 1986; Lewis, Vora, and Haller 1991). An increased capacity of working muscle to utilize available O_2 in oxidative phosphorylation is reflected in an increased peak (a-v)O_2 difference, whereas peak systemic O_2 delivery remains unchanged (see figure 21.3).

At the other extreme, the out-of-wind phenomenon denotes a decrease in exercise and oxidative capacity that accompanies a reduced availability of oxidative fuel. This typically occurs in PFK deficiency after a carbohydrate meal (Haller and Lewis 1991). The mechanism of glucose-induced exertional fatigue in PFK deficiency is apparently an insulin-mediated fall in blood free fatty acids and ketones—the major muscle oxidative fuels in this condition. This leads to a fall in peak rate of oxidative phosphorylation and a decline in the ability of working muscle to extract oxygen from circulating blood (Haller and Lewis 1991).

Substrate-dependent fluctuations in muscle oxidative phosphorylation in McArdle's disease and PFK deficiency are associated with reciprocal fluctuations in circulatory responses to exercise. Increased fuel availability results in a lower heart rate at a given level of exercise, whereas reduced fuel availability is associated with a higher heart rate relative to workload. Correspondingly, cardiac output and the $\Delta Q / \Delta \dot{V}O_2$ ratio is higher at a given work rate when fuel availability is low, while a more normal $\Delta Q / \Delta \dot{V}O_2$ response accompanies exercise when substrate availability is increased (see figure 21.3). These findings indicate that, in these glycolytic defects, fuel delivery limits muscle oxidative phosphorylation. These findings also suggest that muscle metabolic signals responsible for regulating the circulation in exercise reflect the balance between oxidative energy demand and supply in working muscle as determined by the rate of delivery of blood-borne fuels. Furthermore, these findings suggest the hypothesis that the close coupling between O_2 utilization and delivery in healthy subjects is due to the fact that O_2 delivery determines the level of metabolites that signal the balance between energy supply and demand.

Circulatory Regulation in Other Muscle Disorders

A specific link between muscle oxidative phosphorylation and circulatory regulation is suggested both by the presence of abnormal regulation of O_2 transport in exercise in muscle oxidative defects and by

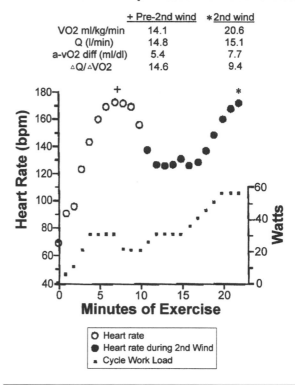

McArdles Disease: Spontaneous Second Wind

	+ Pre-2nd wind	*2nd wind
VO2 ml/kg/min	14.1	20.6
Q (l/min)	14.8	15.1
a-vO2 diff (ml/dl)	5.4	7.7
$\Delta Q/\Delta VO2$	14.6	9.4

○ Heart rate
● Heart rate during 2nd Wind
▪ Cycle Work Load

Figure 21.3 Oxidative and circulatory features of a spontaneous second wind in a 29-year-old woman with muscle phosphorylase deficiency. Cycle exercise was performed after an overnight fast. Note a rapid increase in heart rate as workload is increased to a maximum tolerated level of 30 watts. Subsequently, the workload was transiently lowered; shortly after, the patient noted that she was experiencing a second wind. The workload again was increased to the previous maximum (30 watts) and then to the new maximal work level of 55 watts. Corresponding O_2 consumption and cardiac output are shown above along with calculated systemic (a-v̄) O_2 difference and the ratio of increase in cardiac output relative to the increase in $\dot{V}O_2$ from rest to peak exercise ($\Delta Q / \Delta \dot{V}O_2$) corresponding to the pre-second wind and second wind values.

the absence of significant circulatory abnormalities in muscle disorders that preserve muscle oxidative metabolism. Limited data indicate that partial blocks in muscle glycogenolysis and glycolysis, including deficiency of phosphorylase b kinase, debrancher, phosphoglycerate mutase, and lactate dehydrogenase, preserve pyruvate-dependent oxidative phosphorylation and preserve a normal relationship between O_2 transport and delivery during exercise (Bryan et al. 1990; Haller et al. 1983; Lewis and Haller 1989). Similarly, a deficiency of carnitine palmitoyltransferase (CPT II) preserves muscle oxidative capacity and maintains a normal circulatory regulation in exercise (Lewis and Haller 1989; Lewis, Vora, and Haller 1991).

Matching of O_2 transport and O_2 utilization is also maintained in various forms of muscular dystrophy, including Duchenne, Becker, facioscapulohumeral, and limb-girdle (Haller et al. 1983; Lewis and Haller 1989). Primary diseases of skeletal muscle, including muscular dystrophy, result in a decrease in the number of functional muscle fibers and a decrease in muscle maximal contractile force. As a result, an increased number of motor units is recruited for a given level of muscle work. In other words, for a given work rate, central command increases. Thus, the finding that O_2 transport is not exaggerated in muscular dystrophy implies that exaggerated central command is not the primary mechanism of exaggerated cardiac output in muscle oxidative defects. In patients with exertional muscle pain without clinical, histological, or biochemical evidence of underlying muscle disease, the circulatory response to exercise is normal. This is consistent with the view that pain is not responsible for exaggerated circulatory responses in muscle oxidative defects (Haller et al. 1983; Lewis and Haller 1989).

Exercise Fuel Mobilization in Muscle Oxidative Defects

The increased sympathetic neural activity that underlies the reflex regulation of the circulation in exercise participates in a neurohumoral exercise response that mediates the mobilization of glucose from hepatic glycogen stores and long-chain fatty acids from triglyceride stores in adipose tissue. A general proportionality between fuel-mobilizing neurohumoral responses and relative exercise intensity exists in healthy subjects (Galbo 1983). However, the relative role of central command versus metabolic feedback from working muscle in regulating this response is controversial. Studies of fuel mobilization in patients with defects in muscle oxidative phosphorylation have identified exaggerated substrate mobilization as well as exaggerated circulatory responses during exercise. This suggests that a common link exists between oxidative demand in working muscle and the delivery of both O_2 and oxidizable substrate.

The prime example is muscle phosphorylase deficiency in which the selective block in muscle glycogen availability deprives skeletal muscle of its primary intracellular oxidative substrate. This makes skeletal muscle oxidative metabolism dependent upon the availability of blood-borne fuels, glucose,

and free fatty acids. Compared with healthy subjects exercising at the same absolute workload or a similar relative workload (work level that elicited the same heart rate response to exercise), McArdle's patients have a higher glucose production (from hepatic glycogenolysis) and utilization during exercise (Vissing et al. 1992). Similarly, blood levels of glycerol and free fatty acids increase to a greater extent in patients compared with control subjects. This is consistent with an increased rate of mobilization of fatty acids from adipose tissue. Underlying this exaggerated mobilization of extramuscular fuel was a greater rise in blood levels of growth hormone and cortisol and a greater fall in blood insulin compared with that in healthy subjects exercising at either the same absolute or the same relative workload as patients. Increased fuel mobilization coupled with a greater-than-normal increase in cardiac output augments the delivery of extramuscular fuels that bypass the metabolic block. The resulting substrate-mediated increase in muscle oxidative phosphorylation improves the balance between oxidative energy demand and supply and is responsible for the characteristic second wind (see figure 21.3).

Muscle PFK deficiency blocks the metabolism of both muscle glycogen and blood glucose. As a result, skeletal muscle heavily depends upon the metabolism of lipids to meet muscle oxidative requirements. The ingestion of carbohydrate or infusion of glucose potentiates an energy crisis in working muscle due to insulin-mediated inhibition of lipolysis and a reduction in available fatty acids and ketones (Haller and Lewis 1991). Arguably, therefore, an augmented mobilization of free fatty acids but attenuated hepatic glycogenolysis would be the appropriate response to exercise in PFK deficiency. Paradoxically, glucose mobilization is enhanced and blood glucose levels increase during exercise in PFK deficiency compared with control subjects exercising at the same absolute or same relative (matched heart rate) workload (Vissing, Galbo, and Haller 1996b). Glucose utilization was also high and comparable with that in control subjects exercising at the same heart rate, corresponding to an almost threefold greater workload (Vissing, Galbo, and Haller 1996b). This increase in glucose mobilization and utilization despite the inability of working muscle to utilize glucose as an energy substrate is compatible with the presence of futile glucose cycling.

Exercise in PFK deficiency also results in a greater-than-normal increase in free fatty acids and glycerol, consistent with enhanced lipolysis. Increased lipolysis combined with an exaggerated exercise circulatory response increases the delivery of FFA to working muscle. As in McArdle's disease, increased fuel mobilization results from an exaggerated neurohumoral exercise response. Levels of growth hormone, glucagon, and norepinephrine increased to a greater extent in PFK-deficient patients than in control subjects working at the same heart rate and, thus, presumably at a similar level of central command. These results suggest that fuel mobilization is regulated by muscle oxidative demand and is not directly influenced by the ability of skeletal muscle to oxidize a given fuel. This hypothesis is further supported by findings in patients with mitochondrial myopathy in whom the underlying biochemical abnormality is impaired respiratory chain function.

Exercise in mitochondrial myopathy patients results in a dramatic increase in glucose turnover relative to work rate (Vissing, Galbo, and Haller 1996a) despite the fact that glucose oxidation is blocked by impaired electron transport. Glucose production and utilization and plasma glucose concentrations were higher in patients than in control subjects working at the same heart rate and at an absolute workload that was almost 10-fold higher than that performed by the patients. Exercise increases were twofold (norepinephrine, epinephrine, ACTH) to 10-fold (growth hormone) greater in patients compared with control subjects exercising at the same heart rate. In contrast with glucose turnover, increases in glycerol and free fatty acids were not greater in patients. This possibly occurred due to the steep increase in blood lactate in the mitochondrial myopathy patients, promoting a lactate-mediated inhibition of lipolysis (Rosell and Saltin 1973). This increase in glucose production enhances muscle ATP production via anaerobic glycolysis. However, the benefit to the balance between energy supply and demand is small and achieved at the expense of a progressive accumulation of nonoxidized substrate. These results support the hypothesis that substrate mobilization is linked to oxidative demand rather than to oxidative capacity. The exaggerated sympathetic and hormonal responses in the patients compared with healthy subjects exercising at the same relative workload and, thus, presumably at a similar level of central command suggests reflex activation of neuroendocrine and circulatory centers in the central nervous system in response to stimulation of metaboreceptors in working muscle.

Links Between Muscle Metabolism, O_2, and Fuel Transport in Exercise

The exaggerated delivery of O_2 and fuel during exercise in muscle oxidative disorders strongly suggests that muscle metabolites that accumulate as a consequence of the block in muscle oxidative phosphorylation play a regulatory role in substrate mobilization and circulatory responses to exercise. The nature of these regulatory substances remains uncertain. However, studies of metabolic myopathies have helped to narrow the focus of potential metabolic mechanisms (see table 21.1).

Low PO_2, Increased PCO_2

Falling PO_2 and increased PCO_2 in venous effluent accompanies exercise of increasing intensity and reciprocally correlates with increasing O_2 transport in healthy individuals. However, in muscle oxidative defects, PO_2 remains at near-resting levels. Changes in PCO_2 are attenuated with peak exercise and peak levels of O_2 transport. This indicates that changes in these blood gases are not obligatory for circulatory regulation.

pH and Lactate

Increased levels of lactate and an exaggerated decline in muscle and blood pH relative to workload and metabolic rate accompany exercise in respiratory chain defects and in SDH/aconitase deficiency. However, the opposite is true of muscle glycolytic defects. In both muscle phosphorylase and muscle PFK deficiencies, the complete block in glycogenolysis and glycolysis, respectively, result in no increase in lactate and a slight increase in muscle pH during exercise. All of these conditions are associated with hyperkinetic circulatory and neuroendocrine responses to exercise. They indicate that a fall in muscle or interstitial pH or an increase in lactate is not necessary to mediate the exercise reflex.

Inorganic Phosphate

Increasing exercise intensity is associated with a progressive fall in muscle phosphocreatine and corresponding increases in inorganic phosphate. In addition, the corresponding fall in muscle pH causes a progressive increase in the proportion of diprotonated phosphate (Miller et al. 1988). An exaggerated fall in PCr and an increase in Pi and diprotonated phosphate relative to exercise intensity is typical of mitochondrial myopathies (respiratory chain defects). In McArdle's disease as well, PCr hydrolysis and the increase in Pi is exaggerated in relation to work performed (Ross et al. 1981). However, due to the block in glycogenolysis, pH rises rather than falls in exercise, and no increase of diprotonated

Table 21.1   Substrate Mobilization/Utilization, Muscle Metabolism, and Systemic O$_2$ Transport With Exercise in Muscle Oxidative Defects

| | Substrate utilization | | | | Cellular metabolism | | | | | | | | Systemic/Local O$_2$ Delivery/Uptake | | | |
| | Anaerobic | Aerobic | | | | | | | | | | | | | | |
Defect	Glycogen	Glycogen	Glucose	FFA	Sug phos	Lact	H+	AMP	ADP	Pi	NADH/NAD	NH$_3$	Max a-v O$_2$ diff	ΔQ̇/ΔV̇O$_2$	Muscle blood flow	Muscle O$_2$ uptake
MP	X	X	↑	↑	⇓	⇓	⇓	↑	⇑	±↑	c⇓,m↓	⇑	↓	↑	↑	↓
PFK	X	X	↓	↑	⇑	⇓	⇓	↑	⇑	↓	c⇓,m↓	⇑	↓	↑	↑	↓
TCA cycle	↑	⇓	⇓	⇓	?	⇑	↑	?↑	↑	↑	c↓,m↓	?	⇓	⇑	⇑	⇓
Resp chain	↑	⇓	⇓	⇓	?	⇑	↑ or nl	?↑	↑	↑	c↑,m↑	?	↓	↑	↑	↓

MP = myophosphorylase, PFK = phosphofructokinase, TCA = tricarboxylic acid, resp chain = respiratory chain. FFA = free fatty acids; sug phos = glucose phosphates; lact = lactate; AMP = adenosine monophosphate; ADP = adenosine diphosphate; Pi = inorganic phosphate; NADH/NAD = reduced relative to oxidized nicotinamide adenine dunucleotide, c = cytoplasmic, m= mitochondria; H+ = hydrogen ion accumulation in exercise, (a-v̄) O$_2$ difference = systemic arteriovenous oxygen difference, (↓ = impaired oxygen extraction by working muscle), ΔQ̇/ΔV̇$_2$ = increase in cardiac output relative to increase in oxygen uptake in large muscle exercise. nl = normal, ↑ = increased, ⇑ = greatly increased; ↓ = decreased, ⇓ = greatly decreased. ? = unknown.

phosphate occurs (Argov et al. 1987a). Furthermore, in muscle PFK deficiency, the increase in muscle Pi is markedly attenuated due to phosphate trapping in sugar phosphates (G1P, G6P, F6P) behind the enzymatic block (Argov et al. 1987b; Bertocci et al. 1991). The fact that circulatory and neuroendocrine responses to exercise are similar in PFK and myophosphorylase deficiencies indicates that muscle inorganic phosphate is not the common denominator of exaggerated O$_2$ transport in exercise.

Potassium

Muscle excitation results in a progressive increase in extracellular potassium in proportion to exercise intensity. Potassium is both a potent vasodilator (Kjellmer 1965) and an activator of muscle sympathetic afferents that mediate cardiopulmonary reflexes in exercise (Wildenthal et al. 1968). Blood potassium levels increase to a greater extent in relation to workload in McArdle's patients compared with healthy subjects (Paterson et al. 1990). Possible mechanisms include increased potassium release due to fatigue-related increased motor unit recruitment (Braakhekke et al. 1986), exaggerated

activation of ATP-dependent muscle potassium channels due to the energy deficit (Paterson et al. 1990) muscle fiber injury (Fleckenstein et al. 1989), or limited reuptake due to the energy-limited function of Na$^+$-K$^+$ ATPase (Lewis and Haller 1986) or to low sarcolemmal pump numbers (Haller, Clausen, and Vissing 1998). Proton magnetic resonance imaging indicates that the normal increase in extracellular water is greatly attenuated as a result of the block in glycogenolysis in McArdle's disease (Fleckenstein et al. 1991), thus promoting higher concentrations of extracellular solutes, including potassium, in exercise. In patients with muscle PFK deficiency and mitochondrial myopathy, the exercise increase in blood levels of potassium is also exaggerated relative to workload (Vissing, Galbo, and Haller 1996a, 1996b). However, at similar relative workloads, blood potassium levels are not higher in individuals with these oxidative defects than in healthy subjects (Paterson et al. 1990; Vissing, Galbo, and Haller 1996a, 1996b). Thus, although interstitial potassium may participate in mediating exaggerated stimulation of circulatory- and substrate-mobilizing reflexes in muscle oxidative defects during exercise, potassium is unlikely to be the sole metabolic mechanism.

Adenosine and Related Metabolites

A consistent feature of muscle oxidative defects is an impaired rate of oxidative phosphorylation of ADP. As a result, exercise causes an exaggerated decline in the phosphorylation potential, i.e., [ATP]/[ADP][Pi] or [ATP]/[ADP], in each of the disorders that demonstrate an exaggerated circulatory response to exercise. Experimental inhibition of electron transport causes similarly exaggerated blood flow relative to metabolic rate in which a close inverse relationship occurs between flow and the fall in [ATP]/[ADP][Pi] (Nuutinen et al. 1982). These results suggest that metabolic factors related to the fall in the muscle [ATP]/[ADP] ratio or to increases in hydrolysis products of ATP participate in regulating O_2 delivery to muscle.

In muscle glycolytic defects, an exaggerated exercise increase in muscle ADP is typical (Radda 1986). The mechanism is multifactorial. It relates to impaired oxidative phosphorylation. It also relates to blocked substrate-level phosphorylation and particularly to an absence of the normal fall in muscle pH in heavy exercise that thus shifts the creatine kinase equilibrium (i.e., ADP + PCr + H+ $\leftrightarrow$ ATP + creatine) to the left. Increased ADP results in proportional increases in AMP through the near-equilibrium myokinase reaction (i.e., 2ADP $\leftrightarrow$ ATP + AMP). AMP deamination (via muscle AMP deaminase) results in ammonia and inosine monophosphate (AMP $\rightarrow$ IMP + NH_3) production. Levels of muscle IMP increase excessively in relation to workload in McArdle's patients (Sahlin et al. 1990). Ammonia production is exaggerated in both McArdle's disease and PFK deficiency, consistent with exaggerated AMP deamination. Blood ammonia levels directly correlate with heart rate during cycle exercise in McArdle's patients (Coakley, Wagemnakers, and Edwards 1992). There is also increased availability of blood-borne fuels that bypass the metabolic block and normalize the circulatory response to exercise in McArdle's disease and PFK lower muscle levels of ADP and venous effluent ammonia (Bertocci, Haller, and Lewis 1993; Lewis et al. 1985; Lewis and Haller 1986).

These results reinforce the possibility of a link between oxidative metabolites and circulatory regulation. However, the specific metabolites and mechanisms are not known. One possibility is that an increase in AMP leads to an increase in adenosine, which could promote vasodilation and activate muscle sympathetic afferents (Costa and Biaggioni 1994). Adenosine has not been measured in the metabolic myopathies known to be associated with exaggerated O_2 transport during exercise. Muscle adenosine has been reported to be elevated during exercise in patients with myoadenylate deaminase deficiency (Sabina et al. 1984). However, limited data in two patients with AMP deaminase deficiency revealed a normal circulatory response to exercise (Haller and Lewis, unpublished observation). A patient with a combined deficiency of muscle phosphorylase and myoadenylate deaminase had a circulatory response to exercise that did not differ from typical McArdle's patients (Lewis and Haller 1989). Thus, while a direct role of oxidative phosphorylation in active muscle regulating the circulatory- and substrate-mobilizing response to exercise seems probable, the responsible metabolic mechanism remains obscure.

References

Argov, A., W.J. Bank, J. Maris, J.S. Leigh, and B. Chance. 1987a. Muscle energy metabolism in human phosphofructokinase deficiency as recorded by ^{31}P NMR. *Ann. Neurol.* 22:46–51.

Argov, Z., W.J. Bank, J. Maris, and B. Chance. 1987b. Muscle energy metabolism in McArdle's syndrome by in vivo phosphorus magnetic resonance spectroscopy. *Neurology* 37:1720–1724.

Astrand, P.O., and K. Rodahl. 1986. *Textbook of work physiology: Physiological basis of exercise.* New York: McGraw-Hill.

Bank, W., and B. Chance. 1994. An oxidative defect in metabolic myopathies: Diagnosis by noninvasive tissue oxymetry. *Ann. Neurol.* 36:830–837.

Barcroft, H., B. Greenwood, B. McArdle, R. McSwiney, S.J.G. Semple, R.F. Whelan, and L.J.F. Youlten. 1966. The effect of exercise on forearm blood flow and on venous blood pH, P_cO_2 and lactate in a subject with phosphorylase deficiency in skeletal muscle. *J. Physiol.* 189:44P–46P.

Bertocci, L.A., R.G. Haller, and S.F. Lewis. 1993. Muscle metabolism during lactate infusion in muscle phosphofructokinase deficiency. *J. Appl. Physiol.* 74:1342–1347.

Bertocci, L.A., R.G. Haller, S.F. Lewis, J.L. Fleckenstein, and R.L. Nunnally. 1991. Altered high energy phosphate metabolism during exercise in muscle phosphofructokinase deficiency. *J. Appl. Physiol.* 70:1201–1207.

Braakhekke, J.P., M.I. Debruin, D.F. Stegeman, R.A. Wevers, R.A. Binkhorst, and E.M.G. Joosten. 1986. The second wind phenomenon in McArdle's disease. *Brain* 109:1087–1101.

Brechue, W.F., K.E. Gropp, B.T. Ameredes, D.M. O'Drobinak, W.N. Stainsby, and J.W. Harvey. 1994. Metabolic and work capacity of skeletal muscle of PFK-deficient dogs studied in situ. *J. Appl. Physiol.* 77:2456–2467.

Bryan, W., S.F. Lewis, L. Bertocci, M. Gunder, K. Ayyad, P. Gustafson, and R.G. Haller. 1990. Muscle lactate dehydrogenase deficiency—a disorder of anaerobic glycogenolysis associated with exertional myoglobinuria. *Neurology* 40:203.

Carroll, J.E., J.M. Hagberg, M.H. Brooke, and J.B. Shumate. 1979. Bicycle ergometry and gas exchange measurements in neuromuscular diseases. *Arch. Neurol.* 36:457–461.

Coakley, J.H., A.J. Wagemnakers, and R.H. Edwards. 1992. Relationship between ammonia, heart rate, and exertion in McArdle's disease. *Am. J. Physiol.* 262:E167–72.

Costa, F., and I. Biaggioni. 1994. Role of adenosine in the sympathetic activation produced by isometric exercise in humans. *J. Clin. Invest.* 93:1654–1660.

Drugge, U., M. Holmberg, G. Holmgren, B.G.L. Almay, and H. Linderholm. 1995. Hereditary myopathy with lactic acidosis, succinate dehydrogenase and aconitase deficiency in northern Sweden. *J. Med. Genet.* 32:344–347.

Elliot, D.L., N.R.M. Buist, L. Goldberg, N.G. Kennaway, B.R. Powell, and K.S. Kuehl. 1989. Metabolic myopathies: Evaluation by graded exercise testing. *Medicine* 68:163–172.

Faulkner, J.A., G.F. Heigenhauser, and M.A. Schork. 1977. The cardiac output-oxygen uptake relationship in men during graded bicycle ergometry. *Med. Sci. Sports Exerc.* 9:148–154.

Fleckenstein, J.L., R.G. Haller, S.F. Lewis, L. Bertocci, J. Payne, B. Barker, B. Archer, and R.M. Peshock. 1991. Myophosphorylase deficiency impairs exercise-enhancement on MRI of skeletal muscle. *J. Appl. Physiol.* 71:961–969.

Fleckenstein, J.L., R.M. Peshock, S.F. Lewis, and R.G. Haller. 1989. Focal muscle injury and atrophy in human muscle glycolytic disorders. *Muscle & Nerve* 12:849–855.

Galbo, H. 1983. *Hormonal and metabolic adaptation to exercise.* New York: Thieme-Stratton.

Hagberg, J.M., E.F. Coyle, J.E. Carroll, J.M. Miller, W.H. Martin, and M.H. Brooke. 1982. Exercise hyperventilation in patients with McArdle's disease. *J. Appl. Physiol.* 52:991–994.

Hall, R.E., K.G. Henriksson, S.F. Lewis, R.G. Haller, and N.G. Kennaway. 1993. Mitochondrial myopathy with succinate dehydrogenase and aconitase deficiency: Abnormalities of several iron-sulfur proteins. *J. Clin. Invest.* 92:2660–2666.

Haller, R.G., T. Clausen, and J. Vissing. 1998. Re-duced levels of skeletal muscle Na⁺K⁺- ATPase in McArdle's disease. *Neurology* 50:37–40.

Haller, R.G., K.G. Henriksson, L. Jorfeldt, et al. 1991. Deficiency of skeletal muscle succinate dehydrogenase and aconitase: Pathophysiology of exercise in a novel human muscle oxidative defect. *J. Clin. Invest.* 88:1197–1206.

Haller, R.G., and S.F. Lewis. 1990. Human muscle respiratory chain defects: Metabolic and physiologic implications. In *Biochemistry of exercise V-H,* ed. A.W. Taylor, 251–264. Champaign, IL: Human Kinetics.

Haller, R.G., and S.F. Lewis. 1991. Glucose-induced exertional fatigue in muscle phosphofructokinase deficiency. *N. Engl. J. Med.* 324:364–369.

Haller, R.G., S.F. Lewis, J.D. Cook, and C.G. Blomqvist. 1983. Hyperkinetic circulation during exercise in neuromuscular disease. *Neurology* 33:1283–1287.

Haller, R.G., S.F. Lewis, J.D. Cook, and C.G. Blomqvist. 1985. Myophosphorylase deficiency impairs muscle oxidative metabolism. *Ann. Neurol.* 17:196–199.

Haller, R.G., S.F. Lewis, R.W. Estabrook, S. DiMauro, S. Servidei, and D.W. Foster. 1989. Exercise intolerance, lactic acidosis, and abnormal cardiopulmonary regulation in exercise associated with adult skeletal muscle cytochrome c oxidase deficiency. *J. Clin. Invest.* 84:155–161.

Kissel, J.T., W. Beam, N. Bresolin, G. Gibbona, S. DiMauro, and J.R. Mendell. 1985. Physiologic assessment of phosphoglycerate mutase deficiency. *Neurology* 35:828–833.

Kjellmer, I. 1965. The potassium ion as a vasodilator during muscular exercise. *Acta Physiol. Scand.* 63:460–468.

Lange, A., P. Lund-Johansen, and G. Clausen. 1969. Metabolic and circulatory responses to muscular exercise in a subject with glycogen storage disease. McArdle's disease. *Scand. J. Clin. Lab. Invest.* 24:10–11.

Larsson, L-E, H. Linderholm, R. Muller, T. Ringqvist, and R. Sornas. 1964. Hereditary metabolic myopathy with paroxysmal myoglobinurea due to abnormal glycolysis. *J. Neurol. Neurosurg. Psychiat.* 27:361-380.

Layzer, R. 1985. McArdle's disease in the 1980's. *N. Eng. J. Med.* 312:370–371.

Lewis, S.F., and R.G. Haller. 1986. The pathophysiology of McArdle's disease: Clues to regulation in exercise and fatigue. *J. Appl. Physiol.* 61:391–401.

Lewis, S.F., and R.G. Haller. 1989. Skeletal muscle disorders and associated factors that limit exercise performance. *Ex. Sport Sci. Rev.* 17:67–113.

Lewis, S., R. Haller, J. Cook, and C.G. Blomqvist. 1984. Metabolic control of cardiac output response to exercise in McArdle's disease. *J. Appl. Physiol.* 57:1749–1753.

Lewis, S.F., R.G. Haller, J.D. Cook, and R.L. Nunnally. 1985. Muscle fatigue in McArdle's disease studied by ^{31}P NMR: Effect of glucose infusion. *J. Appl. Physiol.* 59:1991–1994.

Lewis, S.F., S. Vora, and R.G. Haller. 1991. Abnormal oxidative metabolism and O_2 transport in muscle phosphofructokinase deficiency. *J. Appl. Physiol.* 70:391–398.

Linderholm, H., B. Essøn-Gustavsson, and L.-E. Thomell. 1990. Low succinate dehydrogenase. SDH activity in a patient with a hereditary myopathy with paroxysmal myoglobinuria. *J. Int. Med.* 228:43–52.

Linderholm, H., R. Muller, R. Ringqvist, and R. Somas. 1969. Hereditary abnormal muscle metabolism with hyperkinetic circulation during exercise. *Acta Med. Scand.* 185:153–166.

McArdle, B. 1951. Myopathy due to a defect in muscle glycogen breakdown. *Clin. Sci.* 10:13–33.

Miller, R.G., M.D. Boska, R.S. Moussavi, P.J. Carson, and M.W. Weiner. 1988. [31]P nuclear magnetic resonance studies of high-energy phosphates and pH in human muscle fatigue. *J. Clin. Invest.* 31:1190–1196.

Mitchell, J., and C.G. Blomqvist. 1971. Maximal oxygen uptake. *N. Engl. J. Med.* 284:1018–1022.

Mitchell, J.H. 1990. Neural control of the circulation during exercise. *Med. Sci. Sports Exerc.* 22:141–154.

Morgan-Hughes, J. 1994. Mitochondrial diseases. *Myology* 2:1610–1660.

Nuutinen, E.M., K. Nishiki, M. Erecinska, and D.F. Wilson. 1982. Role of mitochondrial oxidative phosphorylation in regulation of coronary blood flow. *Am. J. Physiol.* 243:H159–H169.

Paterson, D.J., J.S. Friedland, D.A. Bascom, I.D. Clement, D.A. Cunningham, R. Painter, and P.A. Robbins. 1990. Changes in arterial K+ and ventilation during exercise in normal subjects and subjects with McArdle's syndrome. *J. Physiol.* 429:339–348.

Pearson, C., D. Rimer, and W.F.H.M. Mommaerts. 1961. A metabolic myopathy due to absence of muscle phosphorylase. *Am. J. Med.* 30:502–517.

Pernow, B.B., R.J. Havel, and D.B. Jennings. 1967. The second wind phenomenon in McArdle's syndrome. *Acta Med. Scand. Suppl.* 472:294–307.

Porte, D., Jr., D.W. Crawford, J.B. Jennings, C. Aber, and M.B. McIlroy. 1966. Cardiovascular and metabolic responses to exercise in a patient with McArdle's syndrome. *N. Eng. J. Med.* 275:406–412.

Radda, G.K. 1986. The use of NMR spectroscopy for the understanding of disease. *Science* 233:640–645.

Rosell, S., and B. Saltin. 1973. Energy need, delivery and utilization in muscular exercise. *The structure and function of muscle Vol. 3,* ed. by G.H. Bourne. New York: Academic Press. 185–221.

Ross, B.D., G.K. Radda, D.G. Gadian, G. Rocker, M. Esiri, and J. Falconer-Smith. 1981. Examination of a case of suspected McArdle's syndrome by ^{31}P nuclear magnetic resonance. *N. Engl. J. Med.* 304:1338–1342.

Sabina, R.L., J.L. Swain, C.W. Olanow, W.G. Bradley, W.N. Fishbein, S. DiMauro, and E.W. Holmes. 1984. Myoadenylate deaminase deficiency. Functional and metabolic abnormalities associated with disruption in the purine nucleotide cycle. *J. Clin. Invest.* 73:720–730.

Sahlin, K., N.-H. Areskog, R.G. Haller, K.G. Henriksson, L. Jorfeldt, and S. F. Lewis. 1990. Impaired oxidative metabolism increases adenine nucleotide breakdown in McArdle's disease. *J. Appl. Physiol.* 69:1231–1235.

Sahlin, K., L. Jorfeldt, K.-G. Henriksson, S.F. Lewis, and R.G. Haller. 1995. Tricarboxylic acid cycle intermediates during incremental exercise: Attenuated increase in McArdle's disease. *Clin. Sci.* 88:687–693.

Vissing, J., H. Galbo, and R.G. Haller. 1996a. Exercise fuel mobilization in mitochondrial myopathy: A metabolic dilemma. *Ann. Neurol.* 40:655–662.

Vissing, J., H. Galbo, and R.G. Haller. 1996b. Paradoxically enhanced glucose production during exercise in humans with blocked glycolysis due to muscle phosphofructokinase deficiency. *Neurology* 47:766–771.

Vissing, J., S.F. Lewis, H. Galbo, and R.G. Haller. 1992. Effect of deficient muscle glycogenolysis on extramuscular fuel production in exercise. *J. Appl. Physiol.* 72:1773–1779.

Wildenthal, K., D.S. Mierzwiak, N.S.J. Skinner, and J.H. Mitchell. 1968. Cardiovascular and ventilatory reflexes from the dog hindlimb. *Am. J. Physiol.* 215:542–548.

Chapter 22

Peripheral Arterial Disease

Simon Green and Jesper Mehlsen

Peripheral arterial disease (PAD) is a debilitating atherosclerotic disease that leads to stenosis and occlusion of the peripheral arteries and restricts the arterial circulation to the limbs. Intermittent claudication is the hallmark of atherosclerosis in large arteries of the lower extremities, and pain in response to walking is the result of an inadequacy of regional blood flow to meet metabolic demands of the exercising muscle. The reported frequency of intermittent claudication varies in different studies from approximately 2% in subjects aged 45 to 55 years up to 10-20 % of individuals over the age of 60 years (Kannel et al. 1970; Donovan 1994). Intermittent claudication most often takes on a benign course leading to critical leg ischemia or amputation in less than 20% of patients within the first decade (Jelnes et al. 1986), but the symptoms of atherosclerosis are associated with a substantial increase in overall mortality, and the physical and social constraints of the disease often constitute a serious and underrated problem for patients. Central to these constraints is the intolerance to common forms of exercise, such as walking.

Like other diseases, PAD is characterised by a deterioration of several physiological systems. The loss of function in these systems could all potentially contribute to exercise intolerance. This chapter focuses on exercise performance and the circulation in PAD. That other systems (e.g., neuromuscular) have not been considered here does not mean that they are not deemed important in limiting exercise performance; in fact, evidence has emerged that suggests that neuromuscular dysfunction also contributes to exercise intolerance in PAD (Brass and Hiatt 1994; Regensteiner and Hiatt 1995).

Exercise Performance

Early work by Hillestad (1963) provided insights into how to approach the problem of establishing what limits exercise performance in PAD. Like some of his Scandinavian contemporaries, Hillestad recognized the value of using both a walking test and an "isolated limb" exercise test to examine factors that limit exercise performance. Walking is the activity most relevant to daily living and requires assessment. Given that claudication in the calf is the primary symptom limiting walking performance in most people with PAD, a test that focuses on the calf's ability to exercise also seems necessary to unravel the physiology that limits exercise performance. To date, very few studies have pursued this theme through the use of both walking and calf exercise tests.

Hillestad (1963) also performed several experiments that focused on the effects that variations in exercise test dimensions exerted on performance. Such a focus in the literature was rare until more recent studies (Hiatt et al. 1988; Gardner et al. 1991). Protocols designed to test walking performance in PAD often require a person to walk until either claudication is first experienced (i.e. pain-onset), "maximal" claudication is reached, or even beyond this point to failure. Different criteria used to terminate the test can influence test variability, with performance to pain-onset being less variable than when continued to failure (Hillestad 1963). When patients use handrails to support their walking performance variability is increased relative to without handrail support (Gardner, Skinner, and Smith 1991). Constant-load or incremental protocols have been used to evaluate performance with the latter type yielding more reliable results (CV = 12% vs. 31%) (Gardner, Skinner, and Smith 1991; Hiatt et al. 1988; Hillestad 1963). Even for incremental protocols, performance variability is initially high (CV = 15–17%) and then reaches a nadir (CV = 7–8%) after two to three tests (Askew et al. 1997). With respect to constant-load tests, differences in walking performance between patients and controls cannot be easily evaluated since the time to failure in the controls can be prohibitively long and/or performance ceilings are often imposed by the investigators. Therefore, close attention should be paid to how exercise performance is evaluated and, given that the effect of various treatments can be small but functionally significant, particular emphasis placed on minimizing performance variability.

In people with PAD who do not experience pain at rest and/or impending gangrene, maximal walking distance on flat ground at 4 km · h^{-1} is, in ~85% of cases, limited to less than 1000 m (Ekroth et al. 1978). Walking performance (i.e., time to failure) determined using an incremental test in PAD is, on average, less than half that (i.e., ~46%) of healthy, age-matched controls (Regensteiner et al. 1993). There is also limited evidence that the capacity to cycle is ~65% lower in PAD compared with age-related normative data (Lundgren et al. 1989). Therefore, the capacity to perform common exercise is severely impaired in PAD.

Depending upon the severity and locii of atherosclerosis, claudication and its severity can vary between the gluteals, thighs, hamstrings, and calves. Although the relationships between the severity and distribution of claudication and walking performance have not been studied, it appears as if the majority of claudicants, who are not limited by angina or other diseases, report calf pain as the symptom that limits performance. The loss in walking performance is generally attributed to the most ischemic or diseased limb and, in particular, the reduced endurance of the calf (i.e., triceps surae). The calf endurance and strength of ischemic limbs are lower than either control or asymptomatic limbs (Bylund-Fellenius et al. 1981; Gerdle et al. 1986; England et al. 1992; Regensteiner et al. 1993). The positive relationship (r = 0.61) between the work capacity of the calf and walking performance (Gerdle et al. 1986) supports the general notion that reduced calf endurance contributes to walking intolerance. That the majority of variance in walking performance is not, however, explained by calf work capacity suggests that other factors also contribute to differences in walking performance. The influence that the loss of calf strength (England et al. 1995) has on the impairment in walking performance is not clear. For example, that resistance training which increased calf strength failed to improve walking performance (Hiatt et al. 1994) suggests that the latter performance is not limited by strength per se. In contrast, oral supplementation with propionyl-L-carnitine improved both calf strength and walking performance in PAD, where these improvements were positively correlated with each other (Barker et al. 1998). In addition to the reduced calf endurance, this evidence suggests that muscle weakness might also contribute to walking intolerance in PAD.

Whole-Body Metabolism

From an energetic perspective, walking performance is determined by five key variables: the peak rate of energy released through aerobic metabolism; the peak rate of energy released through anaerobic metabolism; the time constants for these metabolic processes; the efficiency with which mechanical power is produced by these processes; and the efficiency with which walking speed is produced by these processes (Peronnet and Thibault 1989). There are no estimates of whole-body anaerobic metabolism during walking in PAD; more information is available for aerobic metabolism. Consequently, the relative contributions of aerobic and anaerobic metabolism to whole-body ATP synthesis during exercise in PAD are not known. However, since the durations of maximal walking generally exceed five min, and anaerobic ATP synthesis is relatively moderate even in the most ischemic muscle (Bylund-Fellenius et al. 1981), aerobic metabolism is probably the primary contributor to ATP synthesis during exercise in PAD.

The highest rates of O_2 consumption (i.e., peak VO_2) approximate 12–15 ml · kg^{-1} · min^{-1} in PAD during either maximal, incremental walking (Hiatt et al. 1990; Gardner et al. 1991; Regensteiner et al. 1993; Hiatt et al. 1994), cycling (Askew et al. 1997), or stairclimbing (Gardner et al. 1995). These values are only ~45–50% of those for healthy controls (Regensteiner et al. 1993), but they can vary considerably between individuals, ranging between 10 and 22 ml · kg^{-1} · min^{-1} (Gardner et al. 1992). That the extent of this impairment is similar to that seen for walking performance suggests that factors which limit peak VO_2 mainly limit performance.

Differences in the initial response of VO_2 to exercise can also contribute to differences in performance. During low-intensity exercise, this VO_2 response is normally monoexponential and reaches a steady state within a few minutes. In PAD, however, some subjects do not achieve steady state at low walking speeds on flat ground (Barker et al. 1998), and their VO_2 response displays a second, slow component (Womack et al. 1997). When these responses are treated simply as monoexponential functions, the time constants are considerably longer in PAD than reported for healthy, older adults (Barker et al. 1998). This would impair performance, particularly if there are several increments in workload although the extent of this impairment has yet to be quantified.

Exercise efficiency or economy (i.e., VO_2 at a given intensity) is an important determinant of performance. No data exists on exercise economy for patients and controls. However, short-term training can improve both walking economy and performance in the absence of changes in peak VO_2 (Hiatt et al. 1990; Hiatt et al. 1994). This suggests that factors related to economy but unrelated to peak VO_2 also contribute to differences in performance between people with PAD. Such factors might include alterations in gait or even the temporal response of VO_2, since a reduction in the second, slow component of VO_2 improves economy (Womack et al. 1997).

Cardiac Function

Atherosclerosis is generally found throughout the entire arterial system. For this reason, the majority of people with PAD also have cardiac disease (Barnard 1994) although it was presumably not severe enough to limit exercise performance in participants in most exercise studies. This is supported by an absence of electrocardiograph abnormalities or angina during maximal exercise and by the relatively low peak heart rates that usually lie between 100 and 125 bpm (Gardner et al. 1992; Hiatt et al. 1990; Hiatt et al. 1994). Peak heart rates after training were still much lower than those reported for healthy individuals (mean = 174 bpm) of a similar age (Martin et al. 1990), and claudication still limits exercise performance (Hiatt et al. 1990; Hiatt et al. 1994). This supports the idea that cardiac output is less than maximal and does not limit exercise performance in PAD even after training. Moreover, the failure of PAD patients to achieve a classic plateau in VO_2 during maximal, incremental exercise, which is usually the case with healthy, older people (Martin et al. 1990), provides further evidence of the peripheral, rather than central, hemodynamic limitation to exercise performance in PAD.

Peripheral Arterial Flow and O_2 Extraction

Using the Fick equation, a limitation in VO_2 in PAD can be attributed to either peripheral arterial flow to, and/or O_2 extraction by, active skeletal muscle. Following exercise, muscle blood flow is elevated and remains high for relatively long periods in PAD. In more severe cases, muscle blood flow may only increase after exercise (Alpert, Larsen, and Lassen 1969). The peak calf blood flow, measured using venous occlusion plethysmography, can be reduced to 30–40% (11–16 ml · 100 ml tissue min^{-1}) of that in healthy, older people (45–60 ml · 100 ml tissue^{-1} · min^{-1}) (Zetterquist 1970; Dahllof et al. 1974; Dahllof et al. 1976; Hammarsten et al. 1980; Henriksson et al. 1980; Bylund-Fellenius et al. 1981; Mannarino et al.

1989; Hiatt et al. 1990; Martin et al. 1990). Similarly in unilateral PAD, peak calf blood flow in the symptomatic limb is 25–40% of that observed in the asymptomatic limb (Henriksson et al. 1980; Hillestad 1963). This clearly demonstrates the peripheral limitation to arterial flow and suggests that the reduction in exercise performance and peak VO_2 in PAD can be explained entirely by the decrease in arterial flow to symptomatic limbs.

In contrast, it has been suggested that differences in muscle blood flow do not explain the differences in walking performance in PAD (Regensteiner and Hiatt 1995). This is based on the fact that most studies have not found a significant relationship between walking performance and peak calf blood flow (Andriessen et al. 1989) or their corresponding responses to training (Andriessen et al. 1989; Dahllof et al. 1974; Hiatt et al. 1990; Lundgren et al. 1989; Sørlie and Myhre 1978; Zetterquist 1970), and that there is often no effect of exercise training on peak calf blood flow despite the improvement in exercise performance (Dahllof et al. 1974; Dahllof et al. 1976; Ekroth et al. 1978; Mannarino et al. 1989; Sørlie and Myhre 1978).

The caveat is, however, that none of these studies have accounted for the effects of differences or changes in body weight on performance, which can be significant (Hillestad 1963). When such an influence is accounted for, a much stronger, nonlinear relationship between peak calf blood flow and walking performance has been observed (Hillestad 1963). Even when the influence of body weight is not accounted for, a significant, nonlinear relationship between changes in peak arterial flow and exercise performance have been observed (Alpert, Larsen, and Lassen 1969). The nonlinear nature of these relationships is consistent with the effect that changing the exercise intensity relative to peak VO_2 (and thus peak blood flow) would have on walking performance (Gardner 1993). From the studies cited it also appears that there is considerable individual variability in the response of peak calf blood flow to exercise training, and that other factors also must contribute to performance differences. This suggests that the presence or absence of significant relationships between blood flow and performance also depends upon the nature of the individuals studied.

There is no evidence that peak O_2 extraction across active skeletal muscle is impaired in PAD relative to healthy controls. Exercise training can, however, further increase an already high peak level of O_2 extraction during walking by the calf so that femoral venous O_2 saturation is lower (i.e., 15 to 8%) (Zetterquist 1970) compared with that seen in the quadriceps of younger people performing maximal, knee extensor exercise (i.e., 27%) (Andersen and Saltin 1985). Using the data from Zetterquist

(1970) and assuming normal hemoglobin concentrations (i.e., 15 g · 100 ml⁻¹) and arterial O_2 saturations (i.e., 97%), training increased the arterial-venous O_2 difference across the calf from 17 to 18.5 ml O_2 · 100 ml⁻¹. At a peak arterial flow of 500 ml · min⁻¹ (Sørlie and Myhre 1978), this increase in O_2 extraction would increase calf muscle VO_2 by 7.5 ml · min⁻¹ or 9% which, using economy data for the quadriceps (Andersen and Saltin 1985), approximates a power output of 0.4 W. In the absence of data on calf power output and its relationship to walking performance, it is difficult to quantify the effect that this would have on walking performance and whole-body VO_2. However, quantifying a training-induced increase in peak arterial flow observed (i.e., 38%) (Hiatt et al. 1990) on calf muscle VO_2 using the same analysis as above reveals a relatively larger effect (i.e., 32 ml · min⁻¹) than the scope for further O_2 extraction would allow. Therefore, despite the observation of insignificant relationships between calf blood flow and performance, this demonstrates that changes in calf blood flow that have been observed exert a relatively greater effect on performance than changes in O_2 extraction.

Leg arterial pressures or the ankle-brachial index (ABI) are often used to characterize the hemodynamic status in, and clinical severity of, PAD (Hiatt et al. 1995). The prolonged reduction in systolic ankle blood pressure can be used as an objective indicator of the severity of the disease and may also be used as a guide to the level of obstruction (Wolf, Summer, and Strandness 1972). The ABI in particular has been used in exercise studies, and the recovery of ABI after exercise, termed the *ischemic window*, has been mooted as a method that can quantify the training response (Feinberg et al. 1991). While faster recovery of leg pressures or ABI after training (Skinner and Strandness 1967) is consistent with the reduced duration of postexercise hyperemia after training (Alpert, Larsen, and Lassen 1969), leg arterial pressures, particularly ABI, cannot be used to represent arterial flow, which is the more important variable when trying to explain performance differences. Peak calf blood flow is often not related to resting or postexercise ABI in symptomatic limbs (Barker et al. 1998), and the changes in either leg pressures or ABI with training often do not reflect changes in peak arterial flow (Andriessen et al. 1989; Dahllof et al. 1974; Hiatt et al. 1990).

Physiological Bases of Peripheral Arterial Flow and O_2 Extraction

Differences in arterial flow or leg pressures between people with PAD are usually attributed to the sever-

ity of atherosclerosis in the major feed arteries. Increases in these variables in response to training have usually been attributed to either increased collateralization and/or increased blood flow though collaterals (Skinner and Strandness 1967; Alpert, Larsen, and Lassen 1969). While collateralization is frequently observed in PAD through angiography and might help explain differences in peak arterial flow, there is no evidence in humans that it is increased by exercise training. It is also assumed, but not verified, that training-induced increases in arterial flow occur through increased collateral flow (Alpert, Larsen, and Lassen 1969). There is no evidence that atherosclerosis in peripheral arteries is reduced by training and that increases in arterial flow are created through this process.

Controlling the distribution of arterial blood flow to and within active skeletal muscles during exercise probably is important to exercise performance in PAD. Flow-mediated vasodilation in feed arteries, which depends upon endothelial function, is impaired by atherosclerosis (Celermajer et al. 1992). In addition, the synthesis of nitric oxide, important to this type of flow regulation, is also impaired in PAD to an extent related to the fall in ABI after exercise (Akopov et al. 1997). Plasma levels of endothelin, a vasoconstrictor, are also positively correlated with the severity of atherosclerosis (Lerinan et al. 1991). These findings raise the possibility that factors which help regulate arterial diameter and flow differ between people with PAD and, perhaps in addition to collateral flow, help explain differences in arterial flow and its response to training. In addition, the rheological properties of the blood are also affected by PAD, improve in response to training, and probably increase arterial flow (Ernst and Fialka 1993; Yates et al. 1979). Therefore, many factors could help explain differences in arterial flow but, as yet, their contributions have not been delineated.

Oxygen extraction by skeletal muscle depends critically upon adequate matching of capillary flow and fiber metabolism. This depends upon endothelial function and the competing effects of vasoconstrictors and vasodilators on vascular smooth muscle (Segal and Kurjiaka 1995). Although perfusion-metabolism matching has not been assessed in PAD, it is possible that it differs between individuals, responds to training and, thus, helps explain performance differences.

Oxygen extraction also depends upon the morphology of capillaries and their density in skeletal muscle, since these features are important to distributing blood to active muscle fibers as well as influencing the transit time of erythrocytes through the capillary and, thus, O_2 diffusion into adjacent fibers. One of the mechanisms often used to explain differences in O_2 extraction or exercise performance in PAD is capillarization (Clyne et al. 1985; Henriksson et al. 1980; Lundgren et al. 1989; Zetterquist 1970). Several studies have examined muscle capillarization in human PAD and some have reported higher capillary density in ischemic muscle compared with controls or non-ischemic legs (Clyne et al. 1985; Hammarsten et al. 1980; Henriksson et al. 1980); others have failed to find significant differences in capillary density (Henriksson et al. 1980; Jansson et al. 1988). In contrast, capillary density was slightly lower in PAD than autopsy control muscle, and much lower (22%) in more diseased PAD (Clyne et al. 1982). These data suggest that the capillary response to PAD is not uniform possibly due to differences in the history, lateralization, and/or severity of the disease (Henriksson et al. 1980), or due to the intensity and extent of repetitive activation of ischemic muscle (Hoppeler et al. 1992; Hudlicka et al. 1994). Moreover, in severe PAD there can be extensive capillary disorganization and morphological abnormalities (e.g., endothelial swelling) that tend to reduce luminal diameter and increase capillary size (Makitie 1977; Makitie 1979). Thus, increased capillary density might occur partly in response to these abnormalities and represent a loss rather than a gain of microcirculatory capacity. Therefore, the functional relevance of increased muscle capillarization in PAD is not clear.

The matching of flow and fiber metabolism and optimizing the transit time of the erythrocytes through capillaries also depends upon the rheological properties of the blood, particularly in states of low arterial flow (Chien 1987). Blood viscosity is increased and erythrocyte deformability is reduced in people with atherosclerosis (Dormandy et al. 1973), both of which could compromise microcirculatory flow. Pentoxifylline is a drug that reduces blood and plasma viscosity, reduces the potential for platelet aggregation, and increases erythrocyte deformability. Through some combination of these effects, pentoxifylline can increase peripheral blood flow and increase PO_2 in ischemic tissues (Ward and Clissold 1987). In several placebo-controlled studies, administration of pentoxifylline over an average of 5 months increased walking performance by ~50% in PAD (Bevan, Waller, and Ramsay 1992). A similar ergogenic effect has also been reported for another drug that reduces blood and plasma viscosity, defibrotide (Marrapodi et al. 1994). These findings demonstrate that exercise performance in PAD can be improved through alterations to either blood viscosity or erythrocyte deformability.

In normo-perfused skeletal muscle, a decrease in erythrocyte mass and hematocrit decreases peripheral O_2 flow, muscle VO_2, and performance (Hogan, Bebout, and Wagner 1991). This is not the case in hypoperfused PAD muscle, where both muscle PO_2 (at rest and recovery from exercise) and pain-free walking distance are increased as hematocrit is reduced under normovolemia from ~48% to ~37% (Ernst, Matrai, and Kollar 1986; Hoffkes et al. 1996). This ergogenic effect is probably mediated through either an increase in arterial flow (Yates et al. 1979) and/or improved homogeneity of flow within skeletal muscle (Tyml 1991). These findings demonstrate that the rheological status of the blood contributes to the exercise impairment in PAD, and that its normalization through pharmacotherapy or even exercise training (Ernst and Fialka 1993) contributes to improvements in exercise performance.

Conclusion

Understanding the physiological limits imposed on exercise performance is crucial to designing effective treatment for people with PAD. This chapter has focused only on the circulation and its role in limiting exercise performance. Studying this problem depends upon the reliable measurement of exercise performance and should be of prime importance in all exercise studies. Aerobic metabolism is probably most important to sustained, exercise performance. The reduction in peripheral arterial flow can entirely explain the loss of peak aerobic power and exercise performance in PAD whereas O_2 extraction by active, skeletal muscle is not impaired and its scope for further improvement is relatively small. Therefore, in theory, treatments that focus on restoring blood flow will be most effective in enabling people with PAD. However, clinically significant results have been obtained through therapies (e.g., exercise training, pharmacotherapy) that do not appear to alter peripheral arterial flow, which suggests that other factors contribute to therapeutic effect. Although it has not been addressed here, there is a growing body of evidence which suggests that the pathological changes to skeletal muscle, such as disuse atrophy, could further limit exercise performance and should be considered when designing treatments. Differences in technique will affect exercise efficiency and, thus, performance. Interventions that favorably alter exercise efficiency will improve performance. A simple lesson also can be learned from an earlier study (Hillestad 1963), which demonstrated the importance of considering the influence of body weight on performance and, thus, interventions which assist body weight management in people with PAD.

References

Akopov, S.E., S.S. Pogossian, E.N. Toromanian, G.G. Grigorian, and E.S. Gabrielian. 1997. Increased nitric oxide deactivation by polymorphonuclear leukocytes in patients with intermittent claudication. *J Vasc Surg* 25:704–712.

Alpert, J.S., O.A. Larsen, and N.A. Lassen. 1969. Exercise and intermittent claudication: Blood flow in the calf muscle during walking studied by the xenon-133 clearance method. *Circ* 39:353–359.

Andersen, P., and B. Saltin. 1985. Maximal perfusion of skeletal muscle in man. *J Physiol* 366:233–249.

Andriessen, M.P., G.J. Barendsen, A.A. Wouda, and L. de Pater. 1989. Changes of walking distance in patients with intermittent claudication during six months intensive physical training. *Vasa* 18:63–68.

Askew, C., S. Green, P.J. Walker, and C. Codd. 1997. The reproducibility and stability of walking and cycling capacity in people with intermittent claudication. ANZ Angiology Conference, Brisbane, Australia.

Barker, G., S. Green, C. Askew, and P.W. Walker. 1998. Effect of propionyl-L-carnitine supplementation on exercise performance in peripheral arterial disease. Third Annual Conference of the European College of Sports Science, Manchester.

Barnard, R.J. 1994. Physical activity, fitness, and claudication. In *Physical Activity, Fitness, and Health* edited by C. Bouchard, R.J. Shephard, and T. Stephens. Champaign, Illinois: Human Kinetics. 622–632.

Bevan, E.G., P.C. Waller, and L.E. Ramsay. 1992. Pharmacological approaches to the treatment of intermittent claudication. *Clin Pharmacol* 2(2):125–136.

Brass, E.P., and W.R. Hiatt. 1994. Carnitine metabolism during exercise. *Life Sci* 54(19):1383–1393.

Bylund-Fellenius, P., M. Walker, A. Elander, S. Holm, and T. Schersten. 1981. Energy metabolism in relation to oxygen partial pressure in human skeletal muscle during exercise. *Biochem J* 200:247–255.

Celermajer, D.S., K.E. Sorensen, V.M. Gooch, D.J. Spiegelhalter, O.I. Miller, I.D. Sullivan, et al. 1992. Non-invasive detection of endothelial dysfunction in children and adults at risk of atherosclerosis. *Lancet* 340(8828):1111-1115.

Chien, S. 1987. Physiological and pathophysiological significance of hemorheology. In *Clinical Hemorheology* edited by S. Chien, J. Dormandy, E. Ernst, and A. Matrai. Dordrecht: Martinus Nijhoff. 125–164.

Clyne, C.A., H. Mears, R.O. Weller, and T.F. O'Donnell. 1985. Calf muscle adaptation to peripheral vascular disease. *Cardiovasc Res* 19:507–512.

Clyne, C.A.C., R.O. Weller, W.G. Bradley, D.I. Silber, T.F. O'Donnell, and A.D. Callow. 1982. Ultrastructural and capillary adaptation of gastrocnemius muscle to occlusive peripheral vascular disease. *Surg* 92(2):434–439.

Dahllof, A., J. Holm, T. Schersten, and R. Sivertsson. 1976. Peripheral arterial insufficiency: Effect of physical training on walking distance. *Scand J Rehab Med* 8:19–26.

Dahllof, A., P. Bjorntorp, J. Holm, and T. Schersten. 1974. Metabolic activity of skeletal muscle in patients with peripheral arterial insufficiency. *Eur J Clin Invest* 4:9–15.

Donovan, J. 1994. *Australia's Health 1994*. Canberra: Australian Government Publishing Service.

Dormandy, J.A., E. Hoare, J. Colley, D.E. Arrowsmith, and T.L. Dormandy. 1973. Clinic, hemodynamic, rheological, and biochemical findings in 126 patients with intermittent claudication. *Brit Med J* 4:576–581.

Ekroth, R., A. Dahllof, B. Gundevall, J. Holm, and T. Schersten. 1978. Physical training of patients with intermittent claudication: Indications, methods, and results. *Surg* 84:640–643.

England, J.D., M.A. Ferguson, W.R. Hiatt, and J.G. Regensteiner. 1995. Progression of neuropathy in peripheral arterial disease. *Muscle Nerve* 18:380–387.

England, J.D., J.G. Regensteiner, S.P. Ringel, M.R. Carry, and W.R. Hiatt. 1992. Muscle denervation in peripheral arterial disease. *Neurology* 42:994–999.

Ernst, E., and V. Fialka. 1993. A review of the clinical effectiveness of exercise therapy for intermittent claudication. *Arch Intern Med* 153:2357–2360.

Ernst, E., A. Matrai, and L. Kollar. 1986. A randomized, double-blind, placebo-controlled, crossover trial on isovolemic hemodilution in claudicants: Preliminary results. *Clin Hemorheology* 6:297–301.

Feinberg, R.L., R.T. Gregory, J.R. Wheeler, S.O. Snyder, R.G. Gayle, F.N. Parent, et al. 1991. The ischemic window: A method for the objective quantification of the training effect in exercise therapy for intermittent claudication. *J Vasc Surg* 16:244.

Gardner, A.W. 1993. Dissipation of claudication pain after walking: implications for endurance training. *Med Sci Sports Exer* 25(8):904–910.

Gardner, A.W., J.S. Skinner, C.X. Bryant, and L.K. Smith. 1995. Stair climbing elicits a lower cardiovascular demand than walking in claudication patients. *J Cardiopulmonary Rehab* 15:134–142.

Gardner, A.W., J.S. Skinner, B.W. Cantwell, and L.K. Smith. 1991. Progressive vs. single-stage treadmill tests for evaluation of claudication. *Med Sci Sports Exer* 23(4):402–408.

Gardner, A.W., J.S. Skinner, and L.K. Smith. 1991. Effects of handrail support on claudication and hemodynamic responses to single-stage and progressive treadmill protocols in peripheral vascular occlusive disease. *Am J Cardiol* 68:99–105.

Gardner, A.W., J.S. Skinner, N.R. Vaughan, and C.X. Bryant. 1992. Comparison of three progressive exercise protocols in peripheral vascular occlusive disease. *Angiology* 43:661–671.

Gerdle, B., B. Hedberg, K.-A. Angquist, and A.R. Fugl-Meyer. 1986. Isokinetic strength and endurance in peripheral arterial insufficiency with intermittent claudication. *Scand J Rehab Med* 18:9–15.

Hammarsten, J., A.-C. Bylund-Fellenius, J. Holm, T. Schersten, and M. Krotkiewski. 1980. Capillary supply and muscle fibre types in patients with intermittent claudication: Relationships between morphology and metabolism. *Eur J Clin Invest* 10:301–305.

Henriksson, J., E. Nygaard, J. Andersson, and B. Eklof. 1980. Enzyme activities, fiber types, and capillarization in calf muscles of patients with intermittent claudication. *Scand J Clin Lab Invest* 40:361–369.

Hiatt, W.R., A.T. Hirsch, J.G. Regensteiner, and E.P. Brass. 1995. Clinical trials for claudication: Assessment of exercise performance, functional status, and clinical end points. *Circ* 92:614–621.

Hiatt, W.R., D. Nawaz, J.G. Regensteiner, and K.F. Hossack. 1988. The evaluation of exercise performance in patients with peripheral vascular disease. *J Cardiopulmonary Rehab* 12:525–532.

Hiatt, W.R., J.G. Regensteiner, M.E. Hargarten, E. Wolfel, and E.P. Brass. 1990. Benefit of exercise conditioning for patients with peripheral arterial disease. *Circ* 81:602–609.

Hiatt, W.R., E.E. Wolfel, R.H. Meier, and J.G. Regensteiner. 1994. Superiority of treadmill walking exercise versus strength training for patients with peripheral arterial disease: Implications for the mechanism of the training response. *Circ* 90:1866–1874.

Hillestad, L.K. 1963. The peripheral blood flow in intermittent claudication. IV. Significance of the claudication distance. *Acta Med Scand* 173(4):467–478.

Hoffkes, H.-G., R. Dehn, K. Saeger, A. Franke, H. Landgraf, and A.M. Ehrly. 1996. Effects of normovolemic and hypervolemic hematocrit variations on musle tissue oxygen pressure in patients with chronic ischemia of the calf muscle. *Clin Hemorheology* 16(3):249–265.

Hogan, M.C., D.E. Bebout, and P.D. Wagner. 1991. Effect of hemoglobin concentration on maximal O_2 uptake in canine gastrocnemius muscle in situ. *J Appl Physiol* 70:1105–1112.

Hoppeler, H., O. Hudlicka, E. Uhlmann, and H. Claasen. 1992. Skeletal muscle adaptations to ischemia and severe exercise. *Clin J Sports Med* 2(1):43–51.

Hudlicka, O., M.D. Brown, S. Egginton, and J.M. Dawson. 1994. Effect of long-term electrical stimulation on vascular supply and fatigue in chronically ischemic muscles. *J Appl Physiol* 77(3):1317–1324.

Jansson, E., J. Johansson, C. Sylven, and L. Kaijser. 1988. Calf muscle adaptation in intermittent claudication. Side-differences in muscle metabolic characteristics in patients with unilateral arterial disease. *Clin Physiol* 8:17–29.

Jelnes, R., O. Cmudsting, K.H. Jensen, N. Bøkgaard, K.H. Tonnesen, and T. Schroder. 1986. Fate in intermittent claudication. Outcome and risk factors. *Brit Med J* 293:1137.

Kannel, W.B., J.J. Skinner, M.J. Schwartz, and D. Shurtleff. 1970. Intermittent claudication. Incidence in the Framingham study. *Circ* 41:875.

Lerinan, A., B.S. Edwards, J.W. Hallet, D.M. Heublin, S.M. Søndberg, and J.C. Burnett. 1991. Circulating and tissue endothelin immunoreactivity in advanced atherosclerosis. *New Eng J Med* 325:1997.

Lundgren, F., A.-G. Dahllof, T. Schersten, and A.-C. Bylund-Fellenius. 1989. Muscle enzyme adaptation in patients with peripheral arterial insufficiency: spontaneous adaptation, effect of different treatments and consequences on walking performance. *Clin Sci* 77:485–493.

Makitie, J. 1977. Skeletal muscle capillaries in intermittent claudication. *Arch Path Lab Med* 101:500–503.

Makitie, J. 1979. Peripheral neuromuscular system in peripheral arterial insufficiency. *Scand J Rheumatology* 30:157–162.

Mannarino, E., S. Pasqualini, M. Menna, G. Maragoni, and U. Orlandi. 1989. Effects of physical training on peripheral vascular disease: a controlled study. *Angiology* 40:5–10.

Marrapodi, E., D. Leanza, S. Giordano, M. Nazzari, and C. Corsi. 1994. Effects of defibrotide on physical performance and hemorheologic picture in patients with peripheral arteriopathy. *Clin Trials Meta-Analysis* 29:21–30.

Martin, W.H., W.M. Kohrt, M.T. Malley, E. Korte, and S. Stoltz. 1990. Exercise training enhances leg vasodilatory capacity of 65-year-old men and women. *J Appl Physiol* 69(5):1804–1809.

Peronnet, F., and G. Thibault. 1989. Mathematical analysis of running performance and world running records. *J Appl Physiol* 67(1):453–455.

Regensteiner, J.G., and W.R. Hiatt. 1995. Exercise rehabilitation for patients with peripheral arterial disease. *Exer Sports Sci Rev* 23:1–24.

Regensteiner, J.G., E.E. Wolfel, E.P. Brass, M.R. Carry, S.P. Ringel, M.E. Hargarten, et al. 1993. Chronic changes in skeletal muscle histology and function in peripheral arterial disease. *Circ* 87:413–421.

Segal, S.S., and D.T. Kurjiaka. 1995. Coordination of blood flow control in the resistance vasculature of skeletal muscle. *Med Sci Sports Exer* 27(8):1158–1164.

Skinner, J.S., and D.E. Strandness. 1967. Exercise and intermittent claudication. II. Effect of physical training. *Circ* 36:23–29.

Sørlie, D., and K. Myhre. 1978. Effects of physical training in intermittent claudication. *Scand J Clin Lab Invest* 38:217–222.

Tyml, K. 1991. Heterogeneity of microvascular flow in rat skeletal muscle is reduced by contraction and by hemodilution. *Int J Microcirc Clin Exp* 10:75–86.

Ward, A., and S.P. Clissold. 1987. Pentoxifylline. A review of its pharmacodynamic and pharmacokinetic properties and its therapeutic efficacy. *Drugs* 34:50–97.

Wolf, E.A., D.S. Summer, and D.E. Strandness. 1972. Correlation between nutritive blood flow and pressure in limbs of patients with intermittent claudication. *Surg Forum* 23:238.

Womack, C., D.J. Sieminski, L.I. Katzel, A. Yataco, and A. Gardner. 1997. Improved walking economy in patients with peripheral arterial disease. *Med Sci Sports Exer* 29(10):1286–1290.

Yates, C.J.P., V. Andrews, A. Berent, and J.A. Dormandy. 1979. Increase in leg blood flow by normovolemic hemodilution in intermittent claudication. *Lancet* 11:166–169.

Zetterquist, S. 1970. The effect of active training on the nutritive blood flow in exercising ischemic legs. *Scand J Clin Lab Invest* 25:101–111.

Chapter 23

Hypertension

Anthony F. Lever and Robert Boushel

Hypertension is a major risk factor for the development of heart disease and stroke, two of the leading causes of death in the Western world (National Heart, Lung, and Blood Institute 1997). Furthermore, ~20% of end-stage renal disease is caused by hypertension (United Kingdom Prospective Diabetes Study 1985). Studies in the general population indicate an increased risk of coronary artery disease and stroke with any increase in blood pressure level, and even a small elevation in diastolic blood pressure (5 mmHg) is associated with a 34% increase in the incidence of stroke and a 21% increased incidence of coronary disease (MacMahon et al. 1990). Numerous clinical studies strongly support that early treatment of hypertension has a positive impact on the prevention of cardiovascular disease (CVD) and mortality. It has been proposed that a long-term blood pressure reduction of as little as 5 mmHg would result in a 40% lower risk of stroke and ~15% reduction in risk of myocardial infarction (Collins et al. 1990). Medical treatment of hypertension has been effective in the prevention of stroke and renal disease. Less demonstrable are reductions in manifest CVD morbidity or mortality, perhaps due to the associated metabolic abnormalities contributing to CVD that are not improved by medical management (National Heart, Lung, and Blood Institute 1997; Lip and Li-Saw-Hee 1998). In this regard, exercise training may be an advantageous prophylactic modality because of the favorable effects on metabolic risk factors (Black 1991).

This chapter will review the relation between blood pressure and cardiovascular risk, the biological and environmental factors associated with the etiology of hypertension, and the important role of the sympathetic nervous system in the development of this syndrome. The relevance of hypotension and concomittant rarity of heart disease in individuals with Down's syndrome and Alzheimer's disease will also be discussed in light of the attenuation of sympathetic neural activity in these patients. The hemodynamic responses to acute exercise will be discussed, along with the impact of regular aerobic exercise training as a treatment modality for hypertension.

Hypertension and Cardiovascular Risk

There has been considerable discussion of whether there is a linear and uninterrupted relation between arterial pressure and risk across a wide range of pressures (MacMahon et al. 1990; Glynn et al. 1995) and whether the association of pressure and risk in all or part of the range is causal. It is clear that increased pressure causes stroke; that it also causes CHD is suggested by the ability of hypotensive drugs to reduce CHD in hypertensive patients, particularly in the elderly (Collins et al. 1990; Lever and Ramsay 1995). Some studies, particularly those in elderly subjects, suggest that the relation of pressure and risk is not continuous and that, in the lower part of the range, risk increases with decreasing pressure where J-shaped or U-shaped relations of pressure of risk result (Heikinheimo et al. 1990; Siegel et al. 1987; Taylor et al. 1991). One interpretation is that decreasing blood pressure is advantageous, but only down to a point below which risk is raised by decreasing pressure. Low blood pressure in many old people results from manifest disease such as stroke and coronary heart disease, which are potentially relevent contributors to an association between high mortality and low pressure (Millar, Isles, and Lever 1995). Accordingly, it does not follow that low blood pressure confers negligible risk. Conditions such as postural and postprandial hypotension, and exercised-induced hypotension (Smith and Mathias 1995) are quite common and can become a serious clinical problem, particularly in the elderly.

The Rise of Blood Pressure With Age

Blood pressure rises with age in most Western societies. While the extent of this rise varies from one subject to another, individuals tend to maintain position within the blood pressure distribution (figure 23.1). Those in the upper quintiles of the distribution have the greatest rate of rise, while those in lower quintiles have lower rates, no rise, or even a fall of pressure (figure 23.1). Follow-up studies show that children and young adults in the upper part of the distribution are more likely to develop hypertension and cardiovascular disease in later life (Paffenbarger, Thorne, and Wing 1968; Whincup et al. 1988). Subjects with highest pressure at the outset having the fastest rate of rise will exceed the limit first and in greatest number and become hypertensive.

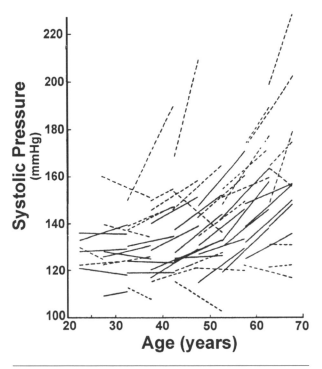

Figure 23.1   Relation of systolic blood pressure and age in groups of less than 10 (- - -) and more than 10 (—) men. Lines join first and subsequent measurement. *Note:* greater rise of pressure in groups with highest initial value. (From Miall and Chinn 1973 with permission of author and publisher.)

Figure 23.2 shows changes of blood pressure and body weight over a wide range of age. Two phases are seen: a rapid increase of blood pressure and weight during childhood, halting abruptly as growth ceases; a plateau is followed by a slowly increasing rate of rise of pressure continuing into late adult life. Longitudinal studies confirm these changes for systolic and diastolic pressures for males and females (figure 23.1; Miall and Chinn 1973; Miall and Oldham 1958; Rosner et al. 1977). Because the rise of blood pressure in children relates so closely to growth, several workers have suggested that a mechanism controlling growth also controls arterial pressure

(Szklo 1986; Lever and Harrap 1992). Thus, an action on cardiovascular maturation or on its programming may be an important factor (Lever and Harrap 1992; van den Bree et al. 1996). Continued activity of the same growth-promoting mechanism is unlikely to explain the second rise of arterial pressure in adults, partly because there is no further growth in height, and partly because factors reducing the rise in adults may not reduce the rise in children. For example, some primitive rural societies show no rise of pressure in adults, but in the children of these societies pressure rises normally by Western standards (Poulter et al. 1990; Marmot 1984).

These observations suggest that two processes govern the rise of pressure with age. While overactivity of the first during childhood may predispose to increasing pressure during the second adult phase, the increase only occurs when environmental factors are present and operating at the time. For example, if adults from a rural and isolated environment, in which blood pressure does not rise with age, migrate from country to town, physical activity and fitness decrease, blood pressure rises, heart rate and body weight increase, and glucose intolerance develops (Poulter et al. 1990; O'Dea 1991). The pattern of disease is similar to that in an industrialized Western society, maybe even an exaggerated version of it.

Mechanisms governing the rise of pressure with age are of great interest currently. There are two lines of inquiry: one suggests that events in fetal life (Barker et al. 1989) or childhood (Lever and Harrap 1992; van den Bree et al. 1996) influence the level of blood pressure in adults and the likelihood that hypertension and cardiovascular disease will develop. The other, with equally strong support, suggests that environmental factors acting during adult life determine blood pressure and the likelihood of hypertension. Although these interpretations are often debated as opposed alternatives, there are data to support involvement of both mechanisms.

Solitude and Isolation: Influence on Blood Pressure in Adults

Patients with psychiatric disease, resident long-term in a hospital, show no rise of blood pressure with age. This is not a feature of a particular psychiatric condition, nor is it a consequence of treatment. Blood pressure on admission is normal (i.e., that of the general population at the same age), and it is only after admission that pressure fails to rise with age (Masterton et al. 1981). Interestingly, the failure

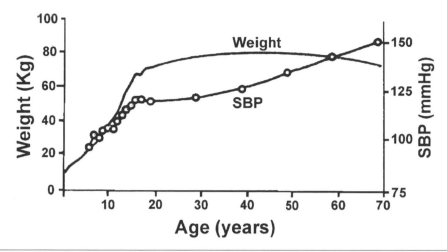

Figure 23.2   Changes of systolic blood pressure (SBP) and body weight in American males. (Reprinted from Lever and Harrap 1992 with permission.)

is clearest in psychiatric wards where isolation is greatest (Main and Masterton 1981). Failure of blood pressure to rise is also seen in patients with physical disease resident long-term in a hospital (Hatama et al. 1993) and in women resident long-term in a nunnery (Timio et al. 1988).

Does the reduction of blood pressure by solitude and isolation produce a commensurate reduction of cardiovascular risk? The studies on urban migration described earlier raise this possibility. An analysis of CHD findings in the study by Timio and colleagues (1988) of nuns isolated in a nunnery is awaited with interest.

Pathophysiology of Hypertension

Hypertension is a complex syndrome often accompanied by multiple underlying pathophysiological abnormalities. Hypertension is frequently associated with, and may be a component of, several metabolic disorders including insulin resistance, glucose intolerance, dyslipidemia, and obesity (e.g., syndrome X) (MacMahon, Cutler, and Stamler 1989; Resnick 1993; Polare, Lithell, and Berne 1990; Reaven, Lithell, and Landsberg 1996; Hardin et al. 1997). For example, ~65% of diabetics have hypertension, and a high coexistence of renal disease exists in this population (American Diabetes Association 1998). Metabolic abnormalities commonly associated with hypertension may play a role in the regulation of blood pressure. For example, elevated sympatho-adrenal activity is a direct correlate of insulin resistance in hypertension (Rowe et al. 1981; Arauz-Pachecho et al. 1996; Dengel et al. 1998; Facchini, Stoohs, and Reaven 1996), and hyperinsulinemia is accompanied by increased sensitivity to potent vaso-

constrictor agents such as norepinephrine and angiotensin II (Tuck, Corry, and Trujillo 1990; Gaboury et al. 1994). Other pathogenic mechanisms affecting vascular tone and blood pressure include underlying atherosclerotic disease, lesions in the brain responsible for blood pressure regulation (Suarez, Pegram, and Frohlich 1981; Burke et al. 1994), nephropathy, neuroendocrine disturbances affecting fluid-electrolyte homeostasis (aldosteronism), ion transport dysregulation (Adragna et al. 1985; Hilton 1986; Rebbeck, Turner, and Sing 1993; Aviv 1994; Picado et al. 1994; Clorius et al. 1997), and mutations of genes coding for vasoactive substances such as angiotensinogen (Caulfield et al. 1994). Advances in molecular biology have focused attention on the important role of the vascular endothelium as a vital organ with autocrine-paracrine systems involved in the pathogenesis of hypertension (Panza et al. 1990; Dzau et al. 1994). Impaired endothelium-mediated vasodilation is a common finding in hypertension, and thus nitric oxide (NO) is implicated as an important factor in the control of blood pressure in hypertension (Luscher and Vanhoutte 1986; Linder, Kiowski, and Buhler 1990; Chen and Sanders 1991; Baylis, Mitruka, and Deng 1992; Kelm et al. 1996; Preik et al. 1996; Franke and Tegeler 1997; Ghiadoni et al. 1998; Palatini 1998). As NO exerts an important systemic vasodilatory influence, impaired NO production leads to increases in vascular resistance and blood pressure (Sander, Hansen, and Victor 1995, 1997). Abnormal production and/or responses to other vascular paracrine system mediators of vascular tone such as type-C natriuretic peptide (Suga et al. 1992; Tomiyama et al. 1997), angiotensin II (Morishita et al. 1992; Naftilan et al. 1991), and endothelin-1 (Clozel et al. 1993; Cottone et al. 1998) may also contribute to hypertension. In addition to the vascular-endothelium, a deficiency in brainstem neuronal NO production has been linked to neurogenic hypertension. NO exerts a modulatory influence on glutamatergic neurotransmission in the brain stem, which mediates sympathetic discharge (Garthwaite and Boulton 1995; Christopherson and Bredt 1997). Accordingly, impaired brainstem NO production causes sympathetic hyperactivity resulting in increases in vascular resistance (Sander, Hansen, and Victor 1995, 1997).

Left ventricular hypertrophy is common in established hypertension and is linked to chronically elevated systolic blood pressure (Roman et al. 1995), and to hormonal abnormalities such as aldosteronism, elevated angiotensin II and urinary C-peptide, and hyperinsulemia (Yoshitomi et al. 1997; Kaneko et al. 1997; Tomiyama et al. 1997; Schlaich et al. 1998). In the large capacitance vessels of the circulation, intimal-medial wall thickening leads to narrowing of vessel lumen cross-sectional area and is associated with a reduction in ventricular diastolic volume (Boutouyrie et al. 1995; Ghiadoni et al. 1998). As hypertrophy develops in hypertension, left ventricular filling becomes progressively impaired (Yoshihara et al. 1996). Vascular medial-layer hypertrophy is prevalent in hypertensives with left ventricular hypertrophy (Conway 1989), but their development may not always occur in parallel (Lucarini et al. 1991; Bergbrant, Hansson, and Jern 1993). Coronary reserve is often reduced in hypertension with and without evidence of ventricular hypertrophy (Lund-Johansen 1994). Impaired coronary vascular reserve is associated with microvascular abnormalities or disordered vascular geometry, which may result in myocardial ischemia and impaired left ventricular function (Grossman, Oren, and Messerli 1994; Takeichi et al. 1997; Virdis et al. 1996; Franz, Tonnesmann, and Erb 1997).

In the peripheral arteries, smooth muscle hypertrophy develops in response to prolonged elevations of blood pressure as well as neurogenic/hormonal factors. Vascular hypertrophy leads to a greater media:lumen ratio and a reduction in vessel compliance, which together with vascular (endothelial) dysfunction cause increases in vascular resistance (Conway 1989; Bergbrant, Hansson, and Jern 1993; Heron et al. 1995; Rosei et al. 1995; Kelm et al. 1996; Ghiadoni et al. 1998; Franke and Tegeler 1997; Palatini 1998). Exacerbating the limited peripheral vasodilatory capacity is the finding that autonomic (cardiopulmonary baroreflex) control of vascular resistance may be impaired in the presence of myocardial hypertrophy (Trimarco et al. 1986; Silva et al. 1997), resulting in an exaggerated vasocontriction response. Furthermore, peripheral chemoreceptor-mediated sympathetic activation is potentiated in hypertensives compared to controls (Somers and Mark 1993).

In virtually all forms of hypertension, including rat models of hypertension (spontaneously hypertensive rats [SHR], Dahl salt-sensitive, Lyon hypertensive, and transgenic renin rats), the final common pathway leading to the long-term elevation of blood pressure is an impaired excretory capacity of the kidney (Cowley et al. 1995). This impairment is characterized by a shift in the relationship between sodium excretion and arterial pressure to higher pressures (Cowley 1992). It has been shown that blood flow to the inner renal medulla is an important determinant of sodium and water excretion. Multiple factors causing increased vascular resistance and reduced perfusion of the renal medulla lead to resetting of the renal pressure-natriuresis relationship and result in a long-term elevation of blood pressure in hypertension (Cowley et al. 1995; Clorius et al. 1997). These factors include enhanced renal sympathetic nerve activity, elevated circulating vaso-active hormones, renovascular disease, as well as altered intrarenal paracrine-autocrine systems (Oparil 1986; Esler et al. 1986; Roman and Zou 1993; Somers and Mark 1993; Kelm et al. 1996; Clorius et al. 1997). There is strong evidence suggesting that the progression of hypertension may involve a cascade of physiological adaptations induced initially by excessive sympathetic activation leading eventually to long-term structural changes in the organs most critical for regulating blood pressure.

Hypertension and Sympathetic Activity

Increased sympathetic activity is a prevalent finding in hypertension, and is especially present in the early phases of the development of borderline and mild hypertension (Julius 1993; Somers and Mark 1993; Lund-Johansen 1994). This has been found in studies using selective receptor blockade revealing increased β-adrenergic (Julius 1993; Chandler and DiCarlo 1998) and α-adrenergic drive (Esler et al. 1977), coupled with elevated plasma norepinephrine levels (Goldstein 1983), norepinephrine spillover (Esler et al. 1986), as well as increases in directly recorded sympathetic nerve activity (Anderson et al. 1989). This pattern is particularly clear in cyclosporine-induced hypertension (Scherrer et al. 1990).

In the early developmental phases of hypertension, the increase in sympathetic drive is accompanied by either higher cardiac output and/or elevated resistance, while in established hypertension, elevated peripheral vascular resistance accounts for the higher arterial pressure at normal heart rates and cardiac output (Julius 1993; Lund-Johansen 1994). Thus, a sympathetically-mediated hyperkinetic circulation is thought to represent an early phase in the development of hypertension, and is a strong predictor of established hypertension (Somers and Mark 1993; Levy et al. 1945; Paffenberger, Thorne, and Wing 1968; Stamler et al. 1975). This pattern may reflect a genetic predisposition to hypertension, as it has been observed in childhood with a strong link to parental hypertension (van den Bree et al. 1996; Nho et al. 1998), and also in SHR rats (Chandler and DiCarlo 1998).

As the duration of hypertension advances, cardiac β-adrenergic responsiveness becomes reduced (Julius 1993; Trimarco et al. 1983) reflecting structural down-regulation of the receptors. This is accompanied by a decrease in resting stroke volume, which is thought to result from a decrease in vascular and cardiac compliance (Julius 1993). The progressive increase in peripheral vascular resistance that occurs with developing hypertension reflects both structural and metabolic adaptations intrinsic to the vasculature (Folkow 1987), resulting in hypersensitivity to adrenergic vasoconstrictors (Amann et al. 1981; Sivertsson, Sannerstedt, and Lundgren 1976; Phillip, Distler, and Cordes 1978; Egan et al. 1987; Lund-Johansen 1994), as well as impaired vasodilation (Preik et al. 1997).

Prolonged exposure of the resistance vessels to sympathetic vasoconstriction and/or circulating norepinephrine and angiotensin II leads to hypertrophy of vascular smooth muscle, which increases vascular resistance (Somers and Mark 1993). Vascular hypertrophy has been proposed as one mechanism contributing to insulin resistance in hypertension (Henrich et al. 1988; Lillioja et al. 1987; Greene et al. 1989). An increase in vessel wall thickness is thought to increase diffusion resistance for insulin uptake by skeletal muscle. Additionally, elevated sympathetic activity, together with altered vascular metabolic vasodilation, contributes to insulin resistance by decreasing muscle blood flow (and its distribution), resulting in lower glucose delivery and uptake (Conway, Julius, and Amery 1968; Kodama et al. 1990; Julius 1993). This is supported by the finding that sympathetic vasoconstriction in response to cardiopulmonary baroreceptor unloading acutely increases insulin resistance (Jamerson et al. 1993). Furthermore, insulin resistance is related more closely to blood pressure and heart rate during mental stress (which induces significant eleva-

tions in sympathetic activity) than to resting blood pressure and heart rate (Moan et al. 1995). The increased resistance to insulin could in turn induce a compensatory elevation of insulin leading to hyperinsulinemia (Julius 1993).

Another mechanism proposed for the link between adrenergic activity, insulin resistance, and hyperinsulinemia in hypertension is that muscle β-receptor stimulation may alter muscle fiber type composition, leading to insulin resistance. Acute β-receptor stimulation by epinephrine infusion decreases glucose uptake in muscle (Diebert and DeFronzo 1980), a response similar to that induced by mental stress (Brod 1963). Prolonged β-adrenergic stimulation may diminish glucose uptake by altering the composition of muscle fibers from type I to type II (Zemen et al. 1988), which are more insulin-resistant. Supporting this is the finding that patients with hypertension have a decreased number of type I fibers (Juhlin-Dannfelt et al. 1979; Karlsson and Smith 1984).

A number of studies have demonstrated a direct correlation between increases in plasma insulin concentration and sympathetic activity, and insulin is known to evoke sympathetic discharge (Rowe et al. 1981; Berne, Fagius, and Niklasson 1989; Arauz-Pachecho et al. 1996; Marks et al. 1996; Manicardi et al. 1986; Ward et al. 1993). Insulin also exerts a local vasodilatory effect, but this appears to be lost or attenuated in hypertensives (Imaizumi 1996). As sympathetic vasoconstriction can increase plasma insulin, and elevated plasma insulin evokes sympathetic discharge, it has been debated which response occurs first in the development of hypertension. Sympathetic activation has been demonstrated in hypertensives in the absence of elevated insulin, and thus it likely precedes the hyperinsulinemic response. Once established, the hyperinsulinemia-sympathetic activity response may represent a feed-forward malignant pattern contributing to hypertension, especially when the peripheral vasculature has chronically adapted with structural and metabolic changes linked to diminished vasodilatory responsiveness.

Virtually opposite to the developmental pathophysiological adaptations in patients with hypertension that stem from elevated sympathetic activation are the responses to exercise training observed in normal healthy adults and animals. The early adaptation to endurance training is characterized by reduced sympathetic activation, resulting in a lowering of vascular resistance and blood pressure, leading to increased tissue perfusion. The long-term adaptations in the vasculature favor enhanced vasodilatory capacity, along with reduced sensitivity to vasoconstrictor agents, enhanced perfusion of muscle and splanchnic region, and increased insulin sensitivity.

Low Blood Pressure in Down's Syndrome

Of considerable interest is the association between impaired sympathetic activation and low blood pressure in individuals with Down's syndrome (Murdoch et al. 1977; Richards and Enver 1979; Lever and Ramsay 1995; Morrison et al. 1996). In fact, hypertension is virtually absent in this population. The reported low blood pressure is not a consequence of isolation or environmental factors as described previously, but is likely related to trisomy 21.

A common mechanistic process underlying this pattern may coexist in patients with Alzheimer's disease, a condition also characterized by low blood pressure—lower than that in other forms of dementia and lower than that in elderly controls (Landin et al. 1993; Kilander, Boberg, and Lithell 1993; Wang, Liao et al. 1994; Burke et al. 1994; Elmstahl et al. 1992; Tresch et al. 1985). By the age of 35, most Down's syndrome patients have the neuropathological changes of Alzheimer's disease; and within a few years a high proportion will develop clinical features as well (Wisniewski, Silverman, and Wegiel 1994). Beta-amyloid precursor protein (APP) and its product β-A_4 protein are likely to cause the neuropathy of Alzheimer's disease in its common sporadic form, in some of its familial forms, and in the form that develops in Down's syndrome (Rossor 1995; Wisniewski, Silverman, and Wegiel 1994; Lantos and Cairns 1994; Hyman, West, and Rebec 1995). In support of this, the APP gene is on chromosome 21 (Wisniewski, Silverman, and Wegiel 1994) and APP plasma levels are raised in Down's syndrome, more so than in Alzheimer's disease (Rumble et al. 1989). Mice transfected with a mutant APP gene from one of the familial forms of Alzheimer's disease develop the same brain lesions seen in Alzheimer's (Games et al. 1995). Centers in the brain regulating blood pressure are affected by the neuropathy, and this is widely believed to explain the reduction of blood pressure in sporadic Alzheimer's (Burke et al. 1994). Morrison and colleagues (1996) suggest that the same neuropathy developing at an earlier age (see Hyman, West, and Rebec 1995) is also responsible for low blood pressure in Down's syndrome. Lowest values are exhibited in patients whose dementia is most marked (Wang, Liao et al. 1994). When serial measurements are made, blood pressure is seen to fall in some patients as dementia worsens (Burke et al. 1994), and hypertension is rare in Alzheimer's disease (Tresch et al. 1985).

Centers in the brain concerned with the regulation of blood pressure are often affected in Alzheimer's disease (reviewed by Burke et al. 1994). Among these are the C_1 adrenergic mechanism in the medulla, a control point for the baroreflex (Burke et al. 1994), the hippocampus, amygdala, caudate nucleus, and locus coeruleus (Yates et al. 1980; Lantos and Cairns 1994). While there is potential in these lesions for a reduction of blood pressure (Burke et al. 1994), the neurological pathways involved cannot be identified from neuropathology alone. Studies of autonomic function may be informative.

If neuropathy lowers blood pressure in Alzheimer's disease, does it also lower blood pressure in Down's syndrome? The Alzheimer's neuropathy that develops in Down's syndrome is indistinguishable in its distribution and staining properties from that which occurs in the sporadic form of Alzheimer's disease (discussed in Hyman, West, and Rebec 1995). The timing of events is compatible with a role for Alzheimer's neuropathy; that is, lowness of blood pressure relative to controls is most apparent in Down's syndrome patients aged 35 years or more (Morrison et al. 1996), and Alzheimer's neuropathy is already established by this age (Wisniewski, Silverman, and Wegiel 1994). Beta-amyloid deposits are seen in the hippocampal region of much younger Down's syndrome patients, and the level reached at ~50 years may be greater than that seen in sporadic Alzheimer's disease (Hyman, West, and Rebec 1995). Amyloid precursor protein, the favored candidate for Alzheimer's disease, is measurable in plasma. Its concentration in Down's syndrome is higher than that in normal subjects and that in patients with sporadic Alzheimer's disease (Rumble et al. 1989).

Another feature shared by Down's syndrome and Alzheimer's disease is that atheroma and coronary heart disease is less frequent than in the general population. For Down's syndrome, this is the conclusion from two studies of pathology (Murdoch et al. 1977; Yla-Herttuala et al. 1989) and from a survey of mortality among Down's syndrome inpatients (Thase 1982). For Alzheimer's disease the conclusion is based on a pathological study (Hughes 1977) and on the frequency of CHD in elderly controls and in a large group of patients with different forms of dementia (Tresch et al. 1985). If atheroma and CHD are low in both diseases, does lowness of blood pressure reduce these risks? This is an intriguing possibility, but it remains to be determined whether β-amyloid found in Alzheimer's disease and Down's syndrome could be responsible by an action of low blood pressure or by a mechanism independent of blood pressure.

Hemodynamic Responses to Dynamic Exercise

Blood pressure is described simply by the product of cardiac output and total peripheral vascular resistance (BP = CO × TPR), and during dynamic exercise both factors contribute to the blood pressure response. During incremental dynamic exercise in normotensive individuals, systolic pressure rises and diastolic pressure tends to fall or remain stable, while mean arterial pressure increases. Both cardiac output and peripheral resistance contribute to the elevation of systolic pressure; however, an inverse relationship exists between cardiac output and peripheral vascular resistance—that is, the higher the cardiac output, the lower the vascular resistance. Diastolic blood pressure reflects the underlying peripheral vascular tone as the arterial pressure component from cardiac ejection is minimized. The stability or lowering of diastolic pressure in normotensive individuals results from vasodilation in active muscle attributed to a release of sympathetic vasocontrictor tone and/or to metabolic sympatholysis (Hansen et al. 1996).

In both Down's syndrome (Pavey 1989) and Alzheimer's disease (Elmstahl et al. 1992; Vitiello et al. 1993), postural hypotension is reported, and the deficit in Alzheimer's disease has features suggesting sympathetic and parasympathetic failure (Yates et al. 1980; Mufson et al. 1993). When this population is studied during exercise, sympathetic activation is impaired, and the rise of heart rate, blood pressure, and exercise capacity are lower than normal (Eberhard, Eterradossi, and Rapacchi 1989; Pitetti et al. 1992). Pitetti and colleagues (1992) compared responses in Down's syndrome and controls with other forms of mental handicap. At maximum exertion, heart rate, minute ventilation, oxygen uptake, and peak work capacity were lower in Down's syndrome. During submaximal exercise at a given oxygen consumption, heart rate and cardiac output were higher in Down's syndrome, and both total peripheral resistance and blood pressure were lower. Interestingly, the arteriovenous oxygen difference was smaller in the Down's syndrome group than in controls. This pattern could be explained by baroreceptor activation of cardiac output as a compensatory response to regulate blood pressure in the face of a lower vascular resistance (Conway 1989). Thus, due to impaired sympathetic control of vascular resistance, an elevation of cardiac output would contribute to maintain blood pressure, and the lower arteriovenous oxygen difference would reflect tissue hyperperfusion.

In hypertensive individuals who also exhibit a hypertensive response to exercise, both hyperkinetic cardiac function and hyper-resistant vascular responses have been observed (Iskandrian and Heo 1988; Montain et al. 1988; Lund-Johansen 1989; Conway 1989; Palatini 1998). The former response is common in early-stage hypertension, while the latter is prevalent in older individuals with more established hypertension and advanced circulatory and metabolic manifestations (Montain et al. 1988). This is consistent with the observation that individuals with established hypertension have both higher systolic and diastolic pressures during acute dynamic exercise in association with elevated vascular resistance (Montain et al. 1988; Iskandrian and Heo 1988; Lund-Johansen 1989, 1994; Colombo et al. 1989). Based on longitudinal data, Lund-Johansen (1989) has proposed that hypertensive individuals progress from a high-output low-resistance phase to a low-output high-resistance phase over time. In both cases, elevated sympathetic activation has been implicated as a contributing factor (Reis, Morrison, and Ruggiero 1988; Tipton 1991; Cottone et al. 1998).

In established hypertension, peak stroke volume, cardiac output, and oxygen uptake are lower, while blood pressure and vascular resistance are higher than in age-matched controls. This response is often accompanied by left ventricular diastolic dysfunction (Fagard et al. 1995; Fagard and Lijnen 1997; Melin et al. 1987; Montain et al. 1988; Missault et al. 1993; Manolis et al. 1997). The hypertensive response to dynamic exercise results in an increase in afterload of the heart, which can lower stroke volume by intensifying asynchronous ventricular wall motion, leading to impaired ventricular relaxation (Miura et al. 1993). Reduced arterial and left ventricular compliance also contribute to impaired ventricular function during exercise, based on the examination of the relation between cardiac filling pressures, ejection fraction, and systolic blood pressures (Fagard, Staessen, and Amery 1988; Melin et al. 1987; Lund-Johansen 1989). This pattern, together with the common manifestation of left ventricular hypertrophy (Fagard and Lijnen 1997; Manolis et al. 1997), also often leads to a decrease in coronary reserve, resulting in myocardial ischemia and impaired inotropic function (Manolis et al. 1997; Ohtsuka et al. 1997). Of interest is that the higher stroke volume and cardiac output observed during exercise in normal adults after training result from virtually the opposite pattern—that is, a lowering of vascular resistance, improved diastolic filling, and greater arterial and ventricular compliance.

Adaptations With Exercise Training

There is general agreement in the literature that aerobic exercise training elicits small reductions in blood pressure in adults classified as normotensive (Fagard 1995; Kelley and Tran 1995). Even in this population, a modest reduction in blood pressure confers prophylaxis against cardiovascular morbidity and mortality as cardiovascular risk is manifest on an approximately linear continuum with blood pressure (MacMahon et al. 1990; Fagard 1995; Arroll and Beaglehole 1992; Tipton 1991; Kelley and Tran 1995). Epidemiological studies estimate that even a 5 mmHg reduction in blood pressure results in a 40% reduction in stroke and an approximately 15% reduction in myocardial infarction (Collins et al. 1990). Moreover, most epidemiological evidence suggests that blood pressure and cardiovascular risk are lower and life span longer in the physically fit (Puddey and Cox 1995; Jennings 1995; Fagard 1995; Paffenbarger et al. 1986).

A recent meta-analysis of clinical studies on the effect of endurance exercise training on blood pressure conducted between 1963 and 1992 involving ~1000 normotensive adults indicated a reduction of 4 and 3 mmHg in systolic and diastolic pressures, respectively (Kelley and Tran 1995). In a 20-year longitudinal study, Kasch and colleagues (1990) followed a group of sedentary and regularly exercising men and found that blood pressure was unchanged in the group engaged in regular exercise, while in the sedentary group, systolic blood pressure rose 15 mmHg and diastolic pressure rose 5 mmHg. A similar effect of exercise was recently reported by Mundal and colleagues (1997). Thus, long-term exercise training has the beneficial effect of preventing the normal age-related increase in blood pressure, and cardiovascular risk can be lowered rapidly as an antihypertensive effect has been shown to occur within 6 days after initiating an exercise program (Meredith et al. 1990).

In hypertensive adults exposed to an endurance exercise training program, reductions of ~5–13 and 3–18 mmHg in resting systolic and diastolic blood pressures, respectively, have been reported (Petrella 1998; Kelley and McClellan 1994; Puddey and Cox 1995; Fagard et al. 1995; Nelson et al. 1986; Urata et al. 1987; Ketelhut, Franz, and Scholze 1997; Zanettini et al. 1997). Similar effects are exhibited in young (Hagberg et al. 1983; Ewart, Young, and Hagberg 1998) and older (Hagberg et al. 1989; Cononie et al. 1991) individuals, in both men and women of different races (Hagberg 1988; Nho et al. 1998; Ewart, Young, and Hagberg 1998), and in those on antihypertensive medication (Motoyama et al. 1998). A recent meta-analysis of 26 training (aerobic) studies conducted between 1980 and 1996 involving 1533 (hypertensive and normotensive) adults also reported blood pressure lowering as an effect of exercise (Halbert et al. 1997).

To what extent regular endurance exercise exerts an antihypertensive effect independent of other beneficial outcomes such as weight loss and dietary and lifestyle changes has not been clearly defined, but evidence in favor of its beneficial role is accumulating. The fact that regular exercise often reduces body-fat weight and improves other metabolic processes such as insulin sensitivity and sympathetic activity, which are causally linked to hypertension, should be considered of critical value. Weight loss reduces basal norepinephrine levels, and a lowering of plasma insulin attenuates sympathetic nerve activity and decreases sodium retention in the kidney (Horton 1981; Lucas et al. 1985; Brown et al. 1997; Uehara and Arakawa 1997). These combined adaptations all contribute to lower blood pressure. In addition, exercise training in obese, hypertensive men has been shown to reduce blood pressure along with triglycerides, total cholesterol, and plasma insulin, and to increase HDL_{2-C} levels, which reduce the risk of atherosclerotic disease (Dengel et al. 1998).

The influence of sympathetic adrenergic activity as a determinant of the (hypotensive) responsiveness to training has received considerable attention. In fact, attenuated sympathetic activity is the most commonly reported training adaptation linked to the lowering of blood pressure in hypertensives (Tipton 1991; Fagard and Tipton 1994). This notion is based on findings of lower plasma norepinephrine (Nomura et al. 1984; Kinoshita et al. 1988; Hagberg et al. 1989; Reiling et al. 1990; Urata et al. 1987; Brown et al. 1997), reduced norepinephrine spillover (Jennings et al. 1986), diminished sympathetic activity, and reduced cardiac β-adrenergic responsiveness (Overton, VanNess, and Takata 1998; Veras-Silva et al. 1997; Uehara and Arakawa 1997). Attenuation of sympathetic activity with training has been linked to weight loss (Landsberg and Young 1984), reduced plasma insulin levels (Horton 1981; Manicardi et al. 1986; Rochini et al. 1987; Brown et al. 1997), and to diminished activation by several other mechanisms, including the baroreflexes, the muscle reflex, the sympathoadrenal system, and central modulatory influences. Even in hypertensives characterized as hyperadrenergic (elevated resting catecholamines), training has been shown to lower blood pressure, and several studies have reported a more pronounced pressure reduction in individuals with pre-existing hyperadrenergic activity (Duncan et al. 1985; Fagard and Tipton 1994).

While there are some conflicting reports in the literature, the general finding is that during exercise blood pressure is lowered in hypertensives after a training program (Fagard and Tipton 1994). Recently, Ketelhu, Franz, and Scholze (1997) found that 18 months of exercise training in hypertensive men resulted in a decrease in arterial pressures not only at rest, but during and after cycle exercise, and also during a cold pressor test. Heart rate was also lower during exercise, contributing to reduced myocardial oxygen consumption. These results conform to the well-established pattern of a reduced pressor and heart rate response to a given absolute exercise load in normotensives after training. Similarly, in SHR rats, 18 weeks of low-intensity (55% max $\dot{V}O_2$) training reduced blood pressure due to a lowering of heart rate and cardiac output (Veras-Silva et al. 1997). This effect has also been demonstrated in humans after swim training (Tanaka et al. 1997). Also in SHR rats, exercise training has been shown to improve arterial and cardiopulmonary baroreflex sensitivity, resulting in a hypotensive and bradycardic effect (Silva et al. 1997), and limited data also supports this pattern in humans (Fagard and Tipton 1994). In normotensive individuals who exhibit a hypertensive response to exercise, 4 months of aerobic exercise training resulted in a reduction in exercise blood pressure and heart rate, suggesting attenuation of sympathetic neural activity (Loma, Herkenhoff, and Vasquez 1998). Together, these findings provide evidence for an attenuation of sympathetic activity contributing to a lower pressor response to exercise after long-term training.

Postexercise hypotension is a common response in both normotensive and hypertensive individuals, and is characterized by a reduction in heart rate and cardiac output. Opiod and serotinergic mechanisms have been implicated in this response (Hoffman et al. 1987; Fagard and Tipton 1994), and it has been proposed that the long-term pressure-lowering effect of exercise training may be mediated in part by these mechanisms. Opiod receptor down-regulation has been proposed as a mechanism contributing to training bradycardia by enhancing the influence of parasympathetic activity, which could account for a hypotensive effect.

The lack of a clearly demonstrable reduction in directly recorded sympathetic nerve activity to rest-ing muscle after training (Somers et al. 1992; Grassi et al. 1994; Svedenhag et al. 1984; Ng et al. 1994; Sheldahl et al. 1994) suggests that sympathetic activity may be reduced to other organs (Jennings et al. 1986; Meredith et al. 1991). The importance of training for reducing renal sympathetic activity in hypertensive individuals has been illustrated by Kohno and colleagues (1997). They found that training reduced blood pressure in parallel with a reduction in renovascular sympathetic tone, lower plasma norepinephrine, lower plasma insulin, and increased sodium excretion. While this adaptation is thought to be of primary importance for long-term improvements in blood pressure control (Julius 1993), exercise training has even produced a reduction in blood pressure in patients with renal failure (Boyce et al. 1997). Training has in some cases been shown to reduce plasma renin activity (Fagard et al. 1985; Geyssant et al. 1981), which is related to lower sympathetic activity (Jennings et al. 1986).

Adaptations in adrenal medullary activity may be of potential importance for the blood pressure-lowering effect of training (Tipton 1991). This notion is supported by the finding that chemically sympathectomized rats exhibit a reduction in resting blood pressure after training, but not if they are also demedullated (Tipton 1984). Thus, an intact adrenal medulla seems important for reducing the adrenergic influence on blood pressure with training.

Changes in renal dopaminergic activity may be a beneficial outcome of exercise training. Sakai and colleagues (1998) found a close inverse relation between increases in urinary free dopamine excretion and decreases in blood pressure after 4 weeks of training in hypertensive individuals. This change was also associated with increased urinary kallikrein activity, sodium excretion, and reduced plasma volume. Dopaminergic stimulation of preganglionic vagal neurons reduces heart rate, and increases in dopaminergic receptors in central and peripheral nerves have been observed after training. This potential bradycardic effect could also exert a hypotensive response.

Recent work has addressed the issue of cardiac and vascular remodeling by pharmacological treatment and exercise regimen in hypertensives. Long-term antihypertensive treatment has been found to normalize left ventricular mass and to reduce vascular resistance and blood pressure at rest (Fagard and Lijnen 1997). This treatment outcome, however, was without improvement in peak exercise performance and may be related to both the lack of improvement in diastolic function, despite normalization of ventricular mass, and to a lack of exercise training.

Accordingly, a reduction in left ventricular mass has been demonstrated with regular exercise training, along with improved fitness and a reduction in blood pressure and plasma fibrinogen (Zanettini et al. 1997). Exercise training may also reverse the negative cardiac inotropic effect of chronic hypertension by enhancing cardiac α-1 receptor responsiveness, as indicated by greater myocardial contractile responses to phenylephrine (Korzick and Moore 1996). Improvements in peripheral vascular function have been observed in hypertensive patients with training. Low-intensity (~45% $\dot{V}O_2$max) walking these times per week over a 6-month period resulted in an increased peak muscle blood flow and vascular conductance, suggesting functional as well as structural improvements in the peripheral vessels (Tanaka, Reiling, and Seals 1998). Pharmacological antihypertensive therapy has proven efficacious for reducing vascular resistance, but endurance training may be unique in altering vascular structure and lumen area (Tipton 1991), as well as functional mechanisms enhancing perfusion.

Resistance training is not a commonly prescribed therapeutic modality in hypertensives, perhaps because this type of training has not consistently produced a blood pressure-lowering effect in normotensive individuals, and also because of safeguard measures against high pressor responses that may accompany static exercise (Borghi et al. 1988). A limited number of studies indicate that strength training elicits little if any change in resting blood pressure in hypertensive individuals as compared to dynamic exercise (Fagard and Tipton 1994).

Despite the generally favorable view that endurance exercise training has an overall pressure-reducing effect and results in reduced cardiovascular risk, some skepticism remains about the benefits of exercise per se as a therapeutic regimen for hypertension (Cox et al. 1993; Gilders, Voner, and Dudley 1989; Blumenthal, Siegel, and Appelbaum 1991). One concern about a generalized prescription of regular exercise in this population pertains to potential risk of myocardial ischemia and infarction, lethal arrhythmias, and sudden death during acute exercise in some hypertensive individuals (Wannamethee et al. 1995; Palatini 1998; Mundal et al. 1997; Manolis et al. 1997; Kohl et al. 1992). Elevated platelet activation and adhesiveness has also been observed during acute exercise in hypertensives (Wang, Jen et al. 1994; Lip and Li-Saw-Hee 1998), which increases the risk of thrombosis. Pathologies associated with long-term hypertension, such as nephropathy, diabetes, ischemic heart disease, and peripheral vascular disease, may also present complications for prescription of regular exercise training. In this regard, long-term cardiovascular risk is greater in those hypertensive patients whose blood pressure rises most during exercise (Mundal et al. 1994). Another basis for caution is that some hypertensive individuals characterized as "non-responders" (Duncan et al. 1985; Attina et al. 1986; Kinoshita et al. 1988) fail to develop a hypotensive adaptation to training. What accounts for this pattern is not well understood, but left ventricular enlargement and a hyperkinetic, noncompliant heart may be important factors (Lund-Johansen 1989; Tipton 1991). There may also be a genetic basis for training responsiveness, as hypertensive individuals with a family history of hypertension show less reduction in blood pressure with training than those without parental history (Nho et al. 1998; Orbach and Lowenthal 1998).

The relation between exercise training, blood pressure, and cardiovascular risk suggests a balance of advantage and disadvantage. For the hypertensive individual with underlying disease, acute vigorous exercise may produce adverse effects apparent during or just after exercise (Puddey and Cox 1995; Manolis et al. 1997). Set against this, the general health benefits produced by exercise training probably offset the short-term disadvantages during acute exercise. In this regard, Fagard, Staessen, and Amery (1991) reported that in 169 patients with essential hypertension (aged 16–66), acute exercise blood pressure did not predispose individuals to any increased risk of manifest organ damage than did resting blood pressure. However, because many hypertensive individuals have other cardiovascular and metabolic complications, the importance of individualized exercise prescription needs to be recognized in order to ensure that hypotensive and other health benefits occur (Orbach and Lowenthal 1998; Puddey and Cox 1995; Jennings 1995; Fagard 1995; Halbert et al. 1997). Also, for meaningful conclusions to be drawn from these trials, improvement is needed in the design of studies on cardiovascular risk in groups receiving and not receiving training.

It has been suggested that low-intensity (40–70% of maximal oxygen uptake) exercise may be more effective for lowering blood pressure in hypertensive individuals than high-intensity training (50–85%), due to diminished sympathetic activation and reduced myocardial stress (Petrella 1998; Tipton 1991; Veras-Silva et al. 1997). Low-intensity training has also been beneficial for attenuating the cardiovascular response to psychological stress (Rogers et al. 1996; Schuler and O'Brien 1997). On the other hand, the reduction in diastolic blood pressure with training is significantly related to the increase in exercise capacity (Fagard and Tipton 1994), which suggests that high-intensity training may be important. Attention is currently focused on determining the effectiveness of various training regimens for both reductions in blood pressure and significant improvements in functional capacity. Special considerations are necessary for hypertensive individuals with underlying disease in order to ensure the appropriate dosage (exercise intensity × duration × frequency) so as to provide safety as well as long-term benefits. More sophisticated methods are being developed to monitor cardiovascular regulatory patterns at rest and during exercise, which may provide better specificity in prescribing and evaluating exercise dosage (Wagner et al. 1998; Manolis et al. 1997; Cody 1997; Orbach and Lowenthal 1998).

Conclusion

While multiple factors contribute to the development of hypertension, there is strong evidence indicating that the sympathetic nervous system plays a significant role in its development, particularly in the early stages. From an epidemiological and physiological viewpoint, the initiation of hypertension seems to relate to autonomic nervous overactivity linked to both genetic and environmental factors. Exercise training in healthy individuals promotes more efficient functioning of most physiological systems; and also reduces cardiovascular morbidity and improves longevity. The efficacy of exercise training as a therapeutic method in patients with established hypertension seems to depend on how advanced the disease process has become and on how well exercise is tolerated. Exercise studies incorporating methods to examine cardiovascular, metabolic, and endocrine responses provide important opportunities for better understanding the mechanisms at work in hypertension.

References

Adragna, N.C., J.L. Chang, M.C. Morey, and R.S. Williams. 1985. Effect of exercise on cation transport in human red cells. *Hypertension* 7:132–139.

Amann, F.W., P. Bolli, W. Kiowski, and F.R. Buhler. 1981. Enhanced alpha-adrenoreceptor-mediated vasoconstriction in essential hypertension. *Hypertension* 3(suppl):I119–I123.

American Diabetes Association. 1998. Position statement: Diabetic nephropathy. *Diabetes Care* 21(suppl 1):S50–S53.

Anderson, E.A., C.A. Sinkey, W.J. Lawton, and A.L. Mark. 1989. Elevated sympathetic nerve activity in borderline hypertensive humans: Evidence from direct intraneural recordings. *Hypertension* 14:177–183.

Arauz-Pacheco, C., D. Lender, P.G. Snell, B. Huet, L.C. Ramirez, L. Breen, P. Mora, and P. Raskin. 1996. Relationship between insulin sensitivity, hyperinsulinemia, and insulin-mediated sympathetic activation in normotensive and hypertensive subjects. *American Journal of Hypertension* 9:1172.

Arroll, B., and R. Beaglehole. 1992. Does physical activity lower blood pressure? A critical review of the clinical trials. *Journal of Clinical Epidemiology* 45(5):439–447.

Attina, D.A., G. Guiliano, G. Arcangeli, R. Musante, and V. Cupelli. 1986. Effects of one year of physical conditioning on borderline hypertension: An evaluation by bicycle ergometer exercise testing. *Journal of Cardiovascular Pharmacology* 8(suppl 5):145–147.

Aviv, A. 1994. Cytosolic Ca^{2+}, Na/H antiport, protein kinase C trio in essential hypertension. *American Journal of Hypertension* 7:205.

Barker, D.J., C. Osmond, J. Golding, D. Kuh, M.E. Wadsworth. 1989. Growth in utero, blood pressure in childhood and adult life, and mortality from cardiovascular disease. *British Medical Journal* 298(6673):564–567.

Baylis, C., B. Mitruka, and A. Deng. 1992. Chronic blockade of nitric oxide synthesis in the rat produces systemic hypertension and glomerular damage. *Journal of Clinical Investigation* 90:278–281.

Bergbrant, A., L. Hansson, and S. Jern. 1993. Interrelation of cardiac and vascular structure in young men with borderline hypertension. *European Heart Journal* 14:1304–1314.

Berne, C., J. Fagius, and F. Niklasson. 1989. Sympathetic response to oral carbohydrate administration. *Journal of Clinical Investigation* 84:1403–1409.

Black, H.R. 1991. Metabolic considerations in the choice of therapy for the patient with hypertension. *American Heart Journal* 121:707–715.

Blumenthal, J.A., W.C. Seigel, and M. Appelbaum. 1991. Failure of exercise to reduce blood pressure in patients with mild hypertension. *Journal of the American Medical Association* 266:2098–2104.

Borghi, C., F.V. Costa, S. Boschi, and E. Ambrosioni. 1988. Impaired vasodilatory capacity and exaggerated pressor response to isometric exercise in subjects with family history of hypertension. *American Journal of Hypertension* 1(3):106S–109S.

Boutouyrie, P., S. Laurent, X. Girerd, A. Benetos, P. Lacolley, E. Abergel, and M. Safar. 1995. Common carotid artery stiffness and patterns of left ventricular hypertrophy in hypertensive patients. *Hypertension* 25(4):651–659.

Boyce, M.L., R.A. Roberms, P.S. Avasthi, C. Roldan, A. Foster, P. Montner, D. Stark, and C. Nelson. 1997. Exercise training by individuals with predialysis renal failure: Cardiorespiratory endurance, hypertension, and renal function. *American Journal of Kidney Disease* 30(2):180–192.

Brod, J. 1963. Hemodynamic basis of acute pressor reactions and hypertension. *British Heart Journal* 25:227–245.

Brown, M.D., G.E. Moore, M.T. Korytkowski, S.D. McCole, and J.M. Hagberg. 1997. Improvement of insulin sensitivity by short term exercise training in hypertensive African American women. *Hypertension* 30(6):1549–1553.

Burke, W.J., P.G. Coronado, C.A. Schmitt, K.M. Gillespie, and H.D. Chung. 1994. Blood pressure regulation in Alzheimer's disease. *Journal of the Autonomic Nervous System* 48(1):65–71.

Caulfield, M., P. Lavender, M. Farrall, P. Munroe, M. Lawson, P. Turner, and A.J. Clark. 1994. Linkage of the angiotensinogen gene to essential hypertension. *New England Journal of Medicine* 330(23):1629–1633.

Chandler, M.P., and S.E. DiCarlo. 1998. Acute exercise and gender alter cardiac autonomic tonus differently in hypertensive and normal rats. *American Journal of Physiology* 274(2):R510–R516.

Chen, P.Y., and P.W. Sanders. 1991. L-arginine abrogates salt-sensitive hypertension in Dahl/Rapp rats. *Journal of Clinical Investigation* 88:1559–1567.

Christopherson, K.S., and D.S. Bredt. 1997. Nitric oxide in excitable tissues: Physiological roles and disease. *Journal of Clinical Investigation* 100(10):2424–2429.

Clorius, J.H., T. Hupp, I. Zuna, P. Schmidlin, S. Denk, and G. van Kaick. 1997. The exercise renogram and its interpretation. *Journal of Nuclear Medicine* 38(7):1146–1151.

Clozel, M., V. Breu, K. Burri, J.M. Cassal, W. Fischli, G.A. Gray, G. Hirth, B.M. Loffer, M. Muller, W. Neldhaart, and H. Ramuz. 1993. Pathophysiological role of endothelin revealed by the first orally active endothelin receptor antagonist. *Nature* 365:759–761.

Cody, R.J. 1997. The sympathetic nervous system and the renin-angiotensin-aldosterone system in cardiovascular disease. *American Journal of Cardiology* 80:9J–14J.

Collins, R., R. Peto, S. MacMahon, P. Hebert, N.H. Fiebach, K.A. Eberlein, J. Godwin, N. Qizilbash, J.O. Taylor, and C.H. Hennekens. 1990. Blood pressure, stroke, and coronary heart disease: Part 2. Short term reductions in blood pressure: Overview of randomised drug trials in their epidemiological context. *Lancet* 335(8693):827–838.

Colombo, F., T. Porro, G.F. del Rosso, P. Bertalero, L. Orlandi, and A. Libretti. 1989. Cardiovascular responses to physical exercise and tyramine infusion in hypertensive and normotensive subjects. *Journal of Human Hypertension* 3:245–249.

Cononie, C.C., J.E. Graves, M.L. Pollock, M.I. Phillips, C. Sumners, and J.M. Hagberg. 1991. Effect of exercise training on BP in 70–79-yr-old men and women. *Medicine and Science in Sports and Exercise* 23:505–511.

Conway, J. 1989. First Bjorn Folkow Award lecture. The role of cardiac output in the control of blood pressure. *Journal of Hypertension* 7(6)(suppl):S3–S7.

Conway, J., S. Julius, and A. Amery. 1968. Effect of blood pressure level on the hemodynamic response to exercise. *Hypertension* 16:79–85.

Cottone, S., M.C. Vella, A. Vadala, A.L. Neri, R. Riccobene, and G. Cerasola. 1998. Influence of vascular load on plasma endothelin-1, cytokines and catecholamine levels in essential hypertensives. *Blood Pressure* 7(3):144–148.

Cowley, A.W. 1992. Long-term control of arterial blood pressure. *Physiological Reviews* 72:231–300.

Cowley, A.W., D.L. Mattson, S. Lu, and R.J. Roman. 1995. The renal medulla and hypertension. *Hypertension* 25(4):663–673.

Cox, K.L., I.B. Puddey, A.R. Morton, L.J. Beilin, R. Vandongen, and J.R. Maserei. 1993. The combined effects of aerobic exercise and alcohol restriction on blood pressure and serum lipids: A two-way factorial study in sedentary men. *Journal of Hypertension* 11:191–201.

Dengel, D.R., J.M. Hagberg, R.E. Pratley, E.M. Rogus, and A.P. Goldberg. 1998. Improvements in blood pressure, glucose metabolism and lipoprotein lipids after aerobic exercise plus weight loss in obese, hypertensive middle-aged men. *Metabolism* 47(9):1075–1082.

Diebert, D.C., and R.A. DeFronzo. 1980. Epinephrine-induced insulin resistance in man. *Journal of Clinical Investigation* 65:717–721.

Duncan, J.J., J.E. Farr, S.J. Upton, R.D. Hagan, M.E. Oglesby, and S.N. Blair. 1985. The effects of aerobic exercise on plasma catecholamines and blood pressure in patients with mild essential hypertension. *Journal of the American Medical Association* 254(18):2609–2613.

Dzau, V.J., G.H. Gibbons, R. Morishita, and R.E.

Pratt. 1994. New perspectives in hypertension research: Potentials of vascular biology. *Hypertension* 23:1132–1140.

Eberhard, Y., J. Eterradossi, and B. Rapacchi. 1989. Physical aptitude to exertion in children with Down's syndrome. *Journal of Mental Deficiency Research* 33:161–174.

Egan, B., R. Panis, A. Hinderliter, N. Schork, and S. Julius. 1987. Mechanism of increased alpha-adrenergic vasoconstriction in human essential hypertension. *Journal of Clinical Investigation* 80:812–817.

Elmstahl, S., M. Petersson, B. Lilja, S.M. Samuelsson, I. Rosen, and L. Bjuno. 1992. Autonomic cardiovascular responses to tilting in patients with Alzheimer's disease and in healthy elderly women. *Age and Ageing* 21(4):301–307.

Esler, M., G. Jennings, B. Biviano, G. Lambert, and G. Hasking. 1986. Mechanism of elevated plasma norepinephrine in the course of essential hypertension. *Journal of Cardiovascular Pharmacology* 8(suppl):S39–S43.

Esler, M., S. Julius, A. Zweifler, O. Randall, E. Harburg, H. Gardiner, and V. DeQuattro. 1977. Mild high-renin essential hypertension: Neurogenic human hypertension? *New England Journal of Medicine* 296:405–411.

Ewart, C.K., D.R. Young, and J.M. Hagberg. 1998. Effects of school-based aerobic exercise on blood pressure in adolescent girls at risk for hypertension. *American Journal of Public Health* 88(6):949–951.

Facchini, F.S., R.A. Stoohs, and G.M. Reaven. 1996. Enhanced sympathetic nervous system activity: The linchpin between insulin resistance, hyperinsulinemia, and heart rate. *American Journal of Hypertension* 9(10):1013–1017.

Fagard, R. 1995. The role of exercise in blood pressure control: Supportive evidence. *Journal of Hypertension* 13:1223–1227.

Fagard, R., R. Grauwels, D. Groeseneken, P. Lijnen, J. Staessen, L. Vanhees, and A. Amery. 1985. Plasma levels of renin, angiotensin II and 6-keto-prostaglandin $F_{1\alpha}$ in endurance athletes. *Journal of Applied Physiology* 59:947–952.

Fagard, R., and P.J. Lijnen. 1997. Reduction of left ventricular mass by antihypertensive treatment does not improve exercise performance in essential hypertension. *Journal of Hypertension* 15(3):309–317.

Fagard, R., J. Staessen, and A. Amery. 1988. Maximal aerobic power in essential hypertension. *Journal of Hypertension* 6:859–865.

Fagard, R., J. Staessen, and A. Amery. 1991. Exercise blood pressure and target organ damage in essential hypertension. *Journal of Human Hypertension* 5:69–75.

Fagard, R., J. Staessen, L. Thijs, and A. Amery. 1995. Relation of left ventricular mass and filling to exercise blood pressure and rest blood pressure. *American Journal of Cardiology* 75(1):53–57.

Fagard, R., and C.M. Tipton. 1994. Physical activity, fitness, and hypertension. In *Physical activity, fitness, and health*, edited by C. Bouchard, R.J. Shephard, and T. Stephens. 633–655. Champaign, IL: Human Kinetics.

Folkow, B. 1987. Structure and function of the arteries in hypertension. *American Heart Journal* 114:938–948.

Franke, W.D., and N.B. Tegeler. 1997. Effects of alpha-1 blockade on maximal vascular conductance in young borderline hypertensives. *Clinical Experimental Hypertension* 19(8):1219–1232.

Franz, I.W., U. Tonnesmann, and D. Erb. 1997. Exercise hemodynamic in hypertension associated with coronary microangiopathy, coronary heart disease and without ischemic syndrome effect of nifedipine. *Z. Kardiol* 86(11):936–944.

Gaboury, C.L., D.C. Simonson, E.W. Seely, N.K. Hollenberg, and G.H. Williams. 1994. Relation of pressor responsiveness to angiotensin II and insulin resistance in hypertension. *Journal of Clinical Investigation* 94(6):2295–2300.

Games, D., D. Adams, R. Allessandrini, R. Barbour, P. Berthelette, C. Blackwell, T. Carr, J. Clemens, T. Donaldson, F. Gillespie, et al. 1995. Alzheimer-type neuropathology in transgenic mice overexpressing V717F beta-amyloid precursor protein. *Nature* 373(6514):523–527.

Garthwaite, J., and C.L. Boulton. 1995. Nitric oxide signaling in the central nervous system. *Annual Review of Physiology* 57:683–706.

Geyssant, A., G. Geelan, C. Denis, A.M. Allevard, M. Vincent, E. Jarsaillon, C.A. Bizollon, J.R. Lacour, and C. Gharib. 1981. Plasma vasopressin, renin activity and aldosterone: Effect of exercise and training. *European Journal of Applied Physiology* 46(1):21–30.

Ghiadoni, L., S. Taddei, A. Virdis, I. Sudano, V. Di Legge, M. Meola, L. Venanzio, and A. Salvetti. 1998. Endothelial function and common carotid artery wall thickening in patients with essential hypertension. *Hypertension* 32(1):25–32.

Gilders, R.M., C. Voner, and G.A. Dudley. 1989. Endurance training and blood pressure in normotensive and hypertensive adults. *Medicine and Science in Sports and Exercise* 21:669–674.

Glynn, R.J., T.S. Field, B. Rosner, P.R. Hebert, J.O. Taylor, and C.H. Hennekens. 1995. Evidence for a positive linear relation between blood pressure and mortality in elderly people. *Lancet* 345(8953):825–829.

Goldstein, D.S. 1983. Plasma catecholamines and essential hypertension: An analytical review. *Hypertension* 5:86–99.

Grassi, G., G. Seravalle, D.A. Calhoun, and G. Mancia. 1994. Physical training and baroreceptor control of sympathetic nerve activity in humans. *Hypertension* 23:294–301.

Greene, A.S., P.J. Tonellato, J. Lui, J.H. Lombard, and A.W. Cowley. 1989. Microvascular rarefaction and tissue vascular resistance in hypertension. *American Journal of Physiology* 256(1 pt 2):H126–H131.

Grossman, E., S. Oren, and F.H. Messerli. 1994. Left ventricular mass and cardiac function in patients with essential hypertension. *Journal of Human Hypertension* 8(6):417–421.

Hagberg, J.M. 1988. Exercise, fitness, and hypertension. In *Exercise, fitness, and health*, edited by C. Bouchard, R.J. Shephard, T. Stephens, J.R. Sutton, and B.D. McPherson. 455–66. Champaign, IL: Human Kinetics.

Hagberg, J.M., D. Goldring, A.A. Ehsani, G.W. Heath, A. Hernandez, K. Schechtman, and J.O. Holloszy. 1983. Effect of exercise training on the blood pressure and hemodynamic features of hypertensive adolescents. *American Journal of Cardiology* 52(7):763–768.

Hagberg, J.M., S.J. Montain, W.H. Martin, and A.A. Ehsani. 1989. Effect of exercise training on 60–69 year old persons with essential hypertension. *American Journal of Cardiology* 64:348–353.

Halbert, J.A., C.A. Silagy, P. Finucane, R.T. Withers, P.A. Hamdorf, and G.R. Andrews. 1997. The effectiveness of exercise training in lowering blood pressure: A meta-analysis of randomized controlled trials of 4 weeks or longer. *Journal of Human Hypertension* 11(10):641–649.

Hansen, J., G.D. Thomas, S.A. Harris, W.J. Parsons, and R.G. Victor. 1996. Differential sympathetic neural control of oxygenation in resting and exercising human skeletal muscle. *Journal of Clinical Investigation* 98:584–596.

Hardin, D.S., J.D. Hebert, T. Bayden, M. Dehart, and L. Mazur. 1997. Treatment of childhood syndrome X. *Pediatrics* 100(2):E5

Hatama, S., T. Tsuchihashi, K. Shida, H. Kawashima, K. Fujii, K. Onoyama, M. Fujishima. 1993. Lack of blood pressure increase with age in long-term hospitalised patients with the sequelae of acute carbon monoxide poisoning. *Journal of Human Hypertension* 7(1):19–23.

Heikinheimo, R.J., M.V. Haavisto, R.H. Kaarela, A.J. Kanto, M.J. Koivunen, and S.A. Rajala. 1990.

Blood pressure in the very old. *Journal of Hypertension* 8(4):361–367.

Henrich, H.A., W. Romen, W. Heimgartener, E. Hartung, and F. Baumer. 1988. Capillary rarefaction characteristic of the skeletal muscle of hypertensive patients. *Klinische Wochenschrift* 66:54–60.

Heron, E., D. Chemla, J.L. Megnien, J.C. Pourny, J. Levenson, Y. Lecarpentier, and A. Simon. 1995. Reactive hyperemia unmasks reduced compliance of cutaneous arteries in essential hypertension. *Journal of Applied Physiology* 79(2):498–505.

Hilton, P.J. 1986. Cellular sodium transport in essential hypertension. *New England Journal of Medicine* 314:22.

Hoffman, P., S. Carlsson, J.O. Skarphedinsson, and P. Thoren. 1987. Role of different serotonergic receptors in the long-lasting blood pressure depression following muscle stimulation in the spontaneously hypertensive rat. *Acta Physiologica Scandinavica* 129:535–542.

Horton, E.S. 1981. The role of exercise in the treatment of hypertension in obesity. *International Journal of Obesity* 5(suppl):165–171.

Hughes, W. 1977. Atherosclerosis, Down's syndrome and Alzheimer's disease. *British Medical Journal* 2(1977):702.

Hyman, B.T., H.L. West, and G.W. Rebec. 1995. Neuropathological changes in Down's syndrome hippocampal formation. *Archives of Neurology* 52:373–378.

Imaizumi, T. 1996. Effects of intra-arterial infusion of insulin on control of forearm vascular resistance in normotensive and hypertensive subjects. *Hypertension Research* 19:S47–S50.

Iskandrian, A.S., and J. Heo. 1988. Exaggerated systolic blood pressure response to exercise: A normal variant or a hyperdynamic phase of essential hypertension. *International Journal of Obesity* 5(suppl):165–171.

Jamerson, K.A., S. Julius, T. Gudbrandsson, O. Andersson, and D.O. Brant. 1993. Reflex sympathetic nervous system activation induces acute insulin resistance in the human forearm. *Hypertension* 21(5):618–623.

Jennings, G. 1995. Mechanisms for reduction of cardiovascular risk by regular exercise. *Clinical and Experimental Pharmacology and Physiology* 22:209–211.

Jennings, G., L. Nelson, P. Nestel, M. Esler, P. Korner, D. Burton, and J. Bazelmans. 1986. The effects of changes in physical activity on major cardiovascular risk factors, hemodynamics, sympathetic function, and glucose utilization in man: A controlled study of four levels of activity. *Circulation* 73(1):30–40.

Juhlin-Dannfelt, A., F. Frisk-Holmberg, J. Karlsson, and P. Tesch. 1979. Central and peripheral circulation in relation to muscle fiber composition in normo- and hypertensive man. *Clinical Science* 56:335–340.

Julius, S. 1993. Sympathetic hyperactivity and coronary risk in hypertension. *Hypertension* 21(6):886–893.

Kaneko, K., D. Susic, E. Nunez, and E.D. Frohlich. 1997. ACE inhibition reduces left ventricular mass independent of pressure without affecting coronary flow and flow reserve in spontaneously hypertensive rats. *American Journal of Medical Science* 314(1):21–27.

Karlsson, J., and H.J. Smith. 1984. Muscle fibers in human skeletal muscle and their metabolic and circulatory significance. In *The peripheral circulation*, edited by S. Hunyor, J. Ludbrook, J. Shaw, and M. McGrath. 67–77. Amsterdam: Elsevier Science Publishers BV.

Kasch, F.W., J.L. Boyer, S.P. Van Camp, L.S. Verity, and J.P. Wallace. 1990. The effect of physical activity and inactivity on aerobic power in older men (a longitudinal study). *Physician and Sportsmedicine* 16:73–83.

Kelley, G., and P.D. McClellan. 1994. Antihypertensive effects of aerobic exercise: A brief meta-analysis. *American Journal of Hypertension* 7:115–119.

Kelley, G., and Z.V. Tran. 1995. Aerobic exercise and normotensive adults: A meta-analysis. *Medicine and Science in Sports and Exercise* 27:1371–1377.

Kelm, M., M. Preik, D.J. Hafner, and B.E. Strauer. 1996. Evidence for a multifactorial process involved in the impaired flow response to nitric oxide in hypertensive patients with endothelial dysfunction. *Hypertension* 27(3):346–353.

Ketelhut, R.G., I.W. Franz, and J. Scholze. 1997. Efficacy and position of endurance training as a non-drug therapy in the treatment of arterial hypertension. *Journal of Human Hypertension* 11(10):651–655.

Kilander, L., M. Boberg, and H. Lithell. 1993. Peripheral glucose-metabolism and insulin sensitivity in Alzheimer's disease. *Acta Neurologica Scandanavica* 87:294–298.

Kinoshita, A., H. Urata, Y. Tanabe, M. Ikeda, H. Tanaka, M. Shindo, and K. Arakawa. 1988. What types of hypertensives respond better to mild exercise therapy? *Journal of Hypertension Supplement* 6(4):S631–S633.

Kodama, J., S. Katayama, K. Tanaka, A. Itabashi, S. Kawazu, and J. Ishii. 1990. Effect of captopril on glucose concentration: Possible role of augmented postprandial forearm blood flow. *Diabetes Care* 13:1109–1111.

Kohl, H.W., K.E. Powell, N.F. Gordon, S.N. Blair, and R.S. Paffenberger. 1992. Physical activity, physical fitness, and sudden cardiac death. *Epidemiol Reviews* 14:37–58.

Kohno, K., H. Matsuoka, K. Takenaka, Y. Miyake, G. Nomura, and T. Imaizumi. 1997. Renal depressor mechanisms of physical training in patients with essential hypertension. *American Journal of Hypertension* 10(8):859–868.

Korzick, D.H., and R.L. Moore. 1996. Chronic exercise enhances cardiac alpha 1-adrenergic inotropic responsiveness in rats with mild hypertension. *American Journal of Physiology* 271(6):H2599–H2608.

Landin, K., K. Blennow, A. Wallin, C.G. Gottfries. 1993. Low blood pressure and blood glucose levels in Alzheimer's disease. Evidence for a hypometabolic disorder? *Journal of Internal Medicine* 233(4):357–363.

Landsberg, L., and J.B. Young. 1984. The role of the sympathoadrenal system in modulating energy expenditure. *Journal of Clinical Endocrinology and Metabolism* 13:475.

Lantos, P.L., and N.J. Cairns. 1994. The neuropathy of Alzheimer's disease. *Dementia* 5:185–207.

Lever, A.F., and S.B. Harrap. 1992. Essential hypertension: A disorder of growth with origins in childhood? *Journal of Hypertension* 10(2):101–120.

Lever, A.F., and L.E. Ramsay. 1995. Treatment of hypertension in the elderly. *Journal of Hypertension* 13:571–579.

Levy, R.L., P.D. White, W.D. Stroud, and C.C. Hillman. 1945. Transient tachycardia: Prognostic significance alone and in association with transient hypertension. *Journal of the American Medical Association* 129:585–588.

Lillioja, S., A.A. Young, C.L. Cutler, J.L. Ivy, W.G. Abbot, J.K. Zawadzki, H. Yki-Jarvinene, L. Christin, T.W. Secomb, and C. Bogardus. 1987. Skeletal muscle capillary density and fiber type are possible determinants of in vivo insulin resistance in man. *Journal of Clinical Investigation* 80:415–424.

Linder, L., W. Kiowski, and F.R. Buhler. 1990. Indirect evidence for release of endothelium-derived relaxing factor in human forearm circulation in vivo: Blunted response in essential hypertension. *Circulation* 81:1762–1767.

Lip, G.Y., and F.L. Li-Saw-Hee. 1998. Does hypertension confer a hypercoagulable state? *Journal of Hypertension* 16(7):913–916.

Loma, E.G., F. Herkenhoff, and E.C. Vasquez. 1998. Ambulatory blood pressure monitoring in individuals with exaggerated blood pressure response to exercise. Influence of physical conditioning. *Arq. Brasilia Cardiologia* 70(4):243–249.

Lucarini, A.R., M. Spessot, E. Picano, C. Marini, F. Lattanzi, R. Pedrinelli, and A. Salvetti. 1991. Lack of correlation between cardiac mass and arteriolar structural changes in mild-to-moderate hypertension. *Journal of Hypertension* 9(12):1187–1191.

Lucas, C.P., J.A. Estigarribia, L.L. Darga, and G.M. Reaven. 1985. Insulin and blood pressure in obesity. *Hypertension* 7:702–706.

Lund-Johansen, P. 1989. Central hemodynamics in essential hypertension at rest and during exercise: A 20-year follow-up study. *Journal of Hypertension* 7(suppl):S52–S55.

Lund-Johansen, P. 1994. Newer thinking on the hemodynamics of hypertension. *Current Opinions in Cardiology* 9(5):505–511.

Luscher, T.F., and P.M. Vanhoutte. 1986. Endothelium-dependent actions to acetylcholine in the aorta of the spontaneously hypertensive rat. *Hypertension* 8:344–348.

MacMahon, S., J.A. Cutler, and J. Stamler. 1989. Antihypertensive drug treatrnent: Potential, expected, and observed effects on stroke and on coronary heart disease. *Hypertension* 13(5 suppl):I45–I50.

MacMahon, S., R. Peto, J. Cutler, R. Collins, P. Sorlie, J. Neaton, R. Abbott, J. Godwin, A. Dyer, and J. Stamler. 1990. Blood pressure, stroke, and coronary heart disease. Part 1. Prolonged difference in blood pressure: Prospective observational studies corrected for the regression dilution bias. *Lancet* 335(8692):765–774.

Main, C.J., and G. Masterton. 1981. The influence of hospital environment on blood pressure in psychiatric inpatients. *Journal of Psychosomatic Research* 25:157–163.

Manicardi, V., L. Camellini, G. Belodi, C. Coscelli, and E. Ferrannini. 1986. Evidence for an association of high blood pressure and hyperinsulinemia in obese man. *Journal of Clinical Endocrinology and Metabolism* 62(6):1302–1304.

Manolis, A.J., D. Beldekos, J. Hatzissavas, S. Foussas, D. Cokkinos, M. Bresnahan, I. Gavras, and H. Gavras. 1997. Hemodynamic and humoral correlates in essential hypertension: Relationship between patterns of LVH and myocardial ischemia. *Hypertension* 30(3):730–734.

Marks, J.B., B.E. Hurwitz, J. Ansley, et al. 1996. Effects of induced hyperinsulinemia on blood pressure and sympathetic tone in healthy volunteers. *Diabetes Care* (S):1465.

Marmot, M.G. 1984. Geography of blood pressure and hypertension. *British Medical Bulletin* 40(4):380–386.

Masterton, G., C.J. Main, A.F. Lever, and R.S. Lever 1981. Low blood pressure in psychiatric inpatients. *British Heart Journal* 45:442–446.

Melin, J.A., W. Wijns, H. Pouleur, A. Robert, M. Nannan, P.M. De Coster, C. Beckers, and J.M. Detry. 1987. Ejection fraction response to upright exercise in hypertension: Relation to loading condition and to contractility. *International Journal of Cardiology* 17(1):37–49.

Meredith, I.T., P. Friberg, G.L. Jennings, E.M. Dewar, V.A. Fazio, G.W. Lambert, and M.D. Esler. 1991. Exercise training lowers resting renal but not cardiac sympathetic activity in humans. *Hypertension* 18(5):575–582.

Meredith, I.T., G.L. Jennings, M.D. Esler, E.M. Dewar, A.M. Bruce, V.A. Fazio, and P.I. Korner. 1990. Time-course of the anti-hypertensive effects of regular endurance exercise in human subjects. *Journal of Hypertension* 8(9):859–866.

Miall, W.E., and S. Chinn. 1973. Blood pressure and results of a 15–17 year follow-up study in South Wales. *Clinical Science* 45(suppl 1):23s–33s.

Miall, W.E., and P.D. Oldham. 1958. Factors influencing arterial blood pressure in the general population. *Clinical Science* 17:409–444.

Millar, J.A., C.G. Isles, and A.F. Lever. 1995. Blood pressure, white coat pressor responses and cardiovascular risk in place: Group patients of the MRC Mild Hypertension Trial. *Journal of Hypertension* 13:175–183.

Missault, L.H., D.A. Duprez, A.A. Brandt, M.L. de Buyzere, L.T. Adang, and D.L. Clement. 1993. Exercise performance and diastolic filling in essential hypertension. *Blood Pressure* 2(4):284–288.

Miura, T., V. Bhargava, B.D. Guth, K.S. Sunnerhagen, S. Miyazaki, C. Indolfi, and K.L. Peterson. 1993. Increased afterload intensifies asynchronous wall motion and impairs ventricular relaxation. *Journal of Applied Physiology* 75(1):389–396.

Moan, A., G. Nordby, M. Rostrup, I. Eide, and E. Kjeldsen. 1995. Insulin sensitivity, sympathetic activity and cardiovascular reactivity in young men. *American Journal of Hypertension* 8:268–275.

Montain, S.J., S.M. Jilka, A.A. Ehsani, and J.M. Hagberg. 1988. Altered hemodynamics during exercise in older essential hypertensive subjects. *Hypertension* 12:479–484.

Morishita, R., J. Higaki, M. Miyazaki, and T. Ogihara. 1992. Possible role of the vascular renin angiotensin system in hypertension and vascular hypertrophy. *Hypertension* 19(2) (suppl II):II62–II67.

Morrison, R.A., A. McGrath, G. Davidson, J.J. Brown, G.D. Murray, and A.F. Lever. 1996. Low blood pressure in Down's syndrome: A link with Alzheimer's disease? *Hypertension* 28(4):569–575.

Motoyama, M., Y. Sunami, F. Kinoshita, A. Kiyonaga, H. Tanaka, M. Shindo, T. Irie, H. Urata, J. Sasaki, and K. Arakawa. 1998. Blood pressure lowering effect of low intensity aerobic training in elderly hypertensive patients. *Medicine and Science in Sports and Exercise* 30(6):818–823.

Mufson, E.J., E. Cochran, W. Benzing, and J.H. Kordower. 1993. Galaninergic innervation of the cholinergic vertical limb of the diagonal band (Ch2) and bed nucleus of the stria terminalis in aging, Alzheimer's disease and Down's syndrome. *Dementia* 4(5):237–250.

Mundal, R., S.E. Kjeldsen, L. Sandvik, G. Erikssen, E. Thaulow, and J. Erikssen. 1994. Exercise blood pressure predicts cardiovascular mortality in middle-aged men. *Hypertension* 24(1):56–62.

Mundal, R., S.E. Kjeldsen, L. Sandvik, G. Erikssen, E. Thaulow, and J. Erikssen. 1997. Predictors of 7-year changes in exercise blood pressure: Effects of smoking, physical fitness and pulmonary function. *Journal of Hypertension* 15:245–249.

Murdoch, J.C., J.C. Rodger, S.S. Rao, C.D. Fletcher, and M.G. Dunnigan. 1977. Down's syndrome: An atheroma-free model? *British Medical Journal* 2:226–228.

Naftilan, A.J., W.M. Zuo, J. Inglefinger, T.J. Ryan, R.E. Pratt, and V.J. Dzay. 1991. Localization and differential regulation of angiotensinogen mRNA expression in the vessel wall. *Journal of Clinical Investigation* 87:1300–1311.

National Heart, Lung, and Blood Institute. 1997. *Fact book fiscal year 1996.* Bethesda, MD: U.S. Department of Health and Human Services, National Institutes of Health.

Nelson, L., G.L. Jennings, M.D. Esler, and P.I. Korner. 1986. Effect of changing levels of physical activity on blood-pressure and haemodynamics in essential hypertension. *Lancet* 2(8505):473–476.

Ng, A.V., R. Callister, D.G. Johnson, and D.R. Seals. 1994. Endurance training is associated with increased basal sympathetic nerve activity in healthy older humans. *Journal of Applied Physiology* 77:1366–1374.

Nho, H., K. Tanaka, H.S. Kim, Y. Watanabe, and T. Hiyama. 1998. Exercise training in female patients with a family history of hypertension. *European Journal of Applied Physiology* 78(1):1–6.

Nomura, G., E. Kumagai, K. Midorikhwa, T. Kitano, H. Tashiro, and H. Toshima. 1984. Physical training in essential hypertension: Alone and in combination with dietary salt restriction. *Journal of Cardiac Rehabilitation* 4:469–475.

O'Dea, K. 1991. Cardiovascular disease risk factors in Australian aborigines. *Clinical and Experimental Pharmacology and Physiology* 18:85–88.

Ohtsuka, S., M. Kakihana, H. Watanabe, R. Ajisaka, and Y. Sugishita. 1997. Relation of a decrease in arterialæ compliance to ST segment depression on exercise electrocardiograms in patients with hypertension. *Hypertension Research* 20(1):11–16.

Oparil, S. 1986. The sympathetic nervous system in clinical and experimental hypertension. *Kidney International* 30:437–452.

Orbach, P., and D.T. Lowenthal. 1998. Evaluation and treatment of hypertension in active individuals. *Medicine and Science in Sports and Exercise* 30(suppl 10):S354–S366.

Overton, J.M., J.M. VanNess, and H.J. Takata. 1998. Effects of chronic exercise on blood pressure in Dahl salt-sensitive rats. *American Journal of Hypertension* 11(1):73–80.

Paffenbarger, R.S., R.T. Hyde, A.L. Wing, and C.C. Hsieh. 1986. Physical activity, all cause mortality, and longevity of college alumni. *New England Journal of Medicine* 314(10):605–613.

Paffenbarger, R.S., M.C. Thorne, and A.L. Wing. 1968. Chronic disease in former college students. VIII. Characteristics in youth predisposing to hypertension in later years. *American Journal of Epidemiology* 88:25–32.

Palatini, P. 1998. Exaggerated blood pressure response to exercise: Pathophysiologic mechanisms and clinical relevance. *Journal of Sports Medicine and Physical Fitness* 38(1):1–9.

Panza, J.A., A.A. Quyyumi, J.E. Brush, and S.E. Epstein. 1990. Abnormal endothelium-dependent vascular relaxation in patients with essential hypertension. *New England Journal of Medicine* 323:22–27.

Pavey, R.J. 1989. Pretreatment systolic orthostatic blood pressure depression in Down's syndrome (letter). *Journal of Clinical Psychopharmacology* 9:146–147.

Petrella, R.J. 1998. How effective is exercise training for the treatment of hypertension? *Clinical Journal of Sports Medicine* 8(3):224–231.

Phillip, T.H., A. Distler, and U. Cordes. 1978. Sympathetic nervous system and blood-pressure control in essential hypertension. *Lancet* 2:959–963.

Picado, M.J., A. de la Sierra, M.T. Aguilera, A. Coca, and A. Urbano-Marquez. 1994. Increased activity of the Mg2+/Na+ exchanger in red blood cells from essential hypertensive patients. *Hypertension* 23(6 pt 2):987–991.

Pitetti, K.H., M. Climstein, K.D. Campbell, P.J. Barrett, and J.A. Jackson. 1992. The cardiovascular capacities of adults with Down syndrome: A comparative study. *Medicine and Science in Sports and Exercise* 24(1):13–19.

Polare, T., H. Lithell, and C. Berne. 1990. Insulin resistance is a characteristic feature of primary hypertension independent of obesity. *Metabolism* 39:167.

Poulter, N.R., K.T. Khaw, B.E. Hopwood, M. Mugambi, W.S. Peart, G. Rose, and P.S. Sever. 1990. The Kenyan Luo migration study: Observations on the initiation of a rise in blood pressure. *British Medical Journal* 300(6730):967–972.

Preik, M., M. Kelm, M. Feelisch, and B.E. Strauer. 1996. Impaired effectiveness of nitric oxide donors in resistance arteries of patients with arterial hypertension. *Journal of Hypertension* 14(7):903–908.

Preik, M., M. Kelm, S. Schafer, and B.E. Strauer. 1997. Impairment of adenosine-induced dilation of forearm resistance arteries in patients with arterial hypertension. *Vasa* 26(2):70–75.

Puddey, I.B., and K. Cox. 1995. Exercise lowers blood pressure—sometimes? Or did Pheidippides have hypertension? *Journal of Hypertension* 13:1229–1233.

Reaven, G.M., H. Lithell, and L. Landsberg. 1996. Hypertension and associated metabolic abnormalities—the role of insulin resistance and the sympathoadrenal system. *New England Journal of Medicine* 334:374.

Rebbeck, T.R., S.T. Turner, and C.F. Sing. 1993. Sodium-lithium counter-transport genotype and the probability of hypertension in adults. *Hypertension* 22:560.

Reiling, M.J., L.A. Bare, P.B. Chase, and D.R. Seals. 1990. Influence of regular exercise on 24-hour blood pressure (BP 24) in middle aged and older persons with essential hypertension (EH) (abstract). *Medicine and Science in Sports and Exercise* 22(suppl):S48.

Reis, D.J., S. Morrison, and D.A. Ruggiero. 1988. The C1 area of the brainstem in tonic and reflex control of blood pressure. *Hypertension* 11(2)(suppl I):I8–I13.

Resnick, L.M. 1993. Ionic basis of hypertension, insulin resistance, vascular disease, and related disorders: The mechanism of syndrome X. *American Journal of Hypertension* 6(suppl 4):123S.

Richards, B.W., and F. Enver. 1979. Blood pressure in Down's syndrome. *Journal of Mental Deficiency Research* 23:123–135.

Rochini, A.P., V. Katch, A. Schork, and R.P. Kelch. 1987. Insulin and blood pressure during weight loss in nonobese adolescents. *Hypertension* 10:267–273.

Rogers, M.W., M.M. Probst, J.J. Gruber, R. Berger, and J.B. Boone. 1996. Differential effects of exercise training intensity on blood pressure and cardiovascular responses to stress in borderline hypertensive humans. *Journal of Hypertension* 14(11):1369–1375.

Roman, R.J., and A-P. Zou. 1993. Influence of renal medullary circulation on the control of sodium excretion. *American Journal of Physiology* 265:R963–R973.

Roman, M.J., T.G. Pickering, R. Pini, J.E. Schwartz, and R.B. Devereux. 1995. Prevalence and determinants of cardiac and vascular hypertrophy in hypertension. *Hypertension* 26(2):369–373.

Rosei, E.A., D. Rizzoni, M. Castellano, E. Porteri, R. Zulli, M.L. Muiesan, G. Bettoni, M. Salvetti, P. Muiesan, and S.M. Guilini. 1995. Media:lumen ratio in human small resistance arteries is related to forearm minimal vascular resistance. *Journal of Hypertension* 13(3):341–347.

Rosner, B., C.H. Hennekens, E.H. Kass, and W.E. Miall. 1977. Age-specific correlation analysis of longitudinal blood pressure data. *American Journal of Epidemiology* 106(4):306–313.

Rossor, M.N. 1995. Catastrophe, chaos and Alzheimer's disease. *Journal of the Royal College of Physicians* 29:412–418.

Rowe, T.W., J.B. Young, K.L. Minaker, et al. 1981. Effect of insulin and glucose infusions on sympathetic nervous system activity in normal man. *Diabetes* 30:219.

Rumble, B., R. Retallack, C. Hilbich, G. Simms, G. Multhaup, R. Martins, A. Hockey, P. Montgomery, K. Beyreuther, and C.L. Masters. 1989. Amyloid A4 protein and its precursor in Down's syndrome and Alzheimer's disease. *New England Journal of Medicine* 320(22):1446–1452.

Sakai, T., M. Ideishi, S. Miura, H. Maeda, E. Tashiro, M. Koga, A. Kinoshita, M. Sasaguri, H. Tanaka, M. Shindo, and K. Arakawa. 1998. Mild exercise activates renal dopamine system in mild hypertensives. *Journal of Human Hypertension* 12(6):355–362.

Sander, M., J. Hansen, and R.G. Victor. 1997. The sympathetic nervous system is involved in the maintenance but not initiation of hypertension induced by N (omega)-nitro-L-arginine methyl ester. *Hypertension* 30:64–70.

Sander, M., P.G. Hansen, and R.G. Victor. 1995. Sympathetically mediated hypertension caused by chronic inhibition of nitric oxide. *Hypertension* 26:691–695.

Scherrer, U., S.F. Vissing, B.J. Morgan, J.A. Rollins, R.S.A. Tindall, A. Ring, P. Hanson, P.K. Mohanty, and R.G. Victor. 1990. Cyclosporine-induced sympathetic activation and hypertension after heart transplantation. *New England Journal of Medicine* 323:693–699.

Schlaich, M.P., H.P. Schobel, M.R. Langenfeld, K. Hilgers, and R.E. Schmieder. 1998. Inadequate suppression of angiotensin II modulates left ventricular structure in humans. *Clinical Nephrology* 49(3):153–159.

Schuler, J.L., and W.H. O'Brien. 1997. Cardiovascular recovery from stress and hypertension risk factors: A meta-analytic review. *Psychophysiology* 34(6):649–659.

Sheldahl, L.M., T.J. Ebert, B. Cox, and F.E. Tristani. 1994. Effect of aerobic training on baroreflex regulation of cardiac and sympathetic function. *Journal of Applied Physiology* 76:158–165.

Siegel, D., L. Kuller, N.B. Lazarus, D. Black, D. Feigal, G. Hughes, J.A. Schoenberger, and S.B. Hulley. 1987. Predictors of cardiovascular events and mortality in the systolic hypertension in the Elderly Program pilot project. *American Journal of Epidemiology* 126(3):385–399.

Silva, G.J., P.C. Brum, C.E. Negrao, and E.M. Krieger. 1997. Acute and chronic effects of exercise on baroreflexes in spontaneously hypertensive rats. *Hypertension* 30(3):714–719.

Sivertsson, R., R. Sannerstedt, and Y. Lundgren. 1976. Evidence for peripheral vascular involvement in mild elevation of blood pressure in man. *Clinical Science and Molecular Medicine* 51:65s–68s.

Smith, G.D.P., and C.J. Mathias. 1995. Postural hypotension enhanced by exercise in patients with chronic autonomic failure. *Quarterly Journal of Medicine* 88:251–256.

Somers, V.K., K.C. Leo, R. Shields, M. Cleary, and A. Mark. 1992. Forearm endurance training attenuates sympathetic nerve response to isometric handgrip in normal humans. *Journal of Applied Physiology* 72:1039–1043.

Somers, V.K., and A.L. Mark. 1993. Sympathetic mechanisms in hypertension. In *Autonomic failure: A textbook of clinical disorders of the autonomic nervous system*, edited by R. Bannister and C. Mathias. 804–821. Oxford: Oxford Medical Publication.

Stamler, J., D.M. Berkson, A. Dyer, M.H. Leper, H.A. Lindberg, O. Paul, H. McKean, P. Rhomberg, J.A. Schoenberger, R.B. Shekelle, and R. Stamler. 1975. Relationship of multiple variables to blood pressure: Findings from four Chicago epidemiologic studies. In *Epidemiology and control of hypertension*, edited by O. Paul. 307–352. Miami, FL: Symposia Specialists.

Suarez, D.H., B.L. Pegram, and E.D. Frohlich. 1981. Systemic and regional hemodynamic changes associated with anterior hypothalamic lesions in conscious rats. *Hypertension* 3(2):245–249.

Suga, S., K. Nakao, H. Itoh, Y. Komatsu, Y. Ogawa, N. Hama, and H. Imura. 1992. Endothelial production of C-type natriuretic peptide and its marked augmentation by transforming growth factor-beta: Possible existence of vascular natriuretic peptide system. *Journal of Clinical Investigation* 90:1145–1149.

Svedenhag, J., B.G. Wallin, G. Sundlof, and J. Henriksson. 1984. Skeletal muscle sympathetic activity at rest in trained and untrained subjects. *Acta Physiologica Scandinavica* 120(4):499–504.

Szklo, M. 1986. Determination of blood pressure in children. *Clinical and Experimental Hypertension* A8(4 and 5):479–493.

Takeichi, Y., K. Nagata, H. Izawa, M. Iwase, T. Sobue, M. Sugawara, and M. Yokota. 1997. Dynamic exercise-induced changes in diastolic properties of the regional myocardium in hypertensive left ventricular hypertrophy. *Heart Vessels* 12:138–141.

Tanaka, H., D.R. Bassett, E.T. Howley, D.L. Thompson, M. Ashraf, and F.L. Rawson. 1997. Swimming training lowers resting blood pressure in individuals with hypertension. *Journal of Hypertension* 15(6):651–657.

Tanaka, H., M.J. Reiling, and D.R. Seals. 1998. Regular walking increases peak limb vasodilatory capacity of older hypertensive humans: Implications for arterial structure. *Journal of Hypertension* 16(4):423–428.

Taylor, J.O., J. Cornoni-Huntley, J.D. Curb, K.G. Manton, A.M. Ostfeld, P. Scherr, and R.B. Wallace. 1991. Blood pressure and mortality risk in the elderly. *American Journal of Epidemiology* 134(5):489–501.

Thase, M.E. 1982. Longevity and mortality in Down's syndrome. *Journal of Mental Deficiency Research* 26:177–192.

Timio, M., P. Verdecchia, S. Venanzi, S. Gentili, M. Ronconi, B. Francucci, M. Montanari, and E. Bichisao. 1988. Age and blood pressure changes. A 20-year follow-up study in nuns in a secluded order. *Hypertension* 12(4):457–461.

Tipton, C.M. 1984. Exercise training and hypertension. In *Exercise and sport sciences*, edited by R.L. Terjung. 245–306. Indianapolis: Benchmark Press.

Tipton, C.M. 1991. Exercise training and hypertension: An update. In *Exercise and sport science reviews,* edited by J.O. Holloszy. 447–505. Baltimore: Waverly Press.

Tomiyama, H., N. Doba, T. Kushiro, M. Yamashita, K. Kanmatsuse, N. Yoshida, and S. Hinohara. 1997. The relationship of hyperinsulinemic state to left ventricular hypertrophy, microalbuminuria, and physical fitness in borderline and mild hypertension. *American Journal of Hypertension* 10(6):587–591.

Tresch, D.D., M.F. Folstein, P.V. Rabins, and W.R. Hazzard. 1985. Prevalence and significance of cardiovascular disease and hypertension in elderly patients with dementia and depression. *Journal of the American Geriatric Society* 33(8):530–537.

Trimarco, B., N. De Luca, B. Ricciardelli, A. Cuocolo, G. Rosiello, G. Lembo, and M. Volpe. 1986. Effects of lower body negative pressure in hypertensive patients with left ventricular hypertrophy. *Journal of Hypertension* 4(suppl 5):S306–S309.

Trimarco, B., M. Volpe, B. Ricciardelli, G.B. Picotti, M.A. Galva, R. Petracca, and M. Condorelli. 1983. Studies on the mechanisms underlying impairment of beta-adrenoreceptor-mediated effects in human hypertension. *Hypertension* 5:584–590.

Tuck, M.L., D. Corry, and A. Trujillo. 1990. Salt-sensitive blood pressure and exaggerated vascular reactivity in the hypertension of diabetes. *American Journal of Medicine* 88:210.

Uehara, Y., and K. Arakawa. 1997. Nonpharmacological therapy in hypertensive patients—effect of physical exercise on hypertension. *Nippon Rinsho* 55:2034–2038.

United Kingdom Prospective Diabetes Study. 1985. III. Prevalence of hypertension and hypotensive therapy in patients with newly diagnosed diabetes. *Hypertension* 7(6)(suppl II):II8.

Urata, H., Y. Tanabe, A. Kiyonaga, M. Ikeda, H. Tanaka, M. Shindo, and K. Arakawa. 1987. Antihypertensive and volume-depleting effects of mild exercise on essential hypertension. *Hypertension* 9(3):245–252.

van den Bree, M.B., R.M. Schieken, W.B. Moskowitz, and L.J. Eaves. 1996. Genetic regulation of hemodynamic variables during dynamic exercise. The MCV twin study. *Circulation* 15(8):1864–1869.

Veras-Silva, A.S., K.C. Mattos, N.S. Gava, P.C. Brum, C.E. Negrao, and E.M. Krieger. 1997. Low-intensity exercise training decreases cardiac output and hypertension in spontaneously hypertensive rats. *American Journal of Physiology* 273(6):H2627–H2631.

Virdis, A., L. Ghiadoni, A. Lucarini, V. Di Legge, S. Taddei, and A. Salvetti. 1996. Presence of cardiovascular structural changes in essential hypertensive patients with coronary microvascular disease and effects of long-term treatment. *American Journal of Hypertension* 9(4):361–369.

Vitiello, B., R.C. Veith, S.E. Molchan, R.A. Martinez, B.A. Lawlor, J. Radcliffe, J.L. Hill, and T. Sunderland. 1993. Autonomic dysfunction in patients with dementia of the Alzheimer-type. *Biological Psychiatry* 34(7):428–433.

Wagner, C.D., H.M. Stauss, P.B. Persson, and K.C. Kregel. 1998. Correlation integral of blood pressure as a marker of exercise intensities. *American Journal of Physiology* 275(5):R1661–1666.

Wang, J.S., C.J. Jen, H.C. Kung, L.J. Lin, T.R. Hsiue, and H.I. Chen. 1994. Different effects of strenuous exercise and moderate exercise on platelet function in men. *Circulation* 90:2877–2885.

Wang, S.J., K.K. Liao, J.L. Fuh, K.N. Lin, Z.A. Wu, C.Y. Liu, and H.C. Liu. 1994. Cardiovascular autonomic functions in Alzheimer's disease. *Age and Ageing* 23(5):400–404.

Wannamethee, G., A.G. Shaper, P.W. Macfarlane, and M. Walker. 1995. Risk factors for sudden cardiac death in middle-aged British men. *Circulation* 91:1749–1756.

Ward, K.D., D. Sparrow, L. Landsberg, et al. 1993. The influence of obesity, insulin, and sympathetic nervous system activity on blood pressure. *Clinical Research* 41:168A.

Whincup, P.H., D.G. Cook, A.G. Shaper, D.J. Macfarlane, and M. Walker. 1988. Blood pressure in British children: Associations with adult blood pressure and cardiovascular mortality. *Lancet* 2(8616):890–893.

Wisniewski, H.M., W. Silverman, and J. Wegiel. 1994. Alzheimer disease and mental retardation. *Journal of Intellectual Disability Research* 38:233–239.

Yates, C.M., J. Simpson, A.F. Maloney, A. Gordon, and A.H. Reid. 1980. Alzheimer-like cholinergic deficiency in Down syndrome. *Lancet* 2(8201):979.

Yla-Herttuala, S., J. Luoma, T. Nikkari, and T. Kivimaki. 1989. Down's syndrome and atherosclerosis. *Atherosclerosis* 76(2–3):269–272.

Yoshihara, F., T. Nishikimi, Y. Yoshitomi, H. Abe, H. Matsuoka, and T. Omae. 1996. Left ventricular structural and functional characteristics in patients with renovascular hypertension, primary aldosteronism and essential hypertension. *American Journal of Hypertension* 9(6):523–528.

Yoshitomi, Y., T. Nishikimi, H. Abe, S. Nagata, M. Kuramochi, H. Matsuoka, and T. Omae. 1997.

Left ventricular systolic and diastolic function and mass before and after antihypertensive treatment in patients with essential hypertension. *Hypertension Research* 20(1):23–28.

Zanettini, R., D. Bettega, O. Agostoni, B. Ballestra, G. del Rosso, R. di Michele, and P.M. Mannucci. 1997. Exercise training in mild hypertension: Effects on blood pressure, left ventricular mass, and coagulation factor VII and fibrinogen. *Cardiology* 88(5):468–473.

Zemen, R.J., R. Ludemann, T.G. Easton, and J.D. Etlinger. 1988. Slow to fast alterations in skeletal muscle fibers caused by clenbuterol, a beta-2-receptor agonist. *American Journal of Physiology* 254:E726–E732.

Chapter 24

Orthostatic Stress and Autonomic Dysfunction

Johannes J. van Lieshout and Niels H. Secher

Assumption of the upright posture causes a vertical displacement of blood away from the heart. In order to maintain cardiac output, the consequent fall in ventricular filling volume must be met by continuous adjustment of arterial and venomotor tone and by regulating cardiac contractility and chronotropy. Humans can stand erect for long periods of time. Their orthostatic circulatory adaptation is provided by the evolution of an effective set of neuromuscular and circulatory mechanisms that are largely involuntary or autonomic. According to Claude Bernard, "Nature thought it provident to remove these important phenomena from the capriciousness of an ignorant will" (Bannister and Mathias 1992a). This chapter is concerned with the circulatory responses to orthostatic stress and the consequences of failing cardiovascular reflexes.

Orthostatic Stress

This first section will deal with orthostatic stress. Specifically, it will discuss passive changes and active changes.

Passive Changes

The first circulatory event upon assumption of the upright position, either passive or active, is a gravitational displacement of blood away from the thorax to the dependent regions of the body with a fall in venous return (Amberson 1943; Hill 1895; Matzen et al. 1991; Piorry 1826) (see figure 24.1). A change of posture to the upright position elicits a shift of 300–800 ml of blood from the chest to the lower parts of the body (Blomqvist and Stone 1984; Sjöstrand 1952). Up to 50% of the total shift takes place in the first few seconds (Brown, Wood, and Lambert 1949; De Mardes, Kunitsch, and Barbey 1973; Kirsch, Merke, and Hinghofer-Szalkay 1980; Smith and Ebert 1990). Yet, gravity does not affect the pressure that drives blood from the arterial side through the venous

system toward the heart because the hydrostatic pressure affects both sides of the circulation. Capillary transmural pressure increases in the dependent parts of the body with continued filtration into the tissue spaces that contributes to the further fall in circulating volume (Smith and Ebert 1990). The hydrostatic load of 5 min quiet standing has been reported to induce a loss of plasma volume of ~400 ml (Lundvall and Bjerkhoel 1994). After standing up, the transcapillary diffusion of fluid approaches stability by ~20–30 min at a net fall in plasma volume of about 15%. However, complete stabilization of plasma volume requires a longer period (Hagan, Diaz, and Horvath 1978; Smith and Ebert 1990; Tarazi et al. 1970; Thompson, Thompson, and Dailey 1928). The decline in plasma volume during passive head-up tilt does not stabilize (Bie et al. 1986; Matzen et al. 1991; Sander-Jensen, Secher, Astrup, et al. 1986).

Assumption of the upright position places the brain above the level of the heart. Approximately 70% of total blood volume shifts below that level, with its largest part stored within the compliant veins. Consequently, intravascular pressure decreases above and increases below the arterial and venous hydrostatic indifference points (HIP), the location in the vascular tree where pressure becomes independent of posture (Blomqvist and Stone 1984; Rowell 1993; Rushmer 1970). The level of arterial and venous HIP depends on intravascular volume, influence of gravity, and vasomotor tone. Blood will pool in the venous system 15–20 cm below the level of the venous HIP (Perko, Payne, and Secher 1993). The pooled volume of blood is not stationary. However, its transit time in regions of the lower part of the body exposed to hydrostatic load is prolonged (Lundvall and Bjerkhoel 1994; Rowell 1993; Wieling and van Lieshout 1993b).

Quietly standing without further movements elicits an increase in skeletal muscle tone that opposes pooling of venous blood (Gauer and Thron 1965; Smith and Ebert 1990). In the upright moving

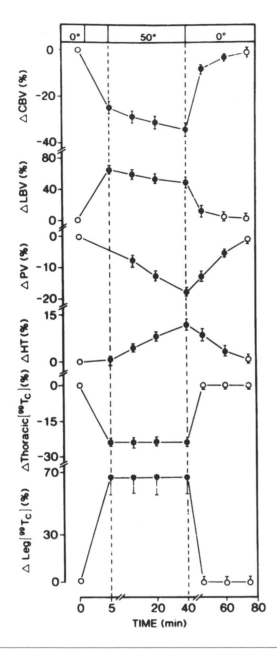

Figure 24.1 Changes in central blood volume (CBV), leg blood volume (LBV), plasma volume (PV), hematocrit (HT), and thoracic 99 Tc° and leg 99 Tc° erythrocytes in eight subjects during head-up tilt. Results are shown for control at supine rest (–30 to 0 min), graded head-up tilt from 0° to 50° (0 to 5 min), sustained 50° head-up tilt (5 to 65 min), and during return to 0° (65 to 95 min). Values are means + SE. Filled symbols: p < 0.05 from rest. (Reproduced from Matzen et al. 1991.)

position, maintenance of sufficient venous return is assisted by the circulatory effects of contracting muscles (skeletal muscle pumps) (Beecher, Field, and Krogh 1935) and the phasic changes in venous flow related to respiration (respiratory pump)

(Rowell 1993). The leg muscle pump is also important during walking and leg exercise. The dynamic increase in muscle tone together with competent venous valves drives blood to the heart and restores cardiac filling. Intra-abdominal pressure increases during deep inspiration. This forces blood into the thorax because venous valves and phasic elevation of extravascular pressure of the iliac and femoral veins prevent retrograde flow into the legs. Since intrathoracic pressure decreases during inspiration, the pressure gradient between the periphery and the right atrium becomes larger (Moreno et al. 1967; Rushmer 1970).

Active Changes

Application of small (~5 mmHg) levels of subatmospheric pressure to the lower part of the body has been proposed to activate cardiopulmonary receptors selectively without altering arterial pressure or arterial baroreceptor activity. Human studies of muscle sympathetic nerve activity in heart transplant recipients during mild lower-body suction suggest that unloading of sinoaortic rather than cardiac and pulmonary baroreceptors produces sympathetic activation during orthostatic stress (Jacobsen et al. 1993). Taylor et al. (1995) recently challenged the concept that small, nonhypotensive reductions of effective blood volume alter neither arterial pressure nor arterial baroreceptor activity. They found that aortic pulse area decreased progressively and significantly during mild lower-body suction, with 50% of the total decline occurring by 5 mmHg. These data suggest that small reductions of effective blood volume reduce aortic baroreceptive areas and trigger hemodynamic adjustments that are so efficient that alterations in arterial pressure escape detection by conventional means. This implies that the separate contributions of cardiopulmonary and arterial baroreceptors to the cardiovascular reflex adjustments to orthostatic stress in humans remain unsolved.

The relative roles of the carotid and aortic mechanoreceptors in baroreflex control of orthostatic blood pressure are likewise difficult to identify. Observations in humans who have undergone bilateral carotid denervation suggest that cardioacceleration during tilt is mediated primarily by carotid hypotension (Eckberg and Sleight 1992). The change in body position during head-up tilt moves the carotid sinus receptors above the level of the heart, and the distending pressure they sense is reduced (Joyner and Shepherd 1993). After carotid deafferentation, the adjustment of arterial pressure to orthostatic stress is impaired both in the initial phase and during the early phases of circulatory stabilization. These observations emphasize that input from carotid baroreceptors dominates the cardiovascular reflex changes involved in the maintenance of postural normotension in humans.

Standing Up

The early circulatory responses to standing are similar but not identical to a passive change induced by head-up tilt (see figure 24.2). The short-term circulatory adaptation to standing up includes an initial response (first 30 s) with marked changes in blood pressure and heart rate. It also includes a relatively steady-state phase attained after being upright for about 1–2 min (Wieling, Ten Harkel, and van Lieshout 1991; Wieling and van Lieshout 1993b). The muscular effort of standing up compresses the venous vessels in the legs and the splanchnic area by contracting leg and abdominal wall muscles (Borst et al. 1984; Sprangers, Wesseling, et al. 1991; Tanaka, Sjöberg, and Thulesius 1996).

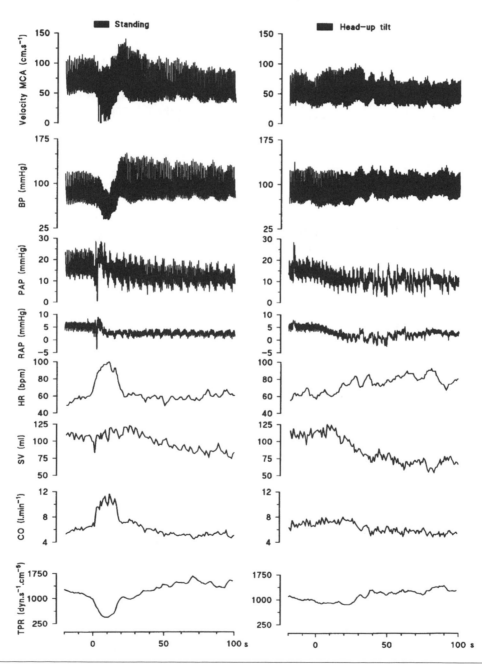

Figure 24.2   Medial cerebral artery velocity and hemodynamic responses to standing and 70° passive head-up tilt in a healthy 28-year-old male subject.

The concomitant abrupt increase in heart rate toward a primary peak at around 3 s results from inhibition of cardiac vagal tone since it is absent after administering intravenous atropine and in many patients with autonomic neuropathy (Wieling et al. 1985). This increase is attributed to a motor signal from higher brain centers to the medullary cardiovascular centers associated with voluntary muscle contraction (central command) (Friedman et al. 1990; Secher 1985;) and by afferent neural input from mechanoreceptors in working muscles (Hollander and Bouman 1975). The more gradual heart rate rise starting at about 5 s after standing up and rising to a peak after about 12 s results from the dual effects of further reflex inhibition of cardiac vagal tone and an increase in sympathetic tone associated with the reduction of central blood volume and drop in blood pressure (Pedersen et al. 1996).

The combination of an instantaneous large heart rate increase and unaltered stroke volume results in a pronounced increase in cardiac output with a maximum level reached ~7 s after standing up (see figure 24.2) (Sprangers, Wesseling, et al. 1991). Nevertheless, blood pressure drops by some 25 mmHg as the result of an ~40% fall in total peripheral resistance unrelated to orthostasis or straining that lasts for 6–8 s (Sprangers, van Lieshout 1991). A similar blood pressure response is observed at the onset of dynamic exercise (Holmgren 1956; Sprangers, Wesseling, et al. 1991). A transient fall in cerebral perfusion likely explains the feelings of dizziness that even healthy humans sometimes experience shortly after standing up (see figure 24.2, left panel).

The initial fall in total peripheral resistance has been attributed to three mechanisms. The first is mechanically induced changes in the conductance of working muscles (Sheriff, Rowell, and Scher 1993; Wieling et al. 1996). The second involves reflex release of vasoconstrictor tone by stimulation of cardiopulmonary afferents through the abrupt increase in right atrial pressure associated with the standing up maneuver (Holmgren 1956; Sprangers, Wesseling, et al. 1991). The third mechanism is cholinergic vasodilatation (Sanders, Mark, and Ferguson 1989) mediated by central command. The readjustment of blood pressure and heart rate after standing is usually completed within 1 min (see figure 24.2) (Wieling and van Lieshout 1993b). The initial and early steady-state circulatory adjustments to the upright posture are neurally mediated. Neuroendocrine control comes into action in a later stage and becomes of importance mainly during prolonged standing (Bakris, Wilson, and Burnett 1986; Cowley 1992; Sancho et al. 1976).

Passive Head-Up Tilt

Passive head-up tilting initiates a rapid decline in central blood volume followed by a continuing decrease in plasma volume by pooling of blood in the legs and splanchnic venous vascular beds (see figure 24.1) (Matzen et al. 1990). Stroke volume will nevertheless decrease only after some six beats of normal cardiac output from the reservoir of blood available in the lungs (see figure 24.2) (Rowell 1993; Sprangers, Wesseling, et al. 1991). The subsequent drop in cardiac filling volume stimulates cardiac and pulmonary receptors in addition to stimulating arterial baroreceptors. This results in an increase in total peripheral resistance with a gradual rise in mean and diastolic pressure but little change in systolic pressure. In addition, an initially fast and subsequently slower rise in heart rate occurs, which then is under sympathetic control (Matzen et al. 1990; Pedersen et al. 1996; Sander-Jensen et al. 1988; Wieling and van Lieshout 1993b). The initial fall in cerebral blood flow upon standing up is absent during passive head-up tilt (see figure 24.2, right panel).

Role of Splanchnic and Leg Capacitance Vessels on Orthostasis

The relative contributions of venous and arterial vasoconstriction in retarding orthostatic volume displacement are not yet fully understood. Active capacitance responses in the highly compliant splanchnic vascular bed that contains a large volume of blood are potentially of great importance in humans. Integrity of splanchnic vascular bed innervation is essential for orthostatic tolerance (Hainsworth 1990; Krogh 1912; Low, Thomas, and Dyck 1978; van Lieshout et al. 1990). Assuming an upright posture involves constriction of splanchnic resistance vessels. When splanchnic arteriolar resistance increases, the consequent fall in venous flow and pressure elicits an elastic recoil of compliant veins. This produces a passive transport of blood out of the splanchnic reservoir (Hainsworth 1990; Krogh 1912; Rothe 1984; Rowell 1993). It implies that passive volume shifts that result from changes in transmural venous pressure are likely to be as important as active capacitance responses. A large fraction of the calf volume change during venous occlusion is attributable to filling of the deep venous spaces (Buckey, Peshock, and Blomqvist 1988). Muscle veins in human limbs have little smooth muscle with a limited supply of sympathetic fibers

(Blomqvist and Stone 1984). The major contribution to leg compliance and degree of orthostatic pooling originates from leg muscle mass and calf intramuscular pressure (Convertino et al. 1988; Mayerson and Burch 1940). In summary, the major rapid adjustment in retarding the gravitational shift of blood seems to be constriction of systemic resistance vessels. Both active and passive capacitance responses in the splanchnic area likely contribute to maintenance of venous return in the upright position.

Local mechanisms support the reflex increase in sympathetic outflow to the vascular wall smooth muscles upon either an active or passive change of posture. An increase in venous transmural pressure by venous distention in the dependent legs stimulates a local axon reflex with the receptor sites located in small veins and the effector sites in the arterioles supplying the dependent tissues. This veno-arteriolar axon reflex probably contributes considerably to the postural increase in limb vascular resistance. The remarkable preservation of orthostatic circulatory adaptation in patients with a transaction transsection of the spinal cord is attributed in part to this neural mechanism (Mathias and Frankel 1992). It thus may be an important adjunct to the cardiovascular reflexes mediated through the central nervous system (Andersen, Boesen, and Henriksen 1991; Henriksen 1977; Henriksen and Skagen 1986; Shepherd 1986; Skagen and Henriksen 1986).

Central Hypovolemia and Vasovagal Syncope

The finding of a low heart rate in air raid casualties with hypovolemic shock (Grant and Reeve 1941; McMichael 1944), in volunteers who bled large amounts (Shenkin et al. 1944), and in patients with serious hemorrhagic shock (Sander-Jensen, Secher, Bie, et al. 1986) was unexpected. Sander-Jensen, Secher, Bie, et al. (1986) showed that patients with serious blood loss initially have a low heart rate that increased only after repletion of the volume deficit. This observation underscores that the traditional teaching that hemorrhage is diagnosed easily by a rapid pulse and a low blood pressure is not legitimate (see figure 24.3).

The complex cardiovascular response to hemorrhage can be divided into three stages (see figure 24.3) (Jacobsen and Secher 1992; Secher et al. 1992). During the first stage, which corresponds to a reduction of the blood volume by approximately 15%, a modest increase in heart rate (< 100 beats/min) and total peripheral resistance compensate for the blood loss with maintenance of a near-normal arte-

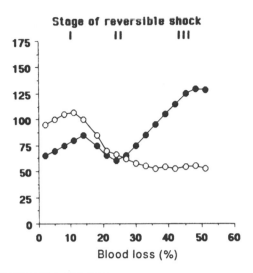

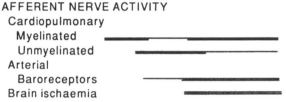

Figure 24.3 Model for changes in heart rate (filled circles = beats/min), mean arterial pressure (open circles = mmHg), and proposed nerve activity from cardiopulmonary and arterial receptor areas and from brain ischemia during three stages of hemorrhage. (Reproduced from Secher et al. 1992.)

rial pressure (Jacobsen and Secher 1992; Schadt and Ludbrook 1991). The second stage is marked by a decrease in heart rate, total peripheral resistance, and blood pressure when the central blood volume becomes reduced by ~30% (Esler, Ludbrook, and Wallin 1995; Matzen et al. 1991; Murray et al. 1968). When blood loss continues, blood pressure falls further, and tachycardia (> 120 beats/min) becomes manifest (see figure 24.3). This third stage probably reflects the transition to irreversible shock (Jacobsen and Secher 1992).

The cardiovascular changes during the second stage of progressive central hypovolemia are thought to reflect the overriding of normal baroreflex circulatory control by an opposing depressor reflex. The result is cardiac slowing and arterial hypotension with development of cerebral hypoperfusion and syncope. Researchers recognize this phenomenon as the vasovagal or vasodepressor response (Lewis 1932). The relative bradycardia results from sudden augmentation of efferent vagal activity. Hypotension follows the abrupt reduction or cessation of sympathetic activity and relaxation of arterial resistance vessels (Burke, Sundløf, and Wallin 1977;

Iversen et al. 1995; van Lieshout, Wieling, et al. 1991; Wallin and Sundløf 1982). This typical response may be evoked by such diverse events like central hypovolemia during hemorrhage or orthostatic venous pooling (Bergenwald, Freyschuss, and Sjøstrand 1977; El-Sayed and Hainsworth 1995), aortic occlusion, emotion, and painful stimuli. The origin of the afferent signal that triggers this depressor reflex has become subject of a continuing debate (Grubb et al. 1993; Waxman, Cameron, and Wald 1993).

Origin of the Depressor Reflex Input

Jarisch (1941), Barcroft et al. (1944), Gauer (1950), and Sharpey-Shafer (1956) proposed that the depressor reflex originates from the heart. Öberg and Thorén (1972) and Thorén (1987; Thorén, Skarphendinsson, and Carlsson 1988) studied this mechanism by using animal experiments. Slowing of heart rate elicited by rapid hemorrhage or vena cava occlusion in anesthetized rats was correlated with and preceded by increased activity in nerve endings of nonmyelinated afferent C fibers from the left ventricle (Thorén, Skarphendinsson, and Carlsson 1988). The finding that adrenergic agonists activate C fibers (Hainsworth 1991) and facilitate the development of vasovagal responses in tilted humans (Waxman et al. 1989) supports an adjunctive role of catecholamines in ventricular C fiber-initiated vasovagal reactions (Waxman, Cameron, and Wald 1993). An increased inotropic stimulus to the heart with a smaller left ventricular volume has been proposed to excite these ventricular mechanoreceptors as a form of the Bezold-Jarisch reflex (Abboud 1989; Blomqvist and Stone 1984; Öberg and Thorén 1972; Thorén 1987).

Some data contest the notion that the signal bringing about the depressor reflex comes solely from cardiac receptors. Morita and Vatner (1985) found that surgical denervation of the heart did not prevent the loss of sympathetic drive to the vessels during hemorrhage in conscious dogs. Aortic occlusion is regarded as a specific stimulus for exciting ventricular mechanoreceptors in the cat. Hainsworth (1991) emphasized that only a minority of afferent fibers identified in that way increased their firing during hemorrhage or caval occlusion (Öberg and Thorén 1972). Furthermore, the postulated decline of the end-systolic left ventricular volume (Shalev, Gal, and Tchou 1991) during central hypovolemia does not surpass ~25% (Jacobsen and Secher 1992; Sander-Jensen et al. 1990). Vasovagal reactions may also happen in cardiac transplant patients. The find-

ing that the associated bradycardia does occur not only in the native but, inconstantly, also in the donor heart (Fitzpatrick et al. 1993) has raised the question of whether the transplanted heart has sufficient ventricular afferent reinnervation to serve as the exclusive sensor in the vasovagal response (Waxman, Cameron, and Wald 1993).

Lewis proposed in 1932 that the carotid sinus baroreceptors act as a possible site of origin of the depressor reflex. Castenfors and Sjöstrand (1972) demonstrated in the rat that hemorrhage-induced hypotension and bradycardia could be prevented by deafferentation of either the heart or the carotid baroreceptors. Landgren (1952) and Angell-James (1971) showed in animal experiments that when arterial baroreceptor areas shrink due to hypovolemia, they may paradoxically increase their firing rate at low blood pressure levels. Dickinson (1995) recently hypothesized a comparable mechanism of collapse and firing at the venous receptor area by sudden invagination of the walls of underfilled atria and great veins. The frequent association of venipuncture with rapid onset of fainting suggests the existence of an afferent pathway from peripheral veins.

Carotid sinus hypersensitivity is a separate cause of syncope mainly in the elderly (Kenny and Traynor 1991). It ranges from a predominantly cardioinhibitory type to a vasodepressor variant (Eckberg and Sleight 1992; Huang et al. 1988). Reflex bradycardia and hypotension in glossopharyngeal neuralgia (Wallin, Westerberg, and Sundløf 1984) is another example of reflex syncope through afferent pathways not originating from the heart.

The role of psychic stress in evoking the vasovagal response is less explicit. L. Hill suggested in 1895 that emotional fainting results from a sudden transient withdrawal of vasomotor neural traffic. In some animals confronted by serious danger, the sympathetic fight-or-flight response (moderate tachycardia, hypertension) shifts to a state analogous to the playing dead reaction (bradycardia, hypotension) (Henry 1984; Roddie 1977). Emotional events presumably modulate the activity of medullary vasomotor centers through output from the limbic system and the cerebral cortex (Marshall 1995; Shepherd and Vanhoutte 1979). Stressful mental tasks and strong emotional stimuli can cause dramatic muscle vasodilatation in humans (Roddie 1977). This is illustrated by the observation that invasive instrumentation may considerably increase the incidence of fainting (Glick and Yu 1963; Jellema et al. 1996; Stevens 1966).

During head-up tilt-induced central hypovolemia in humans, cerebral arterial blood flow falls

(Jørgensen et al. 1993) and brain oxygen saturation drops just prior to the onset of syncope (Madsen, Lyck, Pedersen, et al. 1995). When the central blood volume has become reduced to the degree that cardiac output is too low, cerebral blood flow declines below the critical limit of cerebral perfusion pressure (Njemanze 1992). We therefore suggest that the signal initiating the vasovagal response may originate from different sides of the vascular system, including the brain.

Epinephrine and Opiates

Fainting may be associated with an early failure of sympathetically mediated vasoconstriction (Sneddon et al. 1993; Ten Harkel et al. 1993). This takes place notwithstanding signs of increased sympathetic cardiac and vasomotor activity, e.g., a larger rise in heart rate and a greater variability in the low-frequency band in the power spectrum of blood pressure variability (Karemaker 1993; Ten Harkel et al. 1993). Possible mechanisms involved include reduced vascular responsiveness to α-adrenergic receptor-mediated stimulation and activation of endogenous opioid and serotonergic systems (Kenney and Seals 1993). Epinephrine is suggested as a humoral vasodilating mechanism involved in the vasovagal response. First, plasma levels of epinephrine are elevated in fainting humans (Rowell and Blackmon 1989; Sander-Jensen, Secher, Astrup, et al. 1986). Second, epinephrine produces α-adrenergic dilatation in both skeletal muscle and splanchnic resistance vessels at concentrations measured in humans under stress (Rowell and Blackmon 1989; Rowell and Seals 1990; Shepherd and Vanhoutte 1979).

The finding of less vasodilatation after adrenalectomy is argued to support a contributory role for epinephrine in the mechanism of vasovagal fainting (Barcroft et al. 1960; Robinson and Johnson 1988). However, when the epinephrine response to central hypovolemia is blunted by epidural anesthesia (Jacobsen et al. 1992), by β-adrenoceptor blockade (Pedersen et al. 1996), or by cimetidine (Matzen et al. 1992), it neither improves tolerance to central hypovolemia nor prevents the occurrence of arterial hypotension. The finding that cardioselective α-adrenoceptor blockade diminishes susceptibility to vasovagal responses during head-up tilt in patients with recurrent unexplained syncope (Goldenberg et al. 1987; Waxman et al. 1989) could not be confirmed in healthy humans (Pedersen et al. 1996). Increased epinephrine and decreased norepinephrine levels have been reported in children without

structural heart disease who developed syncope during orthostatic stress testing (Kenny et al. 1986). They were also more sensitive to the vasodilating effect of isoproterenol but not to its chronotropic effects. Some children improve after β-adrenoceptor blocker treatment (Grubb et al. 1992; Kenny et al. 1986; Muller et al. 1993). Theoretically, this might be of help in the avoidance of vasodepressor syncope by adrenergic blockade of skeletal muscle vessels and by modifying the neural traffic from and to the heart (Ferguson, Thames, and Mark 1983; van Lieshout, Wieling, et al. 1991).

The effect of β-adrenoceptor blocking agents on the activity of ventricular mechanoreceptors is, however, controversial (Ferguson, Thames, and Mark 1983; Lipsitz et al. 1989; van Lieshout, Jellema, and Wieling 1993). Interpretation of the reported results of treatment of subjects with unexplained syncope is further hampered since little information is available about the sensitivity and specificity of a positive response to tilt testing as a marker for neurally mediated syncope (Kapoor 1992). In addition, the incidence of positive tilt tests is elevated in healthy young subjects (de Jong-de Vos Van Steenwijk et al. 1995) and decreases over time independent of treatment (Brooks et al. 1993; De Buitleir et al. 1993; Fish, Strasburger, and Benson 1992; Fitzpatrick et al. 1991; Morillo et al. 1993).

Endogenous opiate mechanisms have been suggested as being implicated in the vasodilator response to an acute blood loss. Although the endogenous opiate antagonist naloxone attenuates the fall in blood pressure upon acute blood loss in awake rabbits (Ludbrook and Rutter 1988; Schadt and Ludbrook 1991), the opposite occurs in humans. Naloxone limited the elevation of heart rate and blood pressure in head-up tilt-induced central hypovolemia. It also provoked hypotension and bradycardia at an instant even earlier than in control subjects (Madsen, Klokker, et al. 1995).

Failing Cardiovascular Reflexes—Orthostasis in Sympathetic Failure

Maintenance of arterial pressure in the upright body position depends primarily on an adequate blood volume, an unhindered venous return, and reflex changes in total peripheral resistance through sympathetically mediated adjustments in vasoconstrictor tone of skeletal muscle and splanchnic vascular beds (Rushmer 1970; Shepherd and Mancia 1986). The most important and invalidating manifestation of failure of reflex cardiovascular regulation of arterial pressure is postural hypotension (Bevegård,

Jonsson, and Karløf 1962; Wagner 1959). In patients with failure of autonomic circulatory control, single or multiple lesions are identified in the afferent pathways (carotid sinus syndrome), the vasomotor centers in the brain (tumor or vascular lesion in the medulla, multiple system atrophy) (Shy and Drager 1960), and the efferent sympathetic preganglionic and postganglionic pathways (Bannister and Mathias 1992b).

Bradbury and Eggleston described in 1925 the first cases of idiopathic orthostatic hypotension. Several groups of subjects with generalized or partial primary sympathoneural dysfunction have been since recognized. The group of primary autonomic failure syndromes has two components. The first is pure autonomic failure (PAF) with hypoadrenergic orthostatic hypotension (Bannister 1993) without central neurological signs (peripheral sympathetic failure). The second is multiple system atrophy (MSA) or Shy-Drager syndrome (Shy and Drager 1960). In MSA, autonomic failure is accompanied by overlapping degenerative disorders, strationigral degeneration, and olivopontocerebellar atrophy with a variety of neurological disturbances including Parkinson's disease (central sympathetic failure) (Bannister 1993). In both central and peripheral sympathetic failure, lesions in the intermediolateral columns of the spinal cord are present. In pure autonomic failure, the efferent sympathetic postganglionic pathways have been lost, which are relatively intact in multiple system atrophy (Bannister and Mathias 1992b). Autonomic failure may occur secondary to diseases that affect the autonomic nervous system, e.g., diabetes mellitus and amyloidosis (Bannister and Mathias 1992b). A peculiar form of selective peripheral sympathetic dysfunction is found in patients with an inability to convert dopamine to epinephrine and norepinephrine due to isolated dopamine β-hydroxylase (DBH) deficiency (Man In Et Veld et al. 1987; Mathias and Bannister 1992; Robertson et al. 1986).

The main problem in patients with autonomic cardiovascular dysfunction is that they are unable to modulate sympathetic vasomotor nerve traffic in response to stimuli like standing and exercise that compromise the maintenance of cardiac output at an adequate level. Orthostatic intolerance occurs more commonly in the absence of overt autonomic failure for example during hypovolemia in healthy elderly (Shannon et al. 1986), after physical deconditioning like prolonged recumbency during illness with reduction of muscle mass, or after return from space flight (see chapter 16; Fealey and Robertson 1993; Low et al. 1994; Streeten et al. 1988). Since orthostatic dizziness and syncope are the main clini-

cal problems in patients with disturbances in autonomic circulatory control, the circulatory response to active standing is central in the assessment of autonomic cardiovascular function (see figure 24.4) (van Lieshout and Wieling 1995; Wieling and van Lieshout 1993a). The main types of orthostatic responses can be identified according to the blood pressure and heart rate changes after 1 min upright (Wieling 1992; Wieling, Ten Harkel, and van Lieshout 1991).

Postural tachycardia is a constant finding in the hyperadrenergic orthostatic response and can be regarded as a compensatory response to a variety of conditions. An abnormal degree of central hypovolemia and a strong adrenergic drive in the upright position commonly occur (see figure 24.4c) (Khurana 1995; Wieling, Ten Harkel, and van Lieshout 1991). The postural tachycardia syndrome (POTS) is characterized by orthostatic dizziness, tachycardia, and variable blood pressure changes (Low et al. 1994). It has replaced previous labels such as DaCosta's syndrome, soldier's heart, and neurocirculatory asthenia (Wooley 1976). Tilting in patients with POTS induces a normal-to-excessive increase in total peripheral resistance and an exaggerated decrease in stroke volume (Low et al. 1994). The finding of supersensitive foot vein constrictive responses to noradrenaline infusion in these patients suggests relatively intact sympathetic arteriolar function but selectively impaired sympathetic venomotor function (Streeten and Scullard 1996).

In patients with orthostatic hypotension due to autonomic failure, the initial fall in blood pressure upon standing does not differ in timing or magnitude from healthy humans (see figure 24.5, left panel). The absence of blood pressure recovery between 10 s and 20 s of standing discriminates between the presence or absence of sympathetic vasomotor lesions. In these patients, the decrease in cardiac output after 1 min standing is larger than in healthy humans and is not compensated for by a reflex increase in vascular resistance (see figure 24.5, left panel).

An instantaneous large heart rate increase without relative bradycardia accompanying a progressive fall of arterial pressure can be attributed to correct baroreceptor sensing of an absent recovery of blood pressure. This is indicative for sympathetic vasomotor lesions in the presence of intact parasympathetic heart rate control (see figure 24.4d). Hypoadrenergic orthostatic hypotension in combination with postural tachycardia occurs in some patients with dysautonomia, in quadriplegic patients, and after extensive sympathectomies (van Lieshout et al. 1986, 1990; van Lieshout, Imholz, et

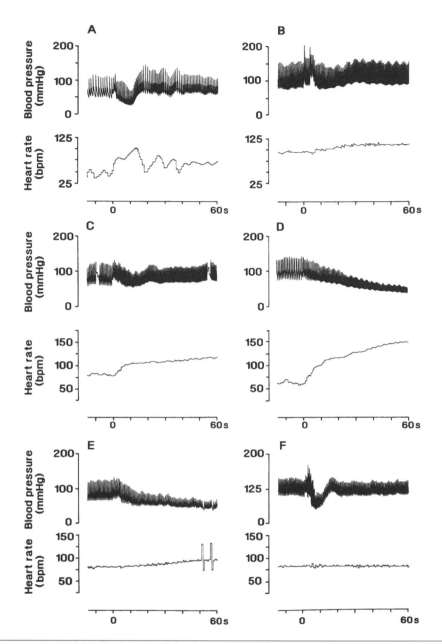

Figure 24.4 Spectrum of normal and abnormal blood pressure and heart rate responses to standing up: *(a)* normal orthostatic circulatory response; *(b)* impairment of parasympathetic cardiac control; *(c)* postural tachycardia with normal blood pressure response; *(d)* orthostatic hypotension (hypoadrenergic) with intact parasympathetic cardiac control; *(e)* orthostatic hypotension (hypoadrenergic) with impairment of sympathetic and parasympathetic cardiac control; and *(f)* total cardiac denervation with intact vasomotor control (cardiac transplant). (Reproduced from Wieling, Ten Harkel, and van Lieshout 1991.)

al. 1991). It implies that a pronounced heart rate increase cannot simply be used as an argument against the diagnosis of orthostatic hypotension. The postural heart rate response in these patients can be reliably interpreted only by monitoring the underlying blood pressure response. The unusual pattern of intact parasympathetic heart rate control and a sympathetic lesion affecting vasomotor control may be interpreted as the mirror image of the

common pattern of autonomic circulatory denervation. For example, this occurs in diabetic autonomic neuropathy where impaired parasympathetic heart rate control precedes overt orthostatic hypotension (Edmonds and Watkins 1992; van Lieshout et al. 1989).

Total parasympathetic denervation is suggested by a normal resting heart rate without sinus arrhythmia and a delayed and sluggish initial heart

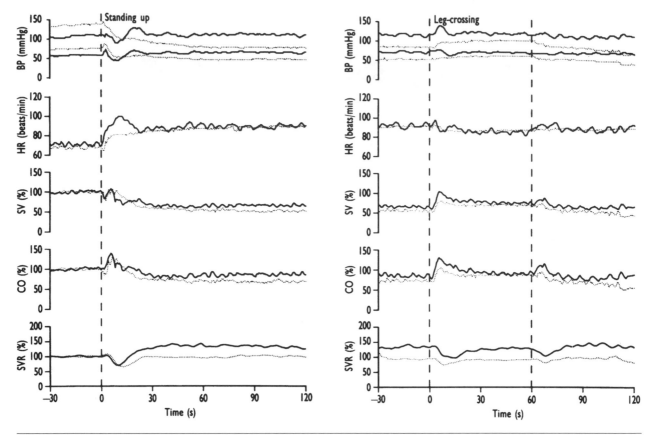

Figure 24.5   Circulatory responses to standing (left panel) and leg crossing (right panel) in healthy subjects (n = 6; complete lines) and patients with orthostatic hypotension (n = 7; dotted lines). (Reproduced from Ten Harkel, Van Lieshout, and Wieling 1994.)

rate response upon standing (see figure 24.4b). The heart rate increase is now of delayed onset but of substantial magnitude afterward. It represents the remaining sympathetic response. This heart rate increase is small in orthostatic hypotension with dual impairment of parasympathetic and sympathetic cardiac control (see figure 24.4e).

Intact vasomotor control with complete cardiac denervation is typical for cardiac transplant patients (see figure 24.4f) (Wieling, Ten Harkel, and van Lieshout 1991). Autonomic blockade of the heart in healthy subjects does not affect orthostatic tolerance; arterial pressure is maintained by greater peripheral vasoconstriction (Rowell 1993). In contrast, pharmacological blockade of sympathetic ganglia in healthy humans elicits orthostatic hypotension despite a normal reflex heart rate increase (Brown, Wood, and Lambert 1949; Schirger et al. 1961). These observations underscore that the integrity of sympathetic nerve outflow to the systemic blood vessels is the key factor in the maintenance of arterial normotension. They also emphasize that peripheral vascular rather than cardiac effector mechanisms are essential for the adjustment of arte-

rial pressure to the upright posture (Wieling and van Lieshout 1993b).

Exercise in Sympathetic Failure

The following discusses exercise in sympathetic failure. Specifically, this section deals with both static and dynamic exercise.

Static Exercise

During both orthostatic stress and exercise, the autonomic nervous system has an important integrating function of local and reflex mechanisms (Joyner and Shepherd 1993) with modulation of sympathetic vasomotor nerve traffic as its effector. In healthy subjects, static exercise like tensing of the leg muscles increases intramuscular pressure (Smith, Hudson, and Raven 1987). This is thought to activate group III and IV muscle mechanosensitive receptors (see figure 24.5) with a reflex augmentation in sympathetic outflow to the heart and vascular wall smooth muscle (Williamson et al. 1994). The result is an increase in heart rate, stroke volume, total peripheral resistance, and arterial pressure in an attempt to perfuse the active muscles in the face of the mechanical obstruction to flow (Joyner and Shepherd 1993; Mitchell and Schmidt 1984).

In patients with autonomic failure, the initial effects on cardiac output and blood pressure of contracting the leg muscles are less pronounced in comparison with healthy subjects. However, the sustained effects on blood pressure are greater. In healthy subjects, the large rise in blood pressure elicited by static exercise of leg muscles is followed by reflex cardio-deceleration and a drop in peripheral vascular resistance (see figure 24.5, right panel) (Ten Harkel, van Lieshout, and Wieling 1994). The result of these reflex adjustments causes the initial blood pressure increase to become offset after 1 minute. In contrast, in patients with sympathetic failure, blood pressure increases steadily during static contraction of leg muscles despite the fact that they remain in a position associated with an otherwise inevitable fall in blood pressure. In these patients, muscle receptor input is not properly translated into reflex changes in vasomotor tone because of peripheral sympathetic dysfunction. Hence, the observed rise in arterial pressure from the onset of leg tensing cannot be explained by a reflex increase in total peripheral resistance. Mechanical compression of the venous vascular beds in the legs into which blood has pooled during standing and translocation of blood to the heart are assumed to be main contributing factors (van Lieshout, Ten Harkel, and Wieling 1992). The sudden blood pressure increase in patients with autonomic failure is not counteracted by a reflex increase in vascular conductance as a consequence of impaired baroreflex vasomotor control. This explains the progressive increase in arterial pressure as seen during sustained leg crossing (see figure 24.5, right panel).

Dynamic Exercise

Dynamic exercise raises arterial pressure in healthy subjects. The magnitude of this increase is slightly larger in the supine position (Bevegård, Holmgren, and Jonsson 1960). The sequence of initial circulatory transients to the onset of upright exercise in humans is remarkably similar to those that occur during standing up when comparable muscle groups come into action (Sprangers, Wesseling, et al. 1991). This suggests that the initial stress posed on the circulation is comparable. The transient drop in systemic vascular resistance associated with a short bout of dynamic leg exercise also occurs in patients

with autonomic failure (see figure 24.6). This drop indicates that an important factor involved appears to be local, nonautonomically mediated vasodilatation in exercising muscles (Wieling et al. 1996).

Marshall, Schirger, and Shepherd (1961) observed in a patient with autonomic failure a reduction in blood pressure during moderate bicycle exercise in the supine position analogous to the drop in blood pressure observed during orthostatic stress. The fall in blood pressure was not prevented by exercising in the 15° head-down position. This observation emphasizes that the blood pressure response to supine exercise in autonomic failure is abnormal during conditions when venous pooling is unlikely to be present. The majority of patients with autonomic failure notice an exacerbation of their postural hypotension symptoms during exercise, even in the supine position (see figure 24.6) (Smith et al. 1995). In these patients, exercise with large muscle groups results in a locally mediated increase in vascular conductance that is not counteracted by reflex vasoconstriction in nonexercising muscles. The impairment of both resistance and capacity vessels results in blood available in the legs and the splanchnic venous reservoir that cannot be recruited in a proper way (Bevegård, Jonsson, and Karløf 1962; Chaudhuri, Thomaides, and Mathias 1992; Marshall, Schirger, and Shepherd 1961; Smith et al. 1995; van Lieshout 1989). The disability of the vascular bed to accommodate its size during an acute fall in blood volume implies that exercise in these patients aggravates the orthostatic decline of venous return by forcing the further loss of blood into an enlarged muscle vascular bed.

Management of Orthostatic Hypotension in Autonomic Failure

Physicians can recognize many of the factors that evoke orthostatic complaints in patients with sympathetic failure by taking into consideration the patients' inability to recruit a proper amount of venous blood when needed. In addition, the unresponsiveness of arterial and venous vascular tone that precludes maintenance of adequate cardiac filling and output when standing up or exercising acts as signals. Since the normal relationship between fall in venous return and sympathetic neuronal activation cannot be reestablished, specific treatment of orthostatic hypotension is not possible. Management should be focused on producing symptomatic improvement (Bannister 1992; Fealey and Robertson 1993). The standing time, defined as the length of time a patient can stand in place before

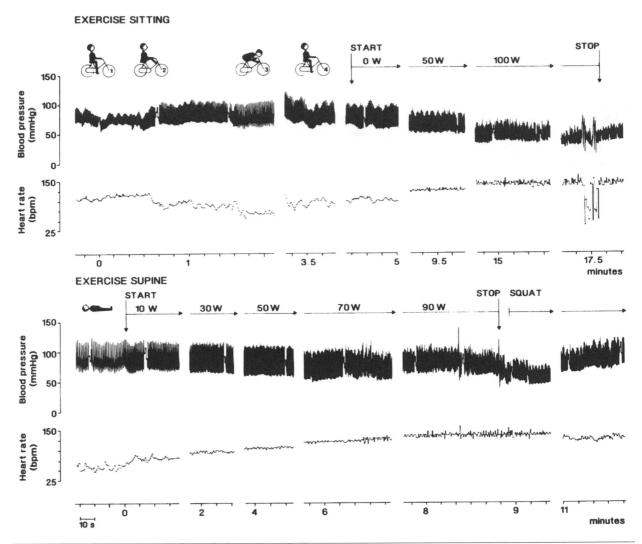

Figure 24.6   Blood pressure and heart rate response to orthostatic maneuvers and both upright and supine bicycle exercise in a patient with a sympathetic postganglionic lesion. Orthostatic maneuvers (upper panel, from left to right): blood pressure drops with the feet on the pedals (bicycle 1) and increases by pulling the legs up to the frame (bicycle 2); an additional increase occurs in pulse pressure by bending over the handle bar (bicycle 3) and again drops when putting the feet down (bicycle 4). Exercise sitting (upper panel, right): note the progressive fall in blood pressure at increasing workloads (10 W every minute up to 100 W). Exercise supine (lower panel): notwithstanding the supine position, blood pressure falls at 90 W during supine exercise. (Reproduced from Van Lieshout 1989.)

experiencing orthostatic symptoms (Fealey and Robertson 1993; Ten Harkel, van Lieshout, and Wieling 1992), is helpful when deciding to treat. In daily life, a prolongation of standing time from only 1 min to ~3–5 min without orthostatic symptoms creates a crucial difference in terms of quality of life (Fealey and Robertson 1993; Wieling and van Lieshout 1993a). When quantifying orthostatic tolerance, one should take into account that the degree of the orthostatic fall in blood pressure may vary substantially during the day and between days. Symptoms tend to be worse in the morning and aggravated after prolonged recumbency, by heat, and postprandially. The development of serious

orthostatic dizziness makes the transition of exercise to rest especially cumbersome, for example, directly after climbing stairs or after stopping bicycle exercise (see figure 24.6, lower panel).

Many of the treatment modalities aim at expansion of the intravascular volume (Bannister 1992; Mehlsen and Boesen 1987; Wieling and van Lieshout 1993a). Chronic expansion of the intravascular volume can be achieved by sleeping in the head-up tilt position, by increasing dietary sodium and water intake, and with mineralocorticoids (Hickler et al. 1959). The therapeutic effect of nocturnal head-up treatment has been attributed to enhancement of endogenous renin production by hypotensive baroreceptor stimulation with increased renal sodium reabsorption and limitation of nocturnal polyuria. In addition, nocturnal head-up treatment moderates the sudden pooling of blood that occurs on arising in the morning, and it reduces supine hypertension (Fealey and Robertson 1993; MacLean and Allen 1940; Ten Harkel, van Lieshout, and Wieling 1992). In patients with severe hypoadrenergic orthostatic hypotension, the level of arterial pressure heavily depends on cardiac filling and output (Mehlsen, Haedersdal, and Stokholm 1994; van Lieshout et al. 1989). This underscores the essential role in treatment of measures aimed at preserving the central blood volume.

Orthostatic tolerance in these patients dramatically increases by physical maneuvers such as leg crossing (cocktail party posture) (see figure 24.5) (Wieling 1996) and bending forward (see figure 24.6, upper panel) (Jeffers, Montgomery, and Burton 1941; Schirger et al. 1961; van Lieshout, Ten Harkel, and Wieling 1992). These measures can prevent orthostatic dizziness while the patient remains upright. We attribute the rise in arterial pressure by leg crossing in autonomic failure patients to be an augmentation of venous return and cardiac output by transporting blood from the veins of the legs and possibly also from the splanchnic vascular bed (Ten Harkel, van Lieshout, and Wieling 1994). Standing time dramatically increases, although the 10–15 mmHg rise in mean arterial pressure observed is relatively modest. Our hypothesis is that such physical maneuvers shift mean arterial pressure from below to just above the critical pressure level of perfusion of the brain. An additional mechanism operative in preventing orthostatic dizziness by bending forward is the shortening of the vertical distance between the heart and the brain.

References

Abboud, F.M. 1989. Ventricular syncope: Is the heart a sensory organ? *New Engl. J. Med.* 320:392–394.

Amberson, W.R. 1943. Physiologic adjustments to the standing posture. *Bull. School Med.* 275:127–145.

Andersen, E.B., F. Boesen, and O. Henriksen. 1991. Local and central sympathetic reflex control of blood flow in skeletal muscle and subcutaneous tissue in normal man. *Clin. Physiol.* 11:451–458.

Angell-James, J.E. et al. 1971. The responses of aortic arch and right subclavian baroreceptors to changes of non-pulsatile pressure and their modification by hypothermia. *J. Physiol. London* 214:201–223.

Bakris, G.L., D.M. Wilson, and J.C. Burnett. 1986. The renal, forearm, and hormonal responses to standing in the presence and absence of propranolol. *Circ.* 74:1061–1065.

Bannister, R. 1992. Treatment of autonomic failure. *Current Opinion Neurol. Neurosurg.* 5:487–491.

Bannister, R. 1993. Multiple-system atrophy and pure autonomic failure. In *Clinical autonomic disorders,* ed. P.A. Low, 517–525. Boston: Little, Brown.

Bannister, R., and C.J. Mathias. 1992a. Introduction and classification of autonomic disorders. In *Autonomic failure. A textbook of clinical disorders of the autonomic nervous system,* ed. R. Bannister and C.J. Mathias, 1–12. Oxford: Oxford University Press.

Bannister, R., and C.J. Mathias. 1992b. Investigation of autonomic disorders. In *Autonomic failure. A textbook of clinical disorders of the autonomic nervous system,* eds. R. Bannister and C.J. Mathias, 255–299. Oxford: Oxford University Press.

Barcroft, H., J. Brod, Z. Hejl, E.A. Hirsjarvi, and A.H. Kitchin. 1960. The mechanism of the vasodilatation in the forearm muscle during stress: Mental arithmetic. *Clin. Sci.* 19:577–586.

Barcroft, H., O.G. Edholm, J. McMichael, and E.P. Sharpey-Schafer. 1944. Posthaemorrhagic fainting. Study by cardiac output and forearm flow. *Lancet* 1:489–490.

Beecher, H.K., M.E. Field, and A. Krogh. 1935. The effect of walking on the venous pressure at the ankle. *Skand. Archiev. Physiol.* 73:133–141.

Bergenwald, L., U. Freyschuss, and T. Sjöstrand. 1977. The mechanism of orthostatic and haemorrhagic fainting. *Scand. J. Clin. Lab. Invest.* 37:209–216.

Bevegård, B.S., A. Holmgren, and B. Jonsson. 1960. The effect of body position on the circulation at rest and during exercise, with special reference to the influence on the stroke volume. *Acta Physiol. Scand.* 49:279–298.

Bevegård, B.S., B. Jonsson, and I. Karlöf. 1962. Circulatory response to recumbent exercise and headup tilting in patients with disturbed sympathetic cardiovascular control postural hypotension. Observations on the effect of norepinephrine infusion and antigravity suit inflation in the head-up tilted position. *Acta Med. Scand.* 172:623–636.

Bie, P., N.H. Secher, A. Astrup, and J. Warberg. 1986. Cardiovascular and endocrine responses to head-up tilt and vasopressin infusion in humans. *Am. J. Physiol.* 251:R735–R741.

Blomqvist, C.G., and H.L. Stone. 1984. Cardiovascular adjustments to gravitational stress. In

Handbook of physiology, eds. J.T. Shepherd and F.M. Abboud, 1025–1063. Washington, D.C.: American Physiological Society.

Borst, C., J.F.M. Van Brederode, W. Wieling, G.A. Van Montfrans, and A.J. Dunning. 1984. Mechanisms of initial blood pressure response to postural stress. *Clin. Sci.* 67:321–327.

Bradbury, S., and C. Eggleston. 1925. Postural hypotension report of 3 cases. *Am. Heart J.* 264–266.

Brooks, R., J.N. Ruskin, A.C. Powell, J. Newell, H. Garan, and B.A. McGovern. 1993. Prospective evaluation of day-to-day reproducibility of upright tilt-table testing in unexplained syncope. *Am. J. Cardiol.* 71:1289–1292.

Brown, G.E., E.H. Wood, and E.H. Lambert. 1949. Effects of tetra-ethyl-ammonium chloride on the cardiovascular reactions in man to changes in posture and exposure to centrifuged force. *J. Appl. Physiol.* 2:117–132.

Buckey, J.C., R.M. Peshock, and C.G. Blomqvist. 1988. Deep venous contribution to hydrostatic blood volume change in the human leg. *Am. J. Cardiol.* 62:449–453.

Burke, D., G. Sundløf, and G. Wallin. 1977. Postural effects on muscle nerve sympathetic activity in man. *J. Physiol. London* 272:399–414.

Castenfors, J., and T. Sjöstrand. 1972. Circulatory control via vagal afferents: 1. Adjustments of heart rate to variations of blood volume in the rat. *Acta Physiol. Scand.* 84:347–354.

Chaudhuri, K.R., T. Thomaides, and C.J. Mathias. 1992. Abnormality of superior mesenteric artery blood flow responses in human sympathetic failure. *J. Physiol. London* 457:477–489.

Convertino, V.A., D.F. Doerr, J.F. Flores, G.W. Hoffler, and P. Buchanan. 1988. Leg size and muscle functions associated with leg compliance. *J. Appl. Physiol.* 64:1017–1021.

Cowley, A.W. 1992. Long-term control of arterial blood pressure. *Physiol. Rev.* 72:231–300.

De Buitleir, M., E.W. Grogan, M.F. Picone, and J.A. Casteen. 1993. Immediate reproducibility of the tilt-table test in adults with unexplained syncope. *American Journal of Cardiology* 71:304–307.

de Jong-de Vos Van Steenwijk, C.C.E., W. Wieling, J.M. Johannes, M.P.M. Harms, W. Kuis, and K.H. Wesseling. 1995. Incidence and hemodynamic characteristics of near-fainting in healthy 6–16 year old subjects. *Journal of the American College of Cardiology* 25:1615–1621.

De Mardes, H., G. Kunitsch, and K. Barbey. 1973. Untersuchung Ilber Kreislaufregulation Wahrend Der Orthostatischen Anpassungsphase. *Basic Res. Cardiol.* 69:462–478.

Dickinson, C.J. 1995. Fainting precipitated by collapse-firing of venous baroreceptors. *Lancet* 342:970–972.

Eckberg, D.L., and P. Sleight. 1992. *Human baroreflexes in health and disease.* Oxford: Oxford University Press.

Edmonds, M.E., and P.J. Watkins. 1992. Clinical presentations of diabetic autonomic failure. In *Autonomic failure. A textbook of clinical disorders of the autonomic nervous system,* eds. R. Bannister and C.J. Mathias, 698–720. Oxford: Oxford University Press.

El-Sayed, H., and R. Hainsworth. 1995. Relationship between plasma volume, carotid baroreceptor sensitivity and ordiostatic tolerance. *Clinical Science* 88:463–470.

Esler, M., J. Ludbrook, and B.G. Wallin. 1995. Cardiovascular regulation: Afferent autonomic pathways and mechanisms. In *Blood loss and shock,* eds. N.H. Secher, J.A. Pawelczyk, and J. Ludbrook, 77–91. London: Edward Arnold.

Fealey, R.D., and D. Robertson. 1993. Management of orthostatic hypotension. In *Clinical autonomic disorders,* ed. P.A. Low, 731–746. Boston, Toronto, London: Little, Brown.

Ferguson, D.W., M.D. Thames, and A.L. Mark. 1983. Effects of propranolol on reflex vascular responses to orthostatic stress in humans. Role of ventricular baroreceptors. *Circulation* 67:802–807.

Fish, F.A., J.F. Strasburger, and D.W. Benson, Jr. 1992. Reproducibility of a symptomatic response to upright tilt in young patients with unexplained syncope. *American Journal of Cardiology* 70:605–609.

Fitzpatrick, A.P., N. Banner, A. Cheng, M. Yacoub, and R. Sutton. 1993. Vasovagal reactions may occur after orthotopic heart transplantation. *Journal of the American College of Cardiology* 21:1132–1137.

Fitzpatrick, A.P., G. Theodorakis, P.E. Vardas, and R. Sutton. 1991. Methodology of head-up tilt testing in patients with unexplained syncope. *Journal of the American College of Cardiology* 17:125–130.

Friedman, D.B., F.B. Jensen, J.H. Mitchell, and N.H. Secher. 1990. Heart rate and arterial blood pressure at the onset of static exercise in man with complete neural blockade. *Journal of Physiology London* 423:543–550.

Gauer, O.H. 1950. Evidence in circulatory shock of an isometric phase of ventricular contraction following ejection. *Federation Proceedings* 9:47.

Gauer, O.H., and H.L. Thron. 1965. Postural changes in the circulation. In *Handbook of physiology. Sec-*

tion 2: The cardiovascular system. Volume 3: The peripheral circulation, eds. J.T. Shepherd and F.M. Abboud. Washington, D.C.: American Physiological Society.

Glick, G., and P.M. Yu. 1963. Hemodynamic changes during spontaneous vasovagal reactions. American Journal of the Medical Sciences 34:42–51.

Goldenberg, I.F., A. Almquist, D.N. Dunbar, S. Milstein, M.R. Pritzker, and D.G. Benditt. 1987. Prevention of neurally-mediated syncope by selective beta-i adrenoceptor blockade. Circulation Suppl. 76:IV:133

Grant, R.T., and E.B. Reeve. 1941. Clinical observations on air-raid casualties. British Medical Journal 2:293–297; 329–332.

Grubb, B.P., A.M. Rubin, D. Wolfe, P. Temesy-Armos, H. Hahn, and L. Elliott. 1992. Head upright tilt-table testing: A useful tool in the evaluation and management of recurrent vertigo of unknown origin associated with near-syncope or syncope. Otolaryngol. Head Neck Surg. 107:570–576.

Grubb, B.P., P.N. Temesy-Armos, D. Samoil, D.A. Wolfe, H. I-Iahn, and L. Elliott. 1993. Tilt table testing in the evaluation and management of athletes with recurrent exercise-induced syncope. Medicine and Science in Sports and Exercise 25:24–28.

Hagan, R.D., F.J. Diaz, and S.M. Horvath. 1978. Plasma volume changes with movement to supine and standing positions. Journal of Applied Physiology 45:414–418.

Hainsworth, R. 1990. The importance of vascular capacitance in cardiovascular control. News in Physiol. Sci. 5:250–254.

Hainsworth, R. 1991. Reflexes from the heart. Physiological Review 3:617–658.

Henriksen, O. 1977. Local sympathetic reflex mechanism in regulation of blood flow in human subcutaneous adipose tissue. Acta Physiologica Scandinavica Suppl. 450:1–48.

Henriksen, O., and K. Skagen. 1986. Local and central sympathetic vasoconstrictor reflexes in human limbs during orthostatic stress. In The sympathoadrenal system. Physiology and pathophysiology. Alfred Benzon symposium, No. 23, eds. N.J. Christensen, O. Henriksen, and N.A. Lassen, 83–94. Copenhagen: Munksgaard.

Henry, J.P. 1984. On the triggering mechanism of vasovagal syncope. Psychosomatic Medicine 46:91–93.

Hickler, R.B., G.R. Thompson, L.M. Fox, and J.T. Hamlin. 1959. Successful treatment of orthostatic hypotension with 9-alpha-fiuorohydrocortisone. New England Journal of Medicine 261:788–791.

Hill, I. 1895. The influence of the force of gravity on the circulation of the blood. Journal of Physiology London 18:15–53.

Hollander, A.P., and L.N. Bouman. 1975. Cardiac acceleration elicited by voluntary muscle contractions of minimal duration. Journal of Applied Physiology 38:70–77.

Holmgren, A. 1956. Circulatory changes during muscular work in man with special reference to arterial and central venous pressures in the systemic circulation. Scandinavian Journal of Clinical and Laboratory Investigation Suppl. 24:1–97.

Huang, S.K.S., M.D. Ezri, R.G. Hauser, and P. Denes. 1988. Carotid sinus hypersensitivity in patients with unexplained syncope: Clinical electrophysiologic and long-term follow-up observations. American Heart Journal 116:989–996.

Iversen, H.K., P. Madsen, S. Matzen, and N.H. Secher. 1995. Arterial diameter during central volume depletion in humans. European Journal of Applied Physiology and Occupational Physiology 72:165–169.

Jacobsen, J., and N.H. Secher. 1992. Heart rate during hemorrhagic shock. Clinical Physiology 12:659–666.

Jacobsen, J., S. Sofelt, V. Brocks, A. Fernandes, J. Warberg, and N.H. Secher. 1992. Reduced left ventricular diameters at onset of bradycardia during epidural anesthesia. Acta Anaesthesiologica Scandinavica 36:831–836.

Jacobsen, T.N., B.J. Morgan, U. Scherrer, S.F. Vissing, R.A. Lange, N. Johnson, W.S. Ring, P.S. Rahko, P. Hanson, and R.G. Victor. 1993. Relative contributions of cardiopulmonary and sinoaortic baroreflexes in causing sympathetic activation in the human skeletal muscle circulation during orthostatic stress. Circulation Research 73:367–378.

Jarisch, A. 1941. Vagovasale Synkope. Zeitschrift für Kreislauf Forschung 23:267–279.

Jeffers, W.A., H. Montgomery, and A.C. Burton. 1941. types of orthostatic hypotension and their treatment. American Journal of the Medical Sciences 202:1–14.

Jellema, W.T., B.P.M. Imholz, J. Van Goudoever, K.H. Wesseling, and J.J. van Lieshout. 1996. Finger arterial versus intrabrachial pressure and continuous cardiac output during head-up tilt testing in healthy subjects. Clinical Science 91:193-200.

Jørgensen, L.G., M. Perko, G. Perko, and N.H. Secher. 1993. Middle cerebral artery velocity during head-up tilt induced hypovolemic shock in humans. Clinical Physiology 13:323–336.

Joyner, M.J., and J.T. Shepherd. 1993. Autonomic control of the circulation. In *Clinical autonomic disorders*, ed. P.A. Low, 55–67. Boston, Toronto, London: Little, Brown.

Kapoor, W.N. 1992. Evaluation and management of the patient with syncope. *JAMA* 268:2553–2560.

Karemaker, J.M. 1993. Analysis of blood pressure and heart rate variability: Theoretical considerations and clinical applicability. In *Clinical autonomic disorders*, ed. P.A. Low, 315–330. Boston: Little, Brown.

Kenney, M.J., and D.R. Seals. 1993. Postexercise hypotension. Key features, mechanisms, and clinical significance. *Hypertension* 22:653–664.

Kenny, R.A., A. Ingram, J. Bayliss, and R. Sutton. 1986. Head-up tilt: A useful test for investigating unexplained syncope. *Lancet* ii:1352–1355.

Kenny, R.A., and G. Traynor. 1991. Carotid sinus syndrome—clinical characteristics in elderly patients. *Age and Aging* 20:449–454.

Khurana, R.K. 1995. Orthostatic intolerance and orthostatic tachycardia: A heterogenous disorder. *Clinical Autonomic Research* 5:12–18.

Kirsch, K.A., J. Merke, and H. Hinghofer-Szalkay. 1980. Fluid volume distribution within superficial shell tissues along body axis during changes of body posture in man: The application of a new plethysmographic method. *Pflügers Archiv.* 383:195–201.

Krogh, A. 1912. The regulation of the supply of blood to the right heart. *Skandinaviske Archiev für Physiologic* 24:229–248.

Landgren, S. 1952. On the excitation mechanism of the carotid baroreceptors. *Acta Physiologica Scandinavica* 26:1–34.

Lewis, T. 1932. Vasovagal syncope and the carotid sinus mechanism with comments on Gowers's and Nothnagel's syndrome. *British Medical Journal* I:873–876.

Lipsitz, L.A., E.R. Marks, J. Koestner, P.V. Jonsson, and J.Y. Wei. 1989. Reduced susceptibility to syncope during postural tilt in old age. *Archives of Internal Medicine* 149:2709–2712.

Low, P.A., T.L. Opfer-Gehrking, S.C. Textor, R. Schondorf, G.A. Suarex, R.D. Fealey, and M. Camilleri. 1994. Comparison of the postural tachycardia syndrome POTS with orthostatic hypotension due to autonomic failure. *Journal of the Autonomic Nervous System* 50:181–188.

Low, P.A., J.E. Thomas, and P.J. Dyck. 1978. The splanchnic autonomic outflow in Shy-Drager syndrome and idiopathic orthostatic hypotension. *Annals of Neurology* 4:511–514.

Ludbrook, J., and P.C. Rutter. 1988. Effect of naloxone on haemodynamic responses to acute blood loss in unanaesthetized rabbits. *Journal of Physiology London* 400:1–14.

Lundvall, J., and P. Bjerkhoel. 1994. Failure of hemoconcentration during standing to reveal plasma volume decline induced in the erect posture. *Journal of Applied Physiology* 77:2155–2162.

MacLean, A.R., and E.V. Allen. 1940. Orthostatic hypotension and orthostatic tachycardia; treatment with the "head-up" bed. *Journal of the American Medical Association* 115:2162–2167.

Madsen, P., M. Klokker, H.L. Olesen, and N.H. Secher. 1995. Naloxone-provoked vaso-vagal response to head-up tilt in men. *European Journal of Applied Physiology and Occupational Physiology* 70:246–251.

Madsen, P., F. Lyck, M. Pedersen, H.L. Olesen, H.B. Nielsen, and N.H. Secher. 1995. Brain and muscle oxygen saturation during head-up tilt induced central hypovolemia in humans. *Clinical Physiology* 153:523–533.

Man In Et Veld, A.J., F. Boomsma, P. Moleman, and M.A.D.H. Schalekamp. 1987. Congenital Dopamine-Beta-Hydroxylase Deficiency. A novel orthostatic syndrome. *Lancet* 1:183–87.

Marshall, J.M. 1995. Cardiovascular changes associated with behavioral alerting. In *Cardiovascular regulation*, eds. D. Jordan and J.M. Marshall, 37–59. Colchester, England: Portland Press.

Marshall, R.J., A. Schirger, and J.R. Shepherd. 1961. Blood pressure during supine exercise in idiopathic orthostatic hypotension. *Circulation* 24:76–81.

Mathias, C.J., and R. Bannister. 1992. Dopamine beta-hydroxylase deficiency and other genetically determined causes of autonomic failure. Clinical features, investigation, and management. In *Autonomic failure. A textbook of clinical disorders of the autonomic nervous system*, eds. R. Bannister and C.J. Mathias, 721–749. Oxford: Oxford University Press.

Mathias, C.J., and H.L. Frankel. 1992. Autonomic disturbances in spinal cord lesions. In *Autonomic failure. A textbook of clinical disorders of the autonomic nervous system*, eds. R. Bannister and C.J. Mathias, 839–811. Oxford: Oxford University Press.

Matzen, S., U. Knigge, H.J. Sci-Rutten, J. Warberg, and N.H. Secher. 1990. Atrial natriuretic peptide during head-up tilt induced hypovolaemic shock in man. *Acta Physiologica Scandinavica* 140:161–166.

Matzen, S., G. Perko, S. Groth, D.B. Friedman, and N.H. Secher. 1991. Blood volume distribution during head-up tilt induced central hypovolaemia in man. *Clinical Physiology* 11:411–422.

Matzen, S., N.H. Secher, U. Knigge, F.W. Back, and J. Warberg. 1992. Pituitary-adrenal responses to head-up tilt in humans: Effects of H1- and H2-receptor blockade. *American Journal of Physiology* 2639:R156–R163.

Mayerson, H.S., and G.E. Burch. 1940. Relationships of tissue subcutaneous and intramuscular and venous pressures to syncope induced in man by gravity. *American Journal of Physiology* 128:258–269.

McMichael, J. 1944. Clinical aspects of shock. *Journal of the American Medical Association* 124:275–281.

Mehlsen, J., and F. Boesen. 1987. Substantial effect of acute hydration on blood pressure in patients with autonomic failure. *Clinical Physiology* 7:243–246.

Mehlsen, J., C. Haedersdal, and K.H. Stokholm. 1994. Dependency of blood pressure upon cardiac filling in patients with severe postural hypotension. *Scandinavian Journal of Clinical and Laboratory Investigation* 54:281–284.

Mitchell, J.H., and R.F. Schmidt. 1984. Cardiovascular control by afferent fibres from skeletal muscle receptors. In *Handbook of physiology*, eds. J.T. Shepherd and F.M. Abboud, 623–658. Washington, D.C.: American Physiological Society.

Moreno, A.H., A.R. Burchell, R. Van Der Woude, and J.H. Burke. 1967. Respiratory regulation of splanchnic and systemic venous return. *American Journal of Physiology* 213:455–465.

Morillo, C.A., J.W. Leitch, R. Yee, and G.J. Klein. 1993. A placebo-controlled trial of intravenous and oral disopyramide for prevention of neurally mediated syncope induced by head-up tilt. *Journal of the American College of Cardiology* 22:1843–1848.

Morita, H., and S.F. Vatner. 1985. Effects of hemorrhage on renal nerve activity in conscious dogs. *Circulation Research* 57:788–793.

Muller, G., B.J. Deal, J.F. Strasburger, and D.W. Benson, Jr. 1993. Usefulness of metoprolol for unexplained syncope and positive response to tilt testing in young persons. *American Journal of Cardiology* 71:592–595.

Murray, R.H., L.J. Thompson, J.A. Bowers, and C.D. Albright. 1968. Hemodynamic effects of graded hypovolemia and vasodepressor syncope induced by lower body negative pressure. *American Heart Journal* 76:799–811.

Njemanze, P.C. 1992. Critical limits of pressure-flow relation in the human brain. *Stroke* 23:1743–1747.

Öberg, B., and P. Thorén. 1972. Increased activity in left ventricular receptors during hemorrhage or occlusion of caval veins in the cat—a possible cause of the vaso-vagal reaction. *Acta Physiologica Scandinavica* 85:164–173.

Pedersen, M., P. Madsen, M. Klokker, H.L. Olesen, and N.H. Secher. 1996. Sympathetic influence on cardiovascular responses to sustained head-up tilt in humans. *Acta Physiologica Scandinavica* 155:435–444.

Perko, G., G. Payne, and N.H. Secher. 1993. An indifference point for electrical impedance in humans. *Acta Physiologica Scandinavica* 148:125–129.

Piorry, P.A. 1826. Recherches Sur l'influence de la pesanteur sur le cours du sang; diagnostic de la syncope et de l'apoplexie; cause et traitement de la syncope. *Archives Génerales De Médecine* 12:527–544.

Robertson, D., M.R. Goldberg, J. Onrot, A.S. Hollister, R. Wiley, J.G. Thompson, and R.M. Robertson. 1986. Isolated failure of autonomic noradrenerge neurotransmission. Evidence for impaired hydroxylation of dopamine. *New England Journal of Medicine* 314:1494–1497.

Robinson, B.J., and R.H. Johnson. 1988. Why does vasodilation occur during syncope? *Clinical Science* 749:347–350.

Roddie, I.C. 1977. Human responses to emotional stress. *Irish Journal of Medical Science* 146:395–417.

Rothe, C.F. 1984. Venous system: Physiology of the capacitance vessels. In *Handbook of physiology. Section 2: The cardiovascular system*, eds. J.T. Shepherd and F.M. Abboud, 397–453. Washington, D.C.: American Physiological Society.

Rowell, L.B. 1993. *Human cardiovascular control*. New York: Oxford University Press.

Rowell, L.B., and J.R. Blackmon. 1989. Hypotension induced by central hypovolemia in hypoxemic humans. *Clinical Physiology* 9:269–277.

Rowell, L.B., and D.R. Seals. 1990. Sympathetic activity during graded central hypovolemia in hypoxemic humans. *American Journal of Physiology* 259:H1197–H1206.

Rushmer, R.F. 1970. Effects of posture. In *Cardiovascular dynamics*, ed. R.F. Rushmer, 192–219. Philadelphia: Saunders.

Sancho, J., R. Re, J. Burton, A.C. Barger, and E. Haber. 1976. The role of the renin-angiotensin-aldosterone system in cardiovascular homeostasis in normal human subjects. *Circulation* 53:400-405.

Sander-Jensen, K., J. Marving, N.H. Secher, I.L. Hansen, J. Giese, J. Warberg, and P. Bie. 1990. Does the decrease in heart rate prevent a detrimental decrease of the end-systolic volume during central hypovolemia in man? *Angiology* 41:687–695.

Sander-Jensen, K., J. Mehlsen, C. Stadeager, N.J. Christensen, J. Fahrenkrug, T.W. Schwartz, J. Warber, and P. Bie. 1988. Increase in vagal activity during hypotensive lower-body negative pressure in humans. *American Journal of Physiology* 253:R149–R156.

Sander-Jensen, K., N.H. Secher, A. Astrup, N.J. Christensen, J. Giese, T.W. Schwartz, J. Warberg, and P. Bie. 1986. Hypotension induced by passive head-up tilt: Endocrine and circulatory mechanisms. *American Journal of Physiology* 251:R742–R748.

Sander-Jensen, K., N.H. Secher, P. Bie, J. Warberg, and T.W. Schwartz. 1986. Vagal slowing of the heart during haemorrhage: Observations from 20 consecutive hypotensive patients. *British Medical Journal* 292:364–366.

Sanders, J.S., A.L. Mark, and D.W. Ferguson. 1989. Evidence for cholinergically mediated vasodilation at the beginning of isometric exercise in humans. *Circulation* 79:815–824.

Schadt, J.C., and J. Ludbrook. 1991. Hemodynamic and neurohumoral responses to acute hypovolemia in conscious mammals. *American Journal of Physiology* 260:H305–H318.

Schirger, A., E.A. Hines, G.D. Molnar, and J.E. Thomas. 1961. Current practices in general medicine. Orthostatic hypotension. *Mayo Clinic Proceedings* 36:239–246.

Secher, N.H. 1985. Heart rate at the onset of static exercise in man with partial neuromuscular blockade. *Journal of Physiology London* 368:481–490.

Secher, N.H., J. Jacobsen, D.B. Friedman, and S. Matzen. 1992. Bradycardia during reversible hypovolaemic shock: Associated neural reflex mechanisms and clinical implications. *Journal of Clinical, Experimental and Pharmacological Physiology* 19:733–743.

Shalev, Y., R. Gal, and P.J. Tchou. 1991. Echocardiographic demonstration of decreased left ventricular dimensions and vigorous myocardial contraction during syncope induced by head-up tilt. *JACC* 38:746–751.

Shannon, R.P., J.Y. Wei, J.Y. Rosa, F.J. Epstein, and J.W. Rowe. 1986. The effect of age and sodium on cardiovascular response to orthostasis. *Hypertension* 8:438–443.

Sharpey-Schafer, E.P. 1956. Emergencies in general practice. Syncope. *British Medical Journal* I:5069.

Shenkin, H.A., R.H. Cheney, S.R. Govons, J.D. Hardy, and A.G. Fletcher. 1944. On the diagnosis of hemorrhage in man. A study of volunteers bled large amounts. *Am. J. Med. Sci.* 208:421–436.

Shepherd, J.T. 1986. Role of venoconstriction for circulatory adjustment to orthostatic stress. In *The sympathoadrenal system. Physiology and pathophysiology. Alfred Benzon symposium 23*, eds. N.J. Christensen, O. Henriksen, and N.A. Lassen, 103–115. Copenhagen: Munksgaard.

Shepherd, J.T., and G. Mancia. 1986. Reflex control of the human cardiovascular system. *Rev. Physiol. Biochem. Pharmacol.* 105:1–99.

Shepherd, J.T., and P.M. Vanhoutte. 1979. *The human cardiovascular system. Facts and concepts.* New York: Raven Press.

Sheriff, D.D., L.B. Rowell, and A.M. Scher. 1993. Is the rapid rise in vascular "conductance" at the onset of dynamic exercise due to the muscle pump? *American Journal of Physiology* 265:H1227–H1234.

Shy, G.M., and G.A. Drager. 1960. A neurological syndrome associated with orthostatic hypotension. *Archives of Neurology* 2:511–527.

Sjöstrand, T. 1952. The regulation of the blood distribution in man. *Acta Physiologica Scandinavica* 26:312–327.

Skagen, K., and O. Henriksen. 1986. Blood pressure regulation during orthostasis: Contribution of local and central sympathetic reflexes. In *The sympathoadrenal system. Alfred Benzon symposium 23*, eds. N.J. Christensen, O. Henriksen, and N.A. Lassen, 95–102. Copenhagen: Munksgaard.

Smith, G.D.P., L.P. Watson, D.V. Pavitt, and C.J. Mathias. 1995. Abnormal cardiovascular and catecholamine responses to supine exercise in human subjects with sympathetic dysfunction. *Journal of Physiology London* 4841:255–265.

Smith, J.J., and J. Ebert. 1990. General response to orthostatic stress. In *Circulatory response to the upright posture*, ed. J.J. Smith, 1–46. Boca Raton, Florida: CRC Press.

Smith, M.L., D.L. Hudson, and P.B. Raven. 1987. Effect of muscle tension on the cardiovascular responses to lower body negative pressure in man. *Medicine and Science in Sports and Exercise* 19:436–442.

Sneddon, J.F., P.J. Counihan, Y. Bashir, G.A. Haywood, D.E. Ward, and A.J. Camn. 1993. Impaired immediate vasoconstrictor responses in patients with recurrent neurally mediated syncope. *American Journal of Cardiology* 71:72–76.

Sprangers, R.L.H., J.J. van Lieshout, J.M. Karemaker, K.H. Wesseling, and W. Wieling. 1991 Circulatory responses to stand up: Discrimination between the effects of respiration, orthostasis and exercise. *Clinical Physiology* 11:221–230.

Sprangers, R.L.H., K.H. Wesseling, A.L. Imholz, B.P.M. Imholz, and W. Wieling. 1991. Initial blood pressure fall on stand up and exercise explained by changes in total peripheral resistance. *Journal of Applied Physiology* 70:523–530.

Stevens, P.M. 1966. Cardiovascular dynamics during orthostatis and the influence of intravascular instrumentation. *American Journal of Cardiology* 17:211–218.

Streeten, D.H.P., G.H. Anderson, R. Richardson, and F. Deaver Thomas. 1988. Abnormal orthostatic changes in blood pressure and heart rate in subjects with intact sympathetic nervous function: Evidence for excessive venous pooling. *Journal of Laboratory and Clinical Medicine* 111:326–335.

Streeten, D.H.P., and T.F. Scullard. 1996. Excessive gravitational blood pooling caused by impaired venous tone is the predominant non-cardiac mechanism of orthostatic intolerance. *Clinical Science* 903:277–285.

Tanaka, H., B.J. Sjöberg, and O. Thulesius. 1996. Cardiac output and blood pressure during active and passive standing. *Clinical Physiology* 16:157–170.

Tarazi, R.C., H.J. Melsher, H.P. Dustan, and E.D. Frohlich. 1970. Plasma volume changes with upright tilt: Studies in hypertension and in syncope. *Journal of Applied Physiology* 28:121–126.

Taylor, J.A., J.R. Halliwill, T.E. Brown, J. Hayano, and D.L. Eckberg. 1995. Nonhypotensive hypovolaemia reduces ascending aortic dimensions in humans. *J. Physiol.* 483:289–298.

Ten Harkel, A.D.J., J.J. van Lieshout, and W. Wieling. 1992. Treatment of orthostatic hypotension with sleeping in the head up tilt position, alone and in combination with fludrocortisone. *J. Int. Med.* 232:139–145.

Ten Harkel, A.D.J., J.J. van Lieshout, and W. Wieling. 1994. Effect of leg muscle pumping and tensing on orthostatic arterial pressure: A study in normal subjects and patients with autonomic failure. *Clin. Sci.* 87:553–558.

Ten Harkel, A.D.J., J.J. van Lieshout, W. Wieling, and J.M. Karemaker. 1993. Differences in circulatory control in normal subjects who faint and who do not faint during orthostatic stress. *Clin. Autonom. Res.* 3:117–124.

Thompson, W.O., P.K. Thompson, and M.E. Dailey. 1928. The effect of posture upon the composition and volume of the blood in man. *J. Clin. Invest.* 5:573–609.

Thorén, P. 1987. Depressor reflex from the heart during severe haemorrhage. In *Cardiogenic reflexes*, eds. R.M. Hainsworth, P.N. McWilliam, and D.A.S.G. Mary, 389–401. Oxford: Oxford University Press.

Thorén, P., J.O. Skarphendinsson, and S. Carlsson. 1988. Sympathetic inhibition from vagal afferents during severe haemorrhage in rats. *Acta Physiol. Scand.* 133 Suppl. 571:97–105.

van Lieshout, J.J. 1989. Cardiovascular reflexes in orthostatic disorders. Ph.D. diss., University of Amsterdam, Rodopi.

van Lieshout, J.J., B.P.M. Imholz, K.H. Wesseling, J.D. Speelman, and W. Wieling. 1991. Singing induced hypotension: A complication of a high spinal cord lesion. *Netherlands J. Med.* 38:75–79.

van Lieshout, J.J., W.T. Jellema, and W. Wieling. 1993. Treatment of neurocardiogenic syncope. *New Engl. J. Med.* 329:969.

van Lieshout, J.J., A.D.J. Ten Harkel, and W. Wieling. 1992. Physical manoeuvres for combatting orthostatic dizziness in autonomic failure. *Lancet* 339:897–898.

van Lieshout, J.J., and W. Wieling. 1995. Circulatory adjustments to orthostatism. *Cardioscopies* 31:533–541.

van Lieshout, J.J., W. Wieling, J.M. Karemaker, and D.L. Eckberg. 1991. The vasovagal response. *Clin. Sci.* 81:575–586.

van Lieshout, J.J., W. Wieling, G.A. Van Montfrans, J.J. Settels, J.D. Speelman, E. Endert, and J.M. Karemaker. 1986. Acute dysautonomia associated with Hodgkin's disease. *J. Neurol. Neurosurg. Psych.* 49:830–832.

van Lieshout, J.J., W. Wieling, K.H. Wesseling, E. Endert, and J.M. Karemaker. 1990. Orthostatic hypotension caused by sympathectomies performed for hyperhidrosis. *Netherlands J. Med.* 36:53–57.

van Lieshout, J.J., W. Wieling, K.H. Wesseling, and J.M. Karemaker. 1989. Pitfalls in the assessment of cardiovascular reflexes in patients with sympathetic failure but intact vagal control. *Clin. Sci.* 765:523–528.

Wagner, H.N. 1959. Orthostatic hypotension. *Bull. Johns Hopkins Hospit.* 105:322–359.

Wallin, B.G., and G. Sundlöf. 1982. Sympathetic outflow to muscles during vasovagal syncope. *J. Auton. Nerv. Syst.* 6:287–291.

Wallin, B.G., C.E. Westerberg, and G. Sundlöf. 1984. Syncope induced by glossopharyngeal neuralgia: Sympathetic outflow to muscle. *Neurology* 34:522–524.

Waxman, M.B., D.A. Cameron, and R.W. Wald. 1993. Role of ventricular vagal afferents in the vasovagal reaction. *J. Am. Coll. Cardiol.* 21:1138–1141.

Waxman, M.B., L. Yao, D.A. Cameron, R.W. Wald, and J. Roseman. 1989. Isoproterenol induction of vasodepressor-type reaction in vasodepressor-prone persons. *Am. J. Cardiol.* 58–65.

Wieling, W. 1992. Non-invasive recording of heart rate and blood pressure in the evaluation of neurocardiovascular control. In *Autonomic failure: A textbook of clinical disorders of the autonomic nervous system*, eds. R. Bannister and C.J. Mathias, 291–311. Oxford: Oxford University Press.

Wieling, W. 1996. External support and physical maneuvers. In *Primer on the autonomic nervous system*, ed. D. Robertson. San Diego: Academic Press.

Wieling, W., C. Borst, J.J. van Lieshout, R.L.H. Sprangers, J.M. Karemaker, J.F.M. Van Brederode, G.A. Van Montfrans, and A.J. Dunning. 1985. Assessment of methods to estimate impairment of vagal and sympathetic innervation of the heart in diabetic autonomic neuropathy. *Netherlands J. Med.* 28:383–392.

Wieling, W., M.P.M. Harms, A.D.J. Ten Harkel, J.J. van Lieshout, and R.L.H. Sprangers. 1996. Circulatory response evoked by a 3s bout of dynamic leg exercise in humans. *J. Physiol.* 494:601-611.

Wieling, W., A.D.J. Ten Harkel, and J.J. van Lieshout. 1991. Spectrum of orthostatic disorders: Classification based on an analysis of the short-term circulatory response upon standing. *Clin. Sci.* 81:241–248.

Wieling, W., and J.J. van Lieshout. 1993a. Investigation and treatment of autonomic circulatory failure. *Current Opinion in Neurol. and Neurosurg.* 6:537–543.

Wieling, W., and J.J. van Lieshout. 1993b. Maintenance of postural normotension in humans. In *Evaluation and management of clinical autonomic failure*, ed. P.A. Low, 69–77. Boston: Little, Brown.

Williamson, J.W., J.H. Mitchell, H.L. Olesen, P.B. Raven, and N.H. Secher. 1994. Reflex increase in blood pressure induced by leg compression in man. *J. Physiol.* 475:351–357.

Wooley, C.F. 1976. Where are the diseases of yesteryear? Dacosta's syndrome, soldier's heart, the effort syndrome, neurocirculatory asthenia—and the mitral valve prolapse syndrome. *Circulation* 53:749–751.

Index

Contributors

Gunnar Blomqvist
Moss Heart Center
University of Texas
Dallas, TX

Robert Boushel
Copenhagen Muscle Research Center
Copenhagen, Denmark

Frederick R. Cobb
Department of Medicine
Duke University Medical Center
Durham, NC

Simon Green
School of Human Movement Studies
Queensland University of Technology
Queensland, Australia

Ronald G. Haller
Institute for Exercise and Environmental Medicine
Presbyterian Hospital of Dallas
Dallas, TX

Michael B. Higginbotham
Department of Medicine
Duke University Medical Center
Durham, NC

Kojiro Ide
Department of Anaesthesia
Rigshospitalet
Copenhagen, Denmark

Bodil Nielsen Johannsen
August Krogh Institute
University of Copenhagen
Copenhagen, Denmark

Lisbeth G. Jørgensen
Department of Vascular Surgery
Rigshospitalet
Copenhagen, Denmark

Atsuko Kagaya
Research Institute of Physical Fitness
Japan Women's College of Physical Education
Tokyo, Japan

Lennart Kaijser
Department of Clinical Physiology
Huddinge Sjukhus
Huddinge, Sweden

Inge-Lis Kanstrup
Department of Clinical Physiology
Herlev Hospital
Herlev, Denmark

Dalane Kitzman
Department of Medicine
Duke University Medical Center
Durham, NC

Maria Koskolou
Department of Physical Education
   and Sport Science
University of Athens
Athens, Greece

Jeffrey Kramer
Physiology, Biophysics and Internal Medicine
University of Illinois
Urbana, IL

Anthony F. Lever
Department of Medicine and Therapeutics
Gardiner Institute
Glasgow, Scotland

Dag Linnarsson
Environmental Physiology Laboratory
Karolinska Institute
Stockholm, Sweden

John Ludbrook
Cardiovascular Research Laboratory
University of Melbourne
Melbourne, Australia

Janice M. Marshall
Department of Physiology
University of Birmingham
Birmingham, United Kingdom

Odile Mathieu-Costello
Department of Medicine
University of California
San Diego, CA

Jesper Mehlsen
Department of Clinical Physiology
Frederiksberg Hospital
Frederiksberg, Denmark

Jere H. Mitchell
Moss Heart Center
University of Texas
Dallas, TX

Markus Nowak
Department of Anaesthesia
Rigshospitalet
Copenhagen, Denmark

James Pawelczyk
Noll Laboratory
Pennsylvania State University
University Park, PA

Jeffrey T. Potts
Department of Physiology
University of Texas
Dallas, TX

Göran Rådegran
Copenhagen Muscle Research Center
Copenhagen, Denmark

Peter B. Raven
Department of Physiology
University of North Texas
Ft. Worth, TX

Chester A. Ray
Milton S. Hershey Medical Center
Pennsylvania State University
Hershey, PA

Robert C. Roach
Department of Life Sciences
Division of Physiology
New Mexico Highlands University
Las Vegas, NM

Mitsuru Saito
Laboratory of Applied Physiology
Toyota Technological Institute
Nagoya, Japan

Bengt Saltin
Copenhagen Muscle Research Center
Copenhagen, Denmark

Niels H. Secher
Copenhagen Muscle Research Center
Copenhagen, Denmark

Steven S. Segal
John B. Pierce Laboratory
Yale University
New Haven, CT

John W. Severinghaus
Department of Anesthesia
University of California
San Francisco, CA

Xiangrong Shi
Department of Physiology
University of North Texas
Ft. Worth, TX

Lawrence I. Sinoway
Milton S. Hershey Medical Center
Pennsylvania State University
Hershey, PA

Peter Snell
Moss Heart Center
University of Texas
Dallas, TX

Martin J. Sullivan
Department of Medicine
Duke University Medical Center
Durham, NC

Johannes J. van Lieshout
Academic Hospital
University of Amsterdam
Amsterdam, Netherlands

John Vissing
Copenhagen Muscle Research Center
Copenhagen, Denmark

Susanne F. Vissing
Copenhagen Muscle Research Center
Copenhagen, Denmark

Tony G. Waldrop
Physiology, Biophysics and Internal Medicine
University of Illinois
Urbana, IL

John B. West
Department of Medicine
University of California
San Diego, CA

About the Editors

Bengt Saltin, MD, PhD, is professor and director of the Copenhagen Muscle Research Centre. He has more than 30 years of cardiovascular research and teaching experience.

Dr. Saltin is a member of the Scandinavian Physiological Society, the American College of Sports Medicine (ACSM), and the Danish Academy of Sciences. He received citation and honor awards from the ACSM and was presented with the Novo Nordisk Award in 1999.

He has received Doctor Honoris Causa from the University of Paris, the University of Athens, the University of Guelph, the Aristotle University, the Norwegian University of Physical Education, and the University of Tartu.

Dr. Saltin obtained his medical degree from the Karolinska Institute in Stockholm. Dr. Saltin earned his PhD in Physiology at the Karolinska Institute in 1964.

Robert Boushel, DSc, is a Research Fellow at the Institute for Sports Medicine Research at Bispebjerg Hospital in Copenhagen.

Dr. Boushel has served in the Division of Cardiology at the University of Texas Southwestern Medical Center, as a lecturer in cardiopulmonary science at Boston University, and as an associate professor/program coordinator in Exercise Sciences at New Hampshire Technical College.

He received his DSc in Applied Anatomy and Physiology at Boston University and an MS in Exercise Science at the University of South Florida.

Niels H. Secher, MD, PhD, is head of the Department of Anesthesia and the Abdominal Center at Rigshospitalet at the University of Copenhagen. A member of the Copenhagen Muscle Research Center, Dr. Secher began his work in cardiovascular research in 1972.

Dr. Secher is a member of the Scandinavian Physiological Society and the ACSM. He received a Young Investigators Award from the University of Copenhagen, where he earned his MD and PhD. He is associate editor of the *European Journal of Applied Physiology*. He is also a former world champion rower.

Jere Mitchell, MD, is Professor of Internal Medicine and Physiology at the University of Texas Southwestern Medical Center. A leading authority in the field of cardiovascular regulation, he has contributed to the basic scientific understanding of how the cardiovascular system is regulated during exercise.

Dr. Mitchell has published 350 papers in the field of cardiovascular regulation and has the distinction of having 40 years of consecutive National Institute of Health funding. His numerous honors include an Established Investigator award from the American Heart Association; Citation, Honor, and Joseph B. Wolffe Distinguished Lecturer awards from the ACSM; the Carl J. Wiggers Award from the Cardiovascular Section of the American Physiological Society; and the Young Investigators' Award and the Distinguished Scientist Award from the American College of Cardiology.

Dr. Mitchell received his medical degree from the University of Texas, Southwestern Medical School.

Other books from Human Kinetics

Guidelines for Cardiac Rehabilitation and Secondary Prevention Programs

(Third Edition)
American Association of Cardiovascular and Pulmonary Rehabilitation
1999 • Paperback • 292 pp • Item BAAC0817
ISBN 0-88011-817-2 • $35.00 ($52.50 Canadian)
Shows health professionals how to gain their patients' full participation in disease management and illustrates how research has shed new light on the vital role prevention efforts play in cardiac rehabilitation programs.

Guidelines for Pulmonary Rehabilitation Programs

(Second Edition)
American Association of Cardiovascular and Pulmonary Rehabilitation
1998 • Paperback • 264 pp • Item BAAC0863
ISBN 0-88011-863-6 • $35.00 ($52.50 Canadian)
Provides direction for existing pulmonary rehabilitation programs and new programs.

Essentials of Cardiopulmonary Exercise Testing

Jonathan N. Myers, PhD
1996 • Hardback • 192 pp • Item BMYE0636
ISBN 0-87322-636-4 • $30.00 ($44.95 Canadian)
The first practical guide to fully explain how to use gas exchange techniques in clinical and research settings.

Heart Disease and Rehabilitation

(Third Edition)
Michael L. Pollock, PhD, and Donald H. Schmidt, MD
1995 • Hardback • 488 pp • Item BPOL0588
ISBN 0-87322-588-0 • $62.00 ($92.95 Canadian)

The most complete and up-to-date reference available on heart disease and cardiac rehabilitation.

To request more information or to order, U.S. customers call 1-800-747-4457, e-mail us at **humank@hkusa.com,** or visit our website at **www.humankinetics.com.** Persons outside the U.S. can contact us via our website or use the appropriate telephone number, postal address, or e-mail address shown in the front of this book.

HUMAN KINETICS
The Information Leader in Physical Activity

Code 2335